Synthesis and Characterization of Glycosides

Synthesis and Characterization of Glycosides

Marco Brito-Arias

Biotechnology Unit
National Polytechnic Institute of Mexico (UPIBI-IPN)

 Springer

Marco Brito-Arias
Biotechnology Unit
National Polytechnic Institute, Mexico
la Laguna Ticomâ
Mexico

Library of Congress Control Number: 2006927420

ISBN 10: 0-387-26251-2
ISBN 13: 978-0-387-26251-2

9 8 7 6 5 4 3 2 1

springer.com

To Carmina Daniela

Contents

Preface

There is no doubt that glycoside chemistry continues to be a dynamic and exciting field related to organic chemistry. Within sugar chemistry, glycosides are of special interest not only because of the challenges represented by their synthesis and structural characterization, but also due to their important biochemical relevance, and hence their applications in a number of essential disciplines, such as pharmaceuticals, food, and biotechnology.

Important biomolecules such as DNA and RNA, or cofactors such as ATP and NAD, are some of the natural glycosidic structures playing key rules at a biochemical level. Also, a considerable number and variety of natural and synthetic glycosides are being extensively used as antibiotics, antiviral, and antineoplasic agents.

There are also a significant number of chromophoric glycosides being used in molecular biology as substrates for detection of enzymatic activity of gene markers.

Solid-phase oligosaccharide synthesis, despite the great progress recently reported by different groups, continues to be a challenging task considering the diversity and complexity of glycosides, especially those present in cellular membranes. However, based on the satisfactory evolution of this approach, it is expected that many complex molecules will be prepared in just in the same way that solid-phase chemistry is currently used to prepare oligopeptides and oligonucleotides.

The aims of this book are to prepare methods and strategies for the formation of glycosides, illustrated by the synthesis of important biologically active glycosides, and also to present an overview of the basic tools needed for the characterization of glycosides through NMR spectroscopy, X-ray diffraction, and mass spectrometry.

From the overwhelming number of excellent articles related to glycoside chemistry, it hasn't been an easy task to select those that are biologically important, and perhaps most importantly serve as didactic models for understanding more about the process of glycoside bond formation.

The text should also serve as a helpful guide to those professionals interested in sugar chemistry, especially the design of synthetic routes, by evaluating suitable protecting and leaving groups, and the best reaction conditions needed for the preparation of glycosides.

The author would like to thank COFAA and SIP-IPN for financial support.

MBA
June 2004

1
Glycosides, Synthesis, and Characterization

1.1   Introduction

Monosaccharides are generally defined as aldoses and ketoses connected to a poly hydroxylated skeleton.[1] In an aqueous solution, the monosaccharides are subject to internal nucleophilic addition to form cyclic hemiacetal structures. When the addition occurs between -OH at C(4) or -OH at C(5) with the carbonyl group, a five- or a six-member ring is formed known as a furanose or a pyranose, respectively. It is also known that equilibrium exists between the open and the cyclic form, being displaced to the latter by more than 90%. Therefore, in aqueous solutions, it is more accurate to consider that most of the sugars are present as cyclic molecules and behave chemically as hemiacetals.

The Haworth structure is a useful way to represent sugars. However, as it is known that for any 6-membered rings, a nonplanar conformation is assumed. The conformation exclusively preferred is called chair and the two possible conformations are $^4C_1$ and $_4C^1$. The first conformation is used for the D enantiomeric form and the second for the L form (Figure 1.1).

On a chair conformation type $^4C_1$, an α anomeric hydroxyl group is positioned in the axial orientation while a β hydroxyl lies equatorial (Figure 1.2).

As a result of this reversible ring formation process, a diastereomer mixture of anomers α and β is produced, as indicated in Table 1.1 for some of the most common monosaccharides.[1,2]

The pioneering work in 1890 by Fischer[3] allowed him to determine the relative configuration and the synthesis of the most known aldohexoses. Based on the assumption that in D-glyceraldehyde, the hydroxyl group was placed to the right, he was able to propose correctly the structure of tetroses, pentoses, and aldohexoses (Figure 1.3). The relative configuration of D-glyceraldehyde was later confirmed by X-ray diffraction by Bijvoet in 1951. Consequently, all the resulting biologically active distereoisomeric aldoses derived from D-glyceraldehyde conserve always the secondary alcohol next to the primary one to the right side in the Fischer projection. Ketoses with three to six carbons are naturally produced from 1,3-Dihydroxyacetone, according to the tree shown in Figure 1.4.

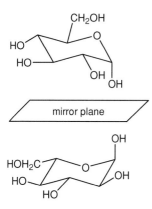

FIGURE 1.1. α-D-glucopyranose-$^4C_1$ and α-L-gluco-pyranose-$^1C_4$.

TABLE 1.1. Distribution of $\alpha\beta$ of some D-monosaccharides in solution at 31°C.

Carbohydrate	% Pyranose		% Furanose	
	α	β	α	β
Glucose	38	62	0.1	<0.2
Galactose	30	64	3	4
Mannose	65.5	34.5	0.6	0.3
Rhamnose	65.5	34.5	0.6	0.3
Fructose	2.5	65.0	6.5	25
Ribose	21.5	58.5	6.4	13.5
Xylose	36.5	63.0	0.3	0.3

1.2   Reactions of Monosaccharides

Carbohydrates own their reactivity to the hemiacetalic centre and to the hydroxyl groups, being primarily more reactive than the secondary. Aldoses and ketoses are susceptible to nucleophilic addition and the latter is less reactive due to steric hindrance. The cyclic forms are adopted when the hydroxyl group positioned at C-5 verifies an intramolecular nucleophilic addition to the carbonyl group producing an anomeric mixture of pyranosides (Figure 1.5).

1.3   Chemical Modifications

The classical reactions on monosaccharides were used initially for identification or sugars or to distinguish between aldoses and ketoses. They have been also very useful for preparing key intermediates in the construction of glycosides. Some of the common reactions used to identify monosaccharides are

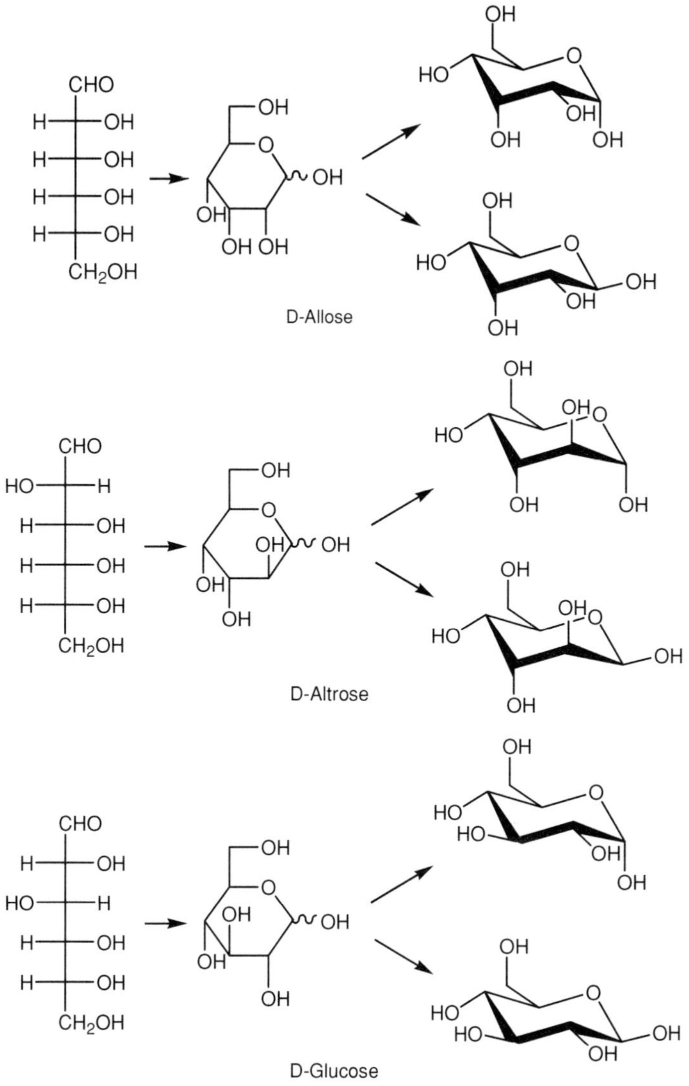

FIGURE 1.2. Fischer projections, Haworth structures, and $^{4}C_{1}$ chair conformation of D-aldohexoses.

1.3.1 Oxidations

The oxidation of nonprotected aldoses may undergo carboxylic acids formation depending on the reaction conditions. Thus, with aqueous bromine the monocarboxylic acid (aldonic acid) is formed, whereas with nitric acid the dicarboxylic acid is favored (aldaric acid) (Figure 1.6).

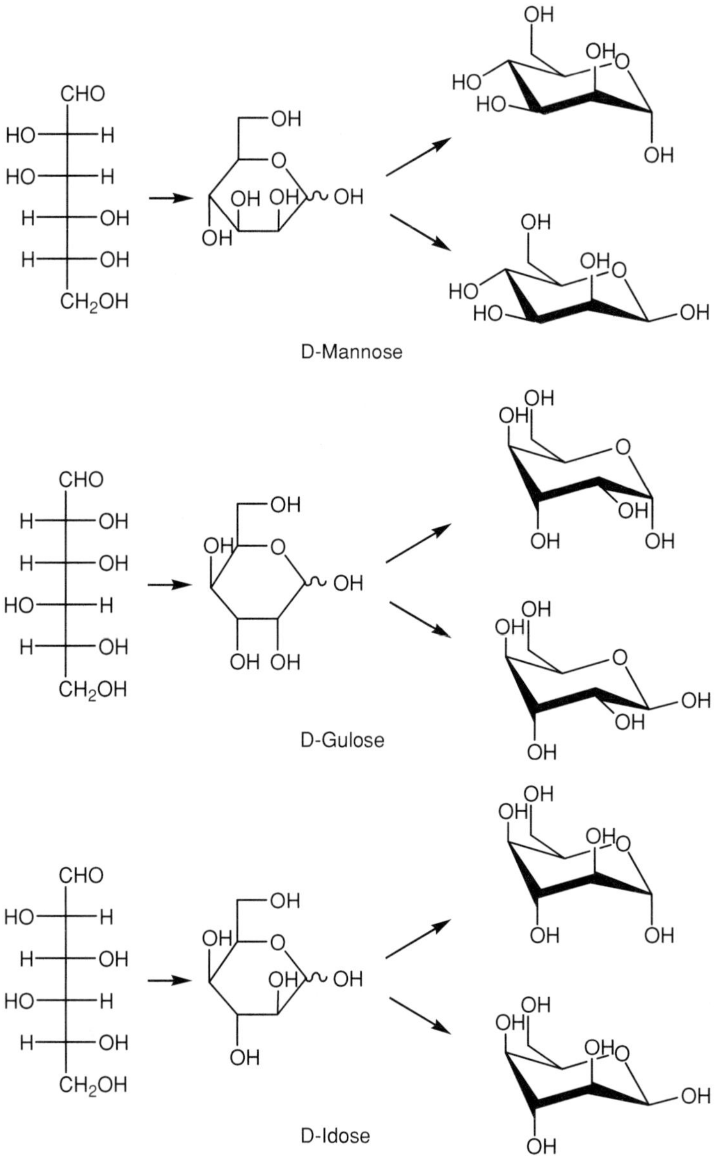

D-Mannose

D-Gulose

D-Idose

FIGURE 1.2. (*continued*)

1.3.2 *Periodate Oxidation*

Periodic acid is a strong oxidizing agent and is capable of breaking 1,2-cis diols to generate after cleavage of the C-C bond carbonyl fragments (Figure 1.7).

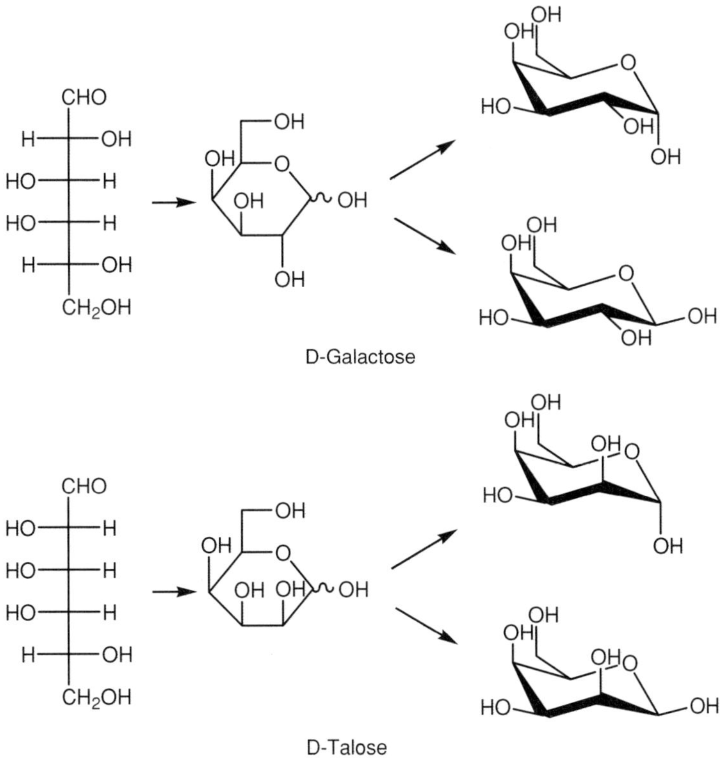

D-Galactose

D-Talose

FIGURE 1.2. (*continued*)

1.3.3   Tollens Reaction

This classical reaction has been very useful for aldose identification and consists of the oxidation of the aldehyde function with a moderate oxidative agent (a silver ammonium salt) to afford the glucuronide ammonium salt and metallic silver, which produce the silver mirror effect (Figure 1.8).

1.3.4   Benedict and Fehling Test

The test consists in the use of a cooper citrate (Benedict reagent) or cooper tartrate complex (Fehling reagent), which upon treatment with the sugar under study produces the glucuronide ion along with cooper (I) oxide which is detected as brick-red precipitate (Figure 1.9).

Based on the Tollens, Benedict or Fehling test, the sugars are classified into reducing when positive or non-reducing sugars if negative. Reducing sugars are hemiacetals in equilibrium with small amounts of the open forms. Under basic conditions, the aldoses and ketoses give positive the Tollens test and/or Benedict/Fehling test as result of equilibrium aldose-ketoses via enediol intermediates.

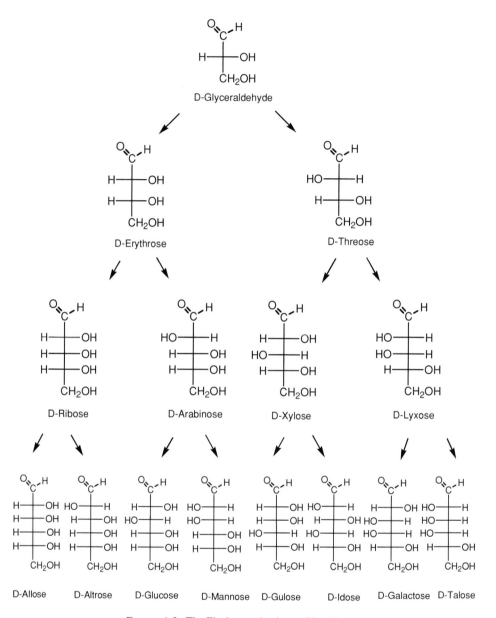

FIGURE 1.3. The Fischer projections of D-aldoses.

1.3.5 Nucleophilic Addition

Aldose and ketone may react with a variety of nucleophiles, giving as results addition/elimination products such as osazones and oximes, or addition products such as reduced derivatives when reacted with hydrides.

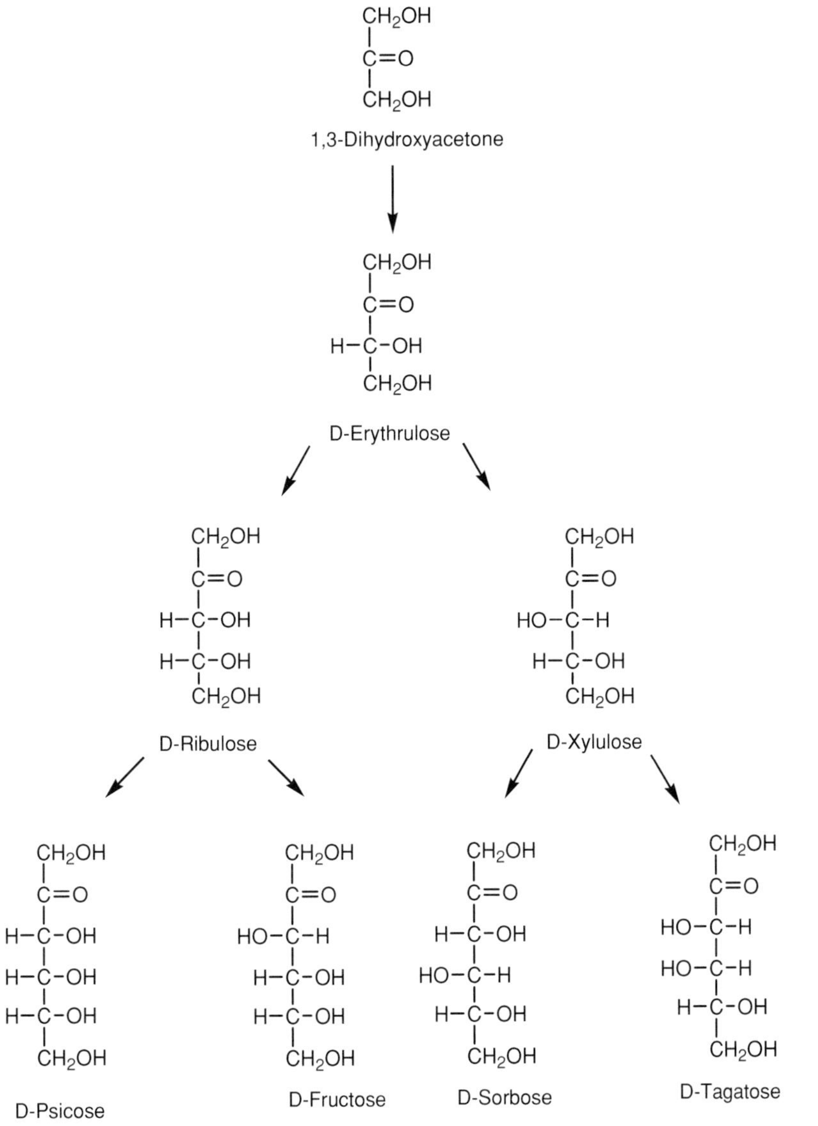

CH$_2$OH
|
C=O
|
CH$_2$OH

1,3-Dihydroxyacetone

↓

CH$_2$OH
|
C=O
|
H−C−OH
|
CH$_2$OH

D-Erythrulose

CH$_2$OH
|
C=O
|
H−C−OH
|
H−C−OH
|
CH$_2$OH

D-Ribulose

CH$_2$OH
|
C=O
|
HO−C−H
|
H−C−OH
|
CH$_2$OH

D-Xylulose

CH$_2$OH
|
C=O
|
H−C−OH
|
H−C−OH
|
H−C−OH
|
CH$_2$OH

D-Psicose

CH$_2$OH
|
C=O
|
HO−C−H
|
H−C−OH
|
H−C−OH
|
CH$_2$OH

D-Fructose

CH$_2$OH
|
C=O
|
H−C−OH
|
HO−C−H
|
H−C−OH
|
CH$_2$OH

D-Sorbose

CH$_2$OH
|
C=O
|
HO−C−H
|
HO−C−H
|
H−C−OH
|
CH$_2$OH

D-Tagatose

FIGURE 1.4. Fischer projections of the 2-ketoses.

The reaction that allowed E. Fischer to determine the structure of the common aldoses was the osazone formation and consisted in the reaction between hydrazine with aldoses (Figure 1.10) to yield crystalline derivatives that can be identified through their melting points values.

The carbonyl group can be reduced by hydrogenation or hydride addition to produce the corresponding alditols (Figure 1.11). These reduced sugars are present

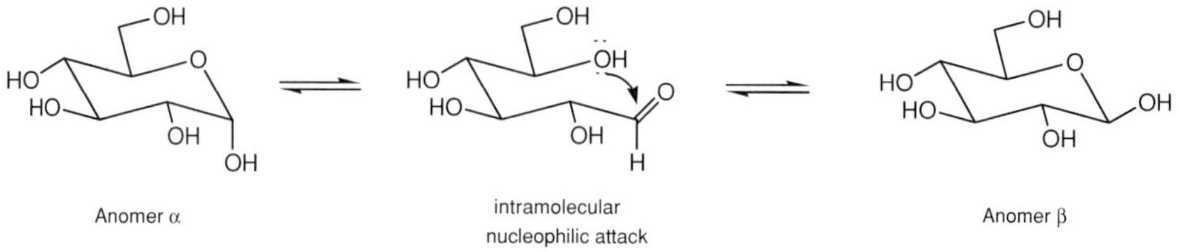

FIGURE 1.5. The pyranose ring formation.

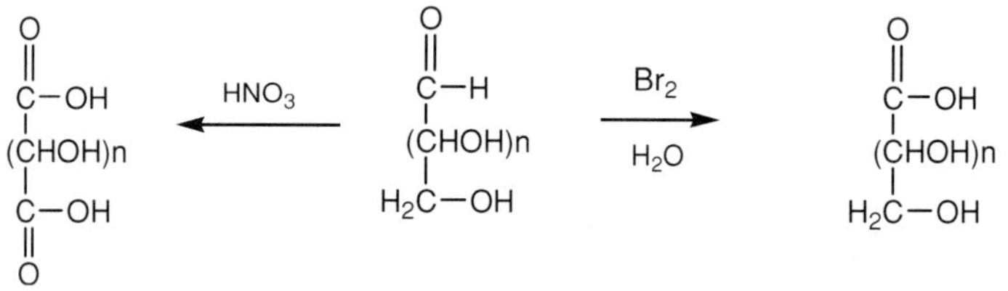

FIGURE 1.6. Oxidative aldose transformation into mono- and dicarboxylic acids.

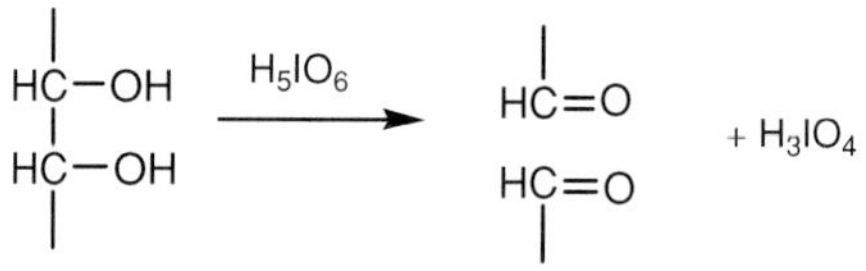

FIGURE 1.7. Oxidative cleavage of diol by periodic acid.

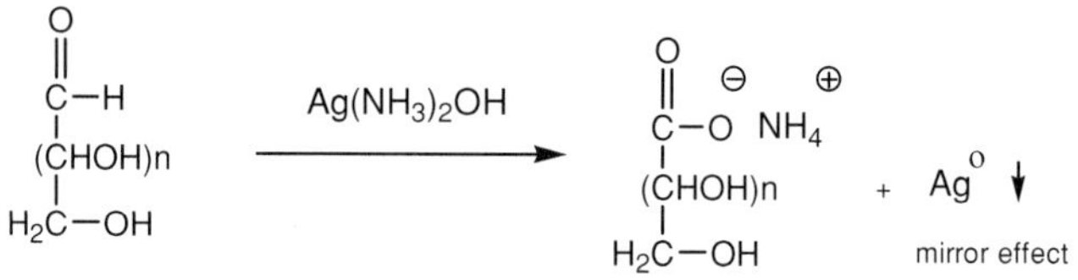

FIGURE 1.8. Tollens reaction.

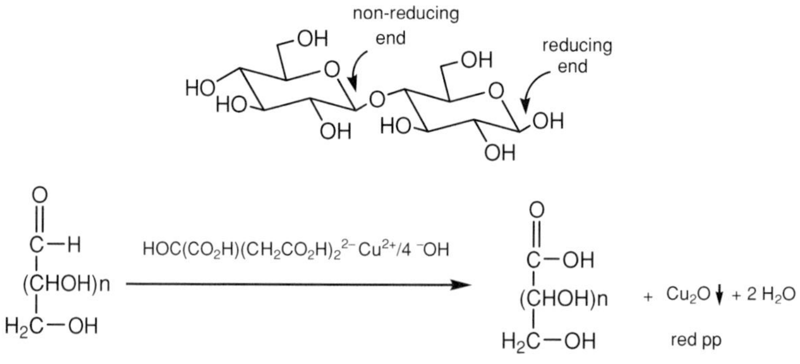

FIGURE 1.9. Benedict and Fehling test.

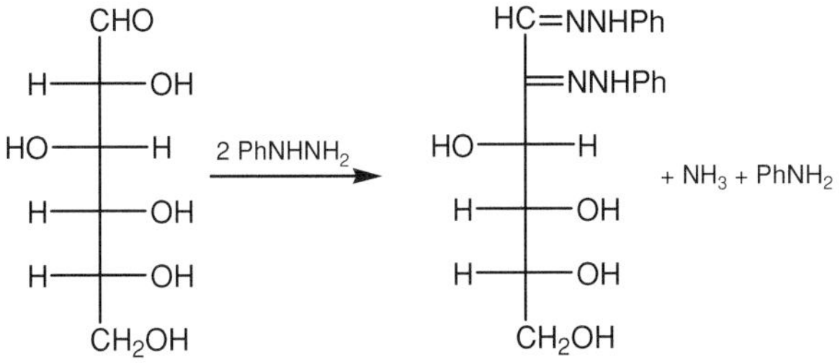

FIGURE 1.10. Osazone formation.

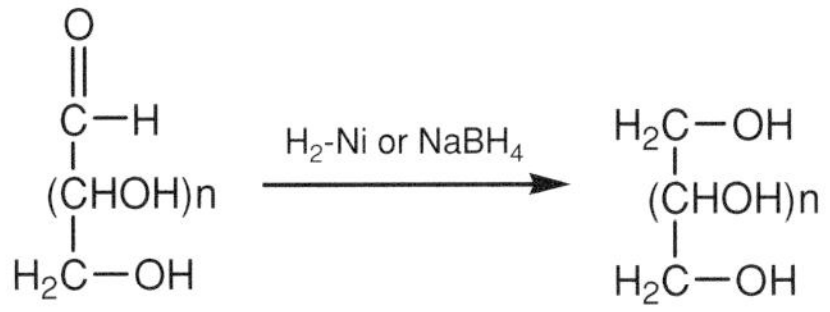

FIGURE 1.11. Carbonyl reduction for the preparation of sorbitols.

in various fruits such as cherries, pies, apples, etc,, and are used as sugar substitute for diabetics.

1.3.6   Enediol Rearrangement

This transformation occurs at basic medium and allows the conversion of epimers, defined as isomeric forms that differ in the position of the hydroxyl group at C-2. In this way it is possible to transform through the enediol intermediate glucose to mannose and vice versa (Figure 1.12).

Another important isomerization process through the enediol rearrangement is the interconversion of glucose and fructose. Thus, the enolization proceeds by migration of proton at position 2, to carbon at 1 (Figure 1.13).

1.3.7   KilianiFischer Synthesis

This sequence was used to increase the number of carbons in a sugar. The reaction involves cyanohydrin formation by nucleofilic addition of cyanide to the aldehyde.

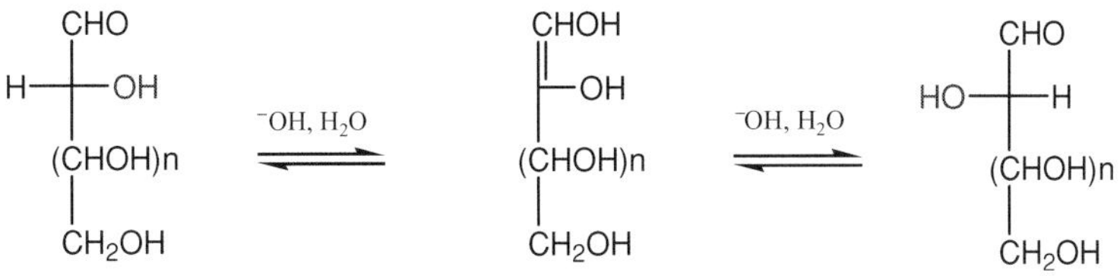

FIGURE 1.12. The enediol rearrangement.

CHOH
‖
—OH   ⁻OH, H₂O
(CHOH)₃   ⇌
|
CH₂OH

CH₂OH
|
=O
|
(CHOH)₃
|
CH₂OH

FIGURE 1.13. The enediol rearrangement.

The diastereoisomeric mixture of cyanohydrins obtained was partially reduced to produce the epimeric forms (Figure 1.14).

1.3.8   The Ruff Degradation

The process of reducing the monosaccharide skeleton in one carbon is known as Ruff degradation and consists in the oxidation of the aldehyde to the carboxylic acid through the use of calcium salt and subsequent peroxide treatment in the presence of ferric salts to produce the aldose reduced in one carbon (Figure 1.15).

1.3.9   Conversion of Pentose to Furfural

Pentoses subjected to high acid concentrations can be transformed to furfural in quantitative yields. The sequence involves a tautomeric keto-enol equilibrium, dehydration and intramolecular nucleophilic addition of the primarily alcohol to the aldehyde to generate furfural (Figure 1.16).

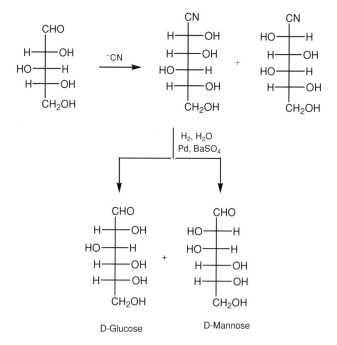

FIGURE 1.14. The Kiliani-Fischer synthesis.

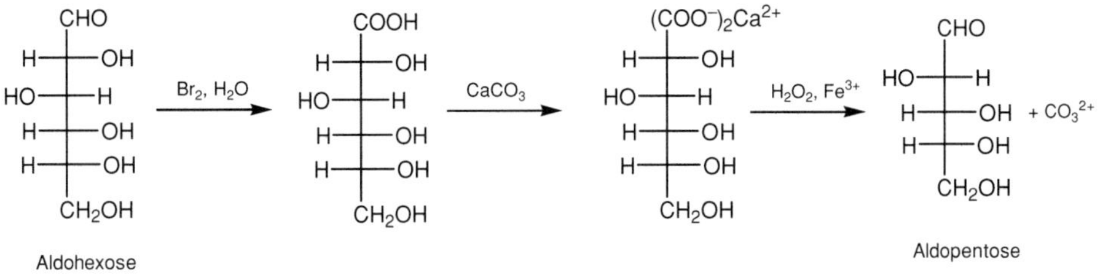

FIGURE 1.15. Ruff degradation.

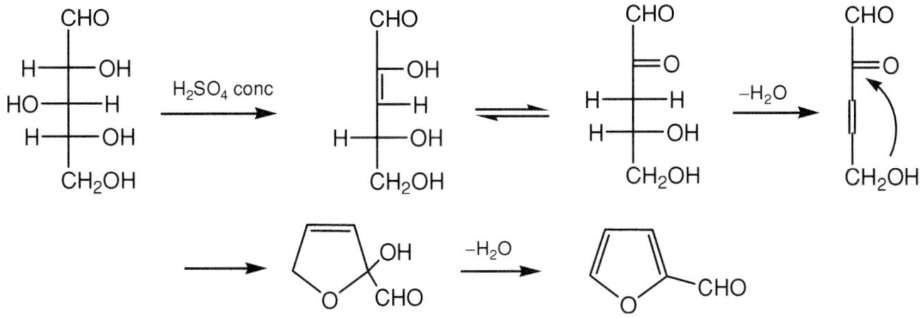

FIGURE 1.16. Conversion of pentoses to furfural.

1.4   Biosynthesis of Sugars

The synthesis of carbohydrates in plants occurs through a mechanism of carbon dioxide fixation, and was understood through the use of long-lived radioactive isotope of carbon ^{14}C. After considerable investigations it was founded that the initial CO_2 acceptor was the five-carbon compound ribulose 1,5-bis-phosphate (RuBP) which after incorporation of carbon dioxide produce a six-carbon molecule. The resulting molecule is fragmented into two molecules of 3-phosphoglycerate (PGA) that is one of the intermediates of glycolysis. This transformations takes place in the chloroplast by a large multisubunit enzyme, ribulose biphosphate carboxylase "Rubisco". The following reaction sequence is cyclic and constitutes what is known as the Calvin cycle which consists after formation of PGA in reduction to glyceraldehydes 3-phosphate (GAP), and regeneration of RuBP. The overall process requires 6 CO_2 molecules fixed, 12 molecules of GAP produced which rearrange to regenerate 6 molecules of the five-carbon CO_2 acceptor RuBP (Figure 1.17).

1.4.1   Sugars as Energy Sources

Metabolically the main monosaccharide useful for the production of energy is glucose. During the glycolysis process glucose is enzymatically transformed and degraded to piruvate and is a preamble which is further introduced into the Krebs cycle.

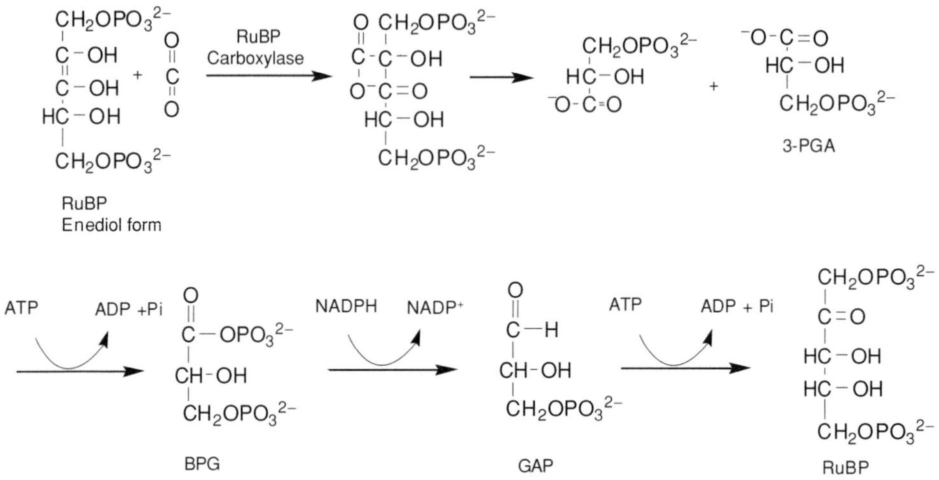

FIGURE 1.17. Carbohydrate synthesis from CO_2 fixation.

Carbohydrates are responsible of several biological events mainly related with the storage and production of energy, as metabolic intermediates and signal molecules. They are also constitutive structural units of essential biomolecules such as polysaccharides (starch, glucogen, and cellulose), glycoproteins, glycolipids, and nucleotides. The process by which glucose is used as energy source, to produce ATP and pyruvate is known as glycolysis and consist in series of events represented in Figure 1.18.

The second cycle of glycolysis is divided in four steps.

1.5   Synthesis of Carbohydrates

The chemical synthesis of carbohydrates can be accomplished by chemical, enzymatic or the combined approach (chemoenzymatic). Their preparation by either of the mentioned methods has received considerable attention especially because they can be used as starting materials for the synthesis of biologically active carbohydrate derivatives known as mimetics or the synthesis of complex molecules such as oligosaccharides or glycopeptides.

1.5.1   Chemical Synthesis

Access to potentially useful sugar or congeners can be obtained from natural sugars such as arabinose and mannose.[4] Thus, convenient routes have been implemented for the preparation of KDN from D-mannose,[5] 3-deoxy-D-manno-

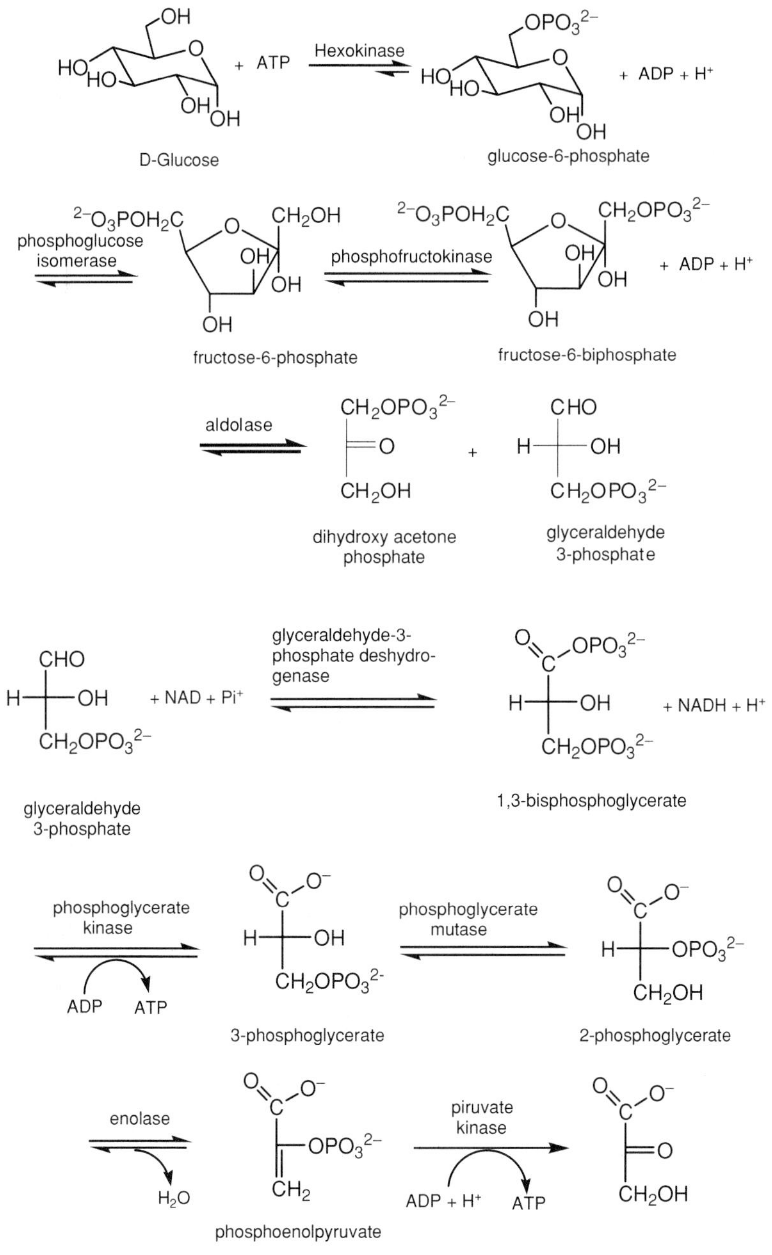

FIGURE 1.18. The glycolysis pathway.

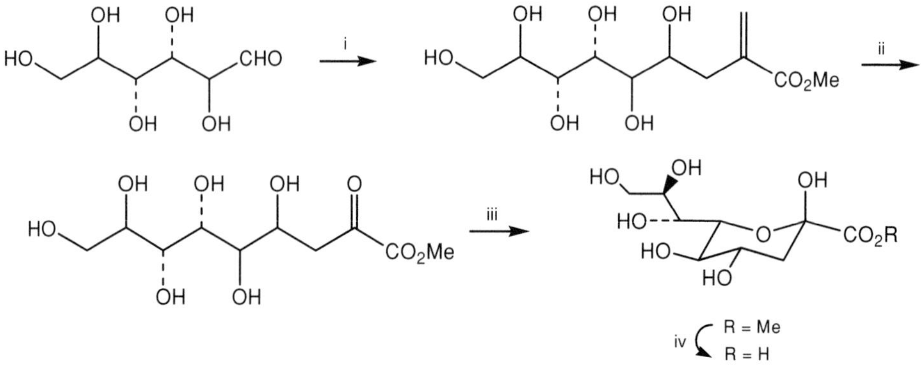

i) methyl 2-(bromomethyl)acrylate, H_2O. ii) O_3, MeOH, −78°C, then Na_2SO_3. iii) spontaneous ciclization. iv) KOH, MeOH.

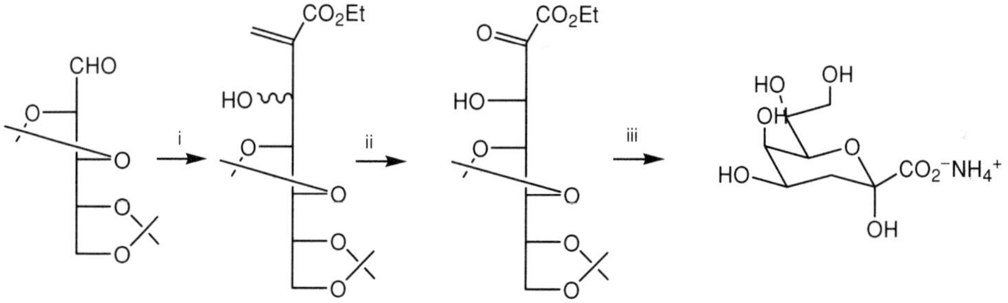

i) ethyl α-(bromomethyl)acrylate,10% formic acid, aq. MeCN. ii) O_3, MeOH, −78°C, then Me_2S, MeOH, −78°C to r.t.. iii) aq. TFA, then NH_4OH.

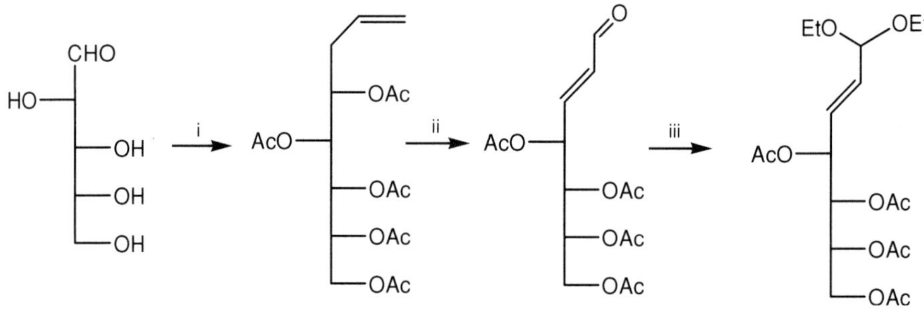

FIGURE 1.19. Chemical synthesis of sugar congeners from natural sugars.

2-octulosonic acid (KDO) from 2,3:4,5-di-O-isopropylidene-D-arabinose,[6] D-*glycero*-D-galacto-heptose from D-arabinose,[7] and KDN from D-mannose[8] (Figure 1.19).

Different approximations for the preparation of monosaccharides from other sources have been reported. One method consists in the asymmetric synthesis of D-galactose via an iterative *syn*-glycolate aldol strategy. The general method is shown in Figure 1.20.[9]

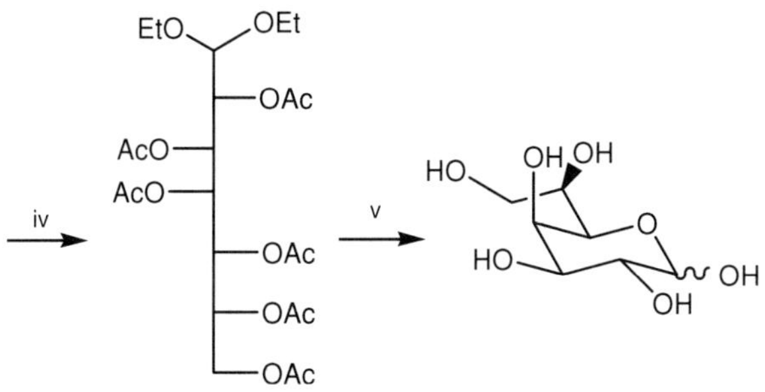

i) allyl bromide, ultrasonication, then Ac$_2$O, Py, DMAP. ii) OsO$_4$, KIO$_4$, then TBAF.
iii) H$_3$O$^+$, (HC(OEt)$_3$. iv) OsO$_4$, NMO, then Ac$_2$O, Py, DMAP. v) NaOMe, MeOH,
then H$_3$O$^+$.

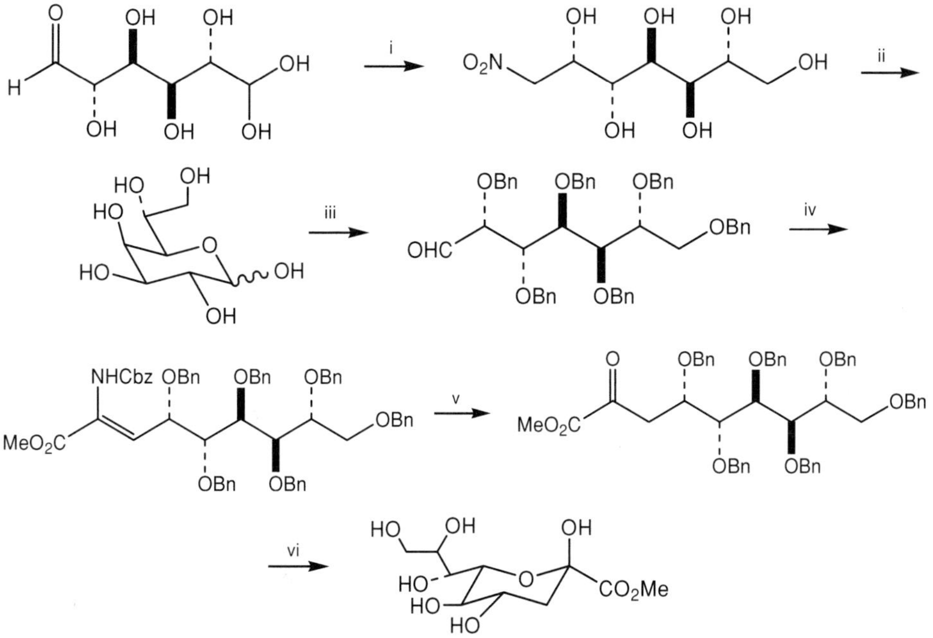

i) MeNO$_2$, DBU. ii) Nef oxidation iii) EtSH, HCl, then NaH, BnBr, DMF, then MeI, Na$_2$CO$_3$. iv) (Et)$_2$P(O)CH(NHCBz)CO$_2$Me,
NaH, CH$_2$Cl$_2$. v) H$_2$, Pd-C. vi) H$_2$, Pd(OH)$_2$, then Dowex H$^+$, MeOH.

FIGURE 1.19. (*continued*)

A promising and simple concept based on a two-step reaction sequence for
preparing monosaccharides via the enantioselective organocatalytic direct aldol
reaction of α-oxyaldehydes is recently described. The summarized sequence is
illustrated in Figure 1.21.[10]

An interesting strategy for preparing KDO and 2-deoxy-KDO from 2,3-O-
isopropylidene-D-glyceraldehyde was reported, based on a hetero Diels-Alder

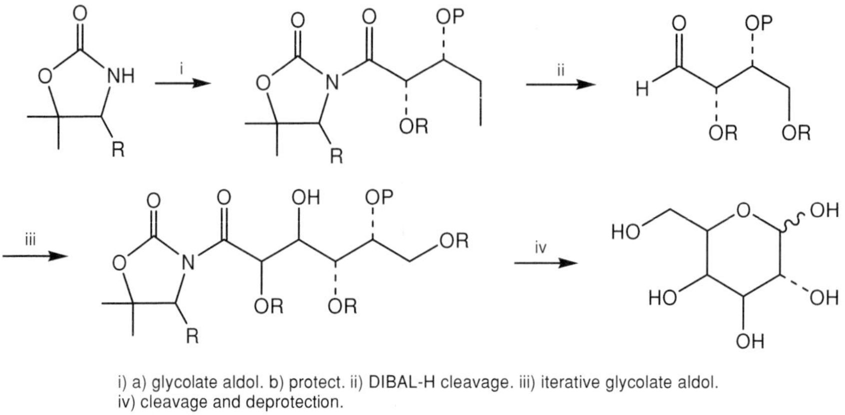
i
ii
iii
iv
NH
O
R
N
OP
OR
H
OH
HO
i) a) glycolate aldol. b) protect. ii) DIBAL-H cleavage. iii) iterative glycolate aldol.
iv) cleavage and deprotection.

FIGURE 1.20. Asymmetric synthesis of D-galactose.

reaction, followed by pyranoside ring formation. Diol formation and double inversion at C-4 and C-5 produced the target molecules (Figure 1.22).[11]

C-methyl heptoses were suitable prepared from nonracemic butenolide as starting material. Asymmetric conjugate addition provided protected lactone which by methylation afforded α-methyl lactone. Each of them under DIBALH treatment, produced C-methylheptoses (Figure 1.23).[12]

Naturally occurring sugar amino acids are another class of interesting modified carbohydrates found as structural components in nucleoside antibiotics. Most of them consist of N- and O-acyl derivatives of neuraminic acids, while other presents a ipso-hidantoin furanosides (Figure 1.24).[13]

Some of these sugars amino acids have been synthesized via azide furanosides,[14,15] as it was the case for β-sugar amino acids shown in Figure 1.25.[13]

1.5.2 C-Glycosyl Amino Acids

It has been mentioned that natural glycopeptides are classified into *O*-glycopeptides when the sugar residue establishes an *O*-glycosyl linkage with *L*-Serine or *L*-Threonine and *N*-glycopeptides if the linkage is with Asparagine. There has been an increasing interest for preparing unnatural *C*-glycol amino acids

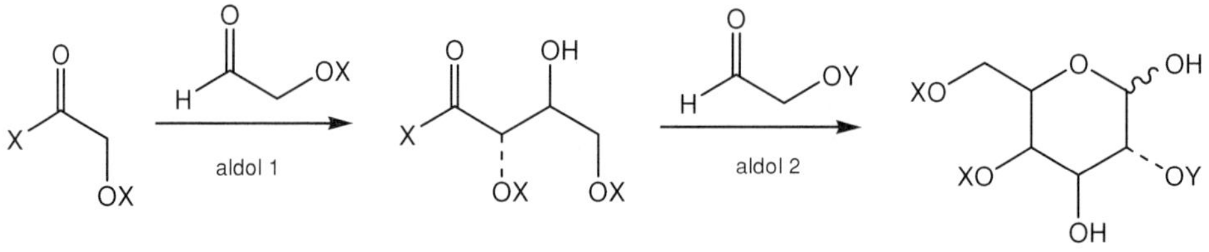
X
O
OX
H
aldol 1
OH
OY
aldol 2
XO
OH

FIGURE 1.21. Two-step carbohydrate synthesis.

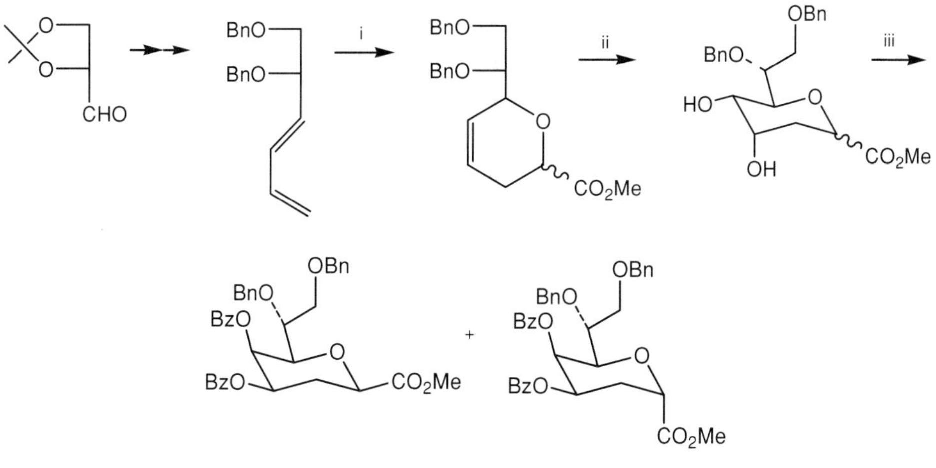

i) HCOCO$_2$Bu, 130°C, then MeOH, TsOH. ii) OsO$_4$, NMO. iii) Tf$_2$O, Py, then PhCO$_2$NBu$_4$.

FIGURE 1.22. Synthesis of protected KDO and 2-deoxy-KDO.

as a potential building block in the assembling of modified glycopeptides that may serve in preparing therapeutically useful mimetics, with higher resistance to hydrolytic enzymes and also displaying superior properties than the natural ones.

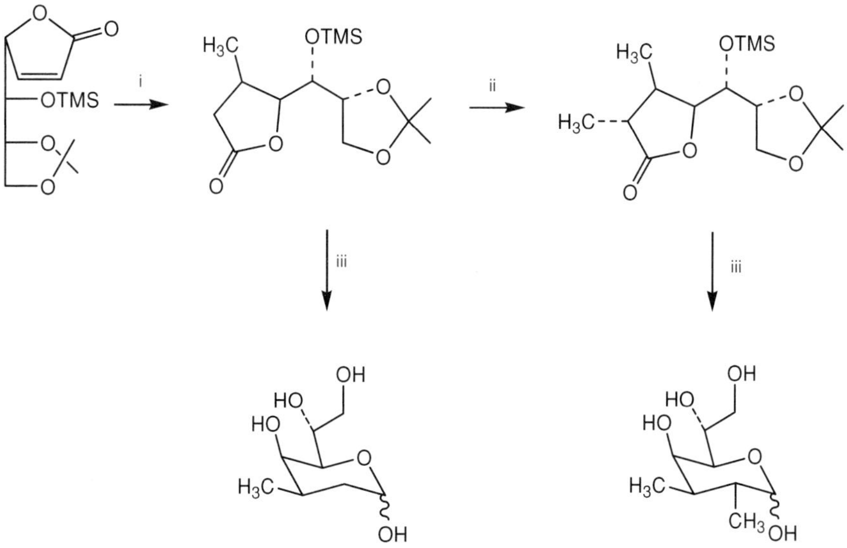

i) Me$_2$CuLi, CH$_2$Cl$_2$. ii) LiHDMS, THF, then MeI. DIBAlH, CH$_2$Cl$_2$, then 3N HCl, THF.

FIGURE 1.23. Synthesis of-methyl heptoses.

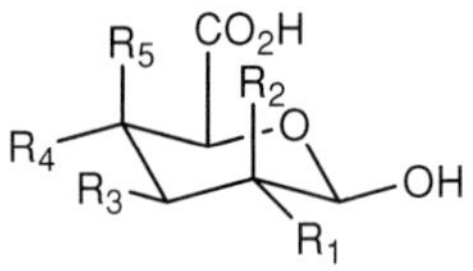

	R^1	R^2	R^3	R^4	R^5
glucosaminuronic acid	NH_2	H	OH	OH	H
galactosaminuronic acid	NH_2	H	OH	H	OH
monnosaminuronic acid	H	NH_2	OH	OH	H
4-amino-4-deoxy-glucuronic acid	OH	H	OH	NH_2	H

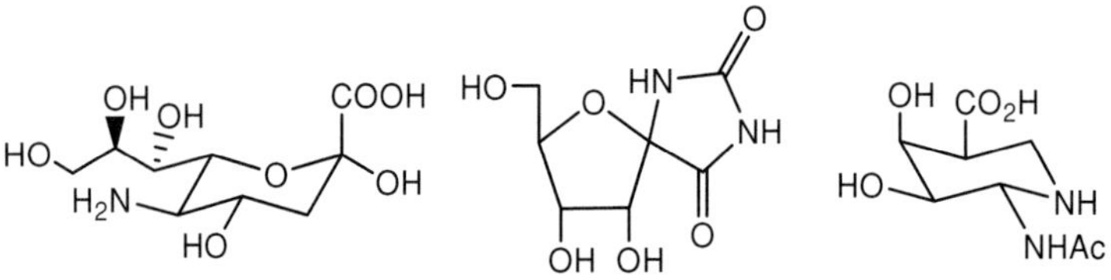

Neuraminic acid   Hydantocidin   Siastatin B

FIGURE 1.24. Naturally occurring sugar amino acids.

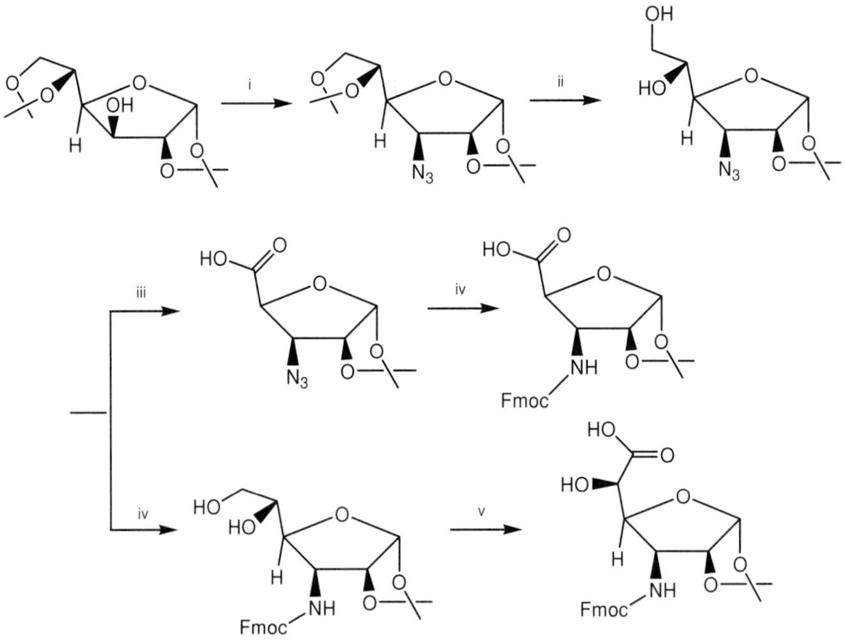

i) a) Tf$_2$O, Py. b) NaN$_3$, Bu$_4$NCl (cat.), 69%. ii) 77% AcOH, quant. iii) a) NaIO$_4$. b) KMnO$_4$, 50% AcOH, 90%. iv) H$_2$, Pd/C, FmocCl, NaHCO$_3$. v) NaOCl, TEMPO (cat.), KBr, NaHCO$_3$, Bu$_4$NCl, 62%.

FIGURE 1.25. Synthesis of protected sugar amino acids.

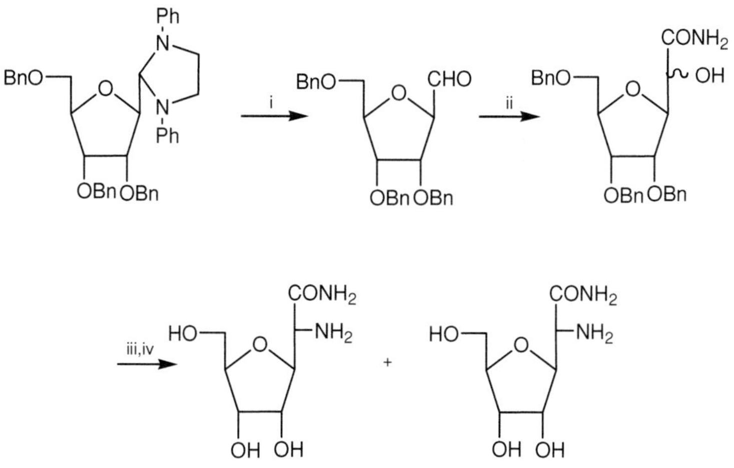

i) TsOH. ii) a) NaCN, K_2CO_3, H_2O. b) H_2O_2. 91%. iii) a) MsCl. b) LiN_3. iv) a) aq. HCl. b) H_2, Pd/C.

FIGURE 1.26. Synthesis of anomeric ribofuranosyl glycines.

A recent review describes methods for the preparation of C-glyosyl glycines, alanines, serines, asparagines tyrosines, and tryptophans.[16]

For instance the synthesis of ribofuranosyl glycine was described under Strecker conditions, starting from 2-(2,3,5-tri-O-benzyl-β-D-ribofuranosyl)-1,3-diphenylimidazolidine, which was after hydrolysis tosylated and reacted with cyanide and peroxide to give the α-hydroxy amide as racemic mixture. The anomers were separated as O-mesyl derivatives which were transformed to the azide and further reduced to the corresponding ribofuranosyl glycines (Figure 1.26).[17]

A method reported for the preparation of C-glycosyl alanines involves the use of (R)-methyleneoxazolidinone, which was linked to the peracetylated iodosugars under promoted radical additions. The α-linked C-glycoside was subjected to hydrogenolysis to give α-D-galactosyl D-alanine and the α-D-glucosyl isomer (Figure 1.27).[18]

C-analogues of glycosyl serines have been prepared by a number of methods and among them Strecker, Witting, and Sharpless asymmetric aminohydroxylation reactions[19]. One of them describes their synthesis via coupling of anomeric pyridyl sulfone with an electrophiles center under samarium catalysis. The resulting C-glycosylation proceeded with α-selectivity (3.3:1). Final deprotection produced the C-glycosyl serine analogue in good yield (Figure 1.28).[20]

More recently, the stereoselective synthesis of a C-glycoside analogue N-fmoc-serine β-N-acetylglucosaminide was described employing the Ramberg-Bäcklung (RB) rearrangement. This procedure involves the coupling reaction between isothiourea and protected iodide to produce thioglycoside in good yield. Oxidation to the sulfone was followed by the RB conditions (KOH/Al_2O_3 in tBuOH/($CBrF_2)_2$

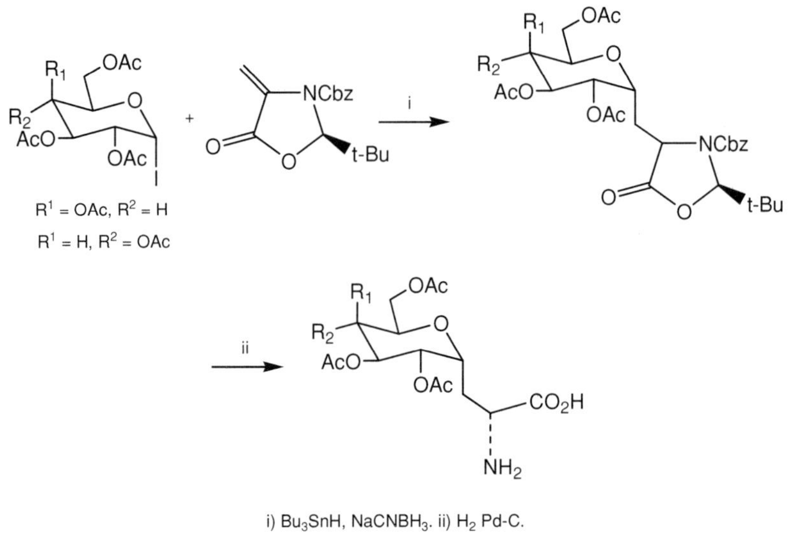

i) Bu$_3$SnH, NaCNBH$_3$. ii) H$_2$ Pd-C.

FIGURE 1.27. Synthesis of α-D-galactosyl D-alanine and the α-D-glucosyl isomers.

at 50°C affording the exoglycal derivative. Final steps which involves hydrogenolysis, deprotection, and oxidation provided the desired C-glycosyl analog (Figure 1.29).[21]

β-2-Deoxy sugar bearing uridin diphosphate (UDP) is a substrate used by a variety of glycosyltransferases. It is described the chemical synthesis of UDP β-2-deoxysugars starting from α-glycosyl chlorides as indicated in Figure 1.30. These

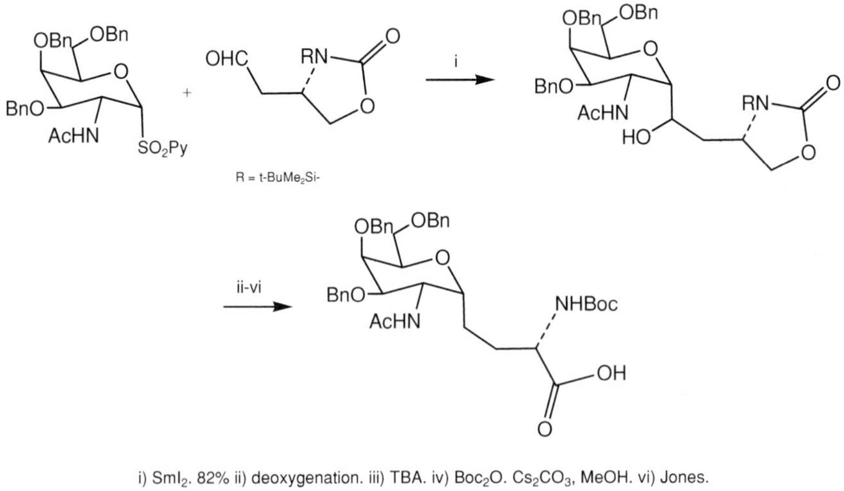

i) SmI$_2$. 82% ii) deoxygenation. iii) TBA. iv) Boc$_2$O. Cs$_2$CO$_3$, MeOH. vi) Jones.

FIGURE 1.28. Synthesis of C-glycosyl serine analogue.

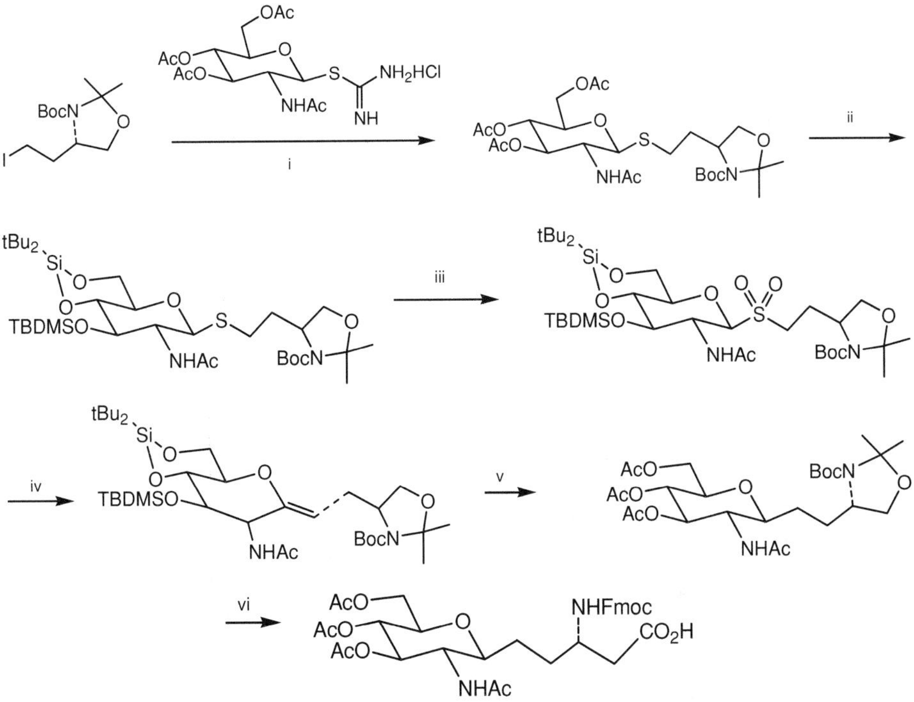

i) K_2CO_3, $Na_2S_2O_5$/acetone-H_2O, 98%. ii) a) Et_3N/MeOH-H_2O. b) $tBu_2Si(OTf)_2$, 2,6-lutidine/DMF, 88%. iii) mCPBA, Na_2HPO_4/CH_2Cl_2, 77%. iv) KOH/Al_2O_3, $CBrF_2CBrF_2$/tBuOH, 50°C, 38%.
v) a) H_2, $Pd(OH)_2$/EtOAc, 78%. b) TBAF/THF. c) Ac_2O/Py, 68% (two steps). vi) a) TFA/$CHCl_3$.
b) FmocCl/iPrNEt/CH_2Cl_2-MeOH, 69% (two steps). c) Jones oxidation/acetone, 77%.

FIGURE 1.29. Synthesis of C-glycoside serine analogue by Ramberg-Bäcklung rearrangement.

useful intermediates are essentials for studying glycosyltransferases involved in the synthesis of biologically active natural products.[22]

1.5.3 Enzymatic Synthesis

The enzymatic synthesis of monosaccharides and carbohydrates mimetics by enzyme catalysts is performed mainly by a group of lyases known as aldolases. This

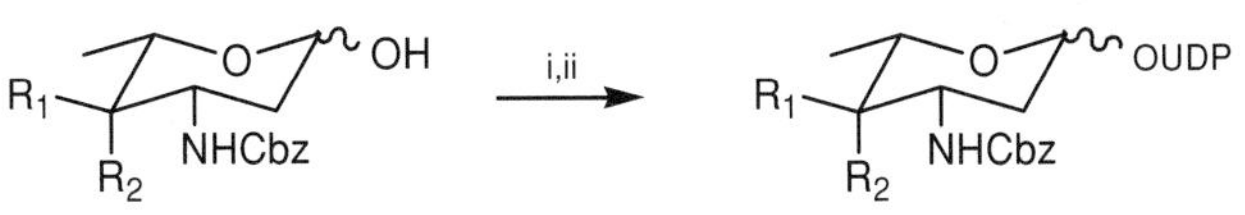

R_1 = OAc, R_2 = H
R_1 = H , R_2 = OAc

α/β 1:2

i) TMS-Cl, TMS-I, –78°C. ii) UDP, Bu_4N salt, 60%.

FIGURE 1.30. Synthesis of protected UDP epi-vancosamine.

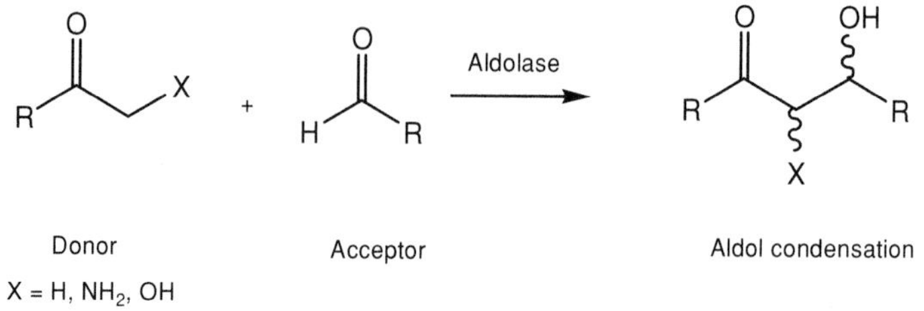

FIGURE 1.31. General scheme of enzymatic-mediated aldol condensation.

enzymes effects the conversion of hexoses from their three-carbon components via an aldol condensation.[23] There are over 30 aldolases identified and isolated, being classified in two types depending on the mechanism involved: aldolase type 1 and type 2, which is Zn-dependent. The general reaction that they catalyze is the stereospecific addition of a ketone donor to an aldehyde acceptor (Figure 1.31).

The aldolases used for synthetic purposes are classified in five groups depending on the ketone donor and the products formed.

Dihydroxyacetone phosphate (DHAP) aldolase
Pyruvate aldolase
2-Deoxyribose 5-phosphate aldolase
Glycine aldolase
Other aldolases

Examples of each of them are indicated in Table 1.2;

Aldolases have been also very useful for the preparation of a variety of common and uncommon monosaccharides. Fructose-1,6-diphosphate (FDP) aldolase effects the conversion of dihydroxyacetone phosphate (DHAP) and glyceraldehyde-3-phosphate (G3P) to D-fructose-1,6-diphosphate (FDP). Table 1.2 summarizes the natural substrates, de donors, and the products obtained through this reaction. Broken lines indicates the bond formed or broken.[24]

DHAP-aldolases catalyze the reversible asymmetrical aldol condensation of DHAP to L-lactaldehyde or to D-glyceraldehyde 3-phosphate (G3P). There are four types of DHAP aldolase, which are classified on the basis of the condensation product formed. D-fructose 1,6-diphosphate (D-FDP) aldolase, which condenses DHP with G3P. D-Tagatose 1,6-diphosphate (TDP) which utilizes the same substrates, Fuculose 1-phosphate, catalyzing the condensation reaction between DHAP and L-lactaldehyde to produce L-fucolose 1-phosphate, and L-rhamnulose 1-phosphate aldolase, which recognizes the same substrates to produce l-rhamnulose 1-phosphate (Figure 1.32).[24]

Likewise, DHAP-dependend aldolases are involved in the incorporation of dihydroxyacetone phosphate (DHAP) on pentose and hexose phosphate introducing consequently three carbons and two quiral centers (Figure 1.33).[25]

Another enzymatic aldol type reaction takes place on N-acetylneuraminic acid also known as sialic acid, which after a reversible aldol reaction of N-acetyl-D-

TABLE 1.2. Natural substrates for aldolases.

rhamnulose-1-P aldolase

sialic acid synthetase

sialic acid aldolase

3-deoxy-2-oxo-6-P-
galactonate aldolase

Donor

Donor

Donor

4-hydroxy-2-oxo-
glutarate aldolase

3-deoxy-2-oxo
-L-arabinoate
aldolase

2-deoxyribose
-5-P-aldolase

D-Thr aldolase

4-hydroxy-4-methyl
-2-oxoglutarate
aldolase

3-deoxy-2-oxo
-D-pentanoate
aldolase

L-Thr aldolase

3-deoxy-2-oxo
-D-glucarate
aldolase

hydroxybutyrate
aldolase

Ser-hydroxymethyl
transferase

mannosamine and pyruvate, produces N-acetyl-5-amino-3,5-dideoxy-D-glycero-
D-galacto-2-nonulosonic acid (NeuAc) (Figure 1.34).[26]

Ketoses can be transformed to aldoses through the use of isomerases.[27] In this
way glucose derivatives can be obtained from fructose as shown in Figure 1.35.[28]

1.5.4 Chemoenzymatic Synthesis

The chemoenzymatic approach is a combination of the chemical and the enzymatic methodologies and intends to explode the versatility and availability of the
chemical reagents with the high stereo- and regioselectivity of the enzymes when
they act as catalysts.

For instance, the enzymatic synthesis of dihydroxyacetone phosphate (DHAP)
is too expensive on large scale, and therefore the combined approach becomes

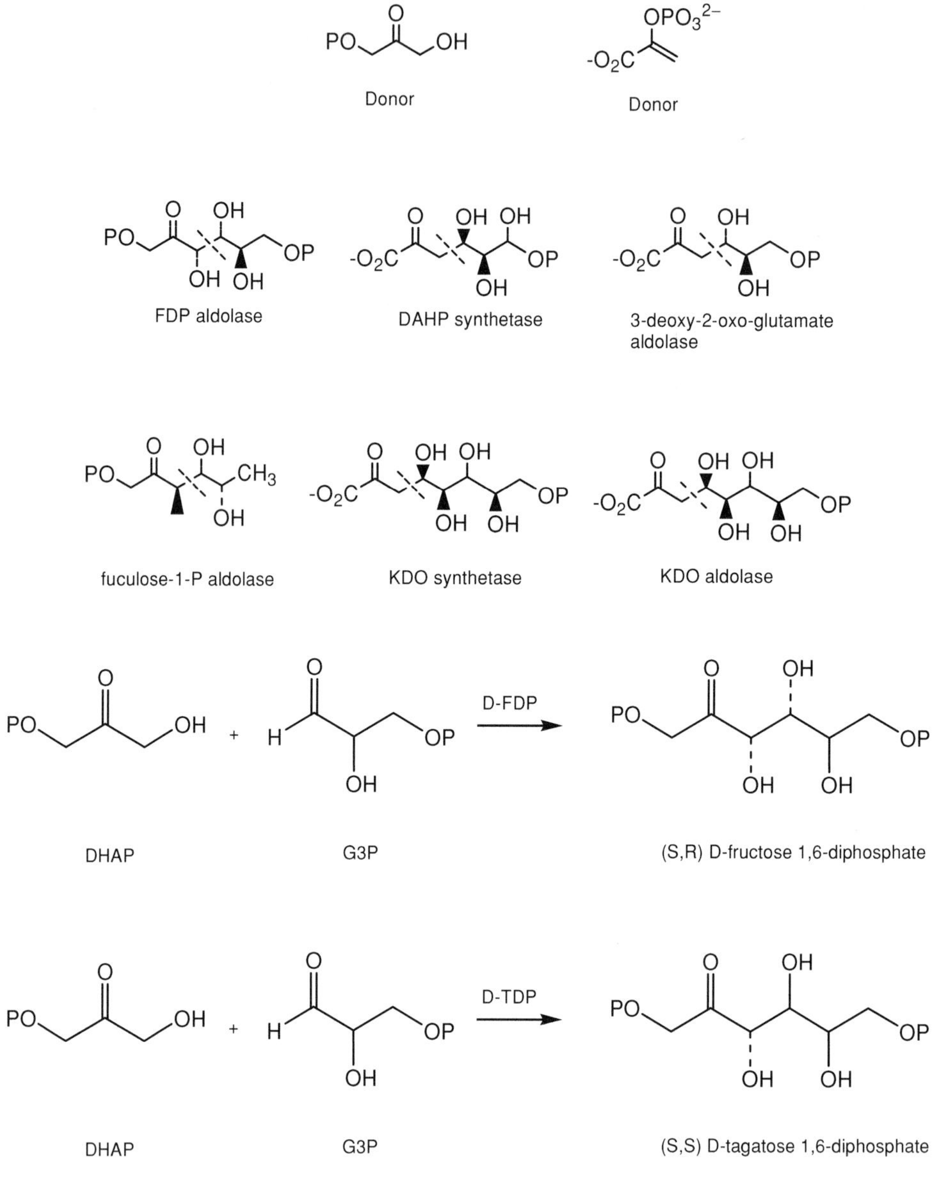

FIGURE 1.32. DHAP-dependent aldolases.

the best choice. The reported procedure consists of the phosphorylation of dihydroxyacetone dimmer with $(PhO)_2POCl$ followed by hydrolysis of the dimmer to generate dihydroxyacetone phosphate in 61% yield.[29] The chemically prepared DHAP is then used as an important material for the synthesis of natural monosaccharides and carbohydrates mimetics (Figure 1.36).

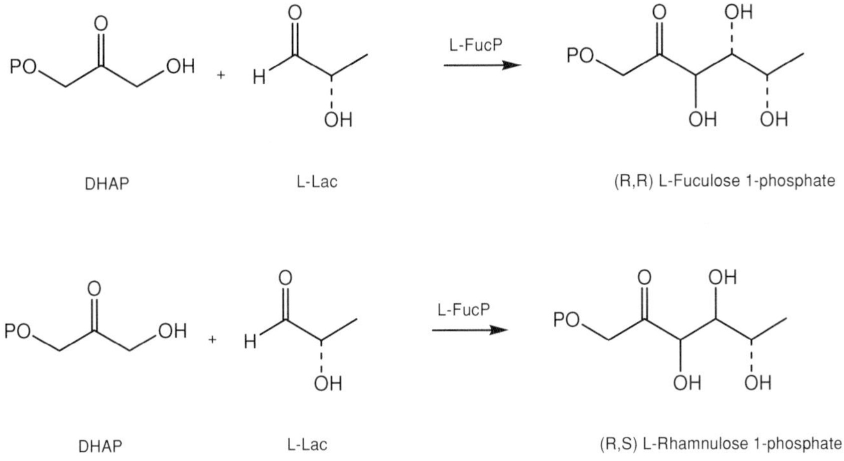

DHAP   L-Lac   (R,R) L-Fuculose 1-phosphate

DHAP   L-Lac   (R,S) L-Rhamnulose 1-phosphate

FIGURE 1.32. (*continued*)

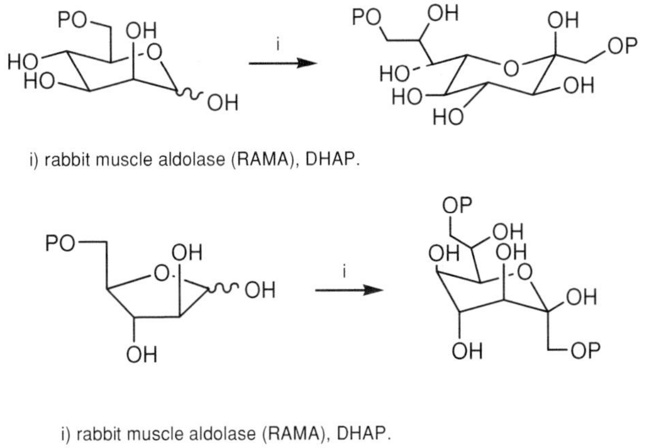

i) rabbit muscle aldolase (RAMA), DHAP.

i) rabbit muscle aldolase (RAMA), DHAP.

FIGURE 1.33. Enzymatic preparation pentose and hexose phosphate.

1.6   Synthesis of Carbohydrates Mimetics

1.6.1   Iminosugars

This class of isosteric sugars also recognized as azasugars have been the subject of intense study because their significant activity as α-glycosidase inhibitors, which is a promising strategy in the treatment against diabetes mellitus type II and other glycosidase associated disorders. It is believed that the mechanism for glycosidase inhibition and to some extent for glycosyltransferases involves the binding of the aza sugars to the active site by charge-charge and hydrogen bond interactions.[30] A significant variety and diversity of either naturally occurring or

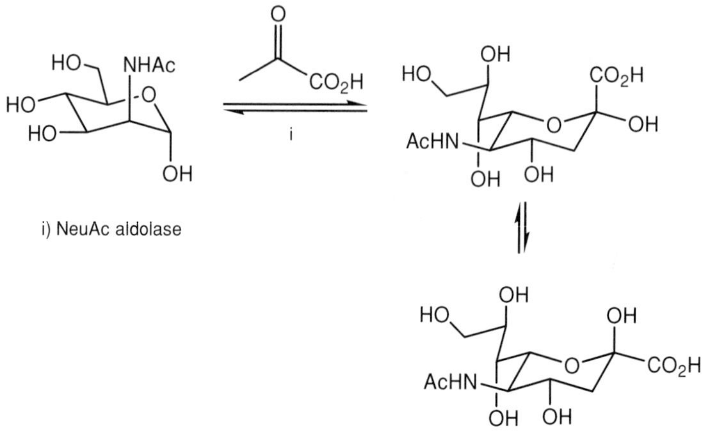

FIGURE 1.34. Enzymatic preparation of sialic acid analogs.

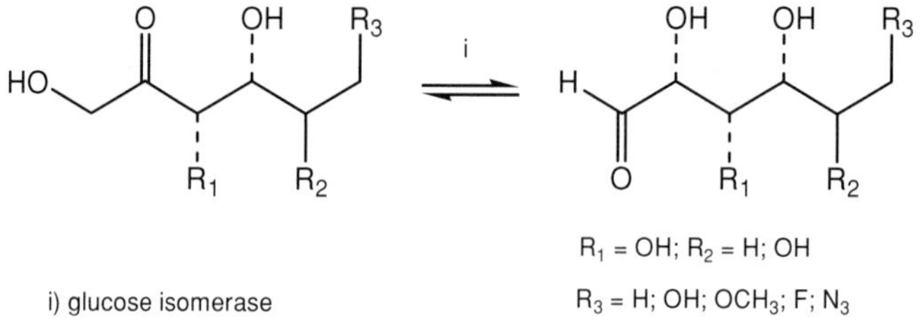

R_1 = OH; R_2 = H; OH

R_3 = H; OH; OCH$_3$; F; N$_3$

i) glucose isomerase

FIGURE 1.35. Enzymatic isomerization of fructose to glucose derivatives.

synthetic aza sugars with glycosidase and glycosyltransferase inhibition activity have been reported.[31–35] The common feature of these derivatives is the replacement by chemical or enzymatic methods of the cyclic oxygen by a nitrogen atom. The representative example is known as deoxynojirimycin (Figure 1.37), which has shown strong inhibition against a variety of α-glycosidases.

Representative examples of natural and synthetic aminoglucosides implicated in inflammation, metastasis, and blocking infection processes are depicted in Figure 1.38.

Series of 1-N-iminosugars including D-glucose-type, D-galactose-type, L-fucose-type, D-glucuronic acid-type, and D-xylose-type were synthesized and evaluated as glycosidase inhibitors (Figure 1.39).

A general procedure for the preparation of 1-N-iminosugars consisted in the azido substitution of a 5-tosyl-1-O-benzoate, followed by aldol reaction, Perlman hydrogenation and cyclization (Figure 1.40).

Another chemical approach described for the preparation of iminosugars consisted in the use of protected L-serinal, which was subjected to Wittig elongation, diol formation, and 2-lithiothiazole treatment, to produce a common thiazole

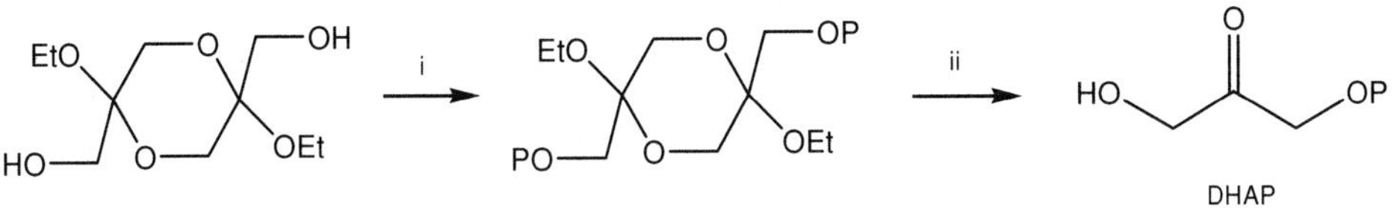

i) a) (PhO)$_2$POCl. b) H$_2$/PtO$_2$. ii) H$_2$O, 65°C.

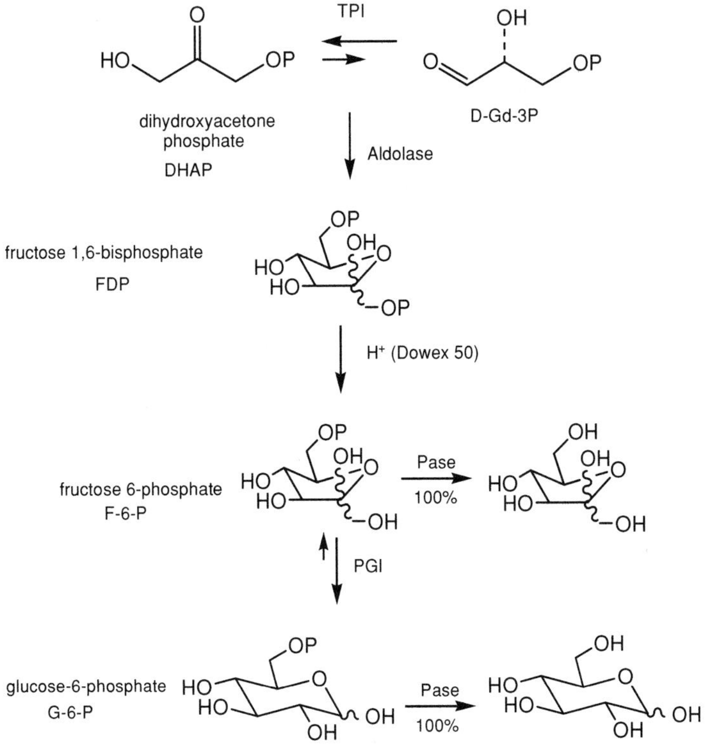

FIGURE 1.36. Chemoenzymatic preparation of glucose.

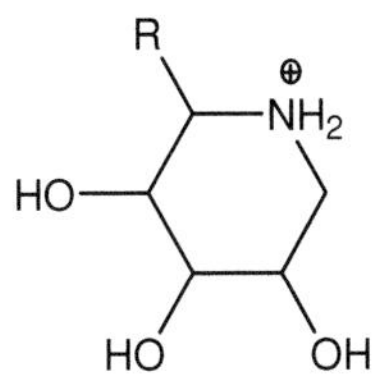

FIGURE 1.37. Iminosugar deoxynojirimycin.

derivative. This intermediate under the appropriate conditions will give access to L-(-)-nojirimycin or L-(-)-mannonojirimycin (Figure 1.41).[36]

Chemoenzymatic preparation of glycosidase inhibitors Deoxynojirimycin and Deoxymannojirimycin was described by using RAMA-aldolase for the aldol

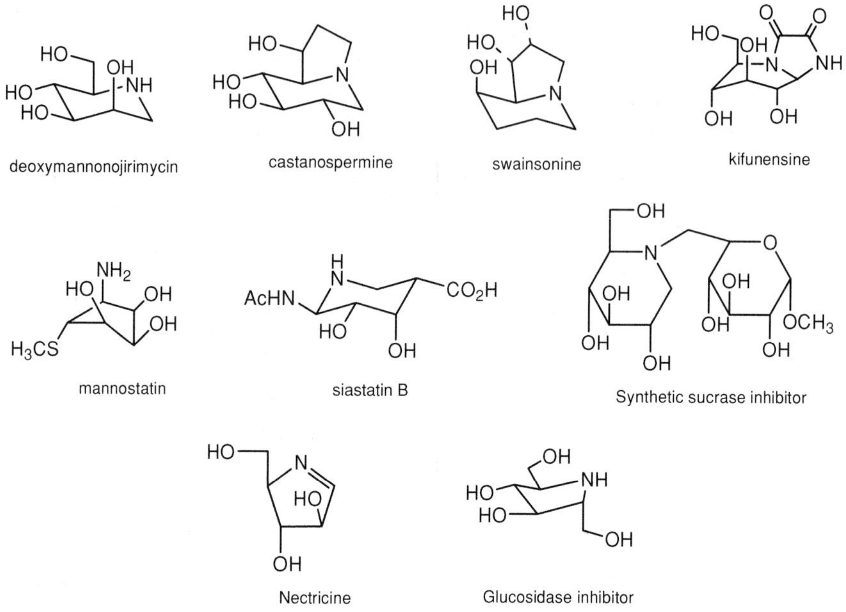

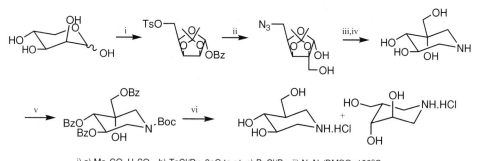

FIGURE 1.38. Representative aminosaccharides.

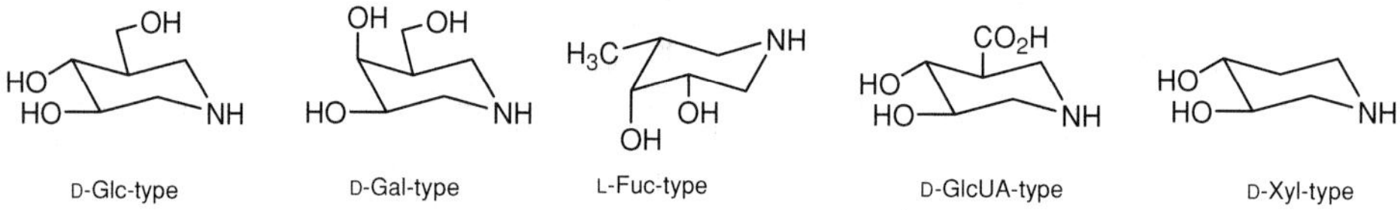

FIGURE 1.39. Series of 1-N-iminosugars.

i) a) Me₂CO, H₂SO₄. b) TsCl/Py, 0oC to r.t. c) BzCl/Py. ii) NaN₃/DMSO, 100°C.
iii) a) NaOMe/MeOH. b) K₂CO₃/HCHO-MeOH. iv) a) H₂Pd(OH)₂/MeOH. b) 1N HCl.
v) a) Boc₂O/Et₃N/MeOH. b) BzCl/Py. vi) a) MeOCOCOCl/DMAP/CH₃CN. b) Bu₃SnH
VAZO/CH₃Ph. c) SiO₂-iPrOH/H₂O/NH₂OH.

FIGURE 1.40. Synthesis of 1-N iminosugars.

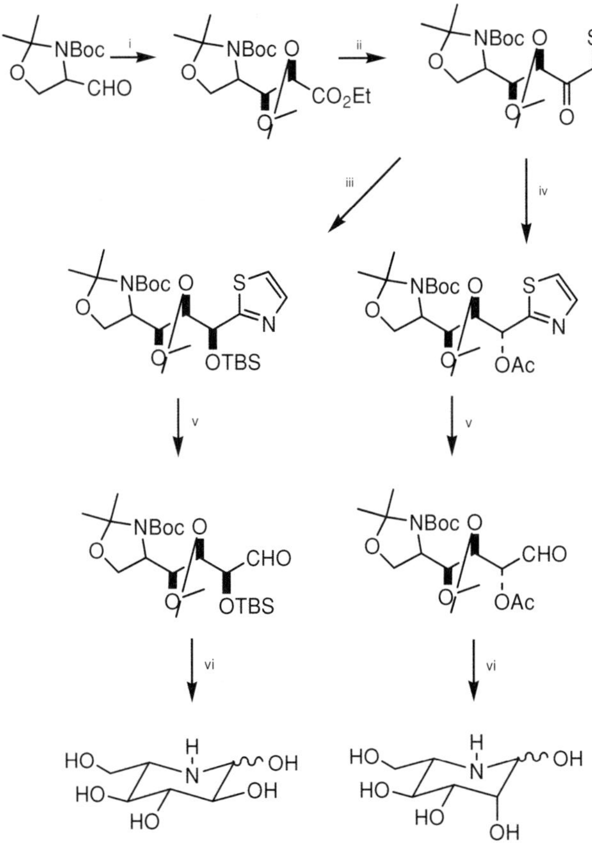

i) a) Ph_3PCHCO_2Et, b) OsO_4, NMO, c) DMP, TsOH. ii) 2-lithiothiazole, Et_2O iii) a) $NaBH_4$. b) TBSCl, imidazole iv) a) Red-Al, toluene, b) Ac_2O, Py, DMAP. v) a) MeI, MeCN. b) $NaBH_4$ c) $HgCl_2$, MeCN, H_2O. vi) TFA, H_2O.

FIGURE 1.41. Chemical synthesis of L-(-)-nojirimycin and L-(-)mannonojirimycin.

condensation and hydrogenolysis for azide reduction and ring formation (Figure 1.42).[37]

Significant achievements have been made for the synthesis of aza sugars based on aldolase reactions particulary fructose-1,6-diphosphate,[30] 2-deoxyribose-5-phosphate,[38] fuculose-1-phosphate,[39] sialic acid aldolase, and Pd/C-mediated reductive amination (Figure 1.43).

1.6.2 Aminosugars

Aminosugars are another class of naturally and nonnaturally sugars which might be considered distinct to the previous in that the nitrogen is exocyclic. Their significance is clearly seen in a family of aminoglycoside antibiotics such as neomycin, kanamycin which are widely used against both gram-positive and gram-negative

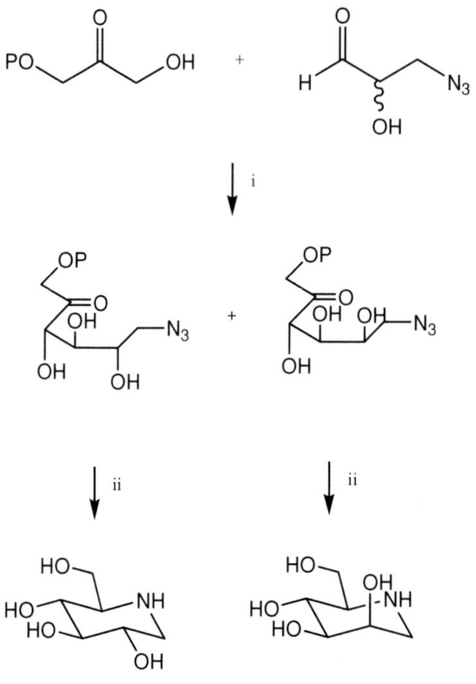

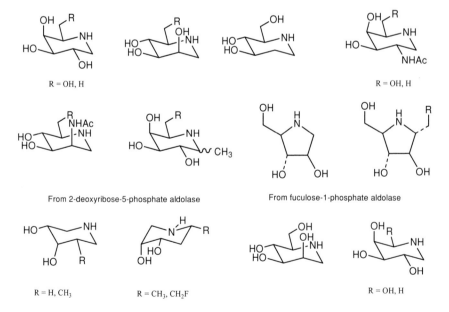

FIGURE 1.42. Chemoenzymatic synthesis of iminosugar.

i) FDP aldolase. ii) a) Pase. b) H₂/Pa

From fructose-1,6-diphosphate aldolase

R = OH, H

R = OH, H

From 2-deoxyribose-5-phosphate aldolase

From fuculose-1-phosphate aldolase

R = H, CH₃

R = CH₃, CH₂F

R = OH, H

FIGURE 1.43. Aza sugars prepared by aldolase reactions.

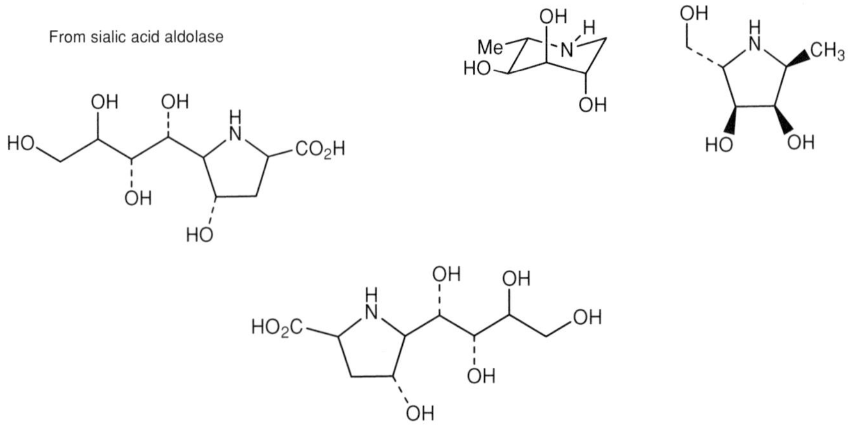

FIGURE 1.43. (*continued*)

bacteria. Although there is no unified protocol for the synthesis of aminosugars, they have been roughly classified in (a) non-azido (Figure 1.44) and (b) azido approaches (Figure 1.45).[40]

(a) The non-azido methodologies usually involves the introduction of an amino group at C-2, and glycals are usually the starting materials.

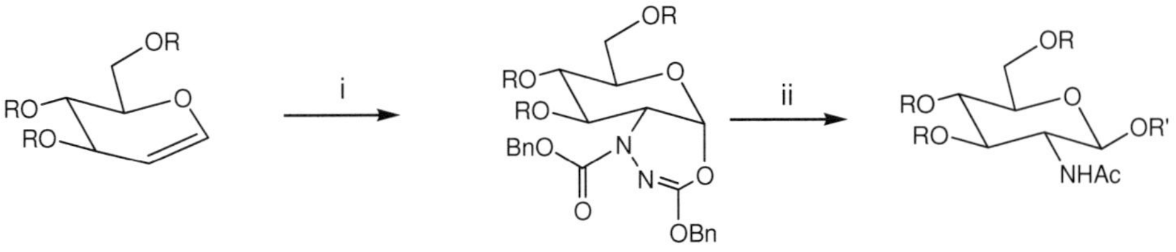

i) $BnO_2C-N{=}N-CO_2Bn$, hv. ii) a) R'OH, Lewis acid b) Raney Ni. 3) Ac_2O, Py

ref.[41]

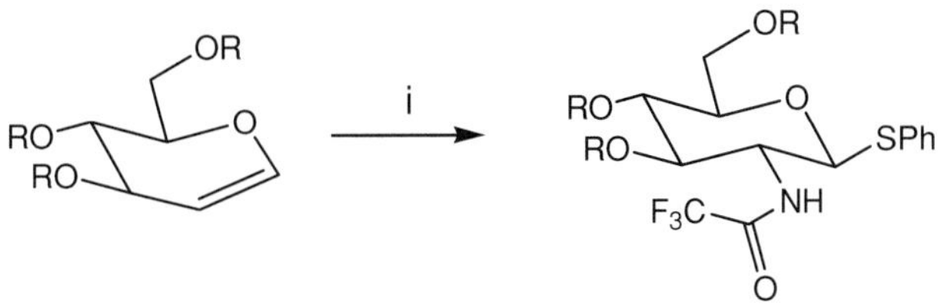

i) $(saltmen)Mn(N)(CF_3CO)_2O$. ii) $PhSH/BF_3\text{-}OEt_2$

ref.[42]

FIGURE 1.44. Non-azido methods for the preparation of aminosacharides.

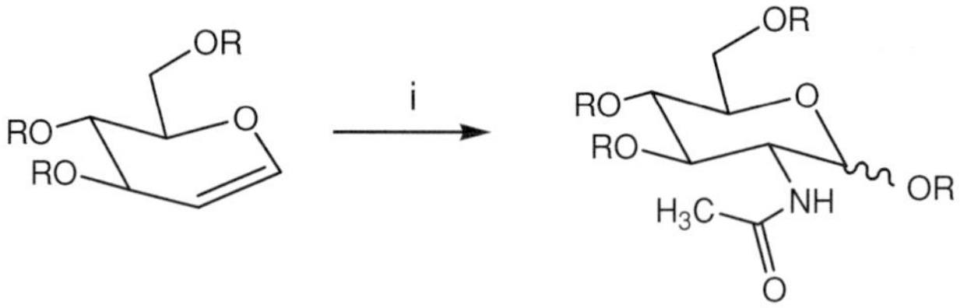

i) a) thianthrene -5-oxide, AcNHSiMe₃, Tf₂O, Et₂NPh. b) Amberlyst -15, HOR.

ref.[43]

FIGURE 1.44. (*continued*)

(b) The azido approach is a more common procedure for amino introduction on sugars due its relative stability, good solubility in organic media, and easy conversion to amines through catalytic hydrogenolysis. Some of the methods reported involves the use of glycals, or protected saccharides containing free primary or secondary alcohols.

Epimerization of hydroxyl groups can be achieved by following an oxidation-reduction sequence in which a secondary alcohol is converted into a keto group, followed by stereoselective hydride reduction and nucleophilic substitution. It has been observed that epimerization by following the Mitsunobu protocol has not been

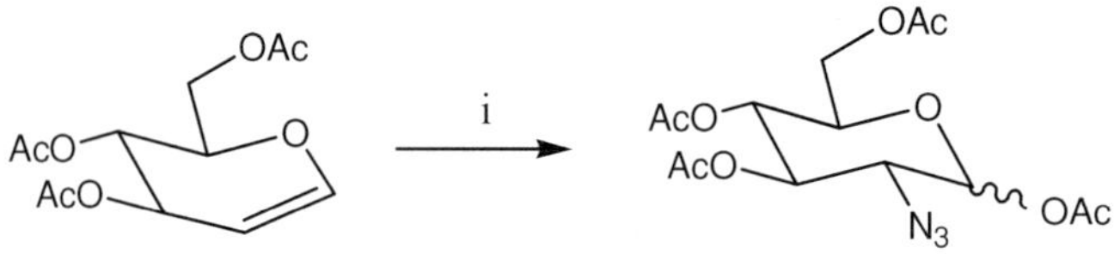

i) ClN₃, CH₂Cl₂ (PhCO₂)₂

ref.[44]

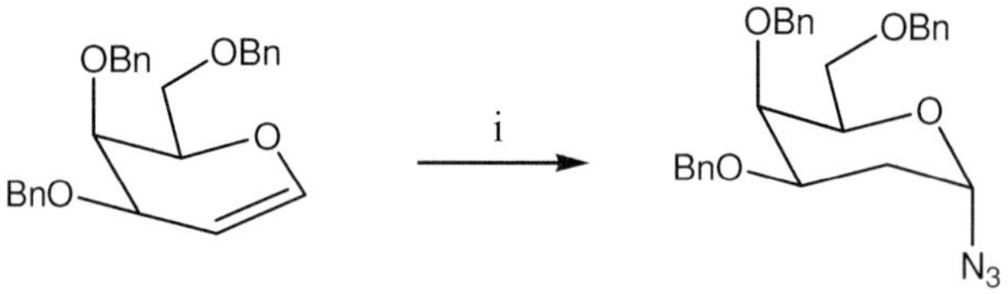

i) Me₃SiN₃, Me₃SiONO₂

ref.[45]

FIGURE 1.45. Azido methods for the preparation of aminosacharides.

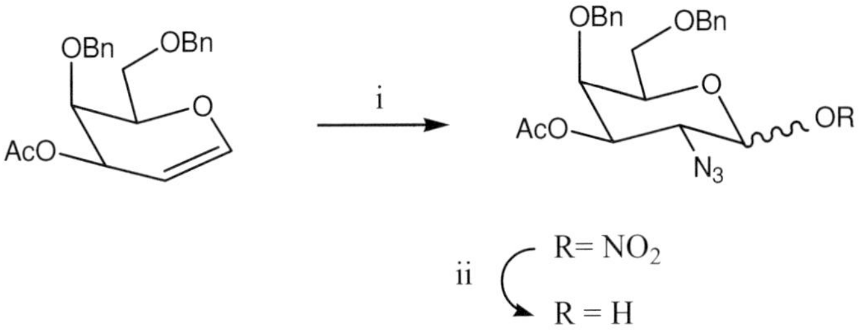

i) CAN, NaN$_3$, CH$_3$CN, −15°C, 45%. PhSH, DIEA, 91%.

ref[64]

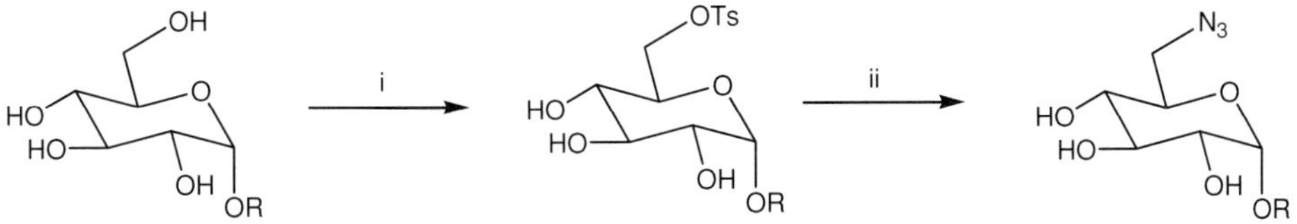

i) TsCl, Py, r.t. ii) NaN$_3$, DMF

ref[1]

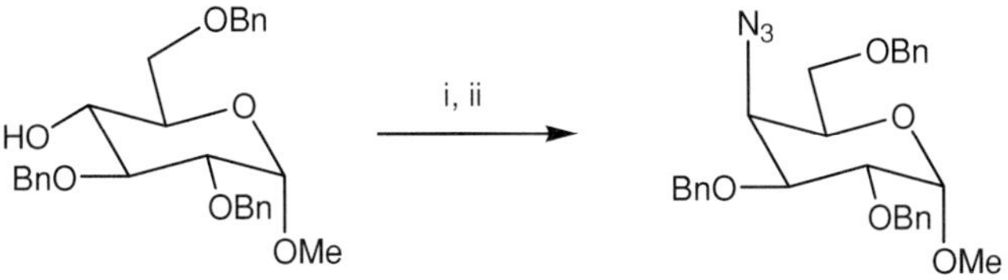

i) Tf$_2$O, Py/ CH$_2$Cl$_2$. ii) NaN$_3$, DMF

ref[46]

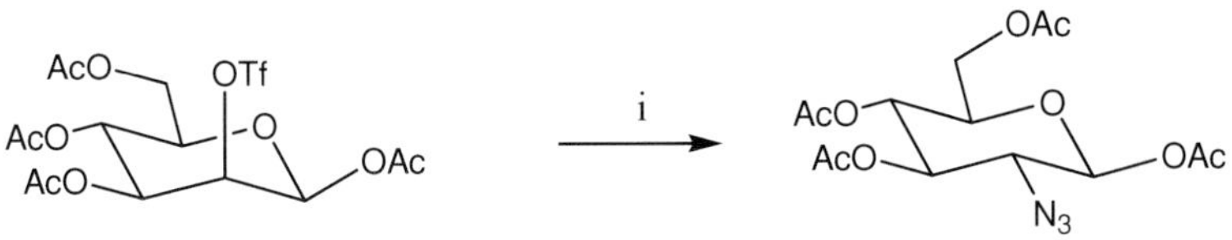

i) NaN$_3$, DMF

ref[47,48]

FIGURE 1.45. (*continued*)

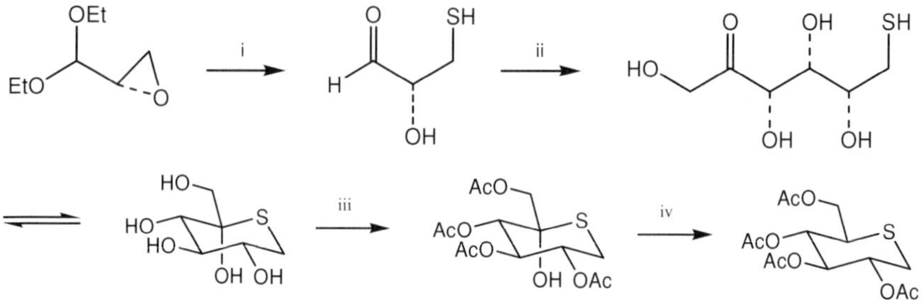

i) a) AcSK, AcSH. b) HCl. ii) a) DHAP FDP A. b) Pase. iii) Ac$_2$O, Py. iv) Et$_3$SiH, BF$_3$.OEt$_2$.

FIGURE 1.46. Synthesis of thiomonosaccharides.

probed satisfactorily due to steric hindrance of the secondary hydroxyl groups on the pyranose ring.[40]

1.6.3   Thiopyranoside Monosaccharides

Thiosugars

Sugars are another class of interesting carbohydrate mimetics. The synthesis of these derivatives can be achieved by using aldolases RAMA for the aldol condensation reaction.. The following reaction sequence was used successfully for the preparation of deoxygluco, manno, galacto, and altropyranosides (Figure 1.46).[49] Another strategy for the synthesis of thiosugars involves the replacement of one of the oxygen atoms at the anomeric carbon of the glycoside by a sulfur atom leading to two distinctly different thiosugars, namely a 5-thio and 1-thioglycosides.[50]

1.6.4   Carbapyranoside-Saccharides

More recently this type of sugar mimics have received increasing attention since some of them present α-glucosidase activity and therefore considered for therapies for non-insulin-dependent diabetes mellitus. Also they have been found to be active as agricultural antibiotics, and because of their recognition by glycosidases and glycosyltransferases as substrates and stability against enzymatic degradation, they have been used also to study oligosaccharide-chain biosynthesis.[51,52] The pseudotetrasaccharide Acarbose (Figure 1.47) has been the first α-glucosidase inhibitor to be explored in humans as an antidiabetic agent along with the amino sugar 1-desoxynojirimycin Miglitol.

The pathway in Figure 1.48[53] has described the chemical synthesis of carbamaltose, carbacellobiose and related carbadisaccharides of biological interest.

Pseudo-N-acetyllactosaminides were found to be acceptors substrates for human-milk α-(1$\rightarrow$ 3/4)-fucosyltransferase. A small scale reaction of the

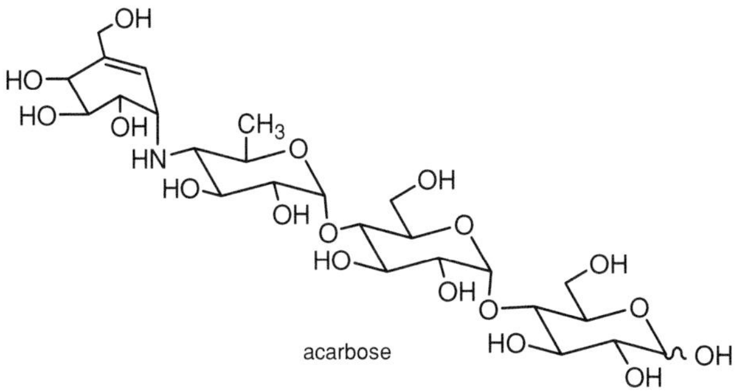

FIGURE 1.47. Pseudotetrasaccharide Acarbose.

mentioned pseudodisaccharides with GDF-fucose resulted in conversion to pseudotrisaccharides (Figure 1.49).[51]

1.7 Glycoside Reactivity

The reactivity for the anomeric carbon C(1) is the typical for acetals, and therefore the nucleophilic addition may occur. On the other hand, the other hydroxyl groups behave typically for alcohols. For coupling reaction with sugars, the anomeric carbon is involved to produce a glycosidic bond, and usually must be activated with a good leaving group in order to form a new linkage (Figure 1.50).

A glycoside is formed when the anomeric carbon of a sugar is connected through a heteroatom (except with C-glycosides) with an aliphatic or aromatic fragment known as aglycon.

The glycosidic bond is formed when a nucleophile (alcohol, amine, thiol, or carbanion) substitutes the hydroxyl group at the anomeric position, which has been previously substituted by a good leaving group. Therefore, when the nucleophiles are an alcohol, amine, or carbanion, O-, N-, or C-glycosides are generated as result, as can be observed in Figure 1.51.

1.8 The Leaving Groups

As mentioned above, the anomeric hydroxyl group can be replaced under suitable conditions with a good leaving group. Initially, the use of halogens such as fluorine, chlorine, and bromine was the strategy of choice, and particulary the last since it presents the best balance between reactivity and stability and this is why it has been extensively used for preparing glycosides. However, halides are in most cases labile and undergo decomposition. Consequently a number of other leaving groups have been designed for glycoside chemistry, and among them, imidates, sulfur, sulfonates, silyl groups, phoshates, and acetates are equally important alternatives.

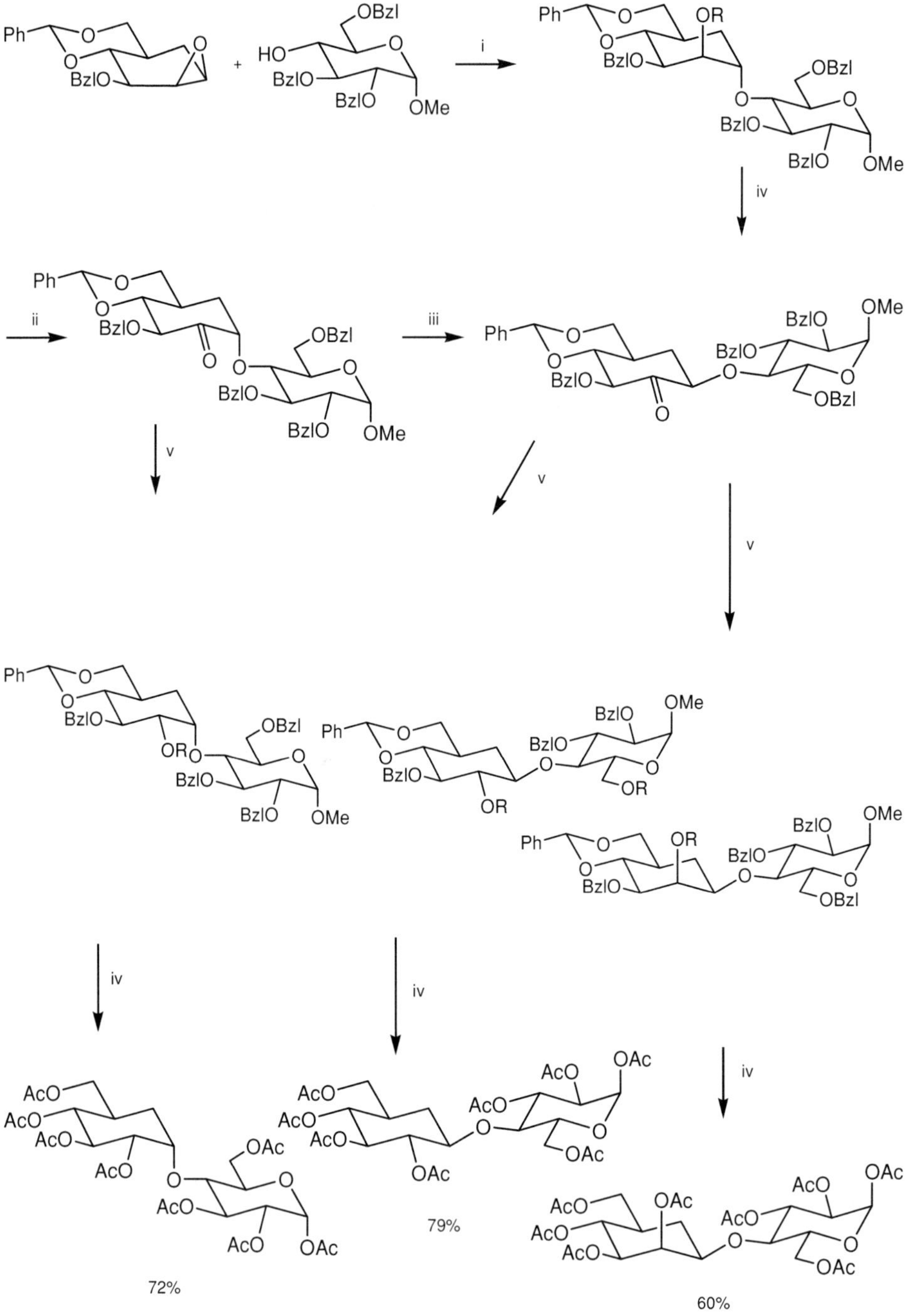

i) DMF, NaH, 15-crown-5 ether, 50°C, 60% ii) DMSO, Ac₂O, r.t., 72% iii) DBU, PhCH3, 70°C, 56%. iv) a) FeCl₃, Ac₂O, −20°C b) H₂, Pd/C, EtOH. c) Ac₂O, Py. v) NaBH₄, CH₂Cl₂/MeOH, O°C.

FIGURE 1.48. Synthesis of carbapyranoside-disaccharides.

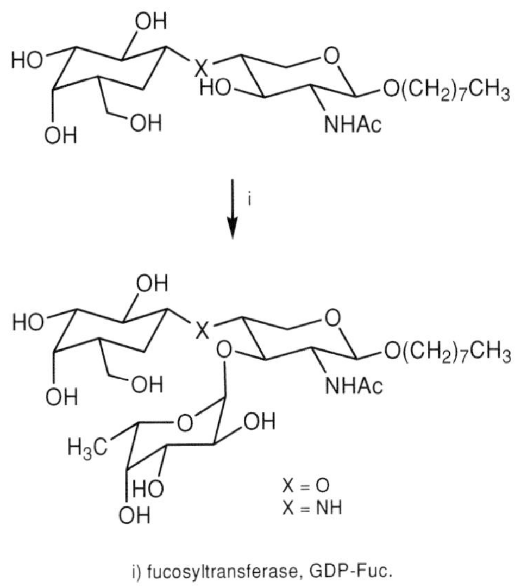

FIGURE 1.49. Chemoenzymatic synthesis of pseudotrisaccharides.

i) fucosyltransferase, GDP-Fuc.

The use of iodide has been restricted due to its low reactivity and fluoride although limitedly has been more used for preparation of some α-glycosides.[54,55] It has been found that in the absence of selective conditions, a leaving group can be found as a mixture of anomers, as in the case of the acetates. However, some others such as bromide and imidate can be introduced preferentially at the α-position (Figure 1.52).

A well-accepted hypothesis that explains the α-stereoselective preference assumed by the leaving group (halogens and imidate) is based on the anomeric effect, consisting in the electronic effect produced by the ring oxygen, which gives rise to a repulsive effect between one of the oxygen lone pairs and the leaving group, forcing the later to assume such position[56] (Figure 1.53).

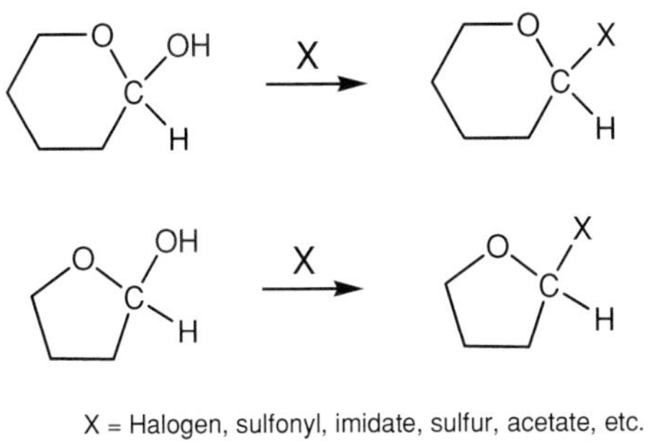

X = Halogen, sulfonyl, imidate, sulfur, acetate, etc.

FIGURE 1.50. Anomeric carbon and activation to a good leaving group.

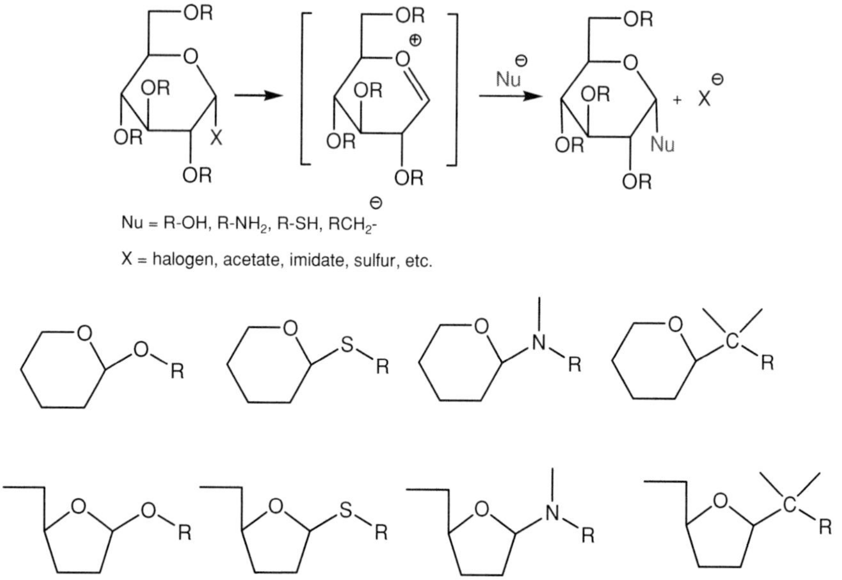

Nu = R-OH, R-NH$_2$, R-SH, RCH$_2$-

X = halogen, acetate, imidate, sulfur, etc.

FIGURE 1.51. Nucleophile displacement on the anomeric carbon and general types of glycosides.

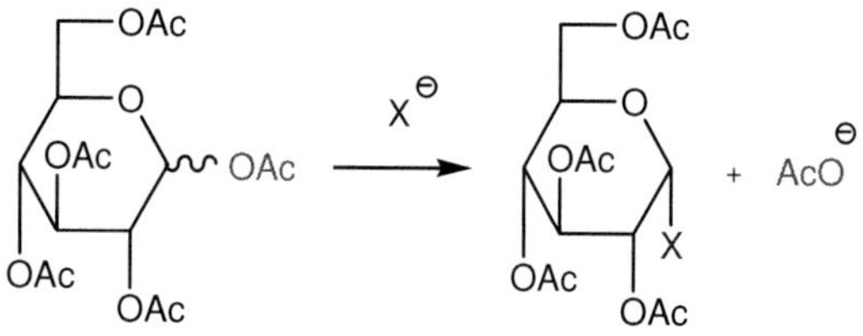

FIGURE 1.52. Stereoselectivity of halides at the anomeric position.

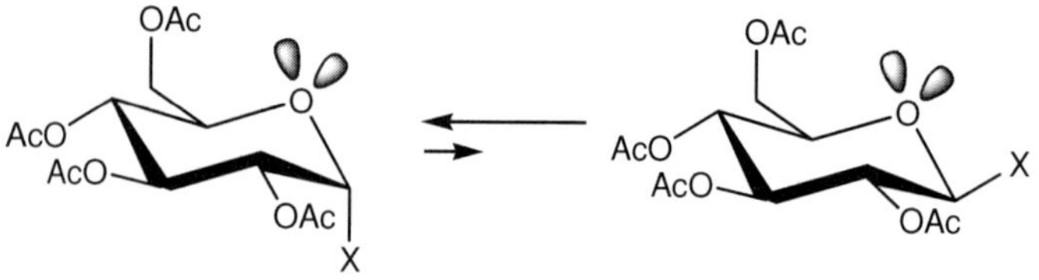

FIGURE 1.53. Anomeric effect on halogens.

1.9   Glycosyl Donors

This term is used to define a glycosidic moiety that contains a leaving group at the anomeric position. When a glycosyl donor is reacted in the presence of a catalyst (also known as promoter) with a free alcohol called glycosyl acceptor, it will produce an O-glycosidic linkage. The first glycosyl donors developed and used specifically for glycoside formation were the glycoyl halides. As mentioned above, glycosyl bromide and chloride are the most widely used halides, and are the glycosyl donors used for the preparation of O-glycosides according to the methods reported by Michael, Koenigs-Knorr and Helferich (see O-glycoside formation).

2,3,4,6-Tetraacetyl-α-D-glucopyranosyl bromide also known as acetobromoglucose is one of the most extensively used sugar intermediates for preparing glycosides derived from glucose.[57] The preparation involves in the initial peracetylation of glucose with acetic anhydride in the presence of a catalyst, commonly pyridine, triethylamine and dimethylaminopyridine, or sodium acetate and zinc chloride, in dichloromethane as solvent.

The resulting 1,2,3,4,6-Pentaacetyl-α,β-D-glucopyranoside (as mixture of anomers) is treated with a 33% solution of HBr-acetic acid in dichlorometane at 5°C during 12 h. The final product is obtained after crystallization from isopropyl ether to yield acetobromoglucose as a white solid.

The chloro derivative could be obtained similarly by using a controlled stream of HCl gas until saturation, or SOCl$_2$ in DMF (Figure 1.54).

When the sugar contains acid sensitive groups such as azide (Figure 1.55) or acid-sensitive protecting groups such as acetonide, or benzylidine (Figure 1.56), alternative milder conditions have been developed. Such is the case of fosgene in DMF and bromotrimethylsilane (TMS-Br) for chlorination and bromination, respectively.[22]

i) Ac$_2$O, DMAP, Et$_3$N, CH$_2$Cl$_2$. ii) HBr/AcOH 33%
iii) HCl (g), CH$_2$Cl$_2$.

FIGURE 1.54. Standard conditions for preparation of acetobromo and acetochloroglucose.

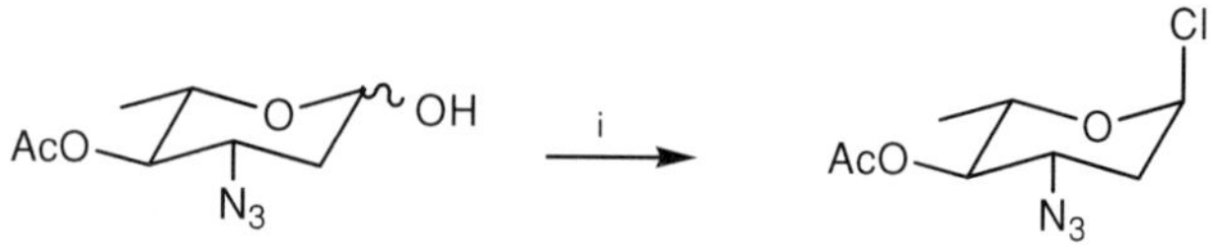

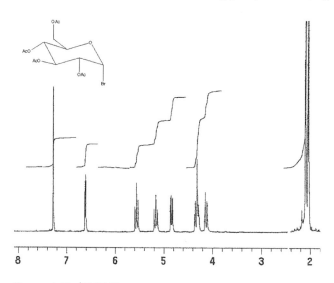

i) (COCl)$_2$, DMF, CH$_2$-CH$_2$

FIGURE 1.55. Preparation of azide glycosyl donors.

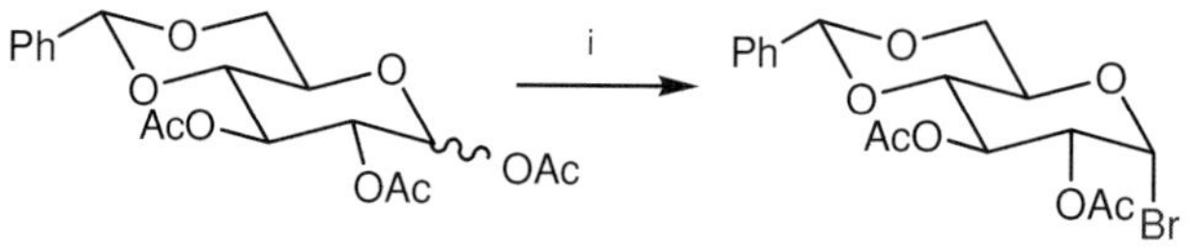

i) TMS-Br, CH$_2$Cl$_2$.

FIGURE 1.56. Anomeric bromination in acid sensitive sugars.

The ^{1}H NMR spectrum of acetobromoglucose shows signals for each of the ring protons, as well as for the primarily alcohol and acetates. The well-defined spectrum allows the net identification of each proton, starting from the anomeric proton at δ 6.60 shifted downfield due to the presence of the halogen, with coupling constant of 4 Hz indicating an equatorial-axial interaction with H-2. Diaxial interactions are evident as triplets for H-3 and H-4, and axial-equatorial as double of double for H-2 (Figure 1.57).

Besides their extensive use in the preparation of glycosides, glycosyl bromide can also be useful for conversion to other suitable glycosyl donors (Figure 1.58),

FIGURE 1.57. ^{1}H NMR spectrum of acetobromoglucose in CDCl$_3$.

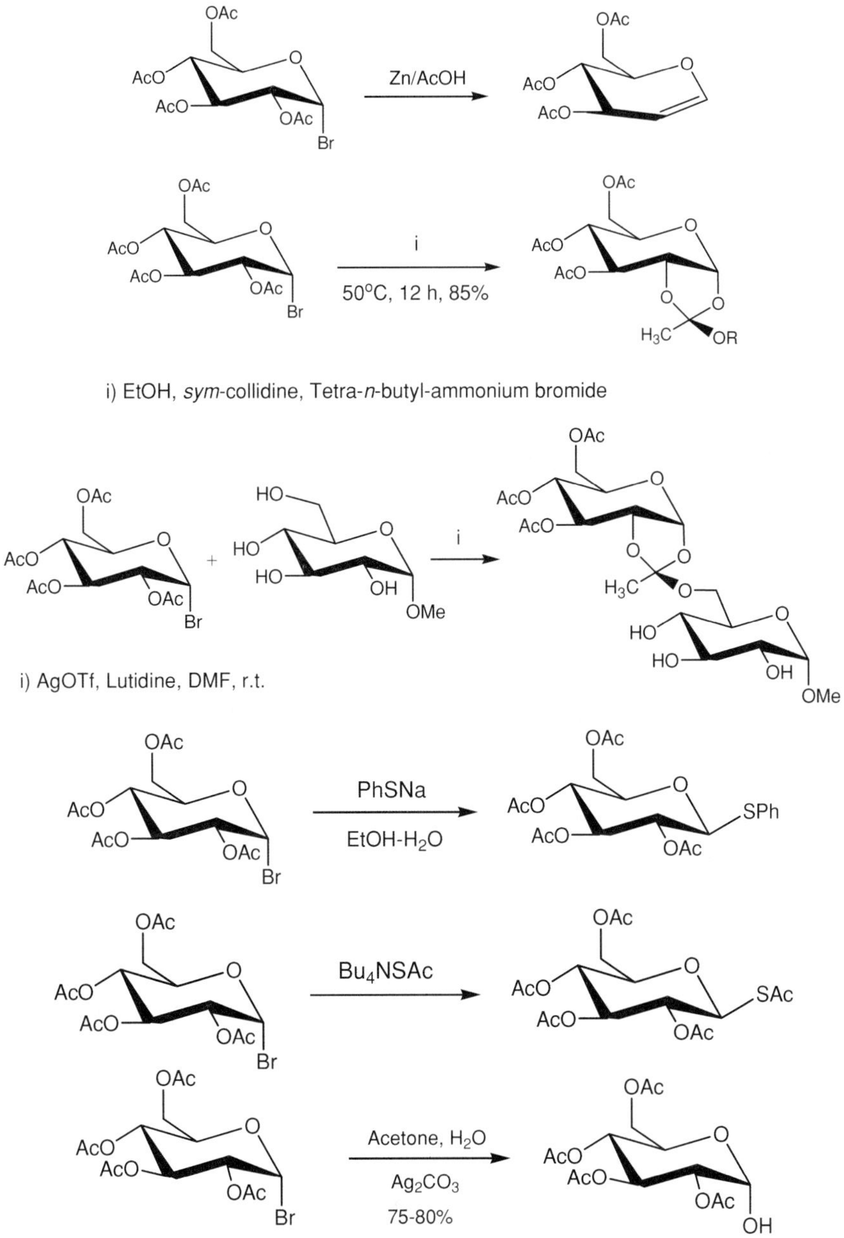

OAc
AcO
AcO
OAc
Br
Zn/AcOH
i
50°C, 12 h, 85%
H3C
OR
i) EtOH, sym-collidine, Tetra-n-butyl-ammonium bromide
HO
OH
OMe
i) AgOTf, Lutidine, DMF, r.t.
PhSNa
EtOH-H2O
SPh
Bu4NSAc
SAc
Acetone, H2O
Ag2CO3
75-80%
OH

FIGURE 1.58. Some glycosyl donors obtained from acetobromoglucose.

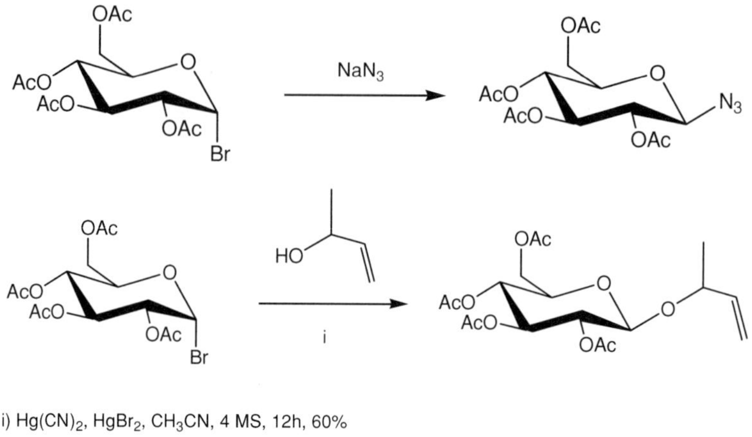

i) Hg(CN)$_2$, HgBr$_2$, CH$_3$CN, 4 MS, 12h, 60%

FIGURE 1.58. (*continued*)

such as glycals,[58] othoesthers,[59] ad thiols.[60] Also, the glycosyl halides can be transformed to glycosyl imidate through the anomeric hydroxyl formation,[61] or to amines via a reaction with azide salt and hydrogenolysis.[52]

More recently a "disarmed" glycosyl iodide derived from glucuronolactone was designed for O-glycosidation. The iodosugar has proved to be a flexible donor, reacting under mild conditions (Figure 1.59).[62]

Glycosyl acetates are also important glycosyl donors and can be used directly under the fusion strategy for the preparation of O- and N-glycosides. The fusion method consists of the reaction between the glycosyl acetate as glycosyl donor with the glycosyl acceptor in the presence of a Lewis acid as a promoter to generate the corresponding glycoside. Likewise, acetates can also be suitable precursors for the preparation of glycosyl donors such halides, thiols.[63] and to imidates, the latter by a two-step process. The first step involves the removal of the anomeric acetate with base, among them hydrazine, benzylamine, ammonia, and piperidine, which are the most preferred. The resulting hydroxyl group is obtained as a mixture of anomers and is subsequently used for the preparation of the glycosyl imidate (see Imidate

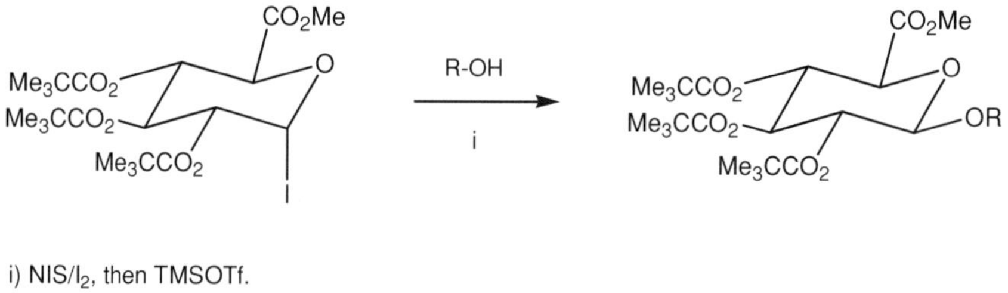

i) NIS/I$_2$, then TMSOTf.

FIGURE 1.59. Glucuronyl iodide as a glycosyl donor.

Method). Another use of glycosyl acetates is the transformation into anomeric amines, through the introduction of the azide group with trimethylsilyl azide under a Lewis acid catalyst, and further hydrogenolysis.[64] This reaction is useful for the preparation of some glycopeptides. Likewise 2-thiophenyl glycosides of Neu5Ac are suitably obtained by treatment of 2-O-acetyl, 2-chloro, or 2-chloro Neu5Ac glycosyl donors with PhSH in the presence of NIS/TfOH as promoter system (Figure 1.60). Other activated agents for preparing S-alkyl and S-aryl glycosyl donors are methyl trifluoromethanesulfonate (MeOTf), dimethyl(methylthio)sulfonium trifluoromethanesulfonate (DMTST), iodo dicollidine perchlorate (IDCP), and phenyl selenyl trifluoromethanesulfonate (PhSeOTf)[65.]

Thioglycosides are stable glycosyl donors widely used for the preparation of glycosides. The usual conditions for achieving this goal are the glycosyl acceptor and N-iodosuccinimide (NIS), or NIS-TfOH as promoter. Thioglycosides are also important starting material for the preparation of other glycosyl donors such as acetates, fluorine,[63] chlorine,[68] sulfoxides,[69] or anomeric alcohols (Figure 1.61).

Glycals are becoming potentially useful glycosyl donors, and an increasing number of simple and complex glycosides have been reported. For this purpose the glycal is usually transformed to the oxirane. and immediately coupled with the glycosyl acceptor in the presence of a Lewis acid (see The Glycal Method). Moreover, glycals are also suitable intermediates for the preparation of a variety of glycosyl donors (Figure 1.62) such as phosphates and thiophosphates,[70] deoxysugars,[71] Diels-Alder adducts,[72] allyl glycosyl donors,[73] and imidates.[74]

1.10   Protecting Groups

An important additional requirement for achieving glycosidic coupling reactions, besides the fact that a good leaving group should be present, is the appropriate use of protecting groups. Their function is to shield those groups (particularly heteroatoms) that are wanted to keep unaltered during the coupling reaction and then release them under mild conditions that do not affect the glycosidic bond (Figure 1.63).

A significant number of protecting groups[75] have been used and combined for pursuing the synthesis of complex natural products including glycosides.

Due to its acetal character, the glycosidic bond is hydrolyzed under acidic conditions, and is significantly more resistant to base, hydride reduction or hydrogenolysis.

The use of ethers such as methyl ether ($-O-CH_3$), methoxymethyl ether ($-O-CH_2OCH_3$, MOM), 2-methoxyethoxymethyl ether ($-O-CH_2OCH_2CH_2OCH_3$, MEM), and tetrahydropyranyl ether ($-O-2-c-C_5H_9O$, THP) have been widely used for protection of alcohols. However, in glycoside synthesis attention has to be paid since deprotection is carry out under acidic conditions, which might be hazardous for the glycosidic bond. Silyl derivatives are also another important choice for protection of hydroxyl group.[65] Some of the most accepted silyl derivatives for carbohydrate hydroxyl protection are *tert*-butyl dimethylsilyl (TBDMS),

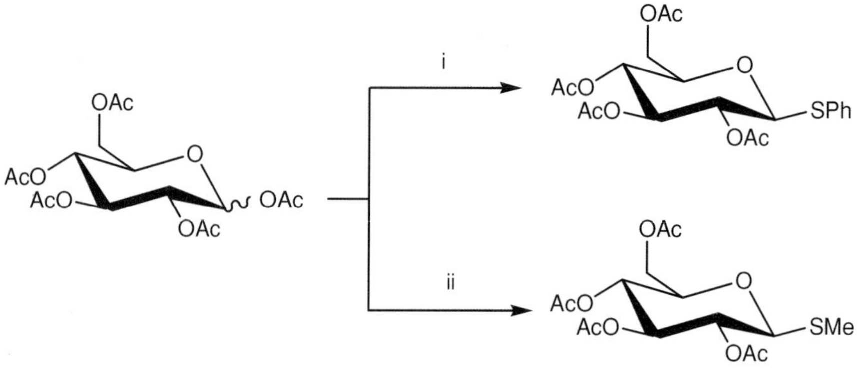

i) PhSH (1.1 eq.), SnCl$_4$ (0.7 eq.), CH$_2$Cl$_2$, 0°C, 4h, 82%. ii) MeSH (excess), SnCl$_4$ (0.7 eq.), CH$_2$Cl$_2$, −20°C, 3h, 85%.

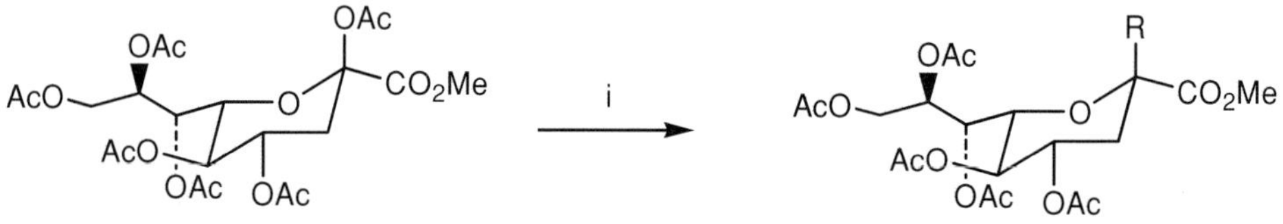

i) TiCl$_4$ or TiBr$_4$

R = Cl, Br

Ref.[66]

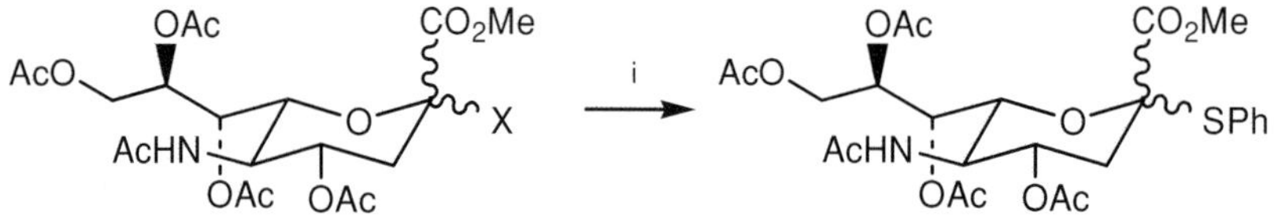

X = OAc, Cl, F.

i) PhSH, NIS/TfOH.

Ref.[67]

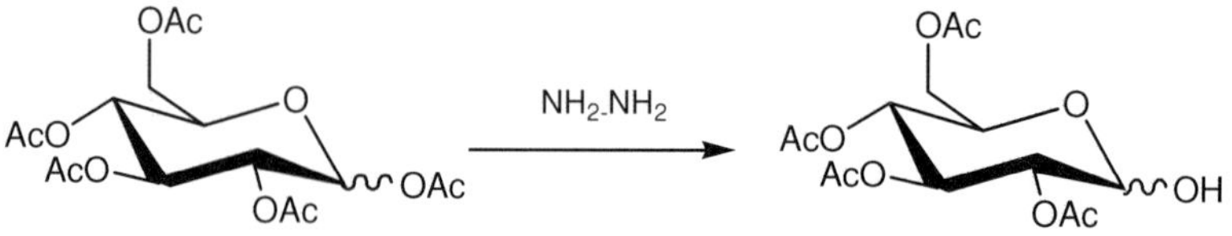

FIGURE 1.60. Preparation of glycosyl donors and precursors from glycosyl acetates.

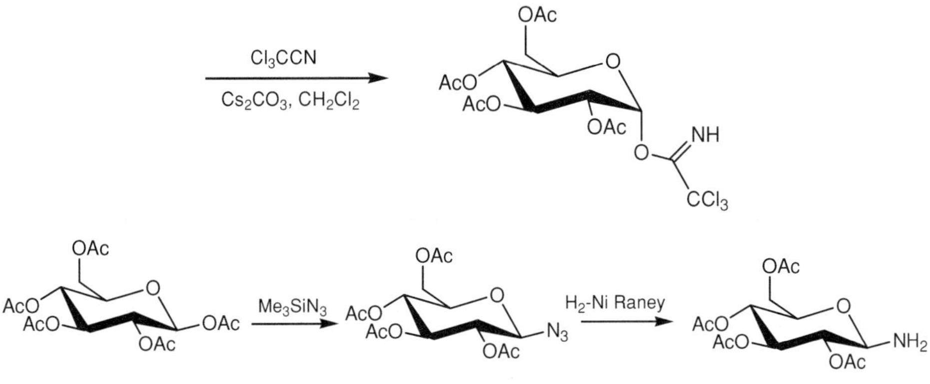

FIGURE 1.60. (*continued*)

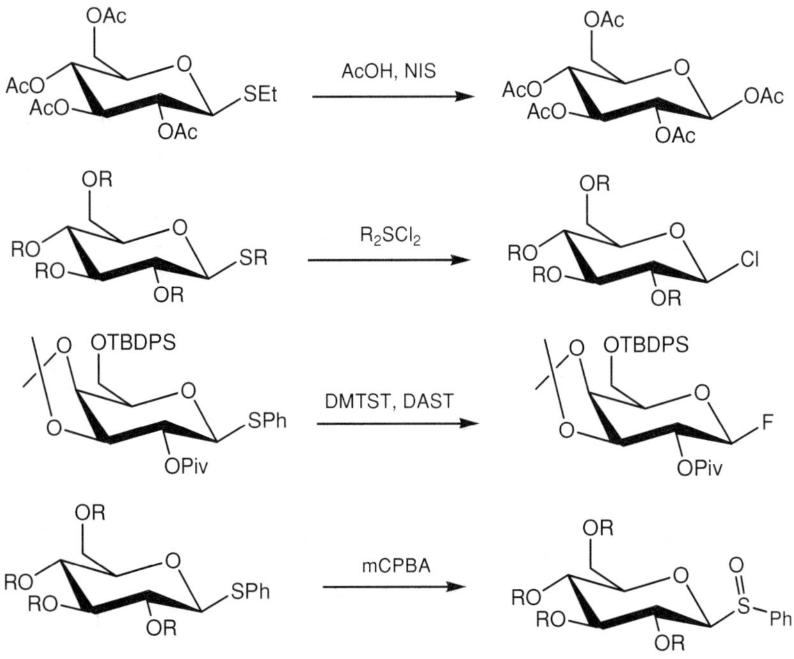

FIGURE 1.61. Modifications of thio glycosyl donors.

triisopropylsilyl (TIPS), *tert*-butyl diphenylsilyl (TBDPS), and triethylsilyl (TES) ethers. Quantitative cleavage is usually achieved upon treatment with tetrabutylammonium fluoride (TBAF) or HF/pyridine.

The most suitable protecting groups for the preparation of glycosides are the affordable acetates, benzoates and benzyl protecting groups since they can be

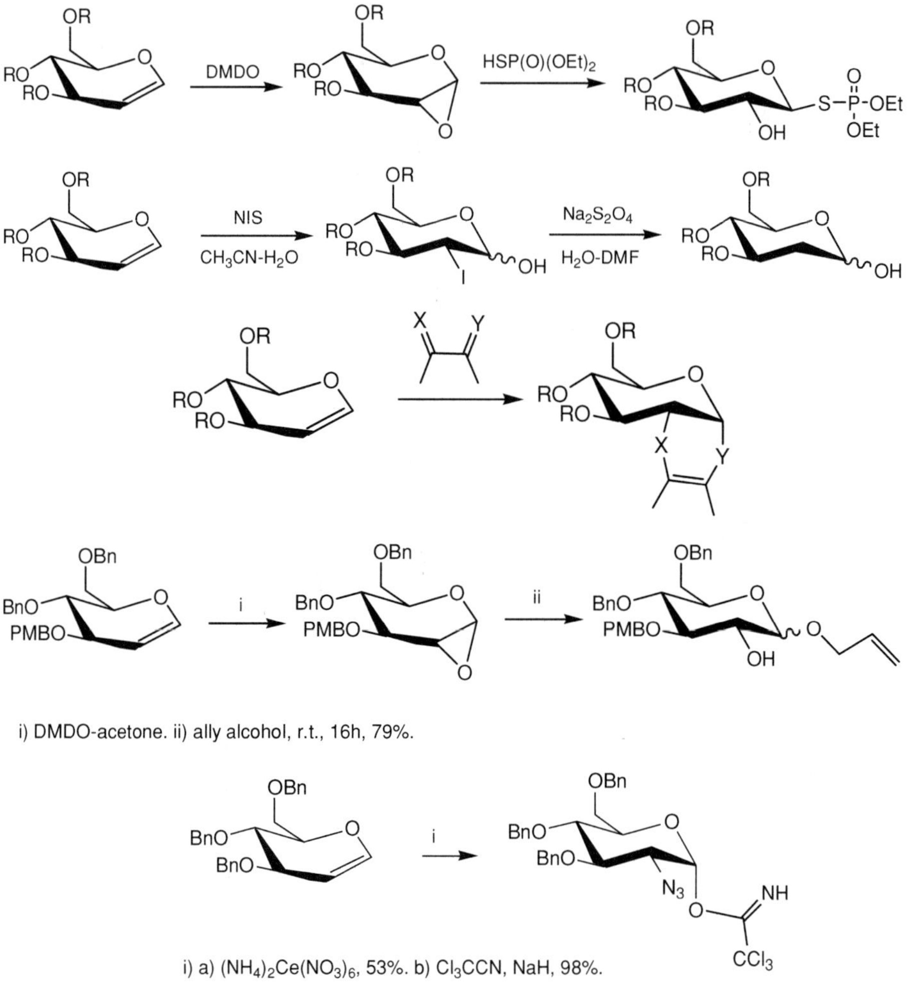
DMDO
HSP(O)(OEt)$_2$
NIS
CH$_3$CN-H$_2$O
Na$_2$S$_2$O$_4$
H$_2$O-DMF
i) DMDO-acetone. ii) ally alcohol, r.t., 16h, 79%.
i) a) (NH$_4$)$_2$Ce(NO$_3$)$_6$, 53%. b) Cl$_3$CCN, NaH, 98%.

FIGURE 1.62. Preparation of glycosyl donors and precursors from glycals.

removed under basic and for the later neutral conditions, being the best conditions for preserving the glycosidic bond. The standard conditions for either installing and removing the most common protecting group described are:

Acetate (Ac-): The standard procedure involves the use of acetic anhydride in the presence of pyridine or triethylamine as acid scavenger, and 4-(dimethylamino) pyridine (DMAP) that improves the rate of reaction. The cleavage of acetates proceeds smoothly with NaOMe solution also known as Zemplen conditions. Acetates are stable at pH from 1-8 and can be cleaved with lithium aluminium hydride (Figure 1.64).

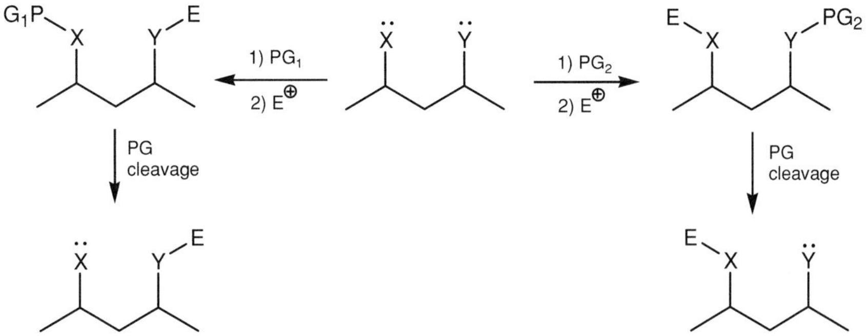

FIGURE 1.63. Schematic representation of protecting group applicability.

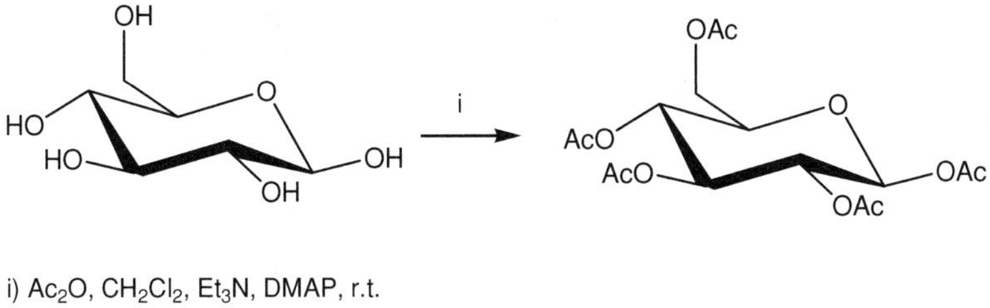

i) Ac$_2$O, CH$_2$Cl$_2$, Et$_3$N, DMAP, r.t.

FIGURE 1.64. Standard protocol for the preparation of peracetylated sugars.

Benzoyl (Bz-): This protecting group is more stable to hydrolysis than acetates and may resist a pH up to 10. The conditions for protection of alcohols are shown in Figure 1.65 and involves the use of benzoyl chloride in pyridine or triethylamine.[76] It is stable to hydrogenolysis and borohydrides but not to lithium aluminium hydride. The cleavage is usually achieved in 1% NaOMe-MeOH solution.

Pivaloyl (Pv-): This protecting group, which is also known as Trimethylacetyl chloride, is used for protection of primary and secondary alcohols in yield. An example of the use of this group is the protection of the hydroxyl group at position 2 of fucose derivative.[77] The standard conditions for protection are pivaloyl chloride in pyridine or DMAP and the cleavage is performed with Bu$_4$N^{+-}OH at 20°C (Figure 1.66).

Trityl (Tr-): This bulky protecting group is selective for primary alcohols (Figure 1.67). The protecting reaction proceeds in pyridine or DMAP-DMF.[78] The cleavage can be performed under neutral conditions with 1% iodide in methanol, or weakly acidic in formic acid-ether solution.

Benzyl (Bn-): This protecting group when attached with alcohol generates an ether (Figure 1.68). However, unlike common ethers, this can be cleaved under neutral condition by hydrogenolysis. The usual conditions for attachment are NaH, THF, and benzyl bromide or chloride.[79] The conditions for removing this group are hydrogen, Pd/C 10% or Pd(OH)$_2$/C 10% in ethanol or ethyl acetate.

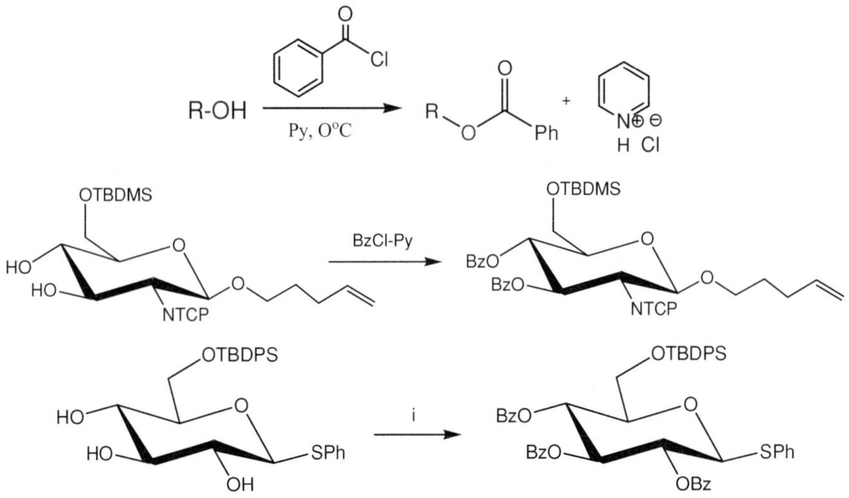

i) PhCOCl (4.0 eq.), Et$_3$N (8.0 eq.), 4-DMAP (0.2 eq), THF, 50°C, 15h, 92%.

FIGURE 1.65. General procedure for the benzoylation of sugars.

p-Methoxybenzyl (PMB-): This benzyl derivative is installed by reacting the free alcohol with PMB-Cl under NaH, DMF conditions at 0°C.[80] An example of its applications can be seen in the protection at the second position of acetonide thioglycoside shown in Figure 1.69. Deprotection is carried out under neutral

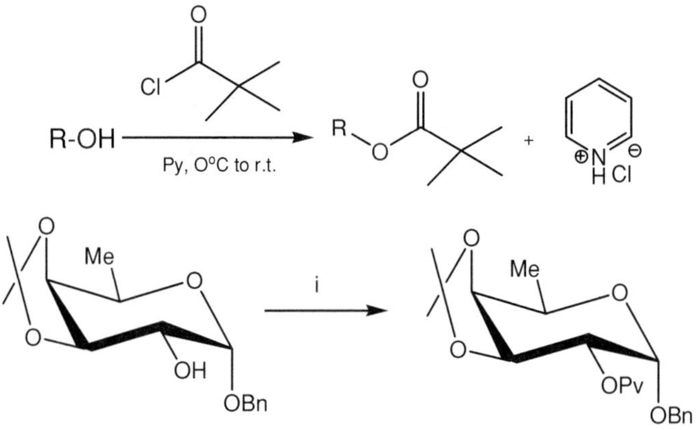

i) PvCl, Py, DMAP, 70°C, 80%.

FIGURE 1.66. Conditions and reagents for protection of alcohols with pivaloyl group.

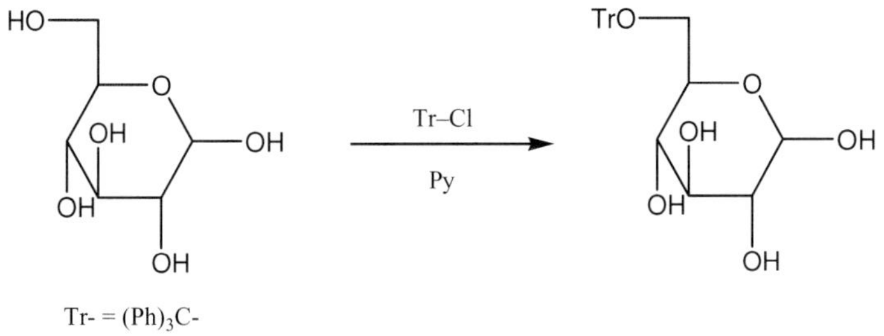
Tr–Cl
Py
Tr- = (Ph)3C-

FIGURE 1.67. Protection of primary alcohol with trityl protecting group.

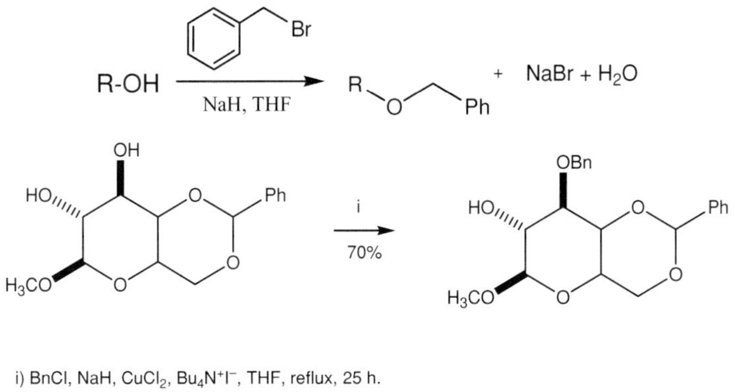
Br
R-OH
NaH, THF
+ NaBr + H2O
Ph
i
70%
OBn
i) BnCl, NaH, CuCl2, Bu4N+I−, THF, reflux, 25 h.

FIGURE 1.68. General conditions for benzylation of carbohydrates.

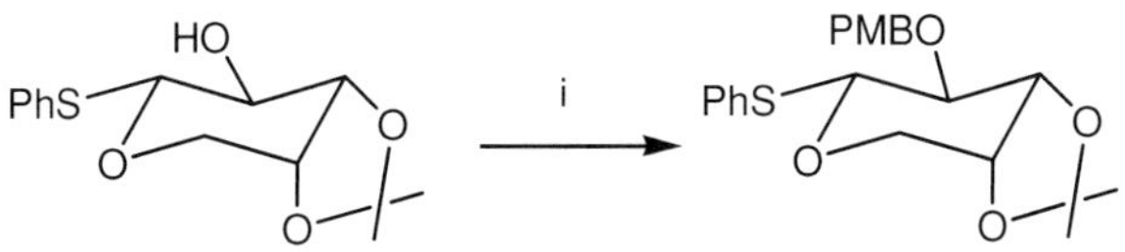
PhS
HO
PMBO
i

i) NaH, DMF, 0°C, 30 min., then 1.2 equiv. PMB-Cl, 2 h, 93%.

FIGURE 1.69. General conditions for benzylation of carbohydrates with PMB.

conditions with 2,3-dichhloro,5,6-dicyano-1,4-benzoquinone (DDQ), in CH_2Cl_2-H_2O (20:1), 1 h, at 25°C, in a 91% deprotection yield.

Acetonide ($(CH_3)_2C(O)_2$-): This protecting group is useful for protection of *cis* diols (Figure 1.70) and the conditions are acetone, 2,2-dimethoxypropane, and *p*-toluenesulfonic acid or camphorsulfonic acid as catalyst.[81] Acetonides are usually stable at a pH between 4 and 12, and the regeneration of the diol can be achieved by treatment with aqueous acid.

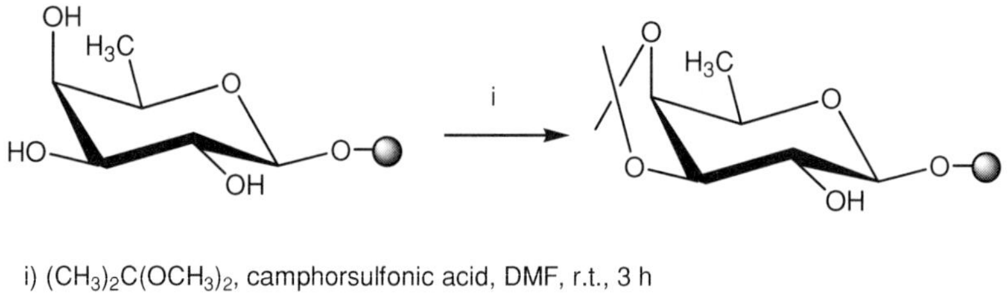

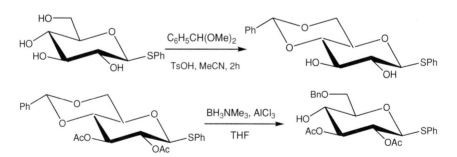

i) (CH₃)₂C(OCH₃)₂, camphorsulfonic acid, DMF, r.t., 3 h

FIGURE 1.70. Acetonide formation for protection of diols.

Benzylidene (PhCH(O)₂-): This classical protecting group is usually selected for protection of position 6 and 4, allowing the remaining positions to be modified. The benzylidene is attached under mild conditions and are useful for either –OH (4) in axial or equatorial positions (Figure 1.71). Deprotection can be effected under different conditions, such as acid conditions, hydrogenolysis, and hydrides such as BH_3NMe_3, $AlCl_3$, THF, 60°C, 1 h.[82]

Carbonate (O=C(O)₂-): This group is suitable for protection of *cis* diols, and it has been used in the synthesis of complex oligosaccharides and also in solid-phase oligosaccharide synthesis. The reagents and conditions used for protection are phosgene in pyridine at 0°C during 1 h (Figure 1.72), and the yield reported is around 70%.[83]

Boronate (PhB(O)₂-): This group has been proposed in solid-phase oligosac-charide synthesis[84] for simultaneous protection of 4,6-OH groups (Figure 1.73). Deprotection is achieved with IRA-743 resin.[85]

Tert-butyldimethylsilyl (TBS)-

More recently introduced for protection of primary and secondary alcohols with reported yield protection around 90% (Figure 1.74). The standard conditions are

FIGURE 1.71. Protection of 4 and 6 hydroxyl groups with benzylidene group and partial removal.

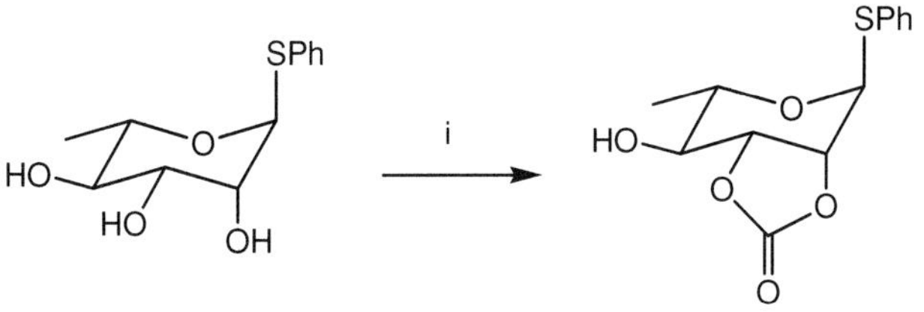

i) (Cl$_2$)C=O, Py, 0°C, 1h, 69%.

FIGURE 1.72. Standard conditions for protection of diols with carbonate protecting group.

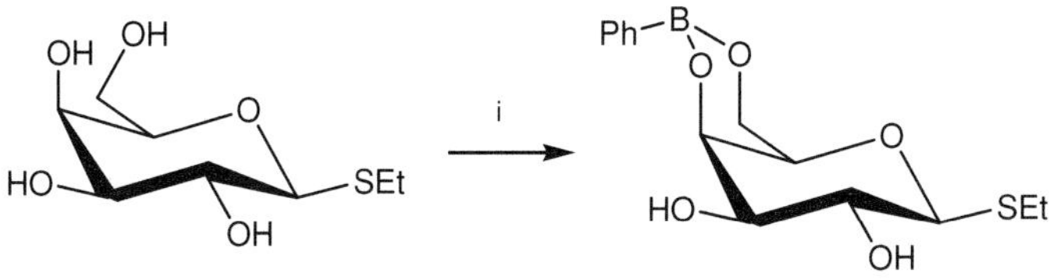

i) PhB(OH)$_2$.

FIGURE 1.73. Protection of 4,6-OH groups with boronate.

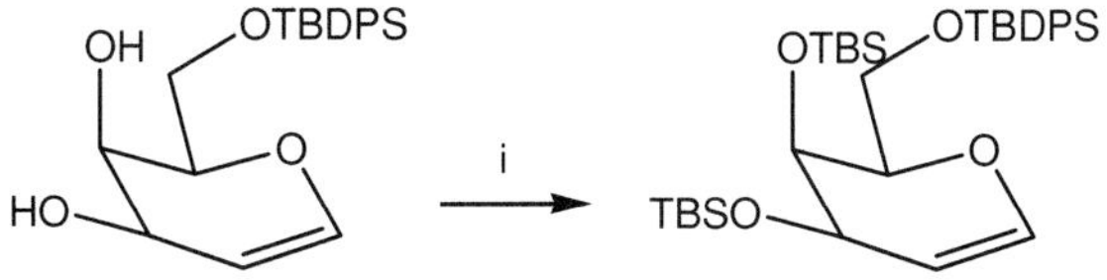

i) TBSOTf, Py.

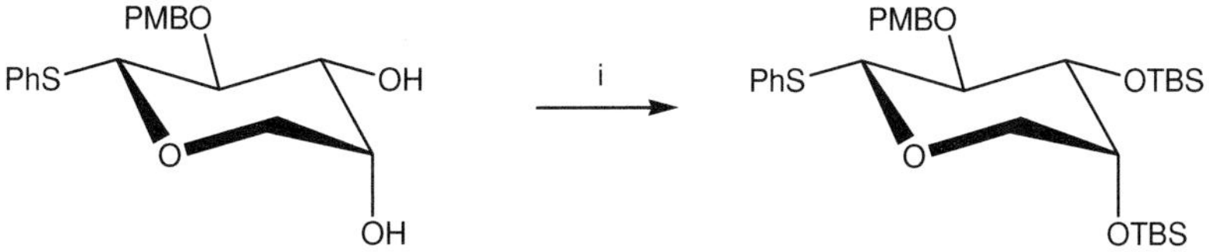

i) TBSOTf (4.0 eq.), Et$_3$N (10.0 eq.), CH$_2$Cl$_2$, 0°C, 2h, 97%.

FIGURE 1.74. Protection of secondary alcohols with TBS protecting group.

tert-butyldimethylsilyltriflate in pyridine,[86] and deprotection is usually achieved with butyl ammonium fluoride (Bu$_4$NF) in THF.

tert-butyldiphenylsilyl (TBDPS-): This protecting group is specific for primary alcohols and the yields reported are quantitatives (Figure 1.75). This bulky silylated

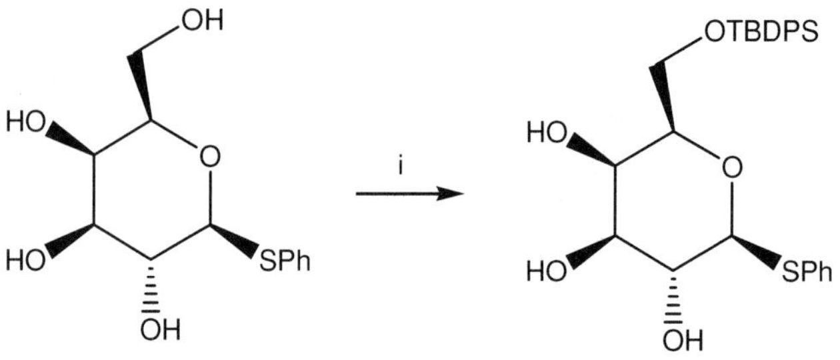

i) t-BuPh₂SiCl, imidazole, DMF, 100%.

FIGURE 1.75. Protection of primary alcohols with TBDPS protecting group.

group has been used for the assembly of oligosaccharide libraries and has been compatible with the use of other highly selective groups.[87] The standard protection conditions are TBDPS-Cl, imidazole, DMF or THF. Deprotection is achieved with hydrogen fluoride-pyridine or TBAF, cat. AcOH, THF, and a yield of 87%.

1.11   Selective Protections

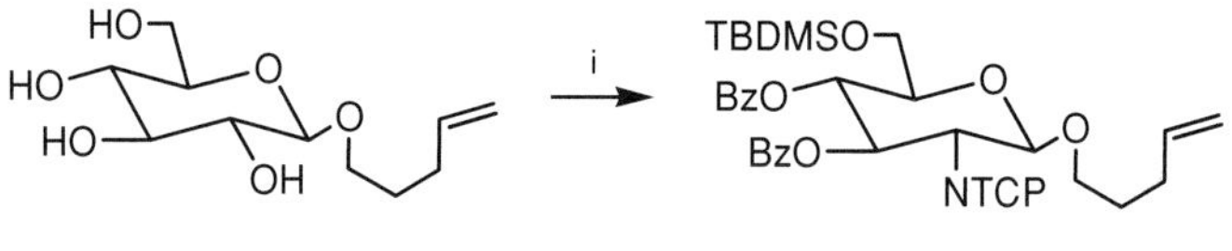

i) TBDMSCl/Py. ii) BzCl/Py.

Ref.[76]

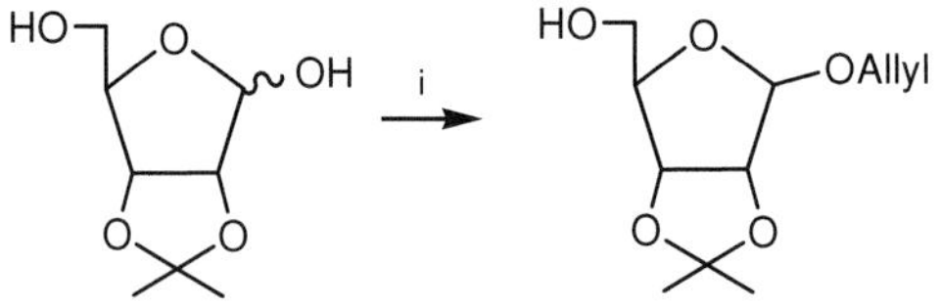

i) CH₂=CHCH₂OCO₂Et, Pd₂(dba)₃, THF, 65°C, 4h, 70%.

Ref.[88]

FIGURE 1.76. Miscellaneous selective protections

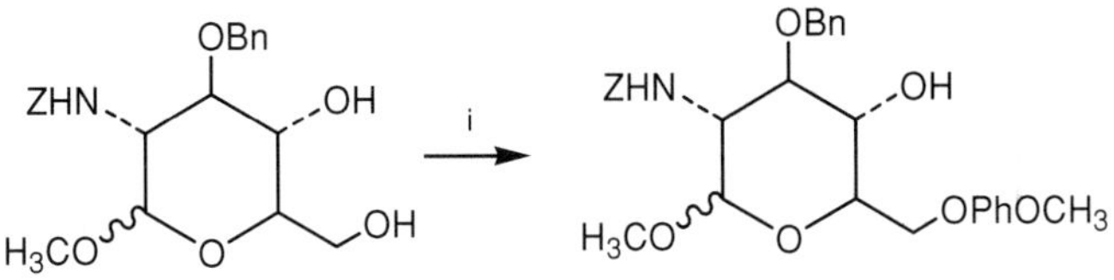

Z = benzyloxycarbonyl

i) 4-CH₃OC₆H₅OH, THF, DEAD, Ph₃P, 80°C, 82%.

Ref.[80]

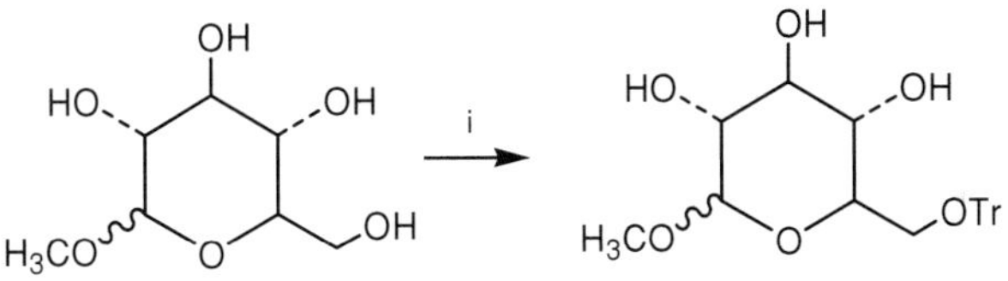

i) Ph₃CCl, DMAP, DMF, 25°C, 12h, 88%.

Ref.[89]

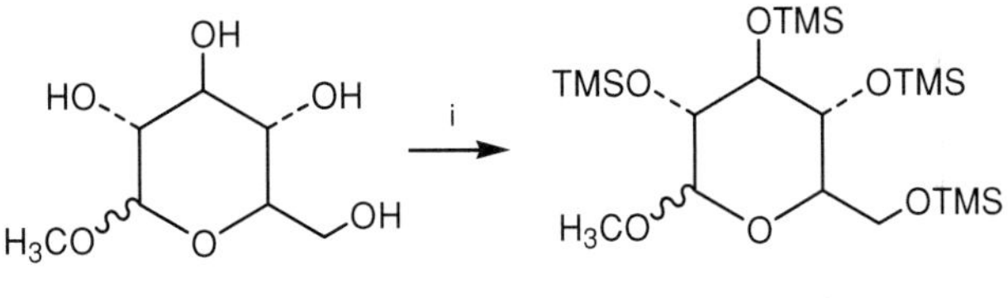

i) Me₃SiCl, Et₃N, THF, 25°C, 8h, 90%.

Ref.[90]

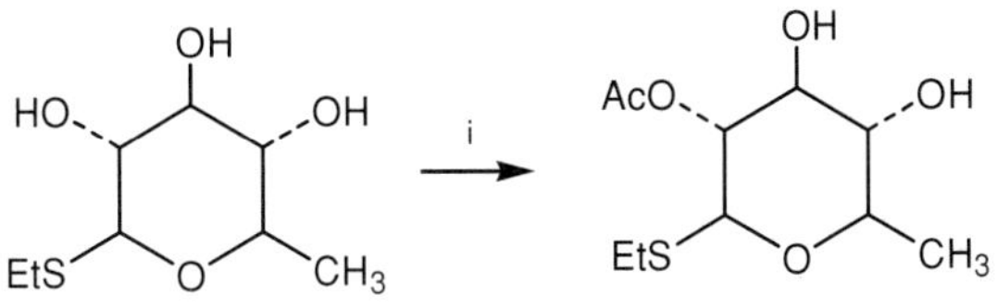

i) MeC(OCH₃)₃, TsOH, DMF, 96%.

Ref.[91]

FIGURE 1.76. (*continued*)

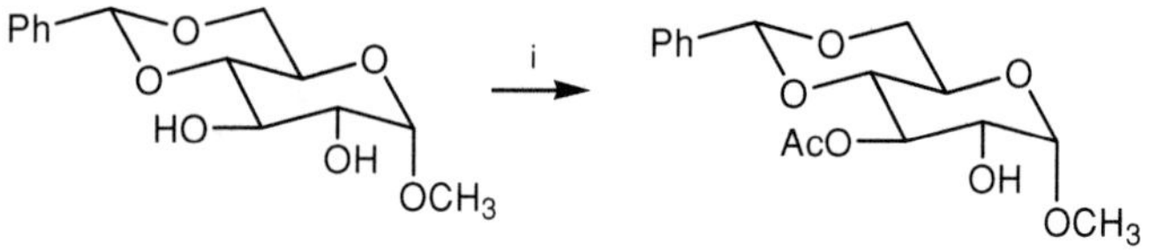

i) Lipase AK, vinyl acetate, 92%

Ref.[92]

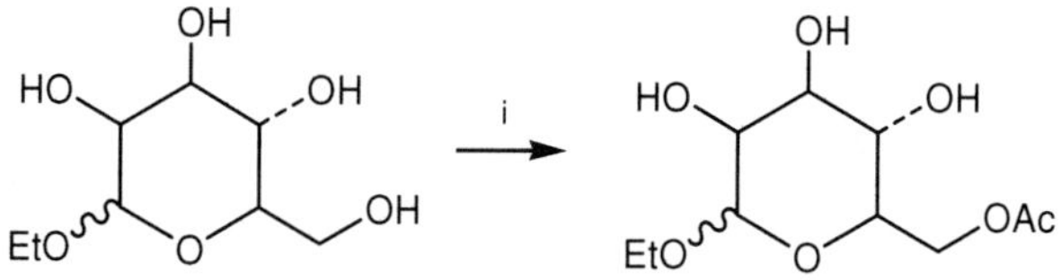

i) Pancreatin, vinyl acetate, THF, TEA 95%.

Ref.[93]

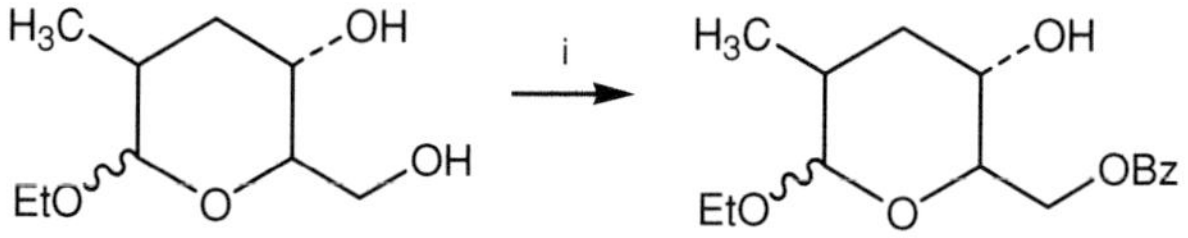

i) Ph₃P, DIAD, PhCO₂H, THF, 84%.

Ref.[94]

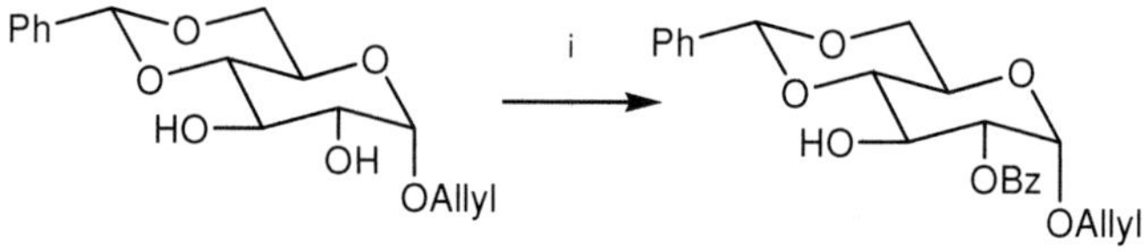

i) BzOBt, 92%

Ref.[95]

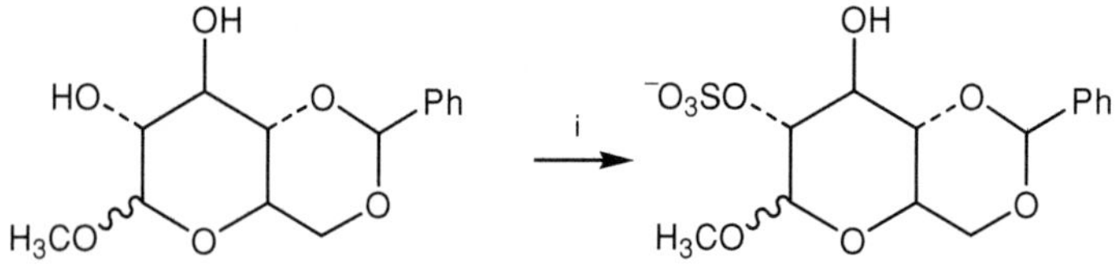

i) Pyr SO₃, Pyr, 69%

Ref.[96]

FIGURE 1.76. (*continued*)

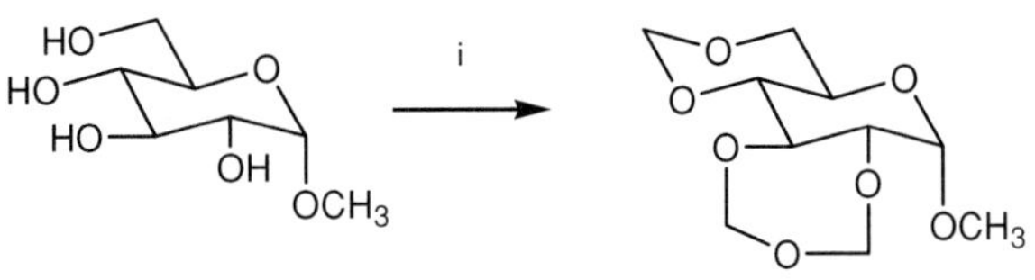

i) DMSO, POCl₃, 85%.

Ref.[97]

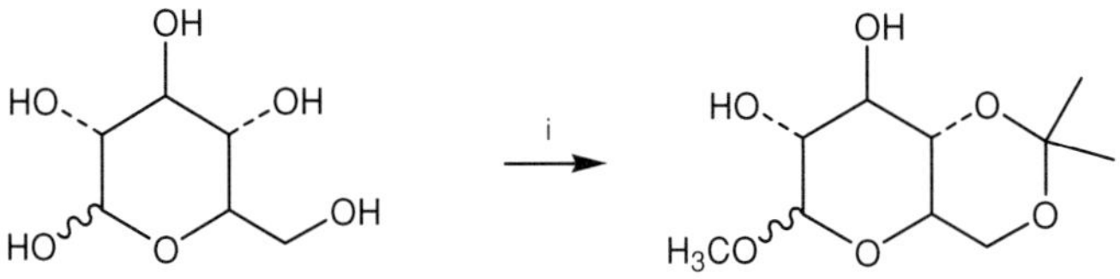

i) CH₂=C(CH₃)OCH₃, DMF, TsOH. 0°C, 95%

Ref.[98]

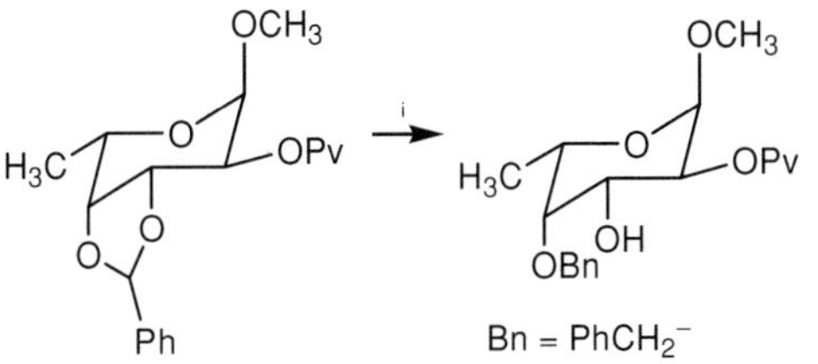

Bn = PhCH₂⁻

i) NBS, CCl₄, 75%.

Ref.[99]

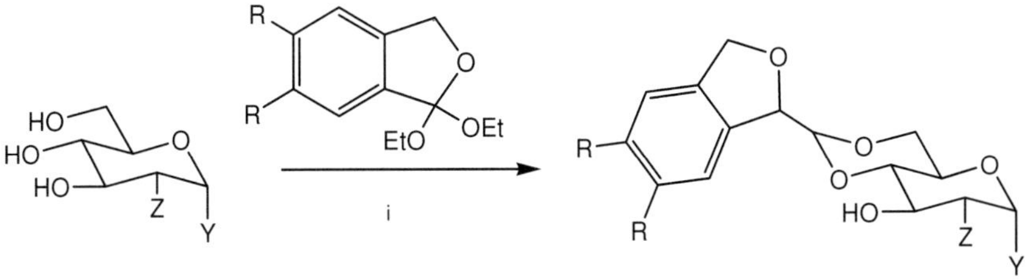

i) PPTS, ACN. 90%

Ref.[100]

FIGURE 1.76. (*continued*)

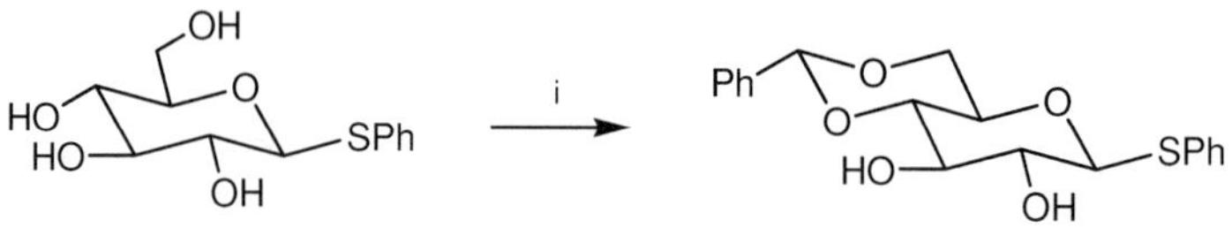

i) C$_6$H$_5$CH(OMe)$_2$, TsOH, MeCN, 2h.

Ref.[82]

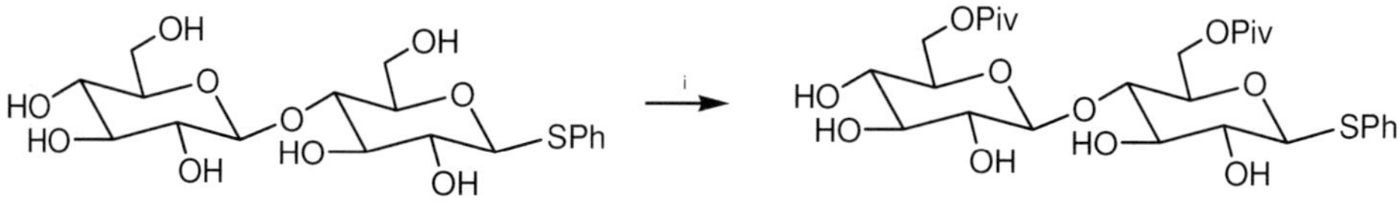

i) Pivaloyl chloride, Py, 91%.

Ref.[101]

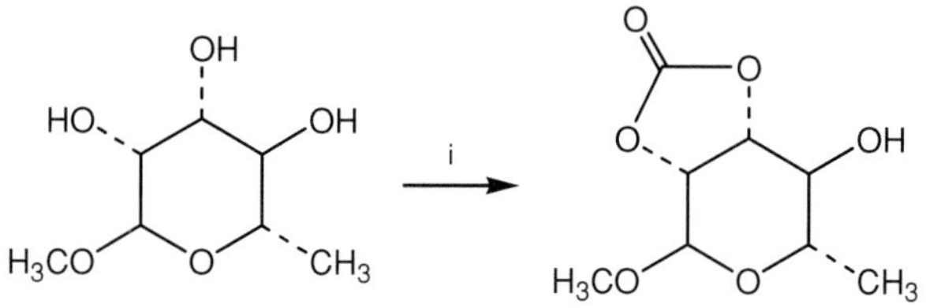

i) Cl$_3$CCOCl, Py, rt, 80%

Ref.[102]

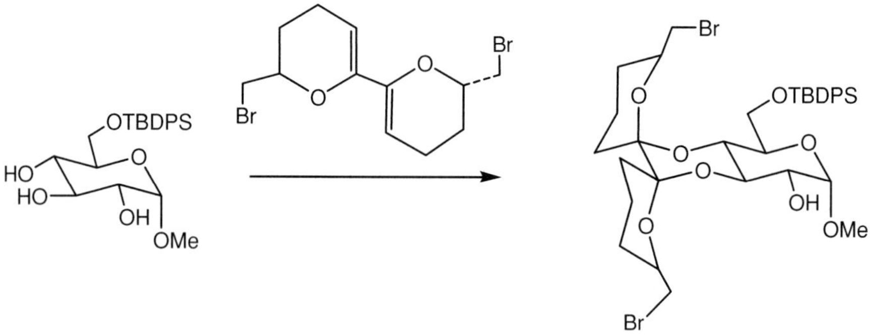

Ref.[103]

FIGURE 1.76. (*continued*)

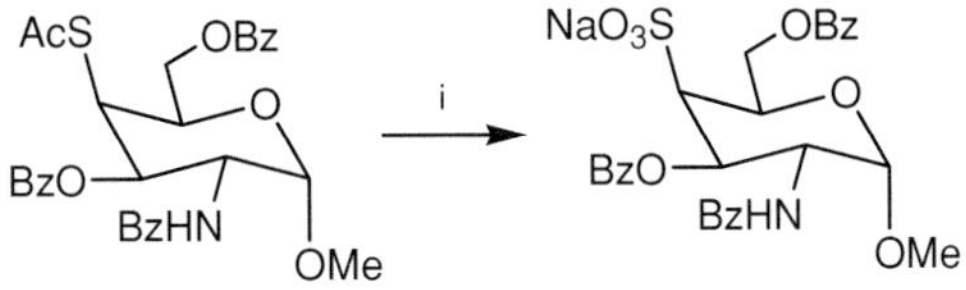

i) aq. H_2O_2 (33%), AcOH, NaOAc, 80°C, 8h.

Ref.[104]

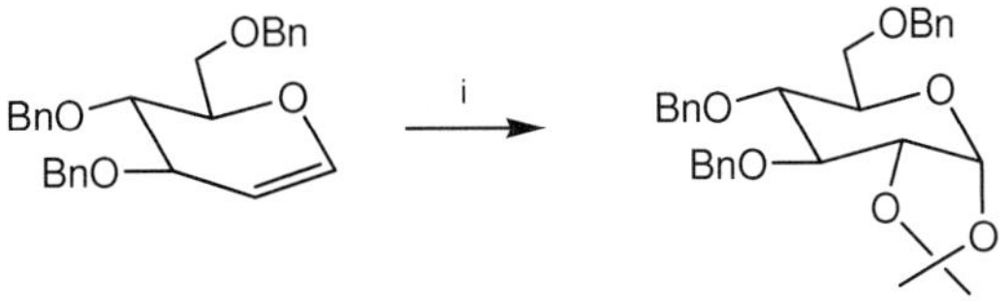

i) DMDO, $ZnCl_2$, acetone.

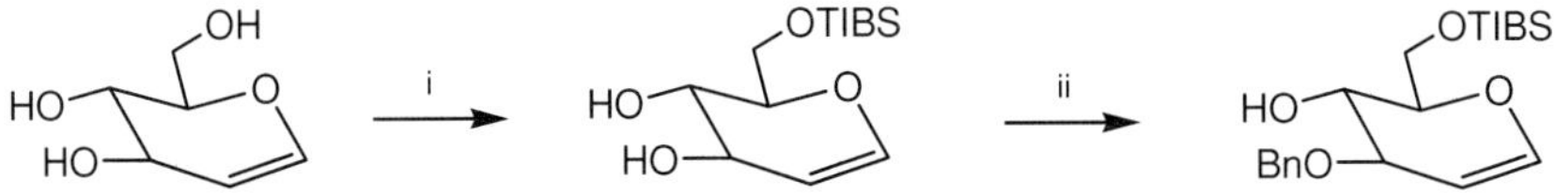

i) TIBS-Cl, Imidazole. ii) Bu_2SnO, BnBr, TBAI, 67%

Ref.[105]

i) a) 1eq. NaOMe/MeOH. b) 1.2 eq. Ac_2O, 12h, 91%.

Ref.[106]

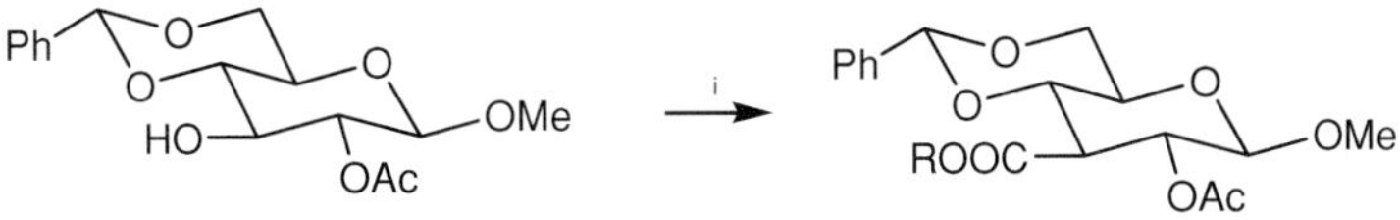

i) TMEDA, $ClCO_2R$, CH_2Cl_2, 0°C.

Ref.[107]

FIGURE 1.76. (*continued*)

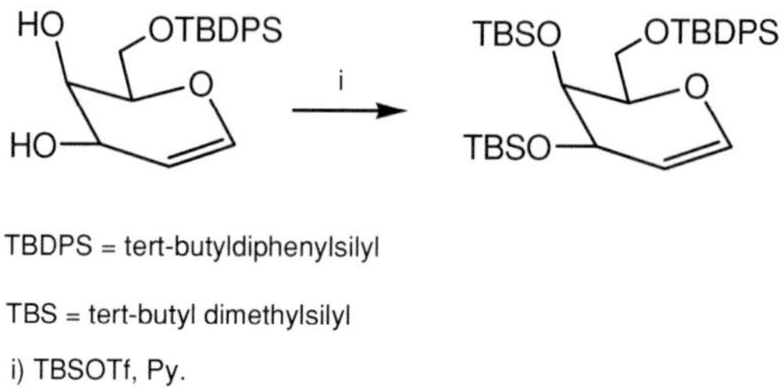

TBDPS = tert-butyldiphenylsilyl

TBS = tert-butyl dimethylsilyl

i) TBSOTf, Py.

Ref.[86]

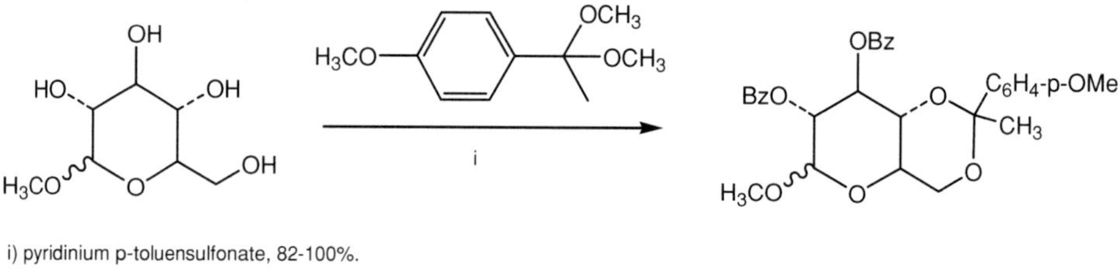

i) pyridinium p-toluensulfonate, 82-100%.

Ref.[108]

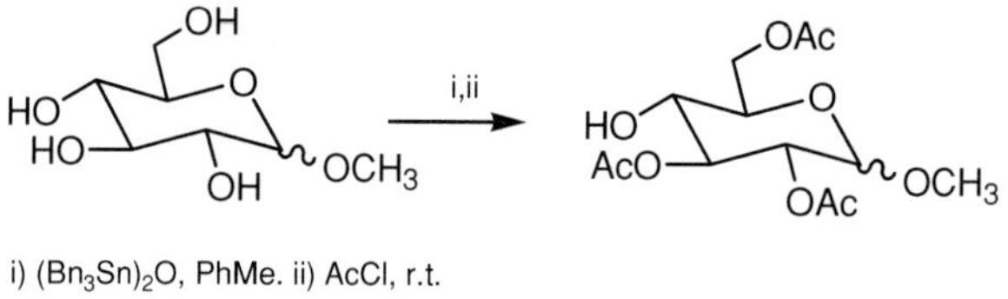

i) (Bn$_3$Sn)$_2$O, PhMe. ii) AcCl, r.t.

Ref.[109]

FIGURE 1.76. (*continued*)

TABLE 1.3. Summary of common protecting and cleavage conditions.

Protecting group	Protection conditions	Cleavage conditions
Acetyl (Ac-)	Ac$_2$O, Et$_3$N, DMAP, CH$_2$Cl$_2$	NaOMe-MeOH
Benzoyl (Bz-)	Bz-Cl, Py	NaOMe-MeOH
Pivaloyl (Pv-)	Pv-Cl, Py, DMAP	Bu$_4$NOH
Trityl (Tr-)	Tr-Cl, DMAP, DMF	1% I$_2$-MeOH
Benzyl (Bn-)	Bn-Br, NaH, THF	H$_2$-Pd(OH)$_2$-EtOH
p-Methoxybenzyl (PMB-)	PMB-Cl, NaH, THF	DDQ, CH$_2$Cl$_2$-H$_2$O
Acetonide ((CH$_3$)$_2$C(O)$_2$-)	(CH$_3$)$_2$CO, 2,2-DMP, p-TSOH	AcOH-H$_2$O
Benzylidene (PhCH(O)$_2$-)	PhCH(OCH$_3$)$_2$, p-TsOH, CH$_3$CN	AcOH-H$_2$O, or H$_2$-Pd(OH)$_2$
tert-butyldimethylsilyl (TBS-)	TBS-OTf-Py	Bu4NF-THF
tert-butyldiphenylsilyl (TBDPS-)	TBDPS-Cl-imidazole, DMF	HF-Py

1.12 Selective Deprotections

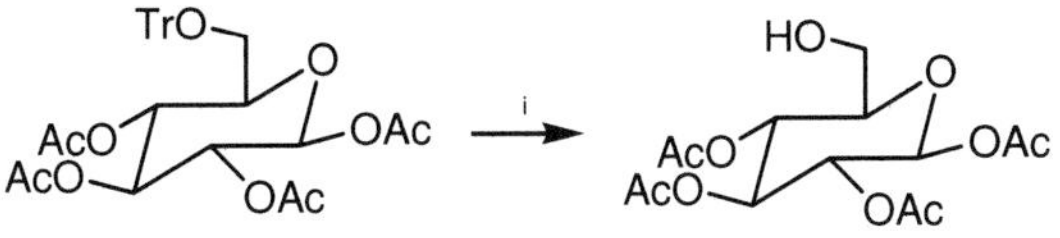

i) 1% I$_2$, MeOH.

Ref.[110]

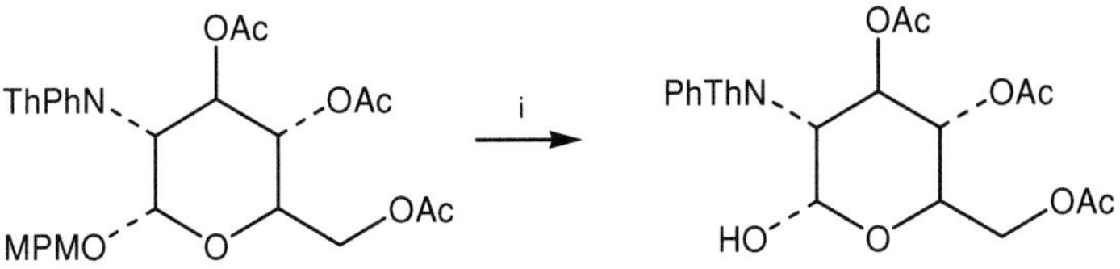

PhTh = phthalimido.
MPM = p-Methoxybenzyl ether p-MeOC$_6$H$_4$CH$_2$OR

i) CAN or NBS, CH$_2$Cl$_2$, H$_2$O.

Ref.[111]

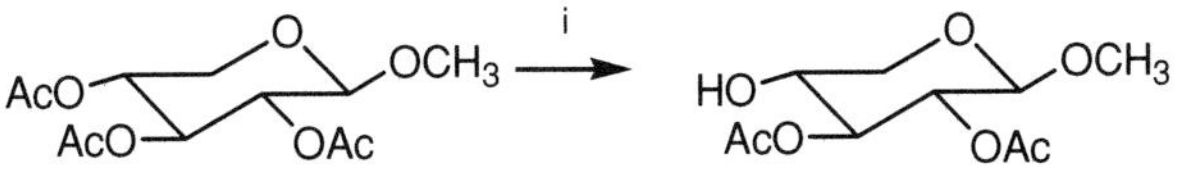

i) Lipase PS, n-pentyl-OH, 93%

Ref.[112]

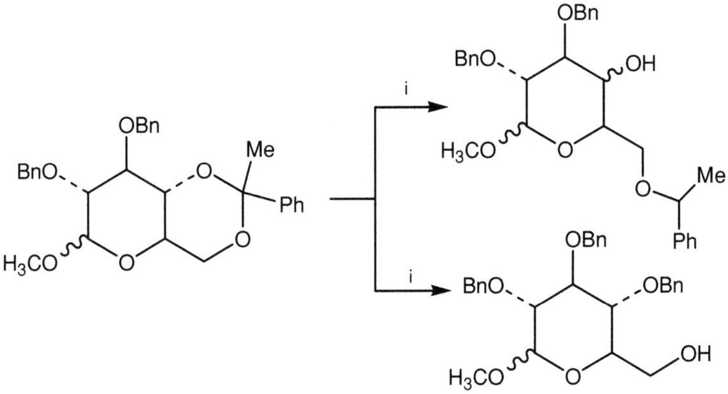

i) LiAlH$_4$, AlCl$_3$, Et$_2$O, CH$_2$Cl$_2$, heat

Ref.[113]

FIGURE 1.77. Miscellaneous selective deprotections.

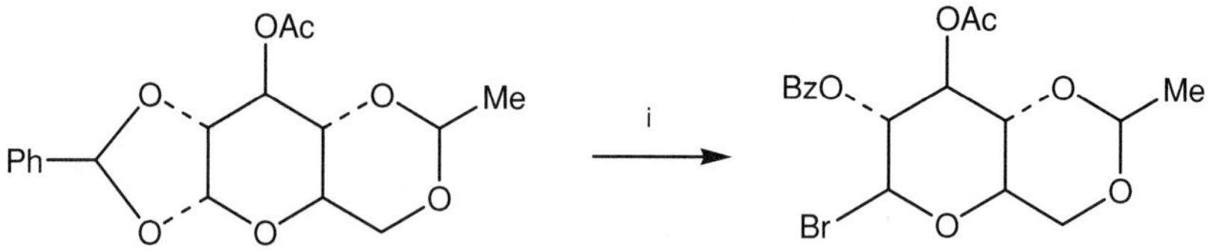

i) BrCCl$_3$, CCl$_4$, hv, 30 min. 100%

Ref.[114]

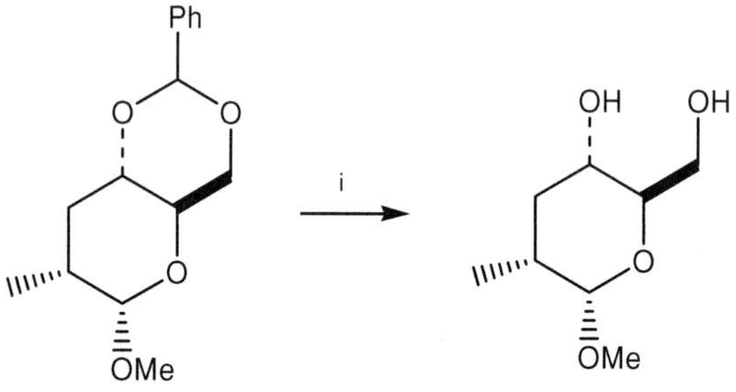

i) H$_2$, Pd(OH)$_2$, EtOH, 92%

Ref.[94]

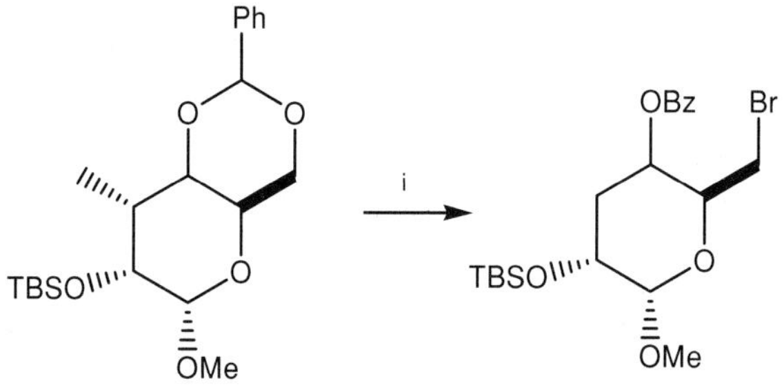

i) NBS, BaCO$_3$, CCl$_4$,Δ

Ref.[115]

FIGURE 1.77. (*continued*)

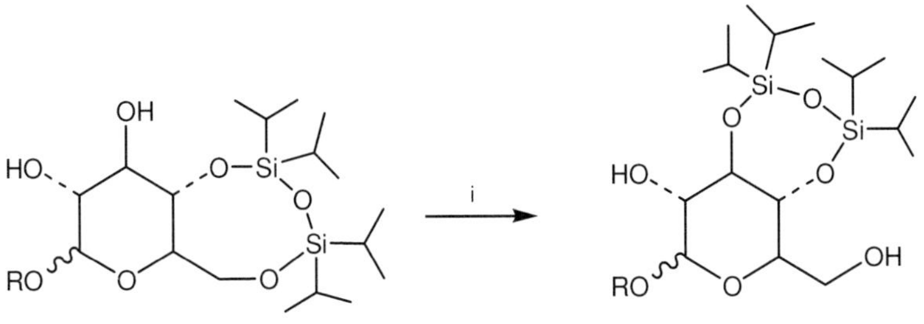

i) DMF, H+, 82%

Ref.[116]

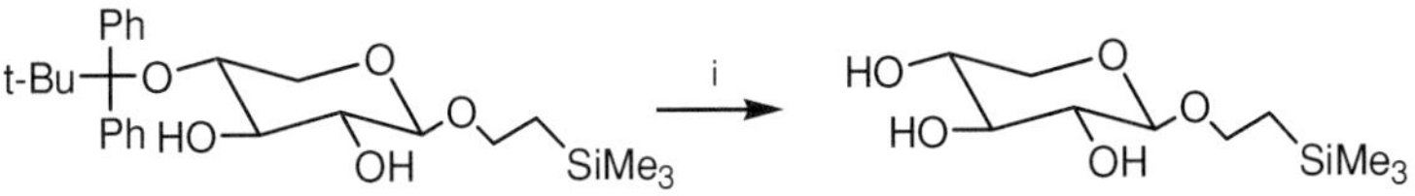

i) Bu$_4$NF.3 H$_2$O, AcOH, THF.

Ref.[117]

i) CF$_3$COOH, CH$_2$Cl$_2$.

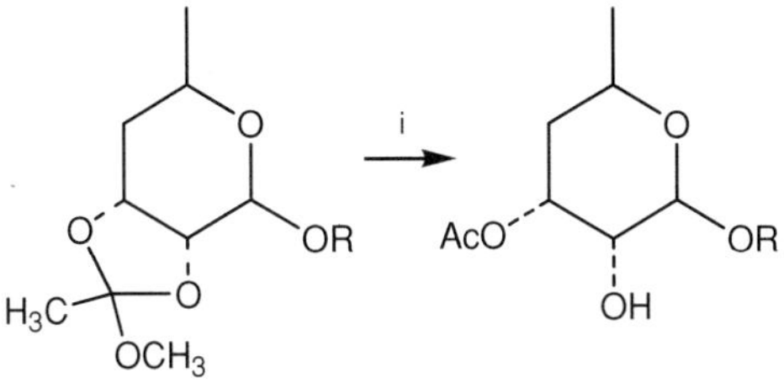

i) AcOH, 100%.

Ref.[118]

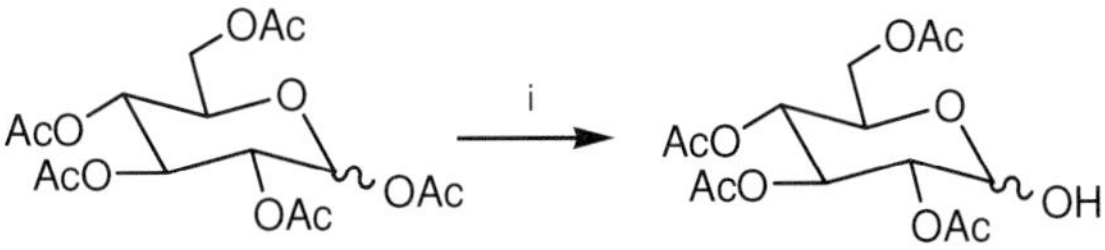

i) a) BnNH$_2$, THF. b) HCl.

Ref.[86]

FIGURE 1.77. (*continued*)

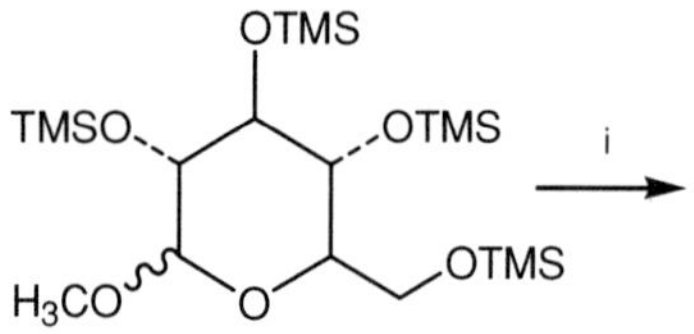

i) K$_2$CO$_3$, MeOH, 0°C, 45 min, 100%

Ref.[119]

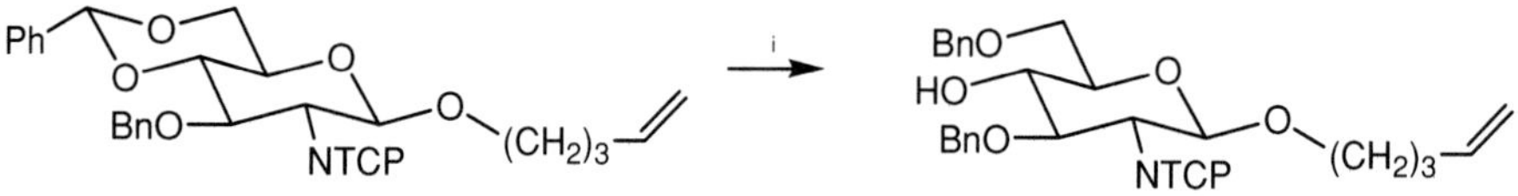

i) a) NaBH$_3$CN. b) HCl, THF, Et$_2$O

Ref.[76]

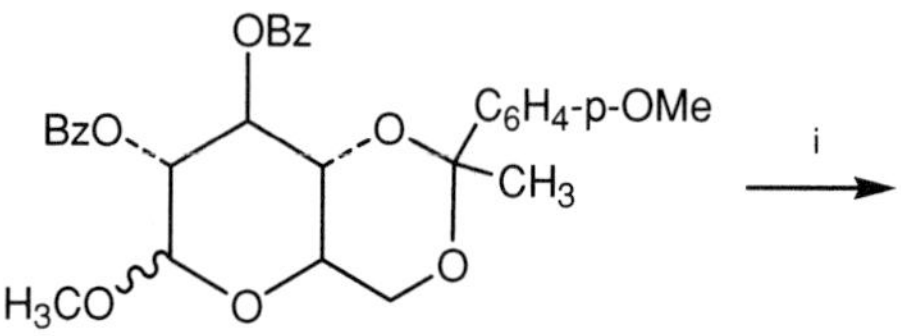

i) a) SnCl$_4$, CH$_2$Cl$_2$, –78°C. b) Bu$_4$NOH, 90%.

Ref.[108]

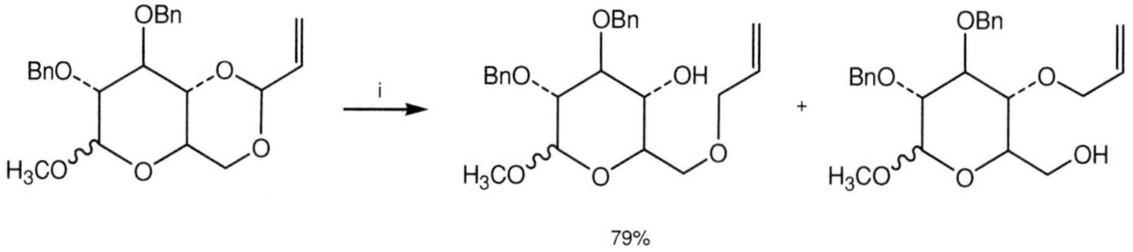

79%

i) a) NaBH$_3$CN. b) HCl, THF.

Ref.[120]

FIGURE 1.77. (continued)

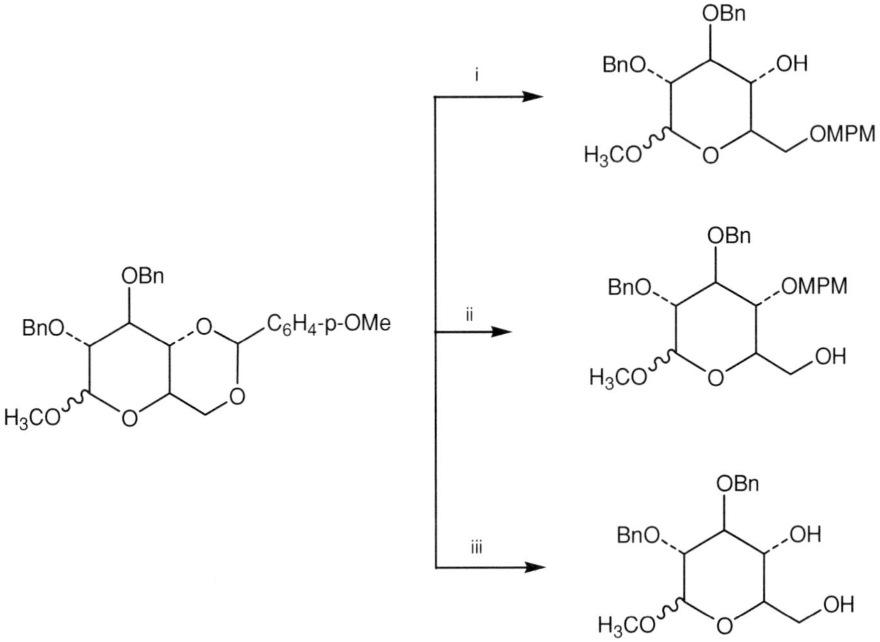

i) NaBH₃CN. b) TFA, DMF. ii) a) NaBH₃CN. b) TMSCl, CH₃CN. iii) CAN, CH₃CN-H₂O (9:1), 95%.

Ref.[121]

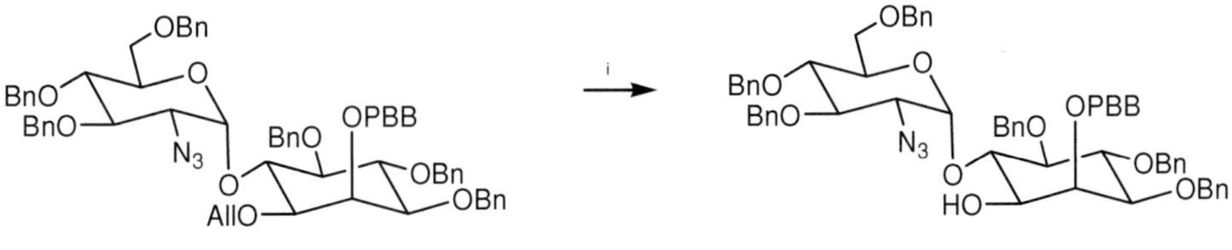

PBB = p-bromobenzyl

i) Pd(OAc)₂, (o-biphenyl)P(ᵗBu)₂, PhN(H)Me, NaOᵗBu, 80°C. ii) SnCl₄, 84%.

Ref.[122]

FIGURE 1.77. (*continued*)

References

1. J.F. Robyt, Essentials of Carbohydrate Chemistry, Springer, NY (1998).
2. H.S. Khadem, Carbohydrate Chemistry, Academic Press, NY (1988).
3. E. Fischer, Ber. **23**, 2114 (1890).
4. G. Casiraghi, F. Zanardi, G. Rassu, and P. Spanu, Chem. Rev. **95**, 1677 (1995)
5. T.-H. Chan, and C-J. Li J. Chem. Soc. Chem. Commun. 747 (1992)

6. J. Gao, R. Härtner, D.M. Gordon, and G.M. Whitesides, J. Org. Chem. **59**, 3714 (1994).

7. R.H. Prenner, W.H. Binder, and W. Schmid, Liebigs Ann. Chem. 73 (1994).

8. K.-I. Sato, T. Miyata, I. Tanai., and Y. Yonezawa, Chem. Lett. 129 (1994).

9. S.G. Davies, R.L. Nicholson., and A.D. Smith, Synlett 1637 (2002).

10. A.B. Northrup, I.K. Mangion, F. Hettche, and D.W.C. MacMillan, Angew. Chem. Int. Ed. **43**, 2152 (2004).

11. A. Lubineau, J. Augé, and N. Lubin, Tetrahedron **49**, 4639 (1993).

12. G. Casiraghi, L. Pinna, G. Rassu, P. Spanu., and F. Ulheri, Tetrahedron: Assimetry **3**, 681 (1993).

13. S.A.W. Gruner, E. Locardi, E. Lohof, and H. Kessler, Chem Rev. **102**, 491 (2002).

14. M.P. Watterson, L. Pickering, M.D. Smith, S.J. Hudson, P.R. Marsh, J.E. Mordaunt, D.J. Watkin, C.J. Newman, and G.W.J. Fleet, Tetrahedron: Asymmetry **10**, 1855 (1999).

15. N.L. Hungerford, and G.W.J. Fleet, J. Chem. Soc. Perkin Trans. 1 3680 (2000).

16. A. Dondoni, A. Marra, Chem Rev. **100**, 4395 (2000).

17. M.J. Robins, and J.M.R. Parker. Can. J. Chem. **61**, 312 (1983).

18. J.R. Axon, and A.L.J. Beckwith, J. Chem. Soc. , Chem. Commun. 549 (1995).

19. G. Li, H.H. Angert, and K.B. Sharpless, Angew. Chem. Int. Ed. Engl. **35**, 2813 (1996).

20. O. Jarreton, T. Skrydstrup, J.-F. Espinosa, J. Jiménez-Barbero, and J.-M. Beau, Chem. Eur. J. **5**, 430 (1999).

21. Y. Ohnishi, and Y. Ichikawa, Bioorg. Med. Chem. Lett. **12**, 997 (2002).

22. M. Oberthur, C. Leimkuhler, and D. Kahne, Org. Lett. **6**, 2873 (2004).

23. O. Meyerhof, and K. Lohmann, Biochem. Z. **271**, 89 (1934).

24. H.M. Gijsen, L. Qiao, W. Fitz, and C.-H. Wong, Chem. Rev. **96**, 443 (1996).

25. M.D. Bednarski, H.J. Waldmann, and G.M. Whitesides, Tetrahedron Lett. **27**, 5807 (1986).

26. W. Baumann, J. Freidenreich, G. Weisshaar, R. Brossmer, and H. Friebolin, Biol. Chem. **370**, 141 (1989).

27. G.J. Boguslawski, J. App. Biochem. **5**, 186 (1983).

28. J.R. Durrwachter, and C.-H. Wong, J. Org. Chem. **53**, 4175 (1988).

29. S.-H. Jung, J.H. Jeong, P. Miller, and C.-H. Wong, J. Org. Chem. **59**, 7182 (1994).

30. G.C. Look, Ch.H. Fotsch, and C.-H. Wong, Acc. Chem. Res. **26**, 182 (1993).

31. T. Aoyagi, T. Yamamoto, K. Kojiri, H. Morishima, M. Nagai, M. Hamada, T. Takechi, and H. Umezawa, J. Antibiot. **42**, 883 (1989).

32. R. Saul, R.J. Moylneux, and A.D. Elbein, Arch. Biochem. Biophys. **230**, 668 (1984).

33. H. Kayakiri, K. Nakamura, S. Takase, H. Setoi, I. Uchida, H. Terano, M. Hashimoto, T. Tada, and S. Koda, Chem. Pharm. Bull. **39**, 2807 (1991).

34. B.C. Baguley, G. Römmele, J. Gruner, and W. Wehrli, Eur. J. Biochem. **97**, 345 (1979).

35. a) J. Swenden, C. Borgmann, G. Legler, and E. Bause, Arch. Biochem. Biophys. **248**, 335 (1986). b) C. McDonnell, L. Cronin, J.L. O'Brien, P.V. Murphy, J. Org. Chem. **69**, 3565 (2004).

36. A. Dondoni, P. Merino, and D. Perrone, Tetrahedron **49**, 2939 (1993).

37. T. Ziegler, A. Straub, and F. Effenberger, Angew. Chem. Int. Ed. Engl. **27**, 716 (1988).

38. C. Augé, and C. Gautheron, Adv. Carbohydr. Chem. **49**, 175 (1991).

39. Y.-F. Wang, D.P. Dumas, and C.-H. Wong, Tetrahedron Lett. **34**, 403 (1993).

40. R. Rai, I. McAlexander, and Ch.-W.T. Chang, Org. Prep. Proced. Int. (OPPI), 37, 339 (2005).

41. Y. LeBlanc, B.J. Fitzsimmons, J.P. Springer and J. Rokach, J. Am. Chem. Soc., 111, 2995 (1989).
42. J. Dubois, C.S., Tomooka, and E.M. Carreira, J. Am. Chem. Soc., 119, 3179 (1997).
43. J. Liu and Y.D. Gin, J. Am. Chem. Soc., 124, 9789 (2002).
44. N.V. Bovin, S.E. Zurabyan and A.Y. Khorlii, Carbohydr. Res., 98, 25 (1981).
45. Reddy et al J. Org. Chem. 69, 2630 (2004).
46. B. Elchert, J. Li, J. Wang, Y. Hui, R. Rai, R. Ptak, P. Ward, J.Y. Takemoto, M. Bensaci, and C.-W. Chang, J. Org. Chem., 69, 1513 (2004).
47. V. Pavliak and P. Kovbk, Carbohydr. Res. 210, 333 (1991)
48. F. Dasgupta and P.J. Garegg, Synthesis, 262 (1988)
49. W.-C. Chou, L. Chen, J.-M. Fang, C.-H. Wong, J. Am. Chem. Soc. 116, 6169 (1994).
50. R.V. Stick, K.A. Stubbs, Tetrahedron Asymmetry 16, 321 (2005).
51. S. Ogawa, N. Matsunaga, H. Li, and M.M. Palcic, Eur. J. Org. Chem. 631 (1999).
52. a) T.D. Heingtmann, and A.T. Vassela, Angew. Chem. Int. Ed. 38, 750 (1999). b) A. Bianchi, A. Russo, A. Bernanrdi, Tetrahedron Asymmetry 16, 381 (2005).
53. S. Ogawa, M. Ohmura, S. Hisamatsu, Synthesis 312 (2001).
54. R.M. van Well, K.P. Kartha, and R.A. Field, J. Carbohydr. Chem. 24, 463 (2005).
55. M. Shimizu, H. Togo and M. Yokohama Synthesis 779 (1998).
56. E. Juaristi, and G. Cuevas, The Anomeric Effect, CRC, Boca Raton 4 (1995).
57. W. Koenigs, and E. Knorr, Chem. Ber. 34, 957, (1901).
58. a) W. Roth, and W. Pigman, in Methods in Carbohydrate Chemistry; Whistler R.L. and Wolfrom, M.L., Eds.; Academic Press: New York, 1963, Vol. 2, 405 (1963).; b) B.K. Shull, Z. Wu, and M. Koreeda, J. Carbohydr. Chem. 15(8), 955 (1996).; A. Fürstner, and H. Weidmann, J. Carbohydr. Chem., 7(4), 773 (1988).
59. R.U. Lemieux, and A.R. Morgan, Can. J. Chem. 43, 2199 (1965).; W. Wang, F. Kong, J. Org. Chem. 63, 5744 (1998).
60. M. Blanc-Muessen, J. Defaye, and H. Driguez, Carbohydr. Res. 67, 305 (1978).
61. Ch. McCloskey, and G.H. Coleman, Org. Synth. 3, 434, (1955).
62. J.A. Perrie, J.H. Harding, C. King, D. Sinnott, and A.V. Stachulski, Org. Lett. 5, 4545 (2003).
63. K.C. Nicolaou, J. Li, and G. Zenke, Helv. Chim. Acta 83, 1977 (2000).
64. Z. Györgydeák, L. Szilágyi, and H. Paulsen, J. Carbohydr. Chem. 12(2), 139 (1993).
65. G.-J. Boons, and A.V. Demchenko, Chem Rev. 100, 4539 (2000).
66. H. Paulsen, and H. Tietz, Carbohydr. Res. 125, 47 (1984).
67. H. Maeda, K. Ito, H. Ishida, M. Kiso., and A. Hasegawa, J. Carbohydr. Res. 14, 387 (1995).
68. S. Sugiyama, and J.M. Diakur, Organic Lett. 2, 2713 (2000).
69. D. Kahne, S. Walker, Y. Cheng, and D. Van Engen, J. Am. Chem. Soc. 111, 6881 (1989).
70. O.J. Plante, and P. Seeberg, J. Org. Chem. 63, 9150 (1998).
71. V. Constantino, C. Imperatore, E. Fattoruso, and A. Magnoni, Tetrahedron Lett. 41, 9177, (2000).
72. A. Dios, A. Geer, C.H. Marzabadi, and W.R. Franck, J. Org. Chem. 63, 6673 (1998).
73. F. Bosse, L.A. Marcaurelle, and P.H. Seeberger, J. Org. Chem. 67, 6659 (2002).
74. R.R. Schmidt, Angew. Chem. Int. Engl. 25, 213 (1986).
75. P.G.M. Guts, and T.W. Greene, Protecting groups in Organic Synthesis; Wiley, New York, (1991).
76. L. Olsson, Z.J. Jia, and B. Fraser-Reid, J. Org. Chem. 63, 3790 (1998).; K.C. Nicolaou, N. Winssinger, J. Pastor, and F. De Roose, J. Am. Chem. Soc. 119, 449 (1997).

77. D.P. Larson, and C.H. Heathcock, J. Org. Chem. **62**, 8406 (1997).
78. L. Liu, and H. Liu, Tetrahedron Lett. **30**, 35 (1989).
79. B. Classon, P.J. Garegg, S. Oscarson, and A.K. Tiden, Carbohydr. Res. **216**, 187 (1991).
80. K.C. Nicolaou, T. Ohshima, F.L. van Delft, D. Vourloumis, J.Y. Xu, J.S. Pfefferkorn, and S. Kim, J. Am. Chem. Soc. **120**, 8674 (1998).
81. I. Kitagawa, K. Ohashi, N.I. Baek, M. Sakagami, M. Yoshikawa, and H. Shibuya, Chem. Pharm., Bull. **45**(5), 786 (1997).
82. V. Ellervik, and G. Magnusson, J. Org. Chem. **63**, 9314 (1998).
83. D. Crich, and H. Li, J. Org. Chem. **67**, 4640 (2002).
84. J.M. Frechét, Polymer-supported Reactions in Organic Synthesis; Hodge, P., Sherrington, D.C., Eds.; Wiley. (1980)
85. G.G. Cross, and D.M. Whitfield, Synlett 487 (1999).
86. J.L. Koviak, M.D. Chapell, and R.L. Halcomb, J. Org. Chem. **66**, 2318 (2001).
87. C.-H. Wong, X.-S. Ye, and Z. Zhang, J. Am. Chem. Soc. **120**, 7173 (1998).
88. R. Lakhmiri, P. Lhoste, and D. Sinou, Tetrahedron Lett. **30**, 4669 (1989).
89. S.K. Chaudhary, and O. Hernández, Tetrahedron Lett. 95 (1979).
90. T.W. Hart, D.A. Metcalfe, and F. Scheinmann, J. Chem. Soc. Chem. Commun. 156 (1979).
91. M. Oikawa, A. Wada, F. Okazaki, and S. Kusumoto, J. Org. Chem. **61**, 4469 (1996).
92. I. Matsuo, M. Isomura, R. Walton, and K. Ajisaka, Tetrahedron Lett. **37**, 8795 (1996).
93. F. Theil, and H. Schick, Synthesis 533 (1991).
94. A.B. Smith III, and K.J. Hale, Tetrahedron Lett. **30**, 1037 (1989).
95. H. Yamada, T. Harada, and T. Takahashi. J. Am. Chem. Soc. **116**, 7919 (1994).
96. L. Leau, N. Goren, Carbohydr. Res. **C8**, 131, (1984).
97. M. Guiso, C. Procaccio, M.R. Fizzano, and F. Piccioni, Tetrahedron Lett. **38**, 4291 (1997).
98. J. Cai, B.E. Davison, C.R. Ganellin, and S. Thaisrivongs, Tetrahedron Lett. **36**, 6535 (1995).
99. R.W. Binkley, G.S. Goewey, and J.C. Johnston, J. Org. Chem. **49**, 992 (1984).
100. A. Arasappan, and P.L. Fuchs, J. Am. Chem. Soc. **117**, 177 (1995).
101. L. Jiang, and T.-H. Chan, J. Org. Chem. **63**, 6035 (1998).
102. Tsatsuda et al., Bull. Chem. Soc. Jpn. **58**, 1699 (1985).
103. S.V. Ley, and S. Mio, Synlett 789 (1996).
104. J.G. Fernandez-Bolaños, J.G. García, J. Fernandez-Bolaños, M.J. Diánez, M.D. Estrada, A. López-Castro, and S. Pérez Garrido, Tetrahedron Assymetry **14**, 3761 (2003).
105. J.G.S. Lohman, and P.H. Seeberger, J. Org. Chem. **68**, 7541 (2003).
106. H. Liang, and T.B. Grindley J. Carbohydr. Chem. **23**, 71 (2004).
107. M. Adinolfi, G. Borone, L. Guariniello, and A. Iadonisi, Tetrahedron Lett. **41**, 9305 (2000).
108. S. Forsén, B. Lindberg, and B.-G. Silvander, Acta Chem. Scand. **19**, 359, (1965).
109. K. Koch, and R. J. Chambers, Carbohydr. Res **241**, 295 (1993).
110. J.L. Wahlstrom, and R.C. Ronald, J. Org. Chem. **63**, 6021 (1998).
111. B. Classon, P.J. Garegg, and B. Samuelson Acta Chem Scan. Ser. B, **38**, 419 (1984).
112. R. López, E. Montero, F. Sanchez, J. Cañada, and A. Fernandez-Mayoralas, J. Org. Chem. 59, 7029 (1994).
113. R. Miethchen, J. Holz, H. Prade, and A. Liptak, Tetrahedron 3061 (1992).

114. P.M. Collins, A. Manro, E.C. Opara-Mottah, and M.H. Ali, J. Chem. Soc. Chem. Commun. 272 (1988).
115. S. Hanessian, and N.R. Plessas, J. Org. Chem. **34**, 1035 (1969).
116. C.A.A. Van Boeckel, and J.H. van Boom, Tetrahedron **41**, 4545 (1985).
117. M. Wilstermann, and G. Magnusson, J. Org. Chem. **62**, 7961 (1997).
118. S. Hanessian, and R. Roy, Can. J. Chem. **63**, 163 (1985).
119. D.T. Hurst, A.G. McIness, Can. J. Chem. **43**, 2004 (1965).
120. P.J. Garegg, Acc. Chem. Res. **25**, 575 (1992).
121. R. Johansson, and B. Samuelson, J. Chem. Soc. Perkin Trans 1 2371 (1984).
122. X. Liu, and P.H. Seeberger, Chem. Commun. 1708 (2004).

2
O-Glycoside Formation

2.1   General Methods

When a monosaccharide (or sugar fragment of any size) is condensed with either an aliphatic or aromatic alcohol, or another sugar moiety through an oxygen, a glycoside bond is formed. General examples of *O*-glycosides are shown in Figure 2.1.

The most common coupling reaction methodologies used for preparing the vast majority of *O*-glycosides known thus far are[1]

The Michael reaction
The Fischer reaction
The Koenigs-Knorr reaction
The Helferich reaction
The Fusion method
The Imidate reaction
The Sulfur reaction
The armed-disarmed approach
The Glycal reaction
The Miscellaneous leaving groups
The solid-phase approach

2.1.1   The Michael Reaction

P = protecting group
X = Br, Cl

Promoter	Conditions
NaH	THF
K$_2$CO$_3$, NaOH	acetone

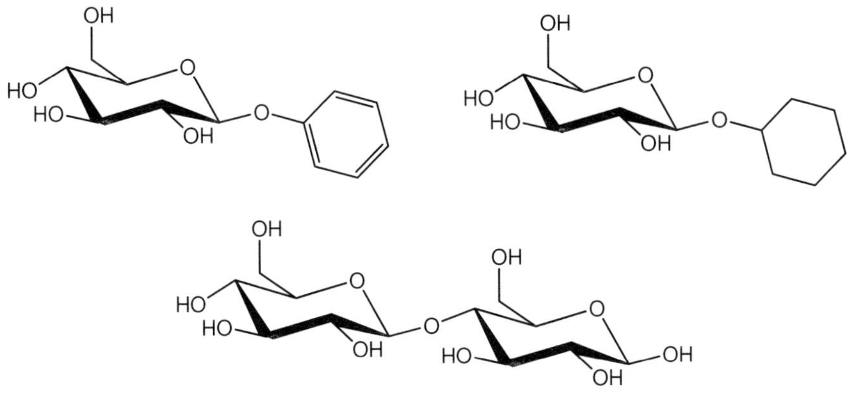

FIGURE 2.1. Examples of O-glycosides.

This pioneering methodology for O-glycosylation consists of the condensation reaction between 2,3,4,6-Tetraacetyl-α-D-glucopyranosyl chloride and potassium phenoxide to generate the acetylated derivate that undergoes basic hydrolysis to give phenyl-β-D-glucopyranoside (Figure 2.2). Since its original methodology, some modifications have been introduced especially for aromatic glycosides.

Some of the main features associated with this methodology are

Preserves the pyranose or furanose ring
Drives the addition of the aromatic aglycon to the anomeric position
Uses protecting groups which are easily removed in basic medium
Produces exclusively the β-O-glycoside as a result of neighboring group participation

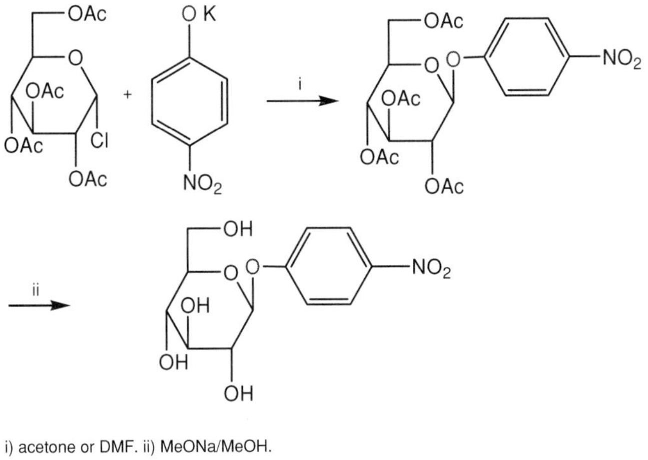

i) acetone or DMF. ii) MeONa/MeOH.

FIGURE 2.2. Synthesis of paranitrophenyl-β-D-glucopyranosyl tetraacetate.

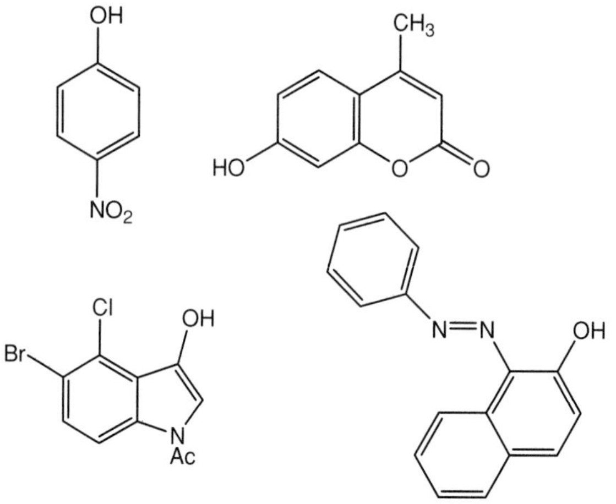

FIGURE 2.3. O-glycoside chromophores used for enzymatic detection.

This reaction has been employed for the preparation of O-glycosides that are used as substrates for detection and measurement of enzymatic activity of most of the known glycosidases.

Using this methodology, several chromophores have been attached to most of the common monosaccharides. After O-glycoside cleavage by the enzyme, the release of the chromophore will indicate the sites and eventually will quantify the enzymatic activity. Some of the chromophores currently used for these purposes are represented in Figure 2.3.

The highly fluorescent O-glycoside substrate 7-hydroxy-4-methylcoumarin-β-D-glucopyranose is prepared by condensation between acetobromoglucose with 4-methylumbelliferone in the presence of potassium carbonate in acetone. The intermediate is deacetylated under basic conditions to afford umbelliferyl β-D-glucopyranoside (Figure 2.4).

Anderson and Leaback[2] were able to prepare 5-Bromo indoxyl-β-D- N-acetylglucopyranoside, a hystochemical substrate for enzymatic detection of quitinase by condensing 3,4,6-triacetyl-β-D N-acetylglucopyranoside chloride with 5-bromo-hydroxy-N acetyl indole at 0°C under nitrogen atmosphere (Figure 2.5).

2.1.2 The Fischer Reaction

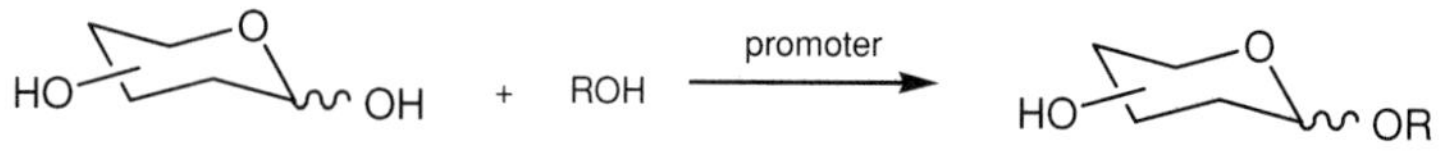

Promoter	Conditions
HCl gas	CH_2Cl_2, r.t.
pTsOH	CH_2Cl_2, r.t.

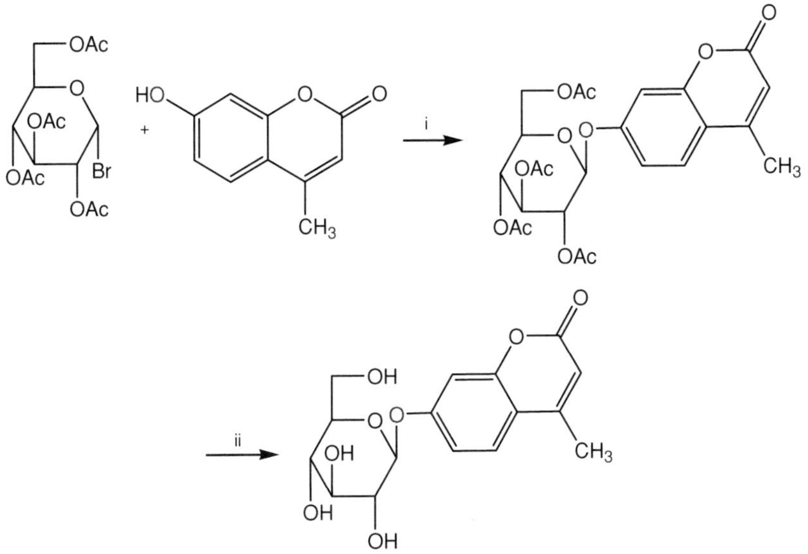

i) K₂CO₃/acetone. ii) MeONa/MeOH.

FIGURE 2.4. Michael approach for preparation umbellyferyl-*O*-glycoside.

This straightforward strategy is used specially for the preparation of simple *O*-glycosides. The advantage of this methodology is that it does not require the use of protecting groups and simply by combining the free sugar with an alcohol under acidic condition we furnish the corresponding *O*-glycoside. However, contrary to the previous method, this procedure is not stereo selective and therefore it provides a mixture of anomers. Also, it has been found satisfactory only for small aliphatic alcohols (Figure 2.6).

The addition of a controlled stream of dry HCl during a period of around 10 min at room temperature generally are the conditions of choice. However, the use

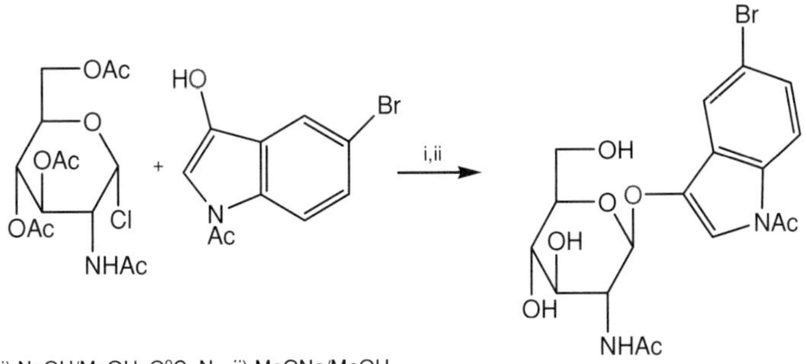

i) NaOH/MeOH, 0°C, N₂. ii) MeONa/MeOH.

FIGURE 2.5. Synthesis of indole *O*-glycoside derivative.

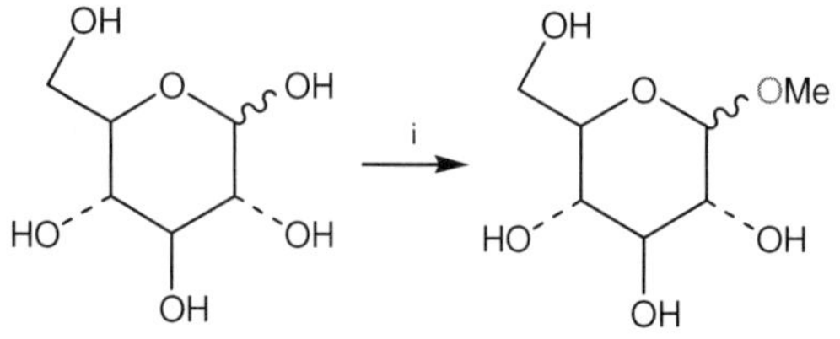

FIGURE 2.6. The Fischer O-glycoside reaction.

i) MeOH-HCl(g).

of Lewis acid, ion exchange resin, and more recently triflic acid have been also reported providing good yields.[3]

It is worth mentioning that besides the main product, a mixture of isomers has been detected, suggesting that a rather complex mechanism is involved. It is also seen that the amount of these isomers depends importantly on the condition reactions employed (Figure 2.7).

The Fischer methodology has been applied successfully for the synthesis of benzyl O-glycosides. L-Fucose was converted into benzyl fucopyranoside[4] by treatment with benzyl alcohol under saturation with HCl at 0°C, to furnish the α and β anomers (ratio 5:1) in 80% yield (Figure 2.8).

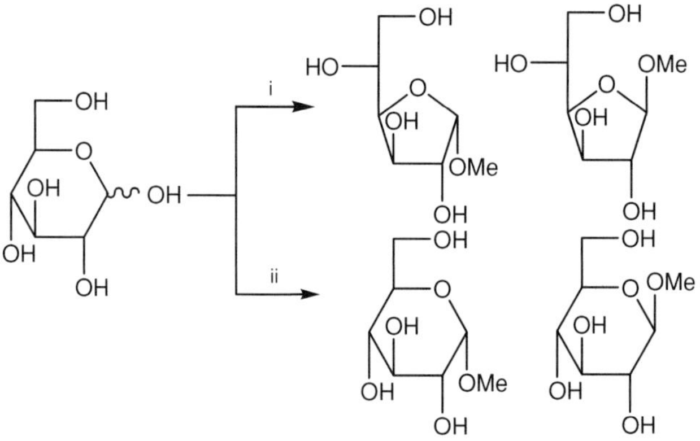

i) MeOH/ 0.7% HCl, 20°C. ii) MeOH/ 4% HCl reflux.

FIGURE 2.7. The Fischer O-glycoside isomers.

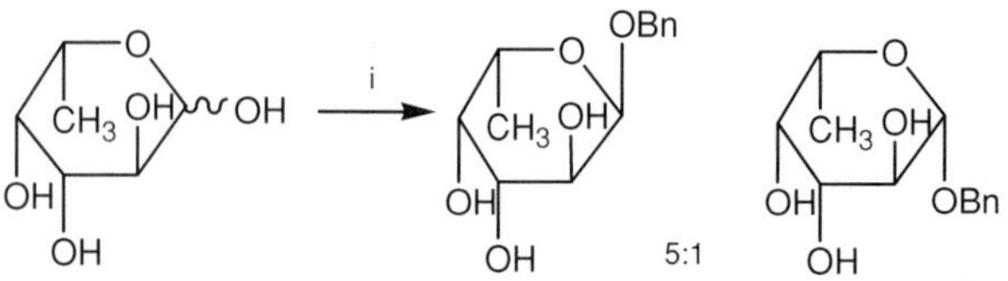

i) BnOH/HCl (g), 10 min. r.t, and O/N at 4°C.

FIGURE 2.8. Fischer conditions for preparation of Benzyl L-fucose.

2.1.3   The Koenigs-Knorr Reaction

X = Br, Cl

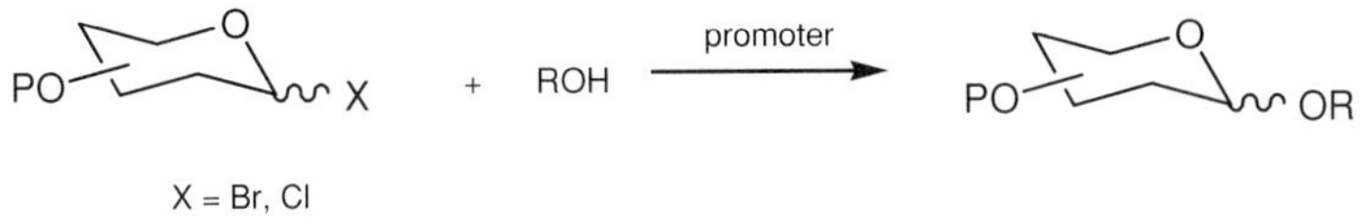

Promoter	Conditions
Ag_2CO_3	PhH, drierite (drying agent), I_2
Ag_2O	s-collidine (acid scavenger)
$AgNO_3$	HgO (acid scavenger)
$AgClO_4$	Ag_2ClO_3 (acid scavenger), THF or toluene, r.t.
AgOTf	CH_2Cl_2, r.t.

This reaction reported in 1901 is still one of the most useful reactions for preparing a wide variety of O-glycosides.[5] It is useful for coupling reactions with either alkyl or aromatic alcohols as well as for coupling between sugars. The methodology requires silver salts as catalyst and among them the oxide, carbonate, nitrate, and triflate silver salts are the most commonly employed (Figure 2.9). Also a drying agent such as calcium sulfate (drierite), calcium chloride, or molecular sieves is recommended. Improved yields are obtained with iodide, vigorous stirring, and protection against light during the course of the reaction.

The stereochemistry observed is 1,2 trans type in most of the cases reported, as a consequence of neighboring group participation. When the protecting group is acetate at C (2), there is an intramolecular nucleophilic displacement of the leaving group, generating an orthoester.[6] This intermediate is responsible for the incorporation of the alcohol on the β-position (Figure 2.10). Only until recently a method for preparing 1,2-cis glycosides has been developed involving the use of (1S)-phenyl-2-(phenylsulfanyl)ethyl moiety at C-2 of a glycosyl donor to give a quasi-stable anomeric sulfonium ion. The sulfonium ion is formed as a trans-decalin ring system. Displacement of the sulfonium ion by a hydroxyl leads to the stereoselective formation of α-glycosides.[7]

This versatile methodology can be applied for preparation of alky, aryl, and oligosaccharide O-glycosides. A steroidal glycoside cholesterol absorption inhibitor was prepared by condensation between acetobromocellobiose and (3β,5α,

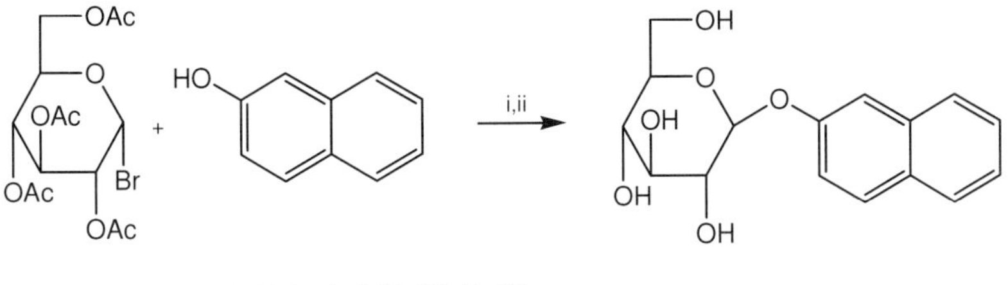

i) Ag_2O or Ag_2CO_3/PhH, drierite, I_2. ii) MeONa/MeOH.

FIGURE 2.9. The Koenigs-Knorr reaction.

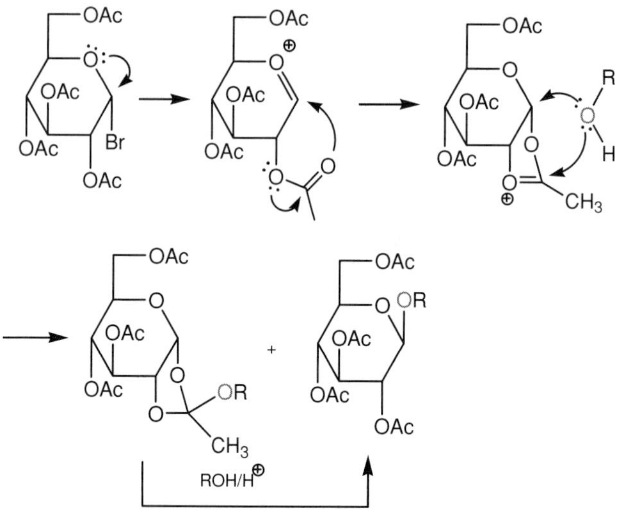

FIGURE 2.10. Proposed mechanism for the Koenigs-Knorr glycosidic reaction.

25R)-3-hydroxyspirostan-11-one with anhydrous ZnF_2 as catalyst in acetonitrile to provide the steroidal glycoside in 93% yield (Figure 2.11).[8]

The steroidal glycoside Estrone-β-D-glucuronide was prepared by condensation between Methyl tri-O-glucopyranosylbromide uronate with estrone, employing cadmium instead of silver carbonate (Figure 2.12).[9] A comprehensive study about methods for the preparation of diverse O-glucuronides has been described.[10]

The syntheses of various disaccharides have been reported under Koenigs-Knorr conditions. Gentobiose octaacetate was prepared through condensation of acetobromoglucose with 1,2,3,4-Tetra-O-acetyl-O-Trityl-β-D-glucopyranose in nitromethane using silver perchlorate as catalyst (Figure 2.13).[11]

Bächli and Percival[12] reported the synthesis of laminaribiose by reacting 1,2,5,6-Diisopropylidenglucose with acetobromoglucose in the presence of silver carbonate, iodine, and drierite to produce an acetonide intermediate, which, upon treatment with oxalic acid and sodium methoxide, furnished the 1,3-disaccharide (Figure 2.14).

The synthesis of various disaccharides containing N-acetylneuraminic acid (Neu5Ac) was achieved by using acetochloro and acetobromo neuraminic acids as glycosyl donors with active glycosyl acceptors under Ag_2CO_3-promoted reactions conditions (Figure 2.15).[13,14]

These conditions are also suitable for preparing short oligosaccharides such as the one presented in Figure 2.16. The donor sugar acetobromogentobiose is coupled to the acceptor intermediate using silver triflate as glycosidation catalyst.[15]

Total synthesis of Bleomycin group antibiotic has been achieved by Katano and Hecht.[16] Thus, glycoside coupling reaction of protected disaccharide glycosyl donor with histidine derivative using silver triflate as glycoside promoter provided Bleomycin key intermediate in 21% (Figure 2.17).

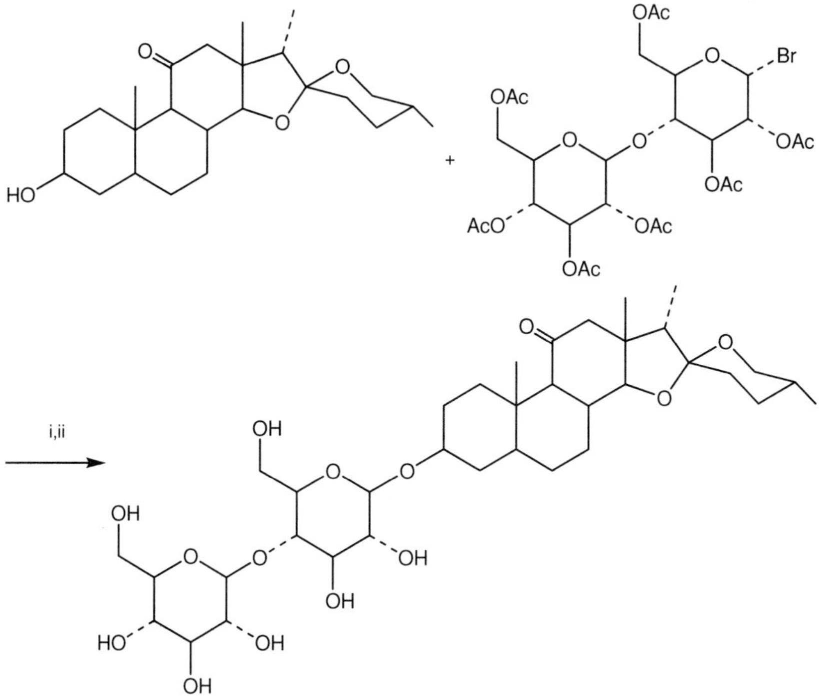

i) ZnF$_2$, CH$_3$CN. ii) NaOMe

FIGURE 2.11. Synthesis of steroidal glycoside.

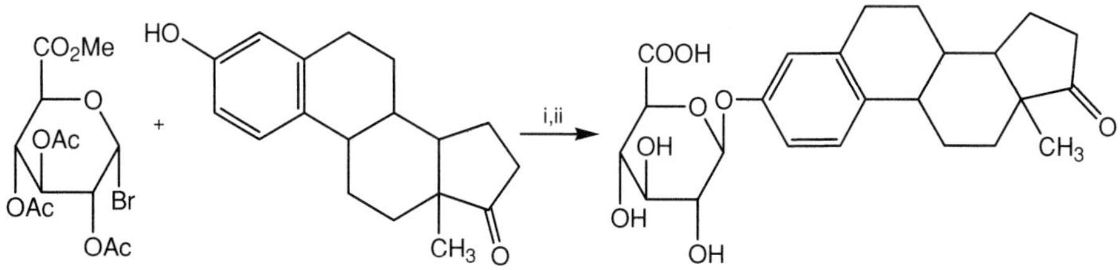

i) Cd$_2$CO$_3$. ii) MeONa/MeOH

FIGURE 2.12. Synthesis of a steroidal O-glycoside.

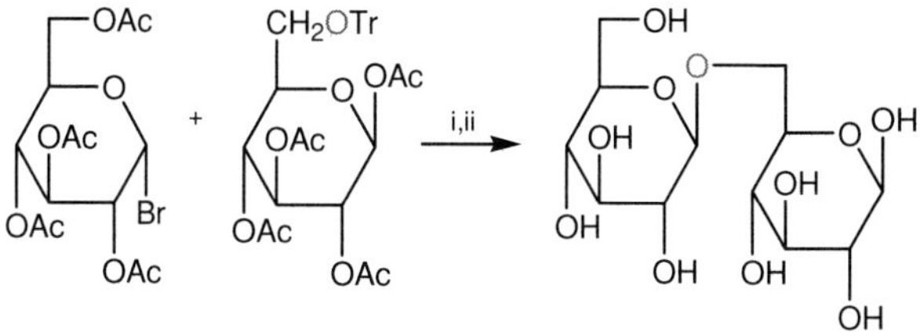

i) AgClO$_4$, CH$_3$NO$_2$. ii) MeONa/MeOH

FIGURE 2.13. Synthesis of gentobiose.

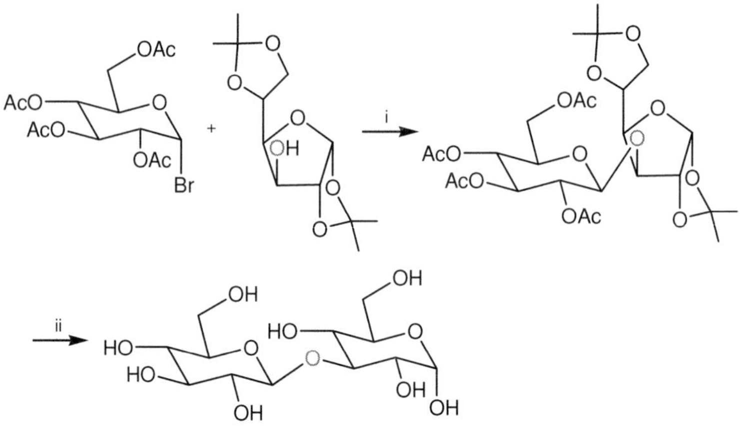

i) Ag$_2$CO$_3$, drierite, I$_2$. ii) a)MeONa/MeOH. b) oxalic acid 0.001 N, 100°C.

FIGURE 2.14. Synthesis of laminaribiose.

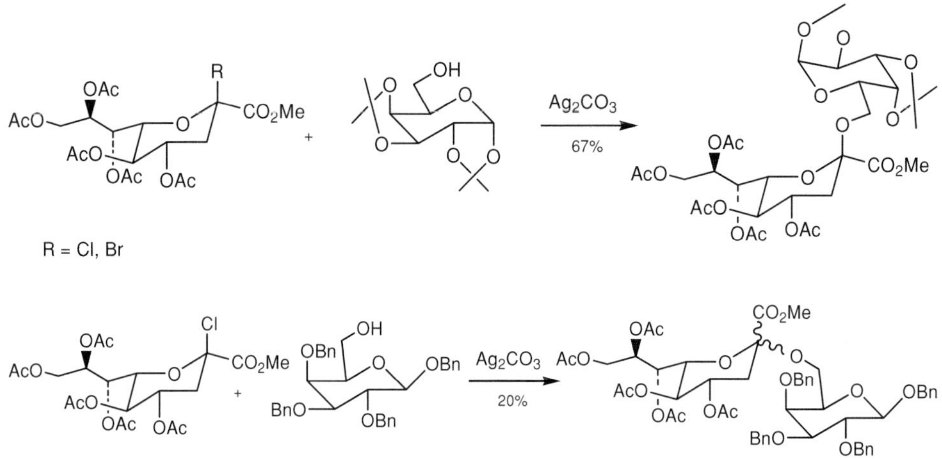

R = Cl, Br

FIGURE 2.15. Silver carbonate promoted synthesis of Neu5Ac(2→6) disaccharides.

2.1.4   The Helferich Reaction

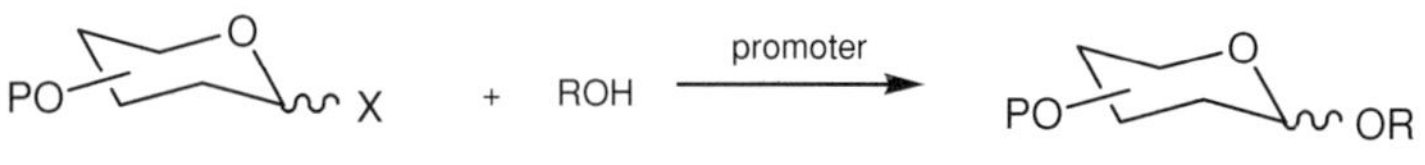

X = Br, Cl

Promoter	Conditions
Hg(CN)$_2$	CH$_3$CN
HgBr$_2$	CH$_3$CN
HgI$_2$	CH$_3$CN

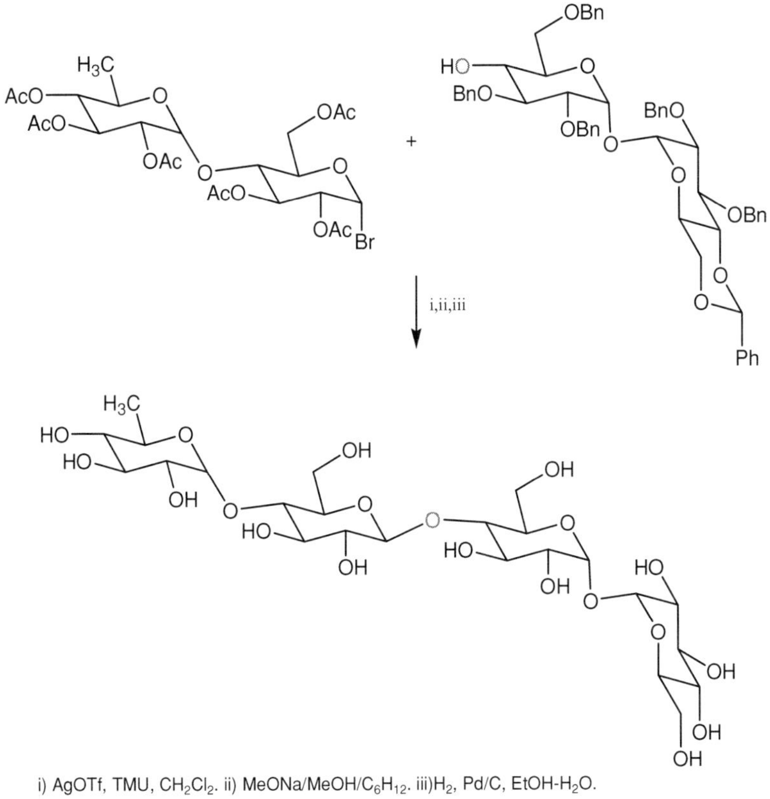

i) AgOTf, TMU, CH$_2$Cl$_2$. ii) MeONa/MeOH/C$_6$H$_{12}$. iii)H$_2$, Pd/C, EtOH-H$_2$O.

FIGURE 2.16. Synthesis of tetrasaccharide.

This methodology is considered a modification of the previous one, and the main change being the use of mercury and zinc salts instead of silver. Also, more polar solvents are used such as acetonitrile or nitromethane (Figure 2.18). The yields reported for this reaction are up to 70%, or higher. However, a mixture of anomers is often observed.

By following this strategy, Umezawa et al.[17] had prepared kanamacin A by condensing 6-O-[2-O-benzyl-3-(benzyloxycarbonylamino)-3-deoxi-4,6-O-isopropylidene-α-D-glucopyranosyl]-N,N'-di(benzoyloxycarbonyl)-2-deoxyestreptamine, as glycosyl acceptor with 2,3,4-tri-O-benzyl-6-(N-benzylacetamido)-6-deoxi-α-D-glycopyranosyl chloride, as glycosyl donor. The catalyst employed was mercury (II) cyanide (Figure 2.19).

The antitumoral O-glycoside Epirubicine was prepared under Helferich conditions[18] using the acetonide form of Adriamicinone and 2,3,6-trideoxi-3-trifluoroacetamido-4-O-trifluoroacetyl-α-L-arabinohexopyranosyl chloride, and a mixture of mercury (II) oxide and bromide as shown in Figure 2.20.

Other coupling reactions between sugars under Helferich conditions have been as well described.[19] For example, the case of trisaccharide Rafinose prepared by

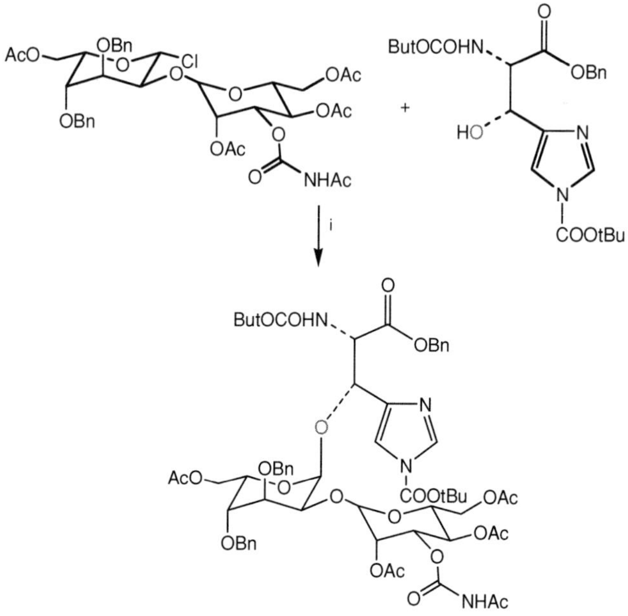

i) AgOTf, tetramethylurea.

FIGURE 2.17. Glycosilation reaction for preparation of Bleomycin precursor.

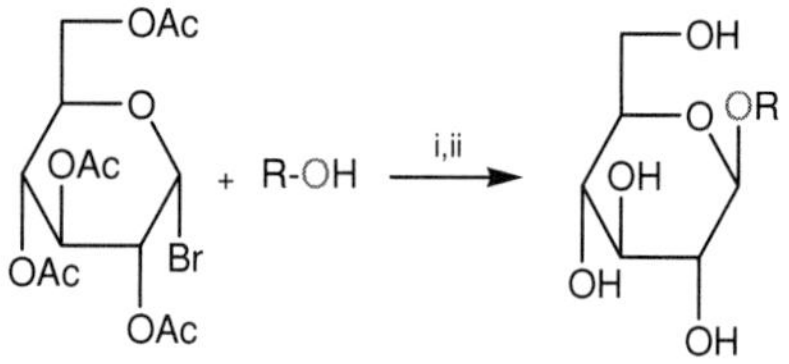

i) HgCN₂, CH₃CN. ii) MeONa/MeOH.

FIGURE 2.18. The Helferich general reaction.

condensation between Tetra-*O*-benzyl-α-D-galactopyranosyl chloride as donor and 2,3,4,1′,3′,4′,6′-hepta-*O*-acetyl sucrose as acceptor (Figure 2.21).

Helferich conditions have been used for preparing disaccharides containing Neu5Ac(2→6)Gal and Glc in good yields, although with low stereocontrol (α:β 3:4) (Figure 2.22).

2.1.5   The Fusion Reaction

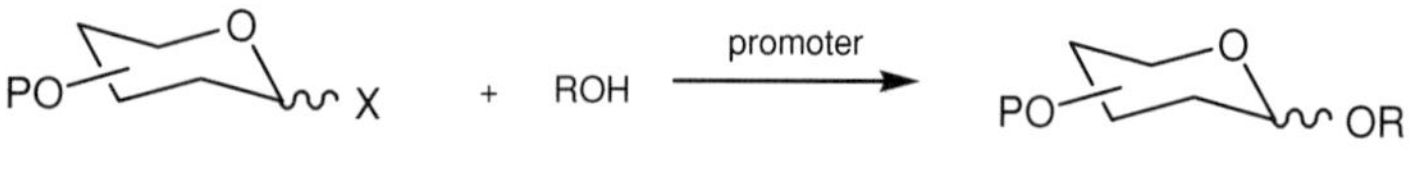

X = OAc

Promoter	Conditions
ZnCl₂	120°C
TsOH	120°C

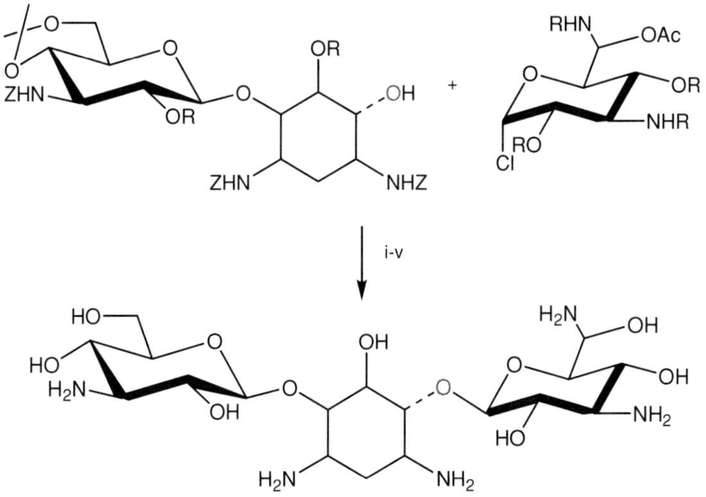

Z = PhCH₂COO-
R = PhCH₂-

i) Hg(II)CN₂, CaSO₄/dioxane, PhH. ii) MeONa/MeOH. iii) AcOH. iv) H₂, Pd-C.

FIGURE 2.19. Synthesis of a kanamacin A derivative.

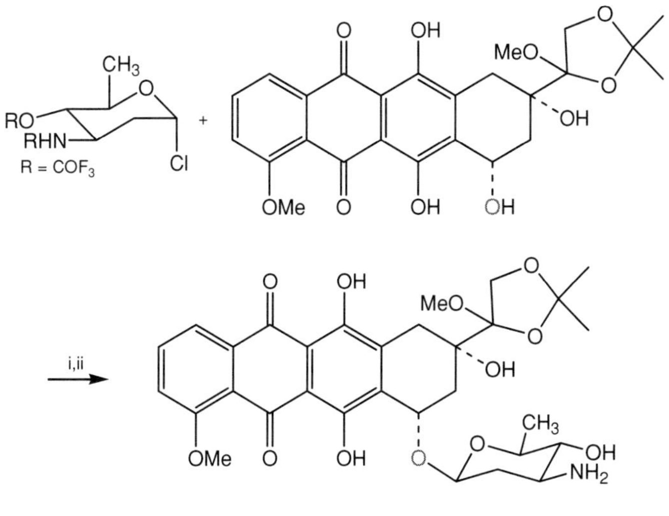

i) HgO-HgBr₂. ii) NaOH

FIGURE 2.20. Synthesis of Epirubicine.

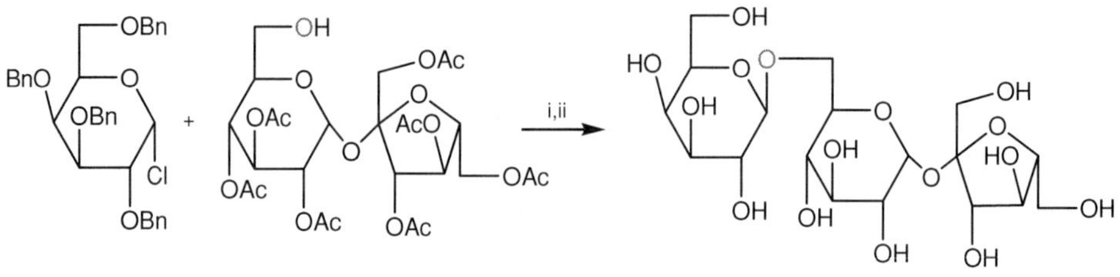

i) Hg(II)CN$_2$, CaSO$_4$, PhH. ii) MeONa/MeOH. iii) H$_2$, Pd-C.

FIGURE 2.21. Synthesis of Rafinose derivative.

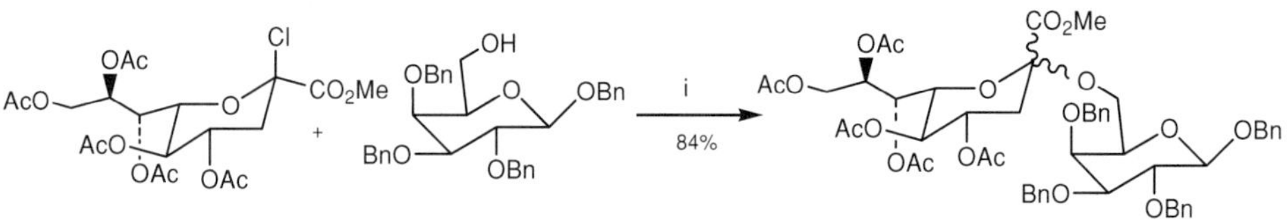

i) Hg(CN)$_2$/HgBr$_2$ (3:1)

FIGURE 2.22. Helferich conditions for the preparation of sialic disaccharide.

This is a process that is mainly used for preparing aromatic glycosides, and generally consists of the reaction between the sugar, having a leaving group either as acetate or bromide with a phenolic aglycon, under Lewis acid conditions, at temperatures above 100°C.

This methodology has been useful to synthesize 1-naphthyl 2,3,4,6-tetra-O-acetyl-α,β-L-idopyranoside by mixing 1,2,3,4,6-penta-O-acetyl-α-L-idopyranose, 1-naphthol, zinc chloride and heating up to 120°C during 1h. Also aromatic S-glycosides could be effectively prepared under the fusion method. Thiophenol 2,3,4,6-tetraacetyl glucopyranose was prepared as a mixture of anomers (40:60, α:β) when thiophenol was combined with ZnO or ZnCO$_3$ and then refluxed with acetobromoglucose in CH$_2$Cl$_2$ (Figure 2.23)[20].

2.1.6   The Imidate Reaction

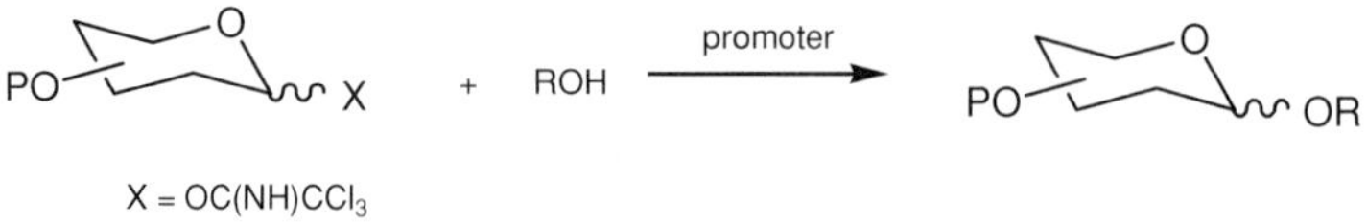

X = OC(NH)CCl$_3$

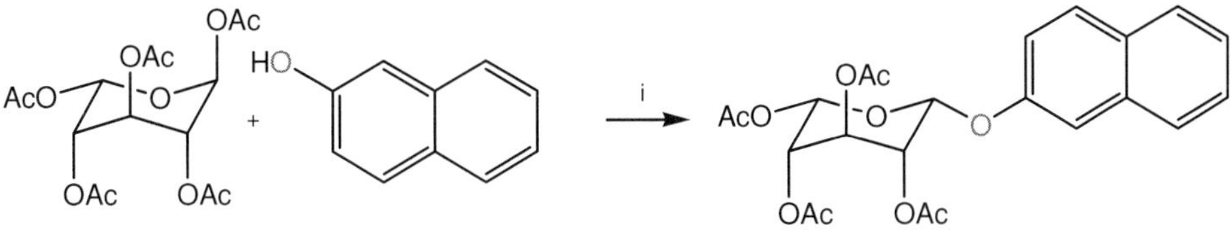

i) ZnCl$_2$, 120°C, 1h.

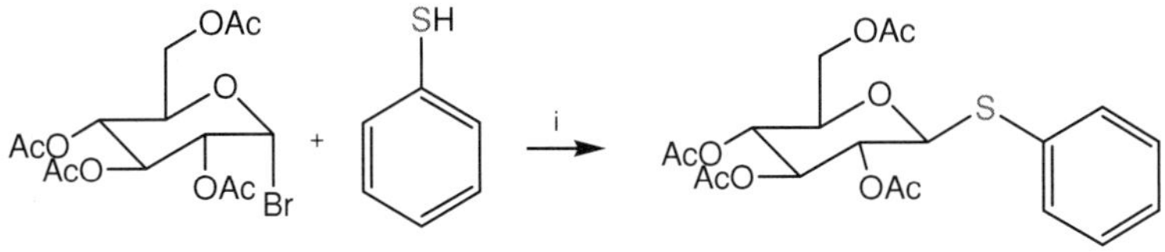

i) ZnO, CH$_2$Cl$_2$ reflux

FIGURE 2.23. Naphtyl O-glycosides and Phenyl S-glycosides.

Promoter	Conditions
AgOTf	CH$_2$Cl$_2$, 0°C→r.t.
TMSOTf	CH$_2$Cl$_2$ or MeCN 0°C
BF$_3$-OEt$_2$	CH$_2$Cl$_2$ or MeCN, −20°C
NaH	CH$_2$Cl$_2$

This is a more recent procedure attributed to Schmidt and coworkers[21a−c] who introduced trichloroacetimidate as a good leaving group for preparation of O-glycosides. A significant number of simple and complex O-glycosides involving the imidate coupling reaction have been described. This strategy involves the use of trichloroacetonitrile that in the presence of a base is incorporated on the anomeric hydroxyl group to generate trichloroacetimidate (Figure 2.24). It should be noted that the resulting imidate derivative is air- sensitive and should be used in coupling reactions immediately following preparation. Imidate formation might be spectroscopically detected by ^{1}H NMR through a signal appearing down field at 6.2 ppm.[22]

Once the imidate is formed, it can be subjected to nucleophilic attack to provide the corresponding S-, N-, C-, or O-glycoside, depending on the chosen nucleophile. The use of a catalyst such as BF$_3$.OEt$_2$, TMSOTf, or AgOTf is necessary to carry out the reaction to completion (Figure 2.25). Although the unquestionable applicability of this approach, an undesirable side reaction has been encountered with glycosyl trichloroacetimidates in the presence of Lewis acid catalysis via the Chapman rearrangement.[21b−c]

Hasegawa et al.[23] have prepared the ganglioside shown in Figure 2.26 using 2,3,4,6-tetrabenzylglucopyranosyl-α-acetimidate with the liphophilic alcohol, to generate a ganglioside.

The total synthesis of calicheamicin α and dynemicin A has been described by Danishefsky's group,[24] and involves glycosilation of calicheamicinone congener

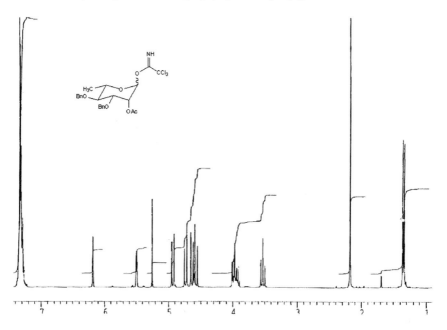

i) Bn-NH₂, HCl, THF. or NH₂NH₂   ii) Cl₃CN, CsCO₃/CH₂Cl₂, r.t.

FIGURE 2.24. Preparation of glycosyl imidate and ^{1}H NMR of imidate ramnosyl derivative.

with the complex glycosyl imidate using $BF_3.OEt_2$ as Lewis acid catalyst (Figure 2.27).

Naturally occurring herbicides known as tricolorin A, F, and G were isolated from the plant *Ipomea tricolor* and since then synthesized involving glycoside coupling reactions. The first total synthesis of tricolorin A was performed by Larson and Heathcock,[25] involving three coupling reactions steps with imidate intermediates used as glycosyl donors (Figure 2.28). The lactonization key step for the preparation of the synthesized tricolorins has been achieved either under macrolactonization conditions reported by Yamaguchi[26,27] and also under ring clousure methathesis conditions.[22]

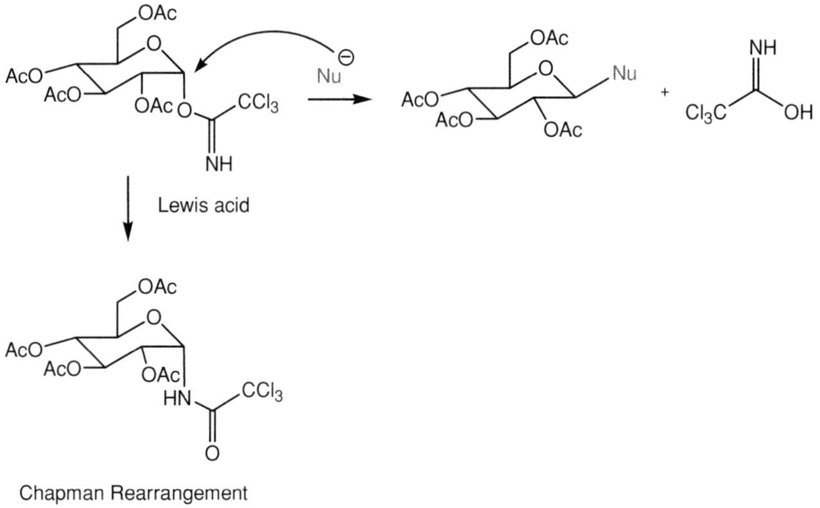

Chapman Rearrangement

FIGURE 2.25. Nucleophilic displacement of imidate leaving group.

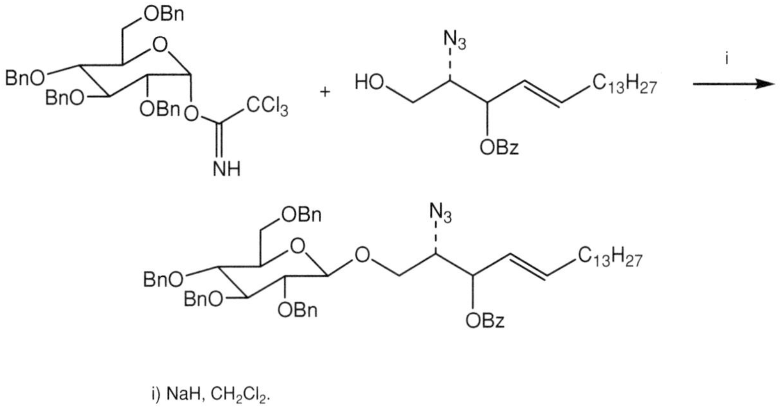

i) NaH, CH$_2$Cl$_2$.

FIGURE 2.26. Coupling reaction for the preparation of ganglioside.

Another hetero-trisaccharide resin glycoside of jalapinolic acid known as tricolorin F has been synthesized involving coupling reactions with imidates as glycosyl donors. In this way disaccharide and trisaccharide were prepared sequentially. The resulting tricoloric acid C derivative was deprotected and subjected to lactonization under Yamaguchi conditions to produce protected macrolactone. Final removal of acetonide and benzyl protecting groups provided Tricolorin F (Figure 2.29).[27]

A convergent approach for obtaining a tumoral antigen fragment of Lewis X has been developed by Boons et al.[28] Condensation of the imidate glycosyl donor and the trisaccharide glycosyl acceptor provided the hexasaccharide, which was

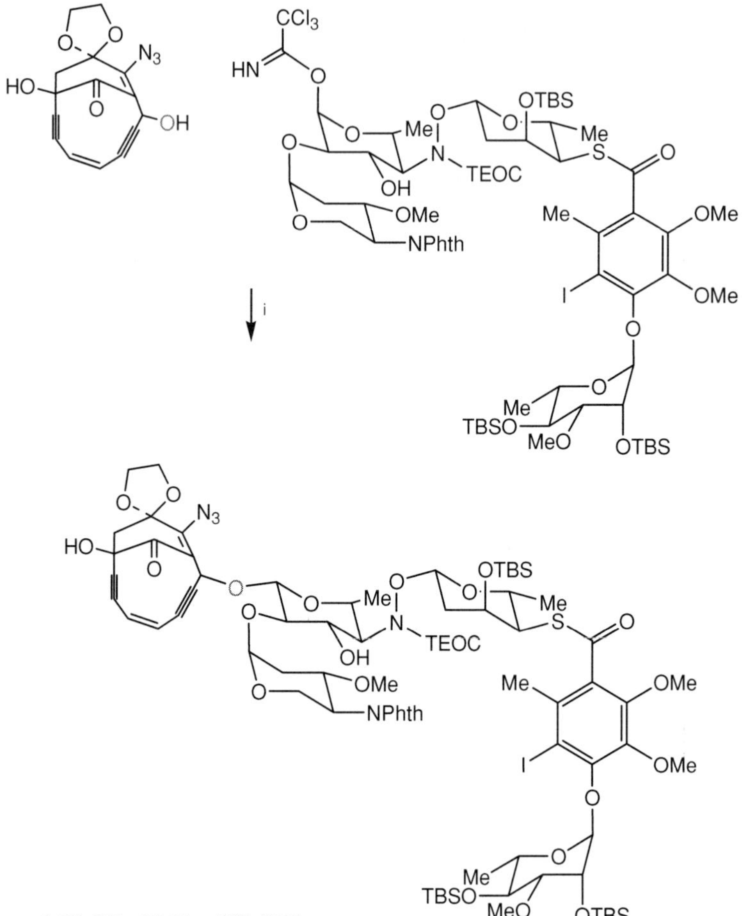

i) BF$_3$.OEt$_2$, CH$_2$Cl$_2$, −78°C (28%).

FIGURE 2.27. Glycosylation of calicheamicinone congener.

further allowed to react with trichloroacetimidate to generate a hexasaccharide glycosyl donor. The final coupling reaction with the disaccharide using BF$_3$.OEt$_2$, furnished the tumoral fragment Lewis X (Figure 2.30).

Selectins (E,P and L) are mammalian C-type lectins involved in the recognition process between blood cells or cancer cells and vascular endothelium. L-selectins play a key role in the initial cell-adhesive phenomena during the inflammatory process, whereas E-selectins bind strongly to sialyl Lewis a and x.[29,30] It has been found that the tetrasaccharide sialyl Lewis x is the recognition molecule and the preparation of sialyl Lewis x confirmed the hypothesis that sulfation increase the affinity for L-selectins.[31] The chemical synthesis of 3e- and 6e-monosulfated and 3e,6e- disulfated Lewis x pentasaccharides has been prepared according to Figure 2.31.

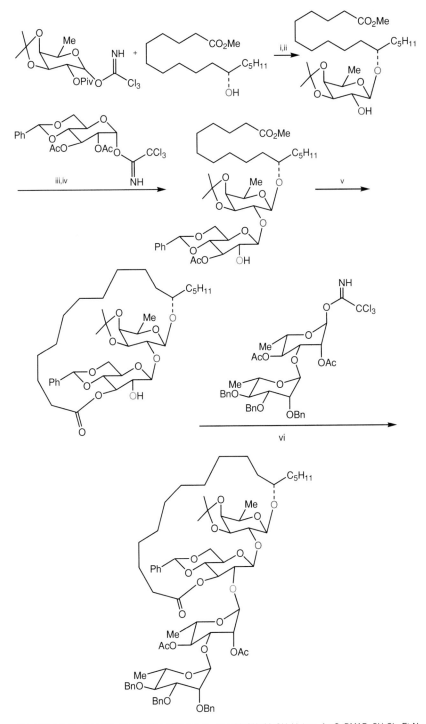

i) AgOTf, CH$_2$Cl$_2$. ii) MeONa/MeOH. iii) AgOTf, CH$_2$Cl$_2$. iv) a)MeONa/MeOH. b) 1 eq. Ac$_2$O, DMAP, CH$_2$Cl$_2$, Et$_3$N. v) a) LiOH, THF, H$_2$O. b) 2,4,6-trichlorobenzoyl chloride, Et$_3$N, MAP, benzene. vi) AgOTf, CH$_2$Cl$_2$.

FIGURE 2.28. Synthesis of tricolorin A precursor.

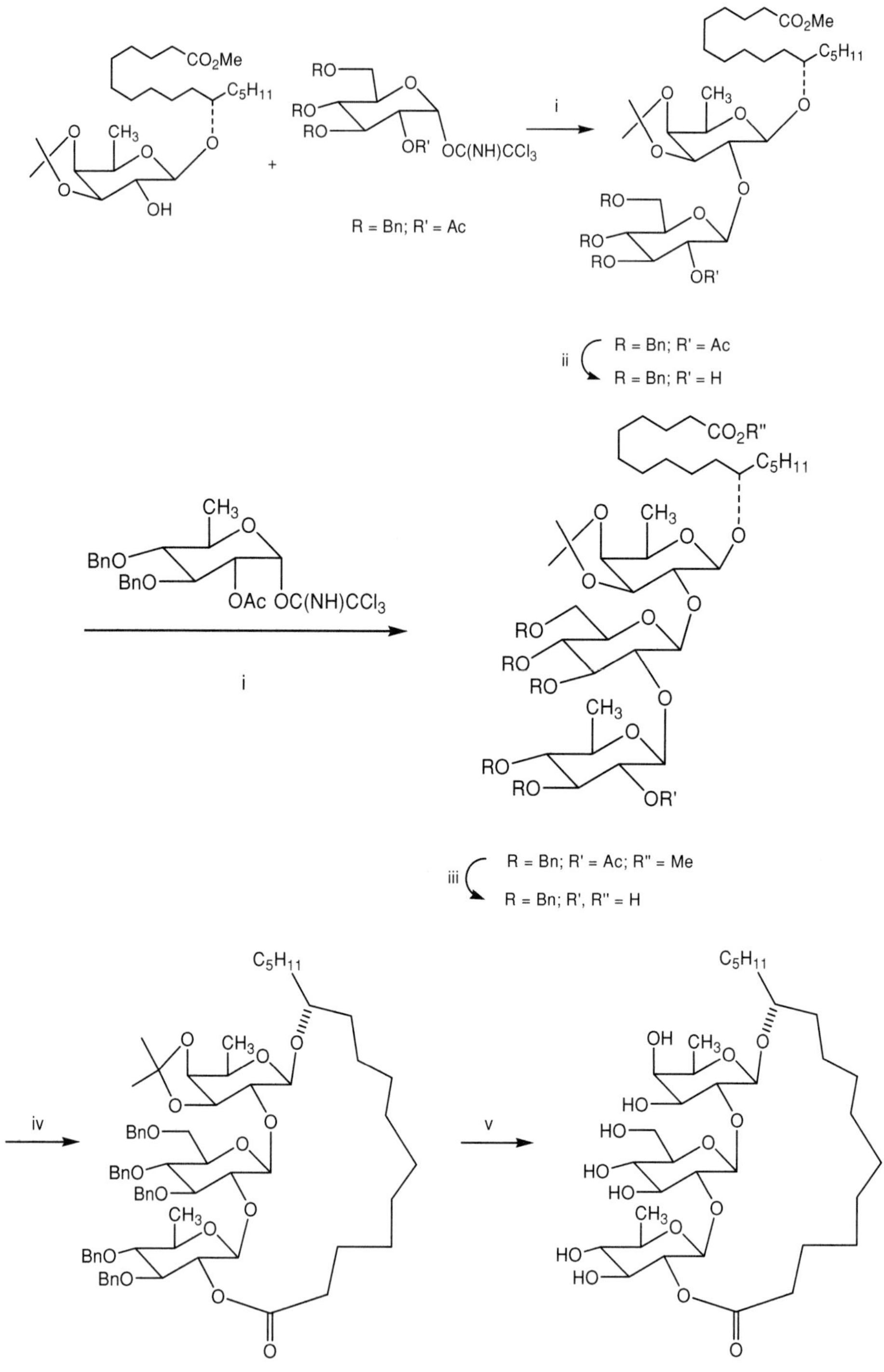

i) $BF_3 \cdot Et_2O$, CH_2Cl_2, -20°C, 1 h; (ii) NaOMe, MeOH, 6h, rt. iii) KOH, MeOH-H_2O, 4 h, reflux. iv) 2,4,6-trichlorobenzoyl chloride, Et_3N, DMAP, PhH. v) 10% HCl-MeOH, $Pd(OH)_2$-C 10%, MeOH.

FIGURE 2.29. Synthesis of Tricolorin F.

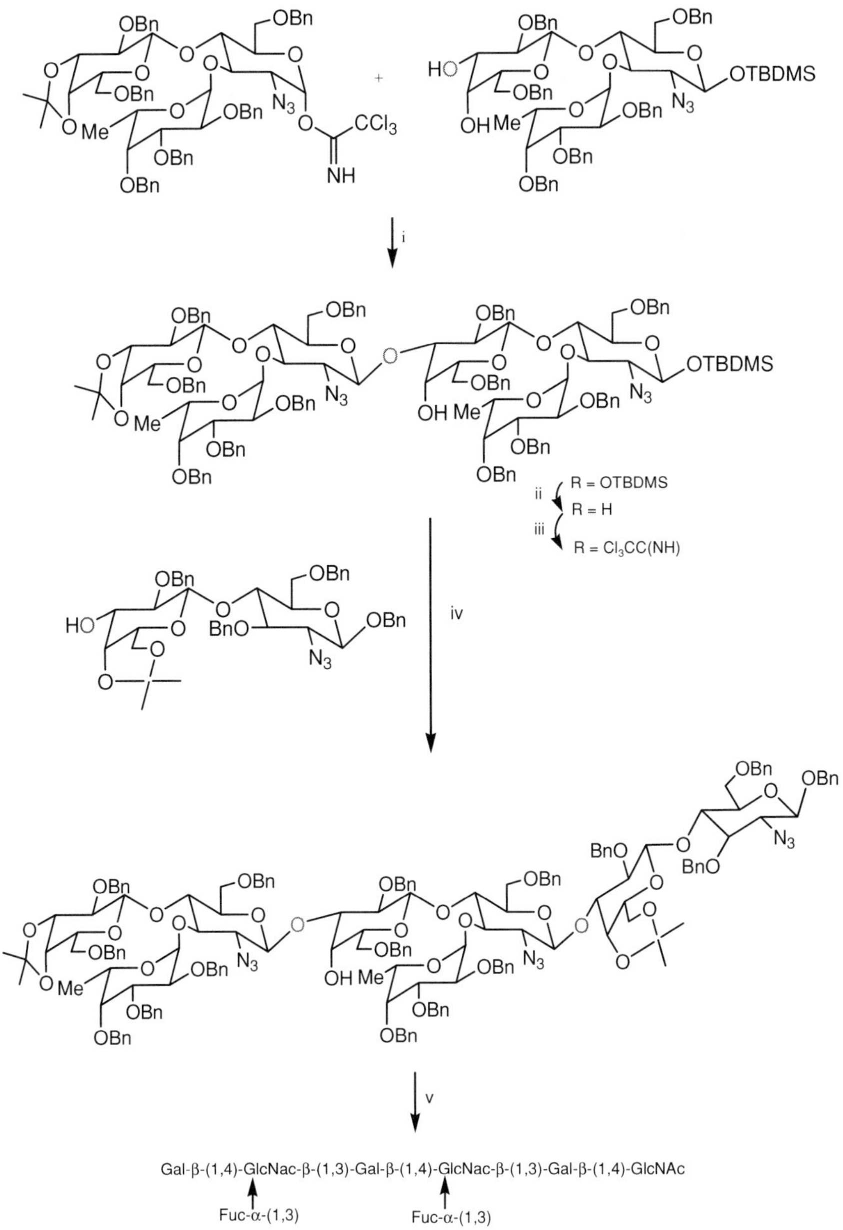

i) BF$_3$.OEt$_2$, CH$_2$Cl$_2$. ii) TBAF, AcOH. iii) Cl$_3$CCN, DBU. iv) BF$_3$.OEt$_2$, CH$_2$Cl$_2$.
v) a) AcOH. b) H$_2$, Pd-C.

FIGURE 2.30. Covergent synthesis of Lewis X fragment.

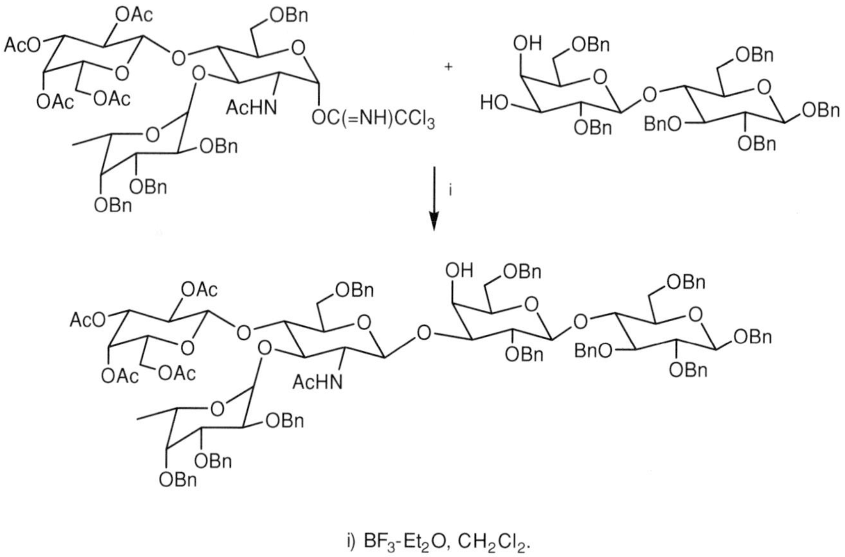

i) BF$_3$-Et$_2$O, CH$_2$Cl$_2$.

FIGURE 2.31. Coupling reaction for the preparation of Lewis x pentasaccharide intermediate.

Likewise, thioaryl donors can also be suitably converted to acetimidates for performing glycoside coupling reactions. This is the case of arabinosyl thio derivative, which is deprotected under NBS-pyridine conditions affording the lactol in 80% yield as a mixture of anomers (2:1). Treatment with NaH, followed by addition of Cl$_3$CCN, provided the desired trichloroacetimidate intermediate. This strategy has been succesfully applied in the syntheses of cytotoxic marine natural products Eleutherobin (Figure 2.32).[32]

2.1.7   The Sulfur Reaction

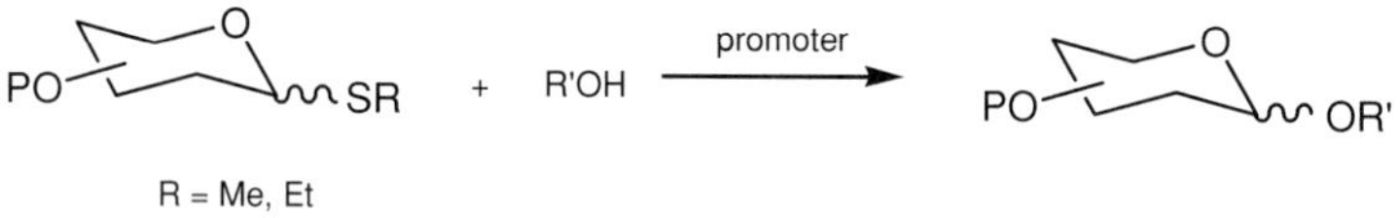

R = Me, Et

Promoter	Conditions
NIS-TfOH	0°C→r.t.
HgCl$_2$	CH$_2$Cl$_2$ or MeCN 0°C
CuBr$_2$-Bu$_4$NBr-AgOTf	CH$_2$Cl$_2$ or MeCN, −20°C
MeOTf	Et$_2$O, r.t.
MeSOTf	Et$_2$O, r.t.
AgOTf-Br$_2$	CH$_2$Cl$_2$
DMTST	MeCN, −15°C
NBS-TfOH	EtCN, −78°C

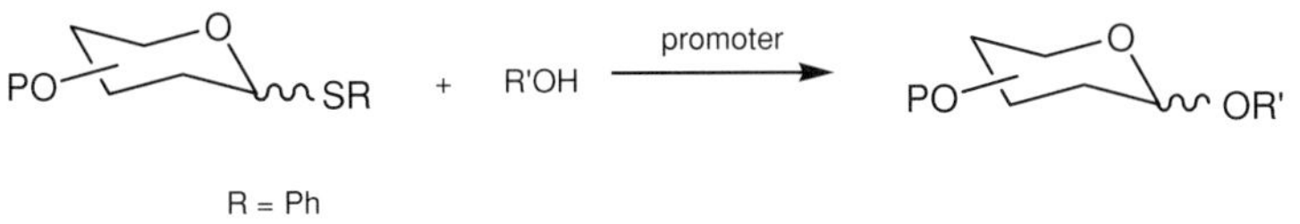

R = Ph

Promoter	Conditions
NBS	CH_2Cl_2, r.t.
BSP	CH_2Cl_2, MS, r.t
DMTST	
MeOTf	
MeSOTf	

Thioglycosides are useful glycosyl donors widely used in the preparation of
O-glycosides. An example of their applicability for the preparation of saccharide synthesis is represented in Figure 2.33. Thus, the synthesis of trisaccharide
intermediate was obtained by combining the thioglycoside donor with a monosaccharide acceptor in the presence of methyltriflate, to provide the target trisaccharide
in 72% yield.[33]

A convergent synthesis of the trisaccharide unit belonging to an antigen polysaccharide from streptococcus has been performed by Ley and Priepke.[34] In this
approach ramnosylalkylsulfur was used as the glycosyl donor, and cyclohexane-
1,2-diacetal as the protecting group (Figure 2.34).

Thioalkyl donor are also useful derivatives for the preparation of biologically
important natural sugars known as Sialic acids.[23] An efficient procedure for introducing thioalkyl groups as leaving groups involves the conversion of acetate into
thiomethyl by treatment with methylthiotrimethylsilane in the presence of TMS-
triflate. O-glycosilation reaction proceeds between the thioglycosylsialic donor
with a glycosyl acceptor (bearing an -OH group available), using a catalyst such
as N-iodosuccinimide-TfOH as promotor (Figure 2.35).

2-Thiophenyl glycosides were used as glycosyl donor for preparing complex oligosaccharides containing sialyl moieties. A remarkable convergent approach was described for preparing a sialyl octasaccharide consisting in the

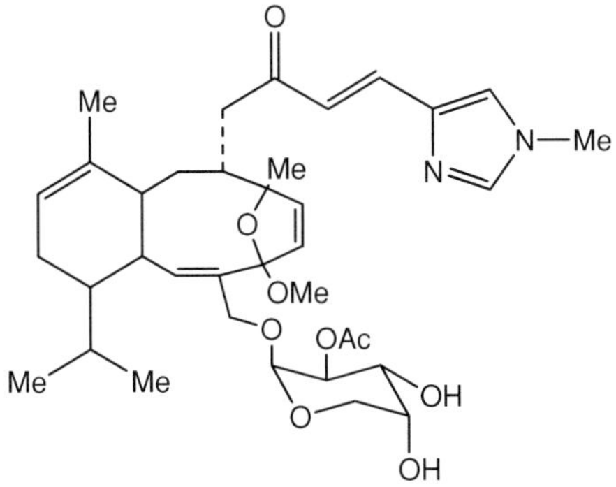

FIGURE 2.32. Citotoxic marine glycoside Eleutherobin.

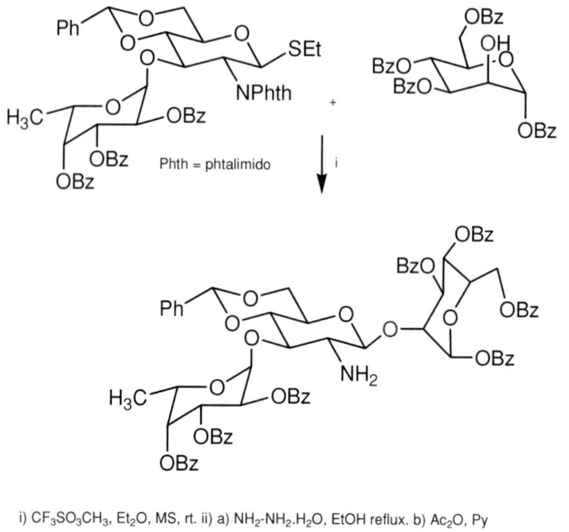

i) CF$_3$SO$_3$CH$_3$, Et$_2$O, MS, rt. ii) a) NH$_2$-NH$_2$.H$_2$O, EtOH reflux. b) Ac$_2$O, Py

FIGURE 2.33. Thioglicoside coupling reaction for preparation of a trisaccharide intermediate.

initial glycosidic reaction between 2-thiophenyl Neu5Ac donor with trisaccharide intermediate to produce the expected tetrasaccharide in 45% having an α(2→6)-linkage. The resulting tetrasaccharide was coupled with dimeric sialyl donor to yield hexasaccharide in 42%. Acetal hydrolysis was followed by coupling reaction

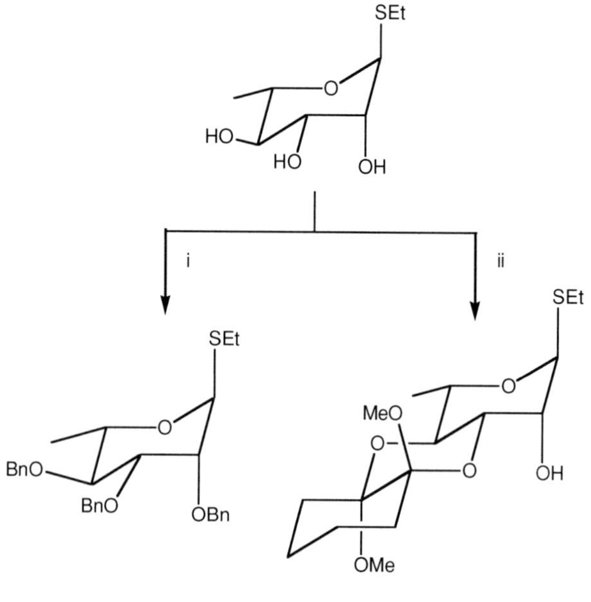

FIGURE 2.34. Synthesis of an antigen polysaccharide fragment.

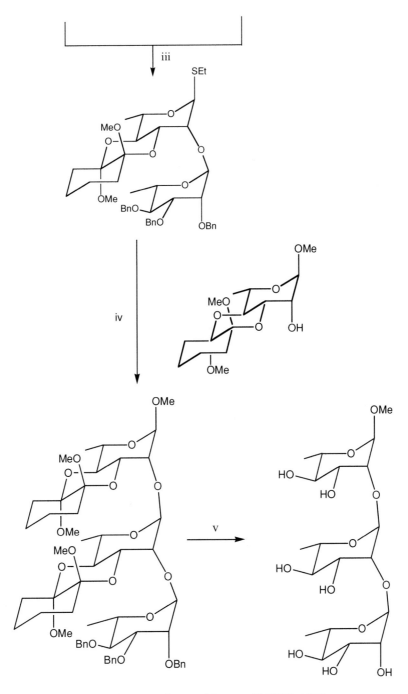

i) BnBr, NaH, DMF. ii) 1,1,2,2-tetramethoxycyclohexane. iii) IDCP, 4 ÅMS.
iv) NIS. v) AcOH-H₂O. vi) H₂, Pd/C, EtOH.

FIGURE 2.34. (*Continued*)

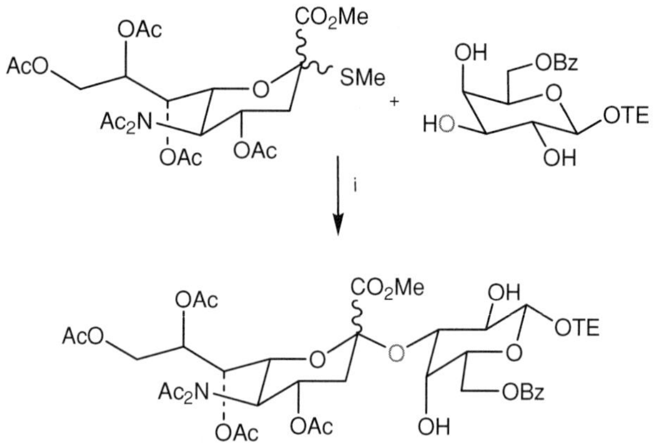

i) NIS/TfOH, MeCN, −40°C.

FIGURE 2.35. Thioalkyl donor for the preparation of sialic acids.

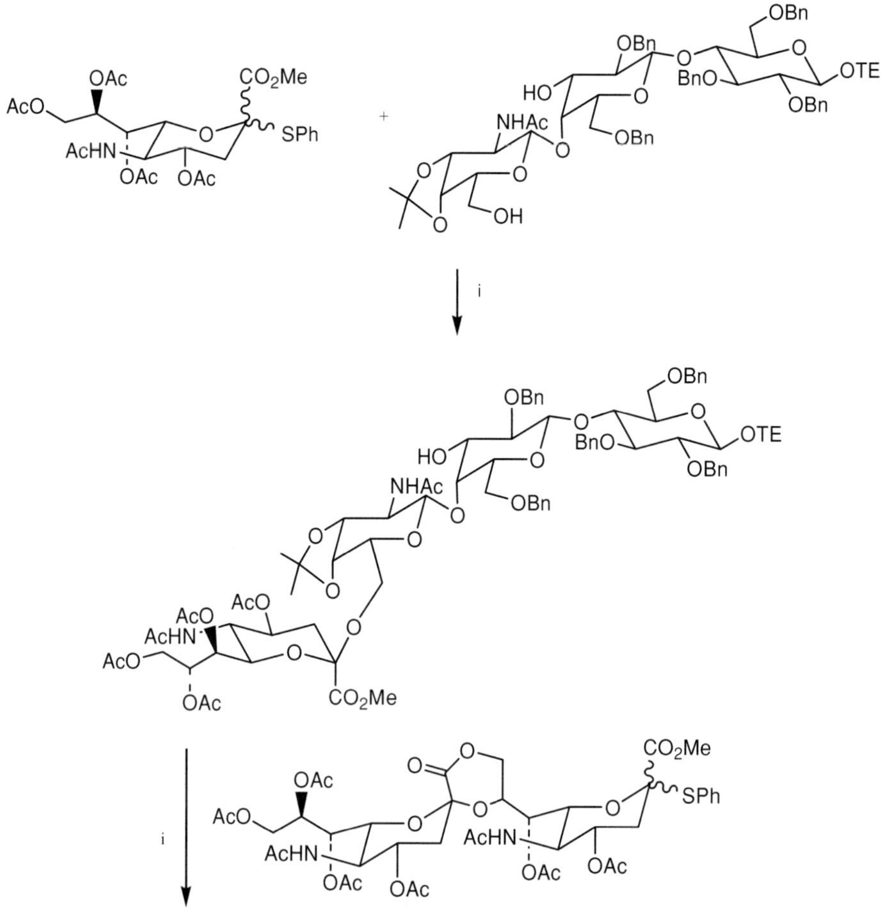

FIGURE 2.36. Convergent synthesis of sialyl oligosaccharide.

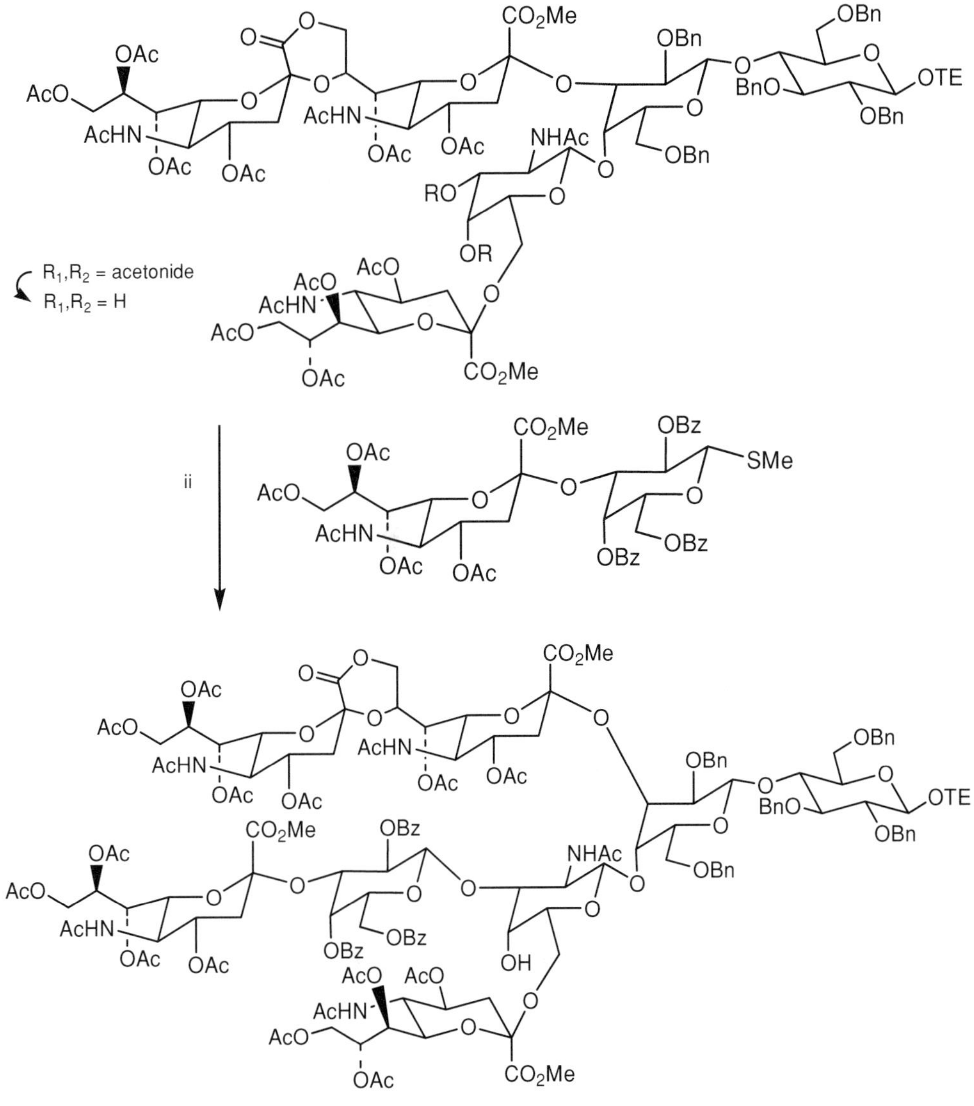

i) NIS/TfOH, 45%. ii) DMTST, 85%.

FIGURE 2.36. (*continued*)

with Neu5Acα(2→3)GalSMe donor to give the octasaccharide in 85% yield (Figure 2.36).[35]

Crich and Li[36] introduced the use of 1-(Benezenesulfinyl)piperidine/triflic anhydride as promoter conditions for preparing O-glycosides from thioglycoside donors. These conditions were applied for preparing Salmonella-type E1 core trisaccharide (Figure 2.37).

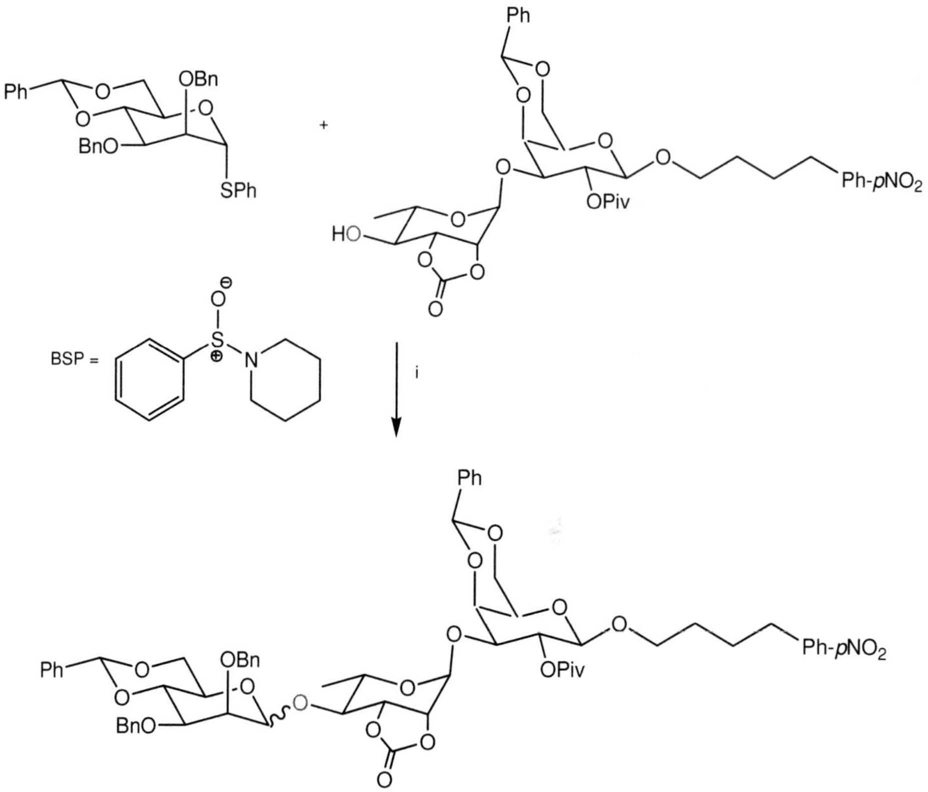

i) a) BSP, m.s., CH$_2$Cl$_2$, r.t. b) Tf$_2$O, –60°C to 0°C 1h.

FIGURE 2.37. Preparation of Salmonella-type E$_1$ core trisaccharide under BSP-Tf$_2$O conditions.

Highly fluorinated thiols have been developed and used as donors in the preparation of disaccharides. The reactivity of these novel fluorinated thiols were examined using different acceptors. Thus, disaccharide formation under glycosidic conditions provided the disaccharides in high yields (Figure 2.38).[37]

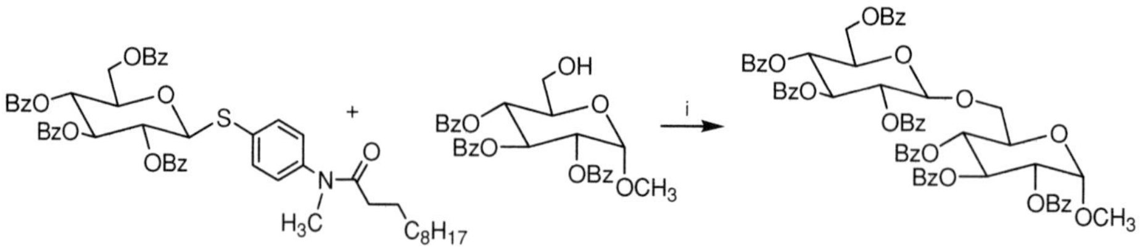

i) NIS (2 eq.), AgOTf (0.2 eq.), CH$_2$Cl$_2$.

FIGURE 2.38. Highly fluorinated thiols glycosyl donor for glycosidation.

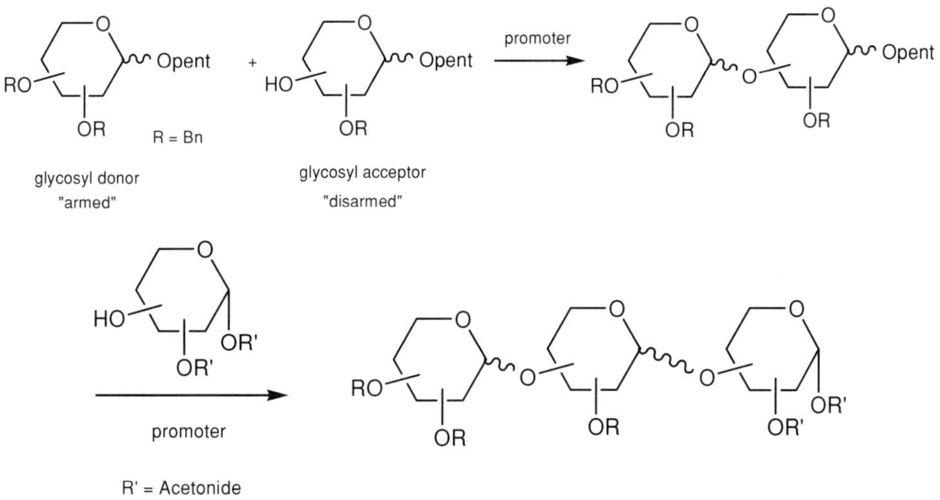

FIGURE 2.39. General scheme for the armed-disarmed approach.

2.1.8   The Armed-Disarmed Method

This versatile approach has been attributed to Mootoo and Fraiser-Reid,[38] and considers the use of a glycosyl donor in the classical sense coined with the term "armed saccharide" (because the reducing end is armed for further coupling reaction), and an acceptor, in this case "disarmed saccharide," which contains both a free alcohol and a leaving group sufficiently resistant for the ongoing coupling reaction. The resulting disaccharide now becomes an armed disaccharide, which in turn is reacted with another glycosyl acceptor or disarmed sugar to produce the oligosaccharide chain elongation (Figure 2.39).

This method was first implemented in the preparation of 1-6 linked trisaccharide shown in Figure 2.40. As one can see, the disarmed sugar intermediates function as glycosyl acceptor bearing the hydroxyl group at position 6 available for establishing a glycosidic linkage with the armed unit.

Despite the usefulness of pentenyl as protecting group, clear preference in the use of thioglycoside donors as armed and disarmed donors is often observed (Figure 2.41).[39]

This concept was applied successfully in the stereocontrolled synthesis of Le^x oligosaccharide derivatives by using two glycosylation steps as described by Yoshida et al.[40] The first coupling between "armed" thiophenyl fucopyranosyl derivative with "disarmed" thiophenyl lactose derivative under NIS-TfOH conditions provided trisacccharide, which was subjected without purification to second condensation with different acceptors, one of which is indicated in Figure 2.42.

The construction of α-linked mannoside disaccharide was achieved under the armed-disarmed approach by using armed thiogalactoside donor activated

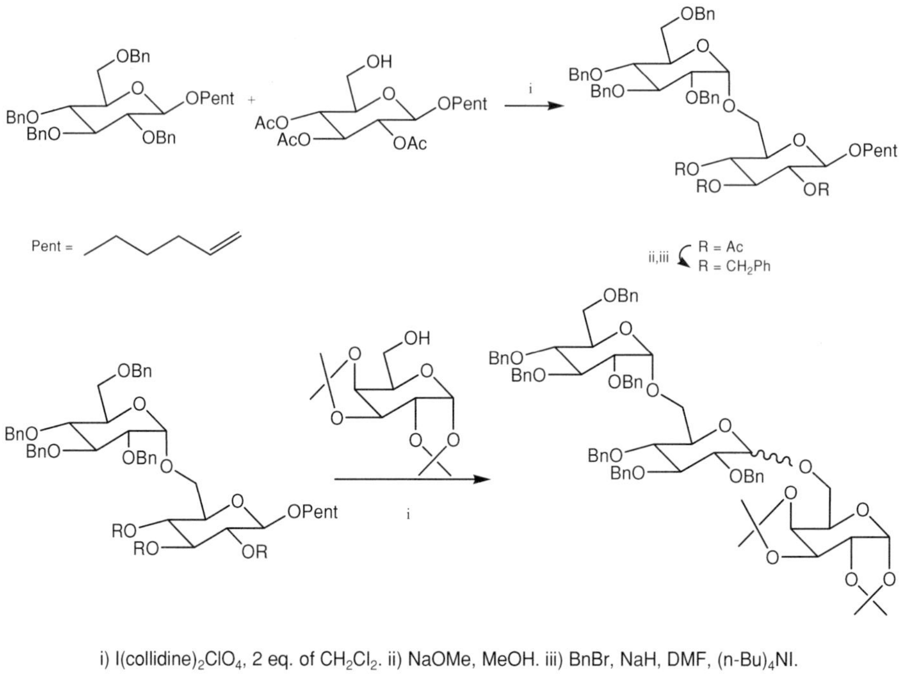

i) I(collidine)$_2$ClO$_4$, 2 eq. of CH$_2$Cl$_2$. ii) NaOMe, MeOH. iii) BnBr, NaH, DMF, (n-Bu)$_4$NI.

FIGURE 2.40. The armed-disarmed approach.

by BSP/Tf$_2$O and condensed with disarmed thiomannoazide intermediate bearing a free hydroxyl group. Addition of triethyl phosphate prior to the aqueous work up led to the generation of the expected α-linked disaccharide in 74% (Figure 2.43).[39]

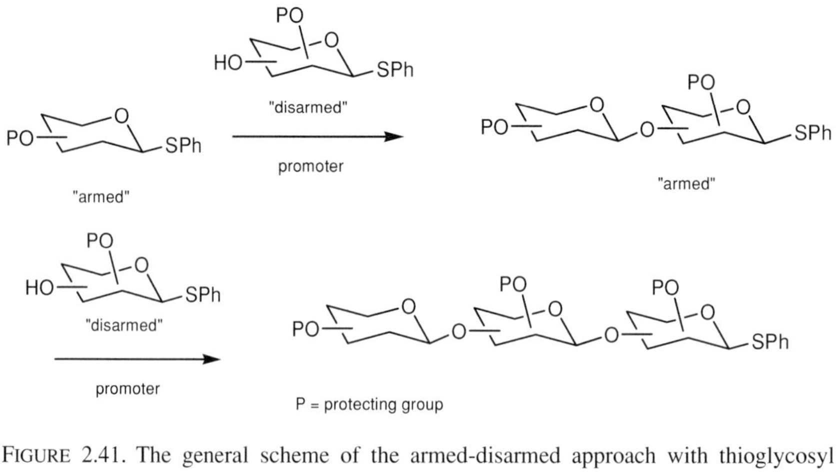

P = protecting group

FIGURE 2.41. The general scheme of the armed-disarmed approach with thioglycosyl sugars.

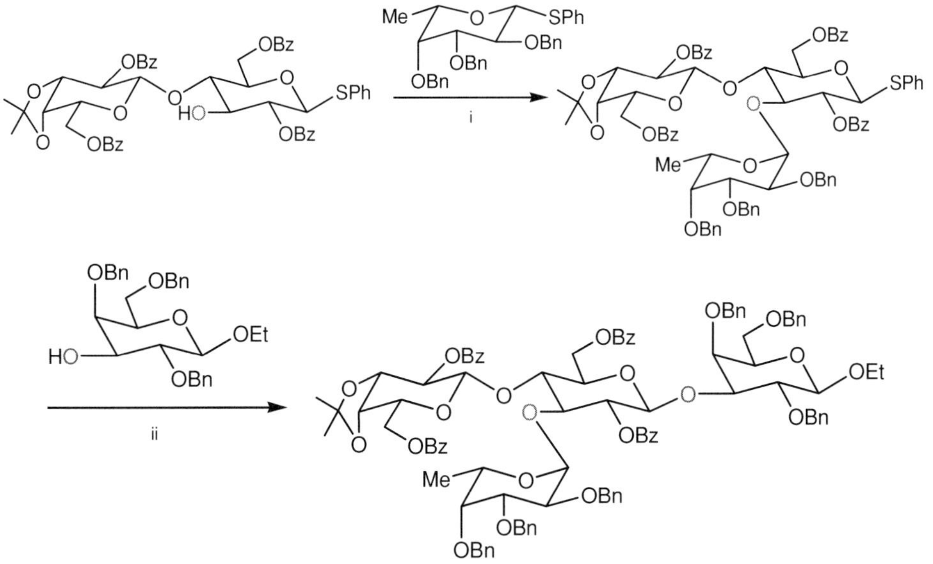

i,ii) CHCl$_3$, 4 Å, NIS-TfOH, −20°C, 1h.

FIGURE 2.42. Preparation of Lewis X tetrasaccharide using armed-disarmed coupling method.

Recently S-benzoxazol thio glycoside (SBox) was synthesized and introduced as alternative glycosyl donor for preparing disaccharides under the armed-disarmed approach. Thus, the SBox glycosyl donor was used as armed donor and condensed with disarmed thioglycoside to provide the target disaccharide (Figure 2.44).[41]

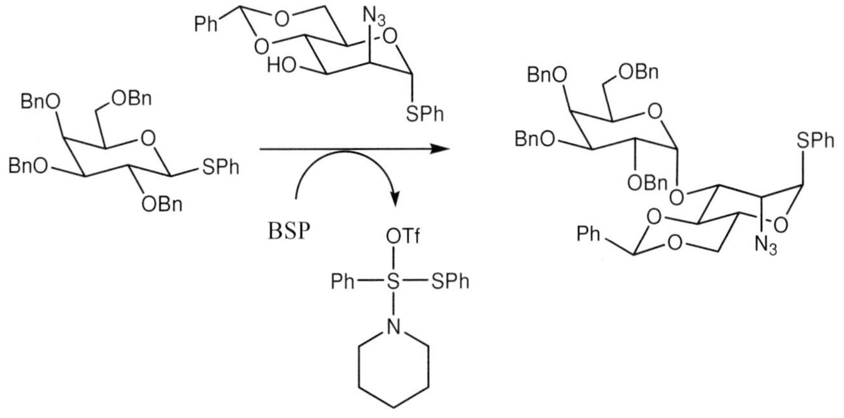

FIGURE 2.43. Synthesis of α-linked mannosyl disaccharide following an armed-disarmed strategy.

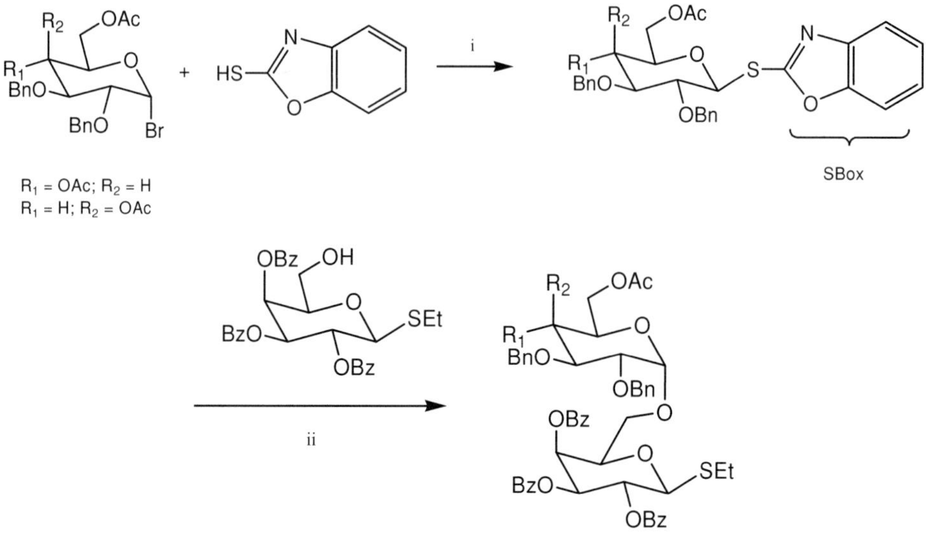

R₁ = OAc; R₂ = H
R₁ = H; R₂ = OAc

i) K₂CO₃, acetone, 90%. AgOTf, CH₂Cl₂.

FIGURE 2.44. Armed-disarmed synthesis using S-benzoxazol (SBox) as disarmed glycosyl donor.

2.1.9   The Glycal Reaction

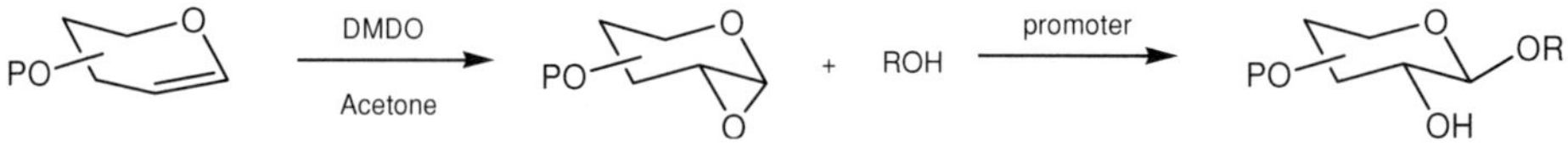

Promoter	Conditions
ZnCl₂	THF

The glycals are unsaturated sugars with a double bond located between C1 and C2. These useful intermediates were discovered by Fischer and Zach in 1913[42] and their utility in the preparation of building blocks for oligosaccharide synthesis is increasingly important. Different routes for the preparation of triacetyl glycals have been examined by Fraser-Reid et al.,[43] involving the Ferrier rearrangement. Moreover, a suitable one-pot preparation of glycals has been more recently described, starting from reducing sugars by Shull et al.[44] The general procedure for preparing these valuable intermediates is based on the reductive removal of a halogen and neighboring acetate group through the use of zinc in acetic acid (Figure 2.45). The completion of this reaction can be followed by ^{1}H NMR, where the presence of a signal around 6.3 ppm as double of double with $J_{1,2} = 6.2$ Hz, $J_{1,3} = 0.3$ Hz is expected for H-1, and a multiple shifted upfield for H-2.

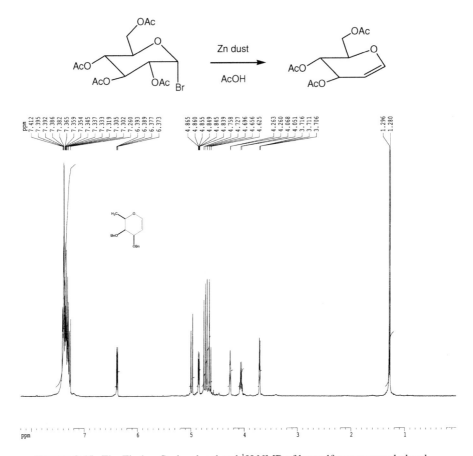

FIGURE 2.45. The Fischer-Sachs glucal and ^{1}H NMR of benzylfucopyranosyl glycal.

More recently the use of alternative catalysts such as titanium complex, Li/NH$_3$, Sodium, Cr (II), and vitamin B-12 as catalysts has been described as improved method, for preparing especially acid sensitive glycals.

As for any double bond, these unsaturated sugars may undergo electrophilic addition, which takes place at the C2 position leaving a positive charge at C1, which instantly reacts with the conjugate base. This reaction is particularly useful for the preparation of 2-deoxypyranosides (Figure 2.46).

FIGURE 2.46. Electrophilic addition.

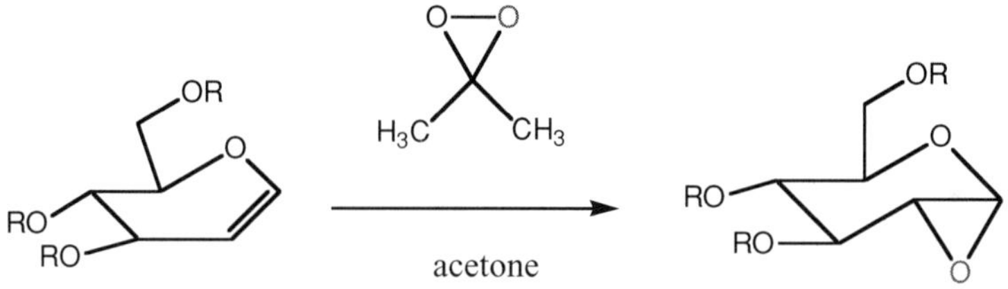

FIGURE 2.47. The Brigl epoxide formation.

A more extended application for glycoside bond formation has been developed recently. Such strategies consist of the conversion of glycals into Brigl's epoxide, and then further treatment with nucleophiles to effect ring opening. The oxidation of the double bond has been successfully achieved with dimethyl dioxirane (DMDO) in acetone (Figure 2.47).

The standard procedure for generation of DMDO was developed by Murray and Jeyaraman[45] and optimized by Adam et al.[46] Such procedure involves the use of potassium monoperoxysulfate as oxidative acetone agent, and the reaction conditions require temperatures below 15°C, an efficient stirring. The DMDO/acetone solution generated must be immediately distilled under moderate vacuum. The concentrations of DMDO are in the order of 0.09–0.11 M (5%), and it is used as acetone solution. The transformation of the glycal to the epoxide can be verified by ^{1}HNMR, where the disappearance of the signal at 6.3 ppm for H-1 double bond is observed, and it is expected the presence of a signal at 5.0, as double for H-1 and at 3.1 as double of double for H-2 (Figure 2.48).

The stereo selectivity of epoxide formation is protecting group-dependent, observing in the case of acetate protecting group a mixture of epoxide anomers, and preferentially the α-anomers if the protecting groups are benzyl, or methyl groups (α:β ratio 20:1). As expected, the epoxide ring opening by nucleophiles occurs with inversion of configuration, providing β-glycosides exclusively (Figure 2.49).

Likewise, alternative epoxide conditions from glycals have been assayed besides DMDO treatment. Among them, cyclization of a bromohydrin,[47] m-chloroperoxybenzoic acid-potassium fluoride complex oxidation of the glycal,[48] and potassium tertbutoxide oxidation of fluoride glycosyl donor[49] has been described (Figure 2.50).

The potential of 1,2-anhydro sugars as glycosyl donor for the preparation of β-linked saccharides was established by Halcomb and Danishefsky[50] and such strategy consist in the treatment of the glucal having available a hydroxyl group at position 6, with the sugar epoxide under Lewis acid conditions (ZnCl$_2$) at low temperature. The resulting glucal disaccharide generated as a single coupling product was further converted to the epoxide, which eventually led to the next coupling reaction with another glucal acceptor (Figure 2.51).

The tetrasaccharide Cap domain of the antigenic lipophosphoglycan of *Leishmania donovani* has been prepared under the glycal approach by Upreti and Vishwakarma.[51] Thus, the preparation of the hexa-O-benzyl-lactal under stardard

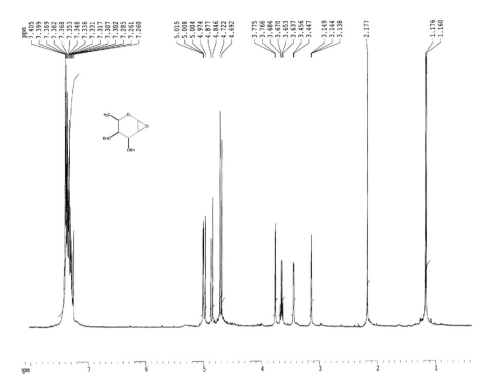

FIGURE 2.48. ^{1}H NMR spectra of 1,2-anhydro-3,4-di-O-benzyl-α-D-fucopyranose (and traces of acetone).

procedures was followed by oxirane formation with dimethyl dioxirane to generate the corresponding oxirane. Methanolysis ring opening and gluco→manno conversion generated the disaccharide intermediate. This was coupled to the mannobiose donor to produce the tetrasaccharide, which after deprotecction led to the tetrasaccharide Cap domain (Figure 2.52).

Brigl's epoxide has been exploited successfully for the preparation of glycosylated peptides such as collagen type II derived glycosides carrying β-Gal and αGlc-1,2-βGal side chains.[52] Galactosyl glycal is reacted with DMDO-acetone solution and the resulting epoxide reacted with hydroxylysine and ZnCl$_2$ as promoter (Figure 2.53). General procedures for preparation of glycosidic bond of glycopeptides can be reviewed in the comprehensive study reported by Kunz.[53]

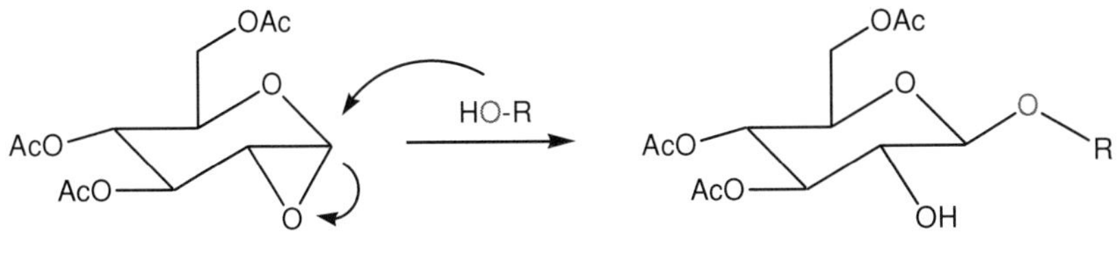

FIGURE 2.49. Ring opening for β-glycoside formation.

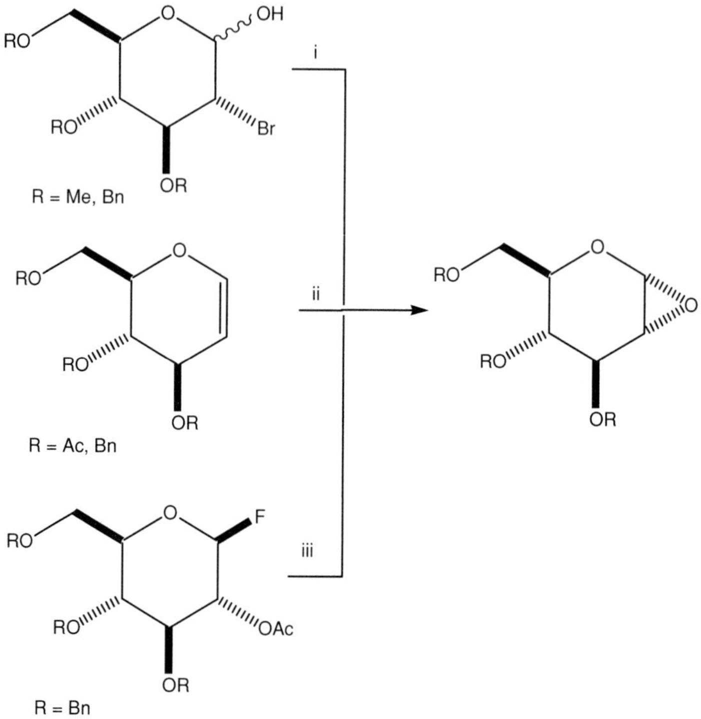

i) KH or, KHMDS, 18-crown-6, −70°C. ii) MCPBA-KF, CH$_2$Cl$_2$, r.t. iii) t-BuOK, THF.

FIGURE 2.50. Alternative glycal-epoxidations.

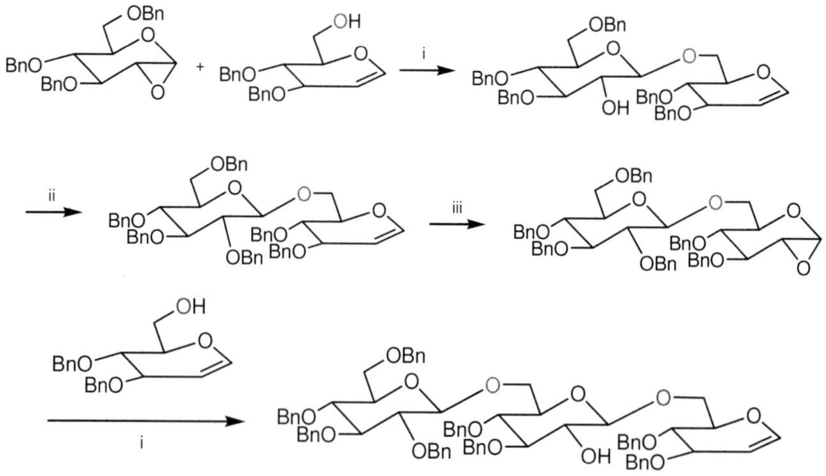

i) ZnCl$_2$/THF, −78°C to r.t. ii) NaH, BnBr. iii) DMDO-acetone.

FIGURE 2.51. Epoxide glycal as glycosil donors.

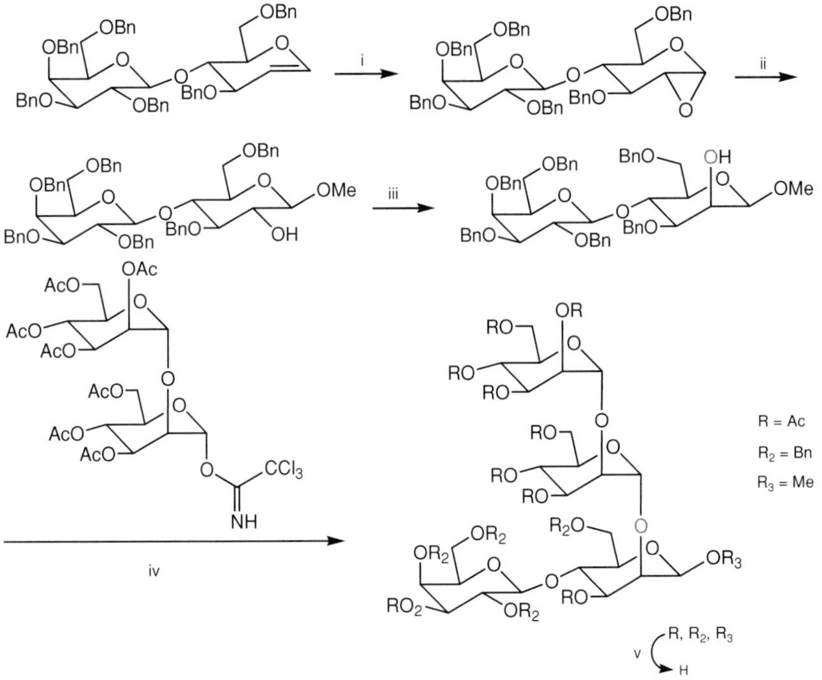

i) DMDO, CH$_2$Cl$_2$, 0°C, 2h. ii) MeOH, rt, 4h. iii) a)(COCl)$_2$, DMSO, CH$_2$Cl$_2$, −78°C. b) NaBH$_4$, rt, 4h.
iv) TMSOTf, CH$_2$Cl$_2$, −30°C, 45 min. v)a) Pd(OH)$_2$, H$_2$, 4h. b) NaOMe, MeOH. c) Ac$_2$O/AcOH/H$_2$SO$_4$.

FIGURE 2.52. Synthesis of a tetrasacharide using an epoxide disaccacharide as glycosyl donor.

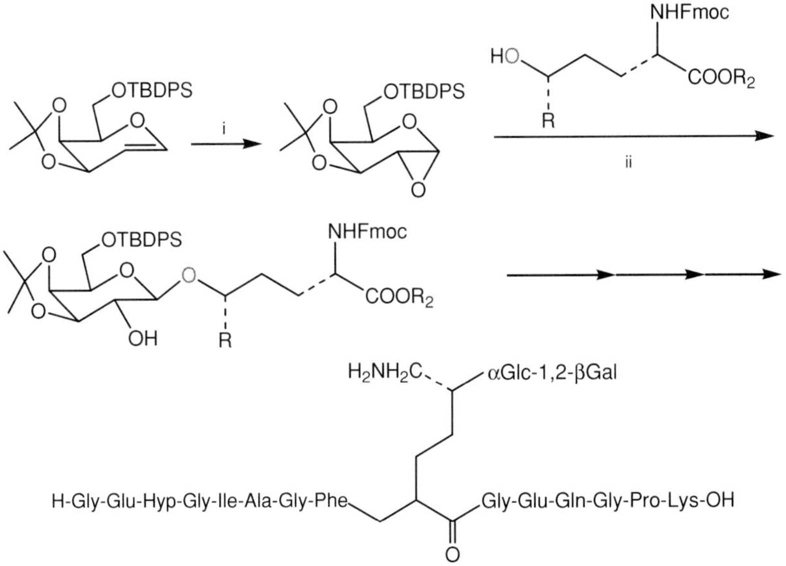

1) DMDO-acetone. ii) ZnCl$_2$, THF

FIGURE 2.53. Amino acids glycosidation.

This methodology has been extended for the preparation of E-selectin ligand tetrasaccharide sialyl Lewis^x (SLe^x), which is located at the terminus of glycolipids present on the surface of neutrophils. The chemoenzymatic sequence consisted in the reaction of the 6-acetylated glucal with β-galactosidase transferase to produce disaccharide, which was subjected to further transformations according to the pathway presented in Figure 2.54.[54]

2.1.10 Miscellaneous Leaving Groups

2.1.10.1   Fluoride Glycosyl Donors

X = F

Ppromoter	Conditions
SnCl$_2$-AgClO$_4$	Et$_2$O, $-15 \rightarrow$ r.t.
Cp$_2$HfCl$_2$-AgOTf	CH$_2$Cl$_2$, $-25°$C
SnCl$_2$-AgOTf	CH$_2$Cl$_2$, $0°$C

Fluorine is considered a poor leaving group, and its use for glycoside bond formation has been more restricted than chlorine and bromine, although display higher thermal and chemical stability. Nonetheless several O-glycoside synthesis involving glycosyl donors with fluorine as leaving group has been described, specially for the preparation of α-O-glycosides with high stereoselectivity.[55]

Based in the use of fluorine glycosyl donors, the synthesis of the marine algae α-agelaspines, was carried out through the condensation of perbenzylated galactopyranosyl fluorine as anomeric mixture with the long chain alcohol in the presence of a mixture of SnCl$_2$-AgClO$_4$ as catalyst (Figure 2.55).[56]

A general procedure for the preparation of ribofuranosyl fluorides and their use as glycosyl donors for O-glycosylation with α-stereocontrol was developed by Mukaiyama et al.,[57] and consists of the conversion of 2,3,5-tri-O-benzyl-D-ribofuranoside that react under mild conditions with 2-fluoro-1-methylpyridinium tosylate at room temperature to give an anomeric mixture (α:β 58:42) in 84% yield. These two fluorines could be either separate or interconverted by treating the α-anomer with boron trifluoride etherate in ether at room temperature (Figure 2.56).

It has been observed that the glycosylation reaction between the glycosyl fluorine with different alcohols under Lewis acid conditions provides mainly α-riboglucosides in high yield as it is shown in Figure 2.57.

Sulfated Le^x and Le^a-type oligosaccharide selectin ligands were synthetically prepared as described below. Thus, glycosyl donor and acceptor were condensed under Mukaiyama conditions (AgClO$_4$-SnCl$_2$) to form the

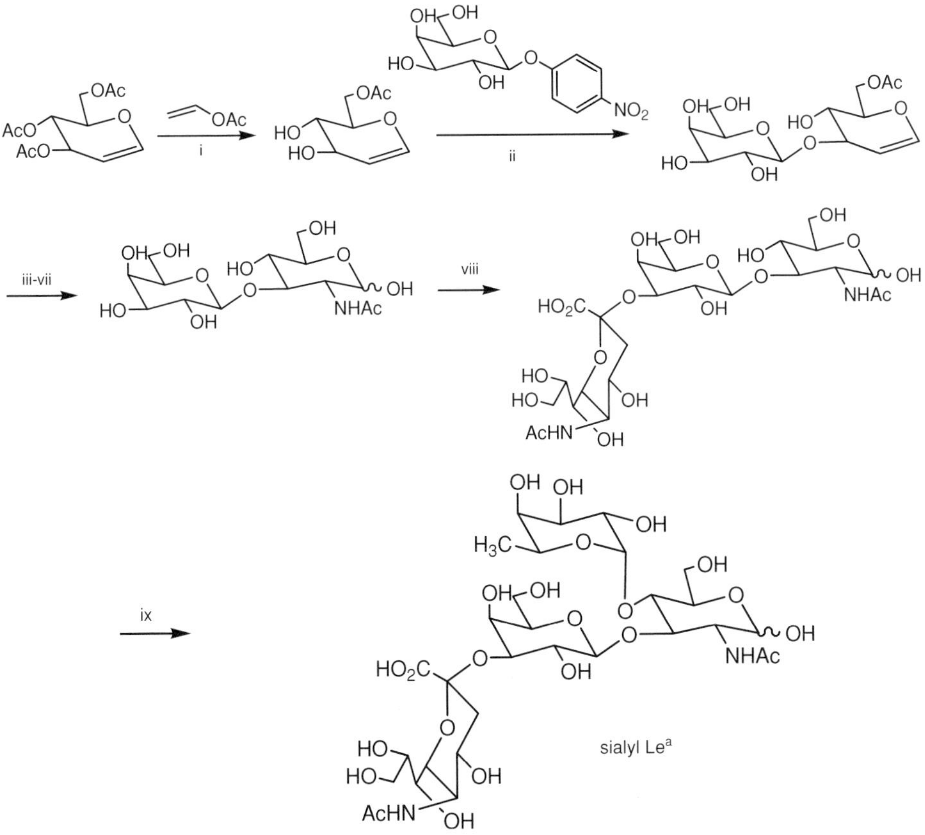

i) subtilisin, DMF. ii) β-galactosidase. iii) Ac$_2$O. iv) NaN$_3$, CAN. v) H$_2$ cat. vi) Ac$_2$O.
vii) saponification. viii) α2-3SiaT, CMP-NeuAc. ix) α1-3/4 FucT, GDP-Fuc.

FIGURE 2.54. Chemoenzymatic synthesis of tetrasaccharide sialyl Le^a.

β-glycoside in 90% yield. The sulphated tetrasaccharide was formed by reaction of tetrasaccharide acceptor with SO$_3$.NM$_3$ complex in anhydrous pyridine (Figure 2.58).[58]

2.1.10.2   Silyl Glycosyl Donors

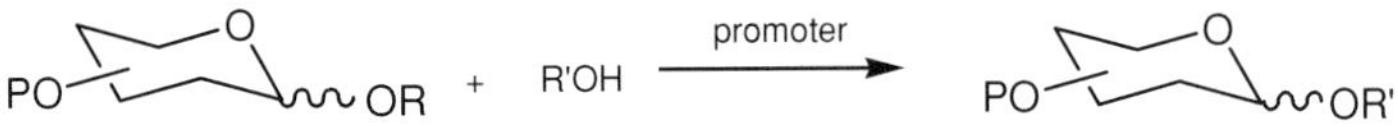

R	Promoter	Conditions
Me$_3$Si	TMSOTf or BF$_3$.Et$_2$O	CH$_2$Cl$_2$, −5°C
^{t}BuMe$_2$Si	TMSOTf	CH$_2$Cl$_2$-acetone, −35°C

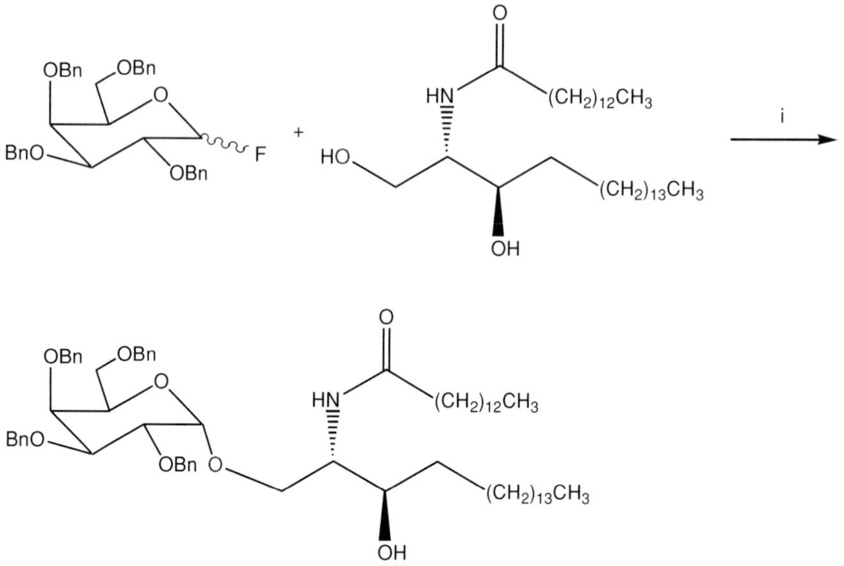

i) SnCl$_2$, AgClO$_4$/THF. ii) H$_2$, Pd-BaSO$_4$/THF.

FIGURE 2.55. Fluorine monosaccharide as glycosyl donor.

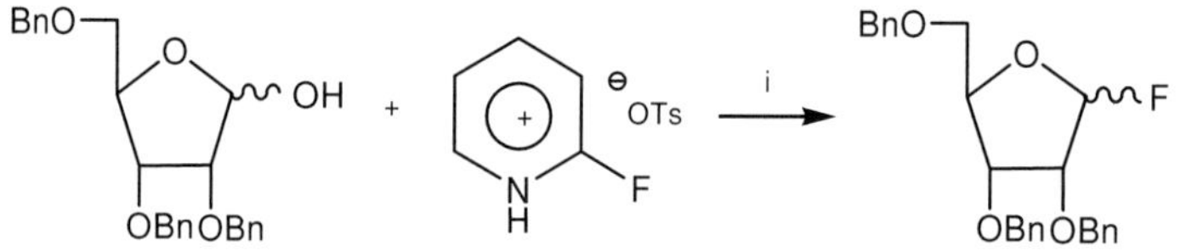

α/β 58/42

i) NEt$_3$, CH$_2$Cl$_2$

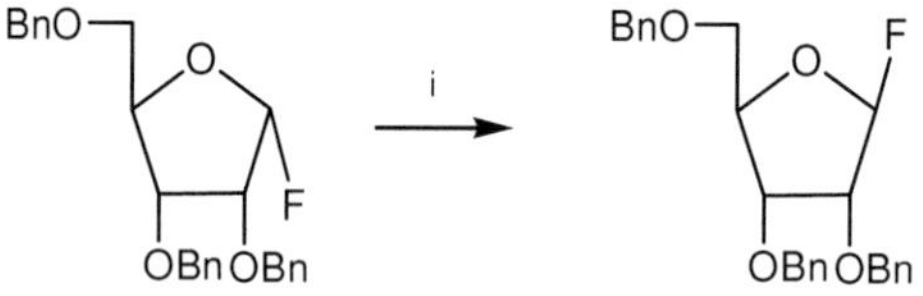

i) BF$_3$.OEt$_2$, Et$_2$O, r.t. 10min, 72%

FIGURE 2.56. The Mukaiyama protocol for preparation of ribofuranosyl fluoride.

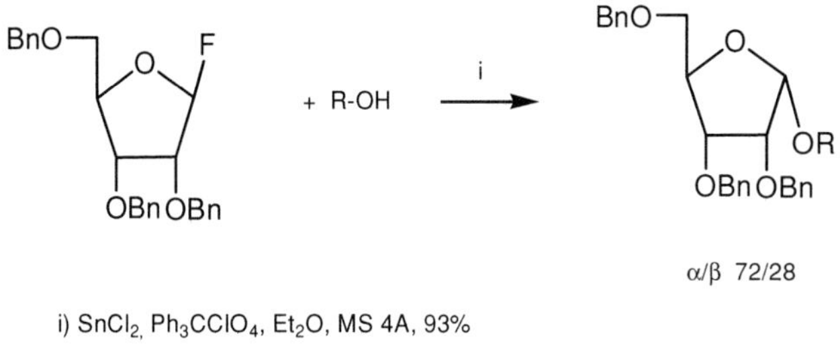

i) SnCl$_2$, Ph$_3$CClO$_4$, Et$_2$O, MS 4A, 93%

FIGURE 2.57. N-glycosylation reaction using ribofuranosyl fluorine.

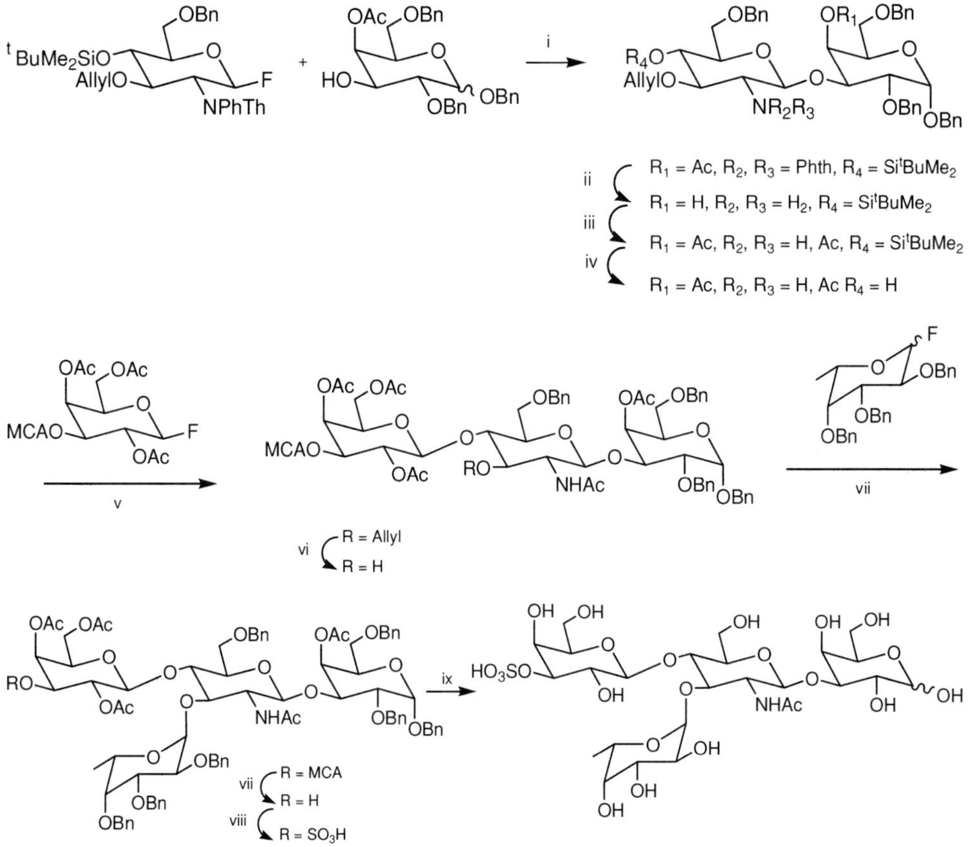

FIGURE 2.58. Total synthesis of sulphated Le^x.

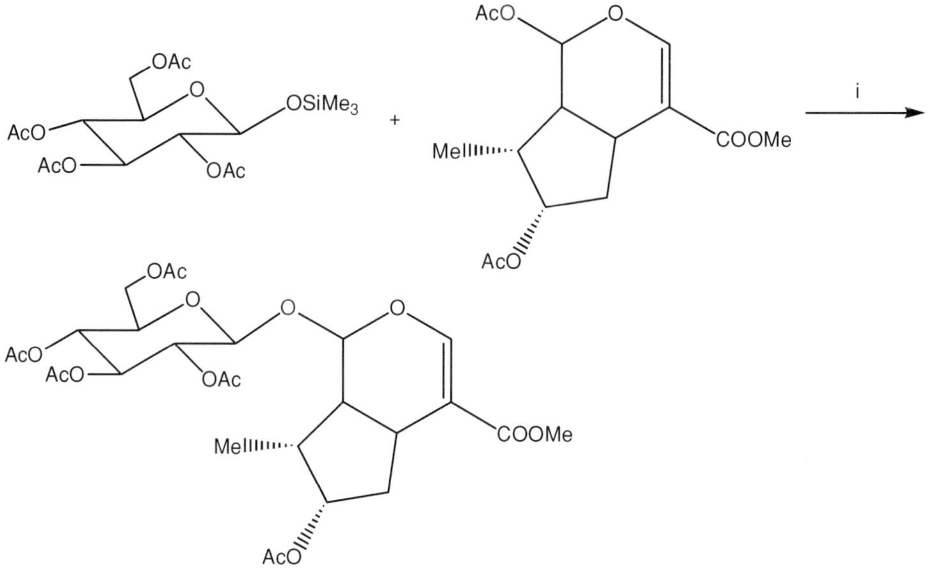

i) TfOSiMe$_3$, −40°C.

FIGURE 2.59. Silyl derivatives as glycosyl donors.

Silyl groups are best known as versatile protecting groups, and their use as leaving groups for glycoside bond formation has been more limited. An example of glycoside formation involving a silyl group as leaving group is reported for the preparation of luganol O-glycoside.[59] In this work, the glycosyl donor is combined with luganine in the presence of trimethylsilyltriflate at low temperature (Figure 2.59). It is worth mentioning that stereoselectivity is dependent on C-2 neighboring group participation. When acetate is the C-2 protecting group, the β-anomer is obtained, while if the protecting group is benzyl, the α-anomer is preferred.

The use of selenoglycosides as glycosyl donors and acceptor in glycosilation reactions has also been described by Metha and Pinto.[60] A typical glycosidation procedure with phenylselenoglycoside donors involves the glycosyl acceptor, 4-Å molecular sieves, silver triflate, and potassium carbonate in dichloromethane (Figure 2.60).

Tetrazol has also been tested as a leaving group for the preparation of an antibiotic fragment.[61] A coupling reaction with the methoxyphenyl glycosyl acceptor was catalyzed with (Me$_3$)$_3$OBF$_4$ as shown in Figure 2.61.

2-Aminodisaccharides were obtained by an elegant [3,3] sigmatropic rearrangement, by Takeda et al.[62] The addition of thiophenol to an unsaturated C-1 in the presence of Lewis acid was followed by a sigmatropic rearrangement with an imidate group that migrates from C-4 to C-2. Disaccharide formation was catalyzed with Pd(CH$_3$CN)$_2$-AgOTf complex in dichloromethane (Figure 2.62).

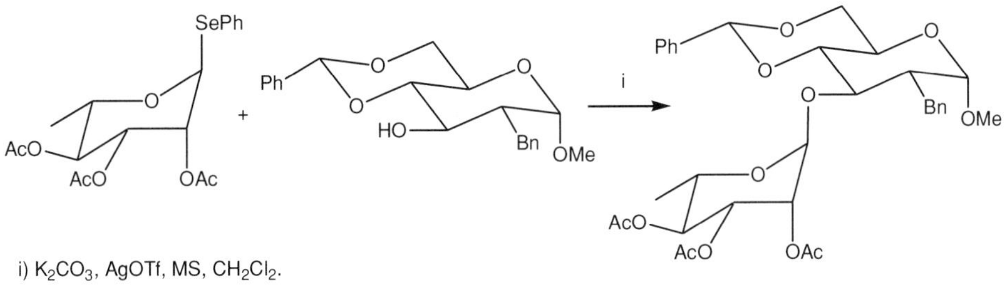

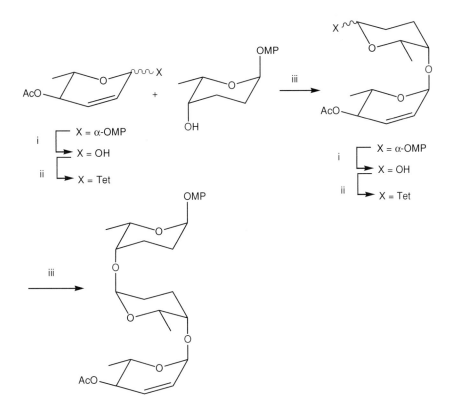

i) K$_2$CO$_3$, AgOTf, MS, CH$_2$Cl$_2$.

FIGURE 2.60. Phenylselenosugars as glycosyl donors.

The total synthesis of the cyclic glycolipid Arthrobacilin A, a cell growth inhibitor was achieved by Garcia and Nishikawa,[63] under zinc *p-tert*-butylbenzoate salt as glycoside catalyst, obtaining the β-galactoside glycoside in 73% along with α-isomer in 27% (Figure 2.63).

i) CAN. ii) 1H-tetrazole. iii) (CH$_3$)$_3$OBF$_4$, MS.

FIGURE 2.61. The use of tetrazol as a leaving group.

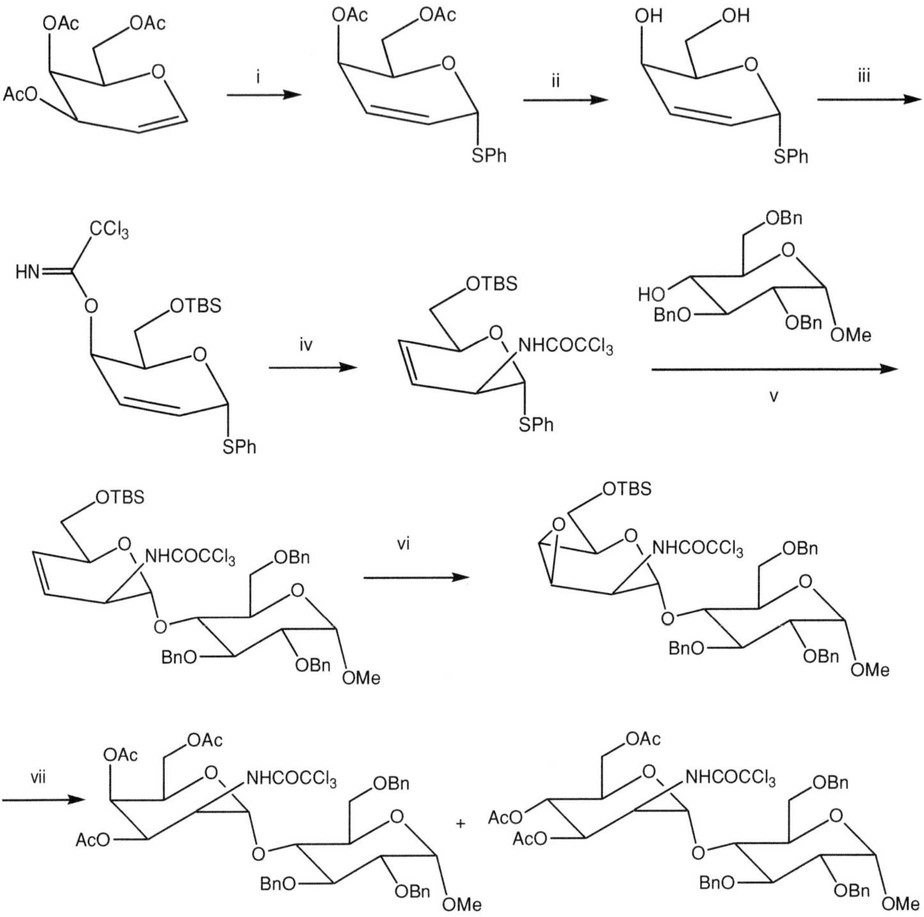

i) PhSH, SnCl$_4$/CH$_2$Cl$_2$. ii) MeONa/MeOH. iii) a) TBSCl-imidazol/CH$_2$Cl$_2$. b) Cl$_3$CCN-NaH/CH$_2$Cl$_2$.
iv) xylene, reflux. v) Pd(CH$_3$CN)$_2$-AgOTf, MS 4A/CH$_2$Cl$_2$. vi) m-CPBA/CH$_2$Cl$_2$. viii) Ac$_2$O-AcOH/
BF$_3$.OEt$_2$.

FIGURE 2.62. Sigmatropic rearrangement.

2.1.10.3   Heterogenous Catalysis

Stereocontrolled α- and β-glycosylations by using environmentally benign heterogenous catalyst has been developed as a novel approach for stereoselective formation of β-O-glycosidic linkages. Polymeric materials such as montmorillonite K-10,[64] heteropoly acid (H$_4$SiW$_{12}$O$_{40}$),[65] sulphated zirconia (SO$_4$/ZrO$_2$),[66] and perfluorinated solid-supported sulfonic acids (Nafion resins)[67] have been assayed successfully providing series of stereocontrolled O-glycosides in high yield (Figure 2.64).

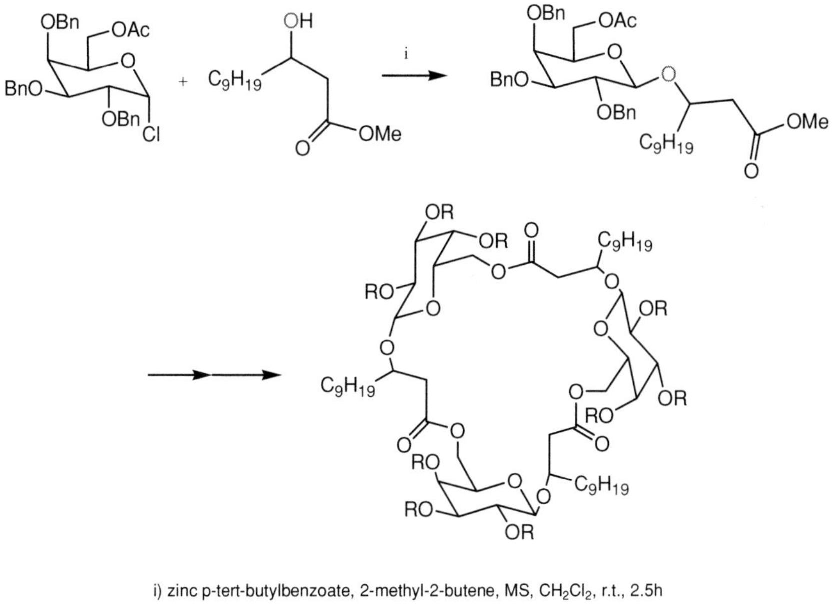

i) zinc p-tert-butylbenzoate, 2-methyl-2-butene, MS, CH$_2$Cl$_2$, r.t., 2.5h

FIGURE 2.63. Glycosylation reaction for preparation of Arthrobacilin A.

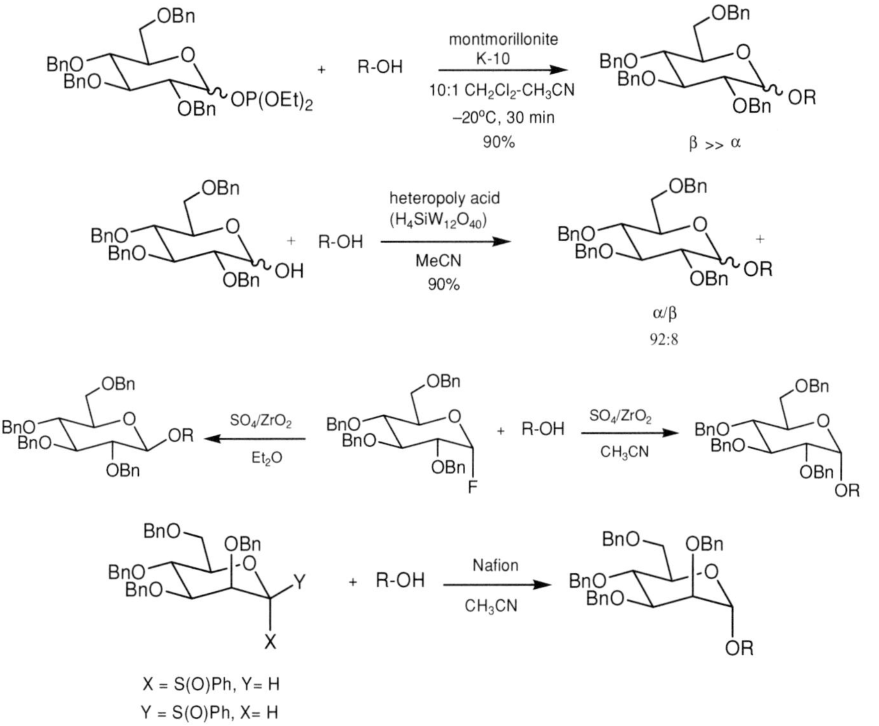

FIGURE 2.64. Stereocontrolled O-glycosidations using heterogeneous polymeric materials.

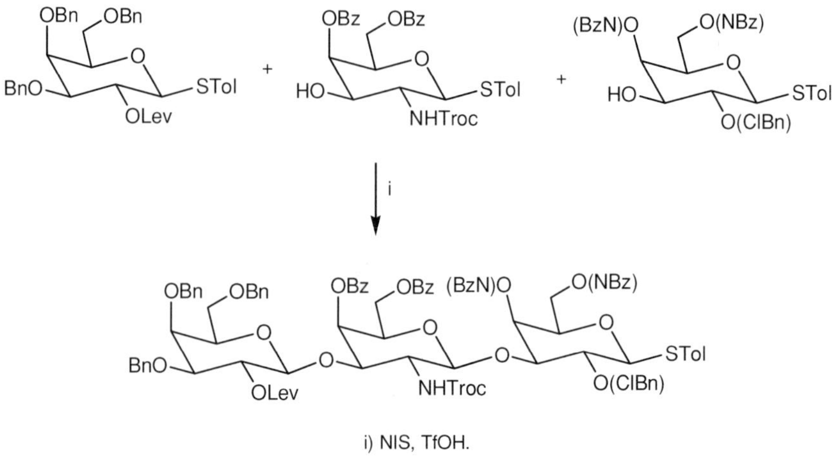

i) NIS, TfOH.

FIGURE 2.65. One-pot reaction for two β-linkages formation.

2.1.10.4   The Pool Strategy

This term applies to define a one-step reaction used to build up two β-linkages simultaneously from 3 sugar intermediates.[68] This approach has been described for the preparation of the glycosyl ceramide Globo H hexasaccharide identified as an antigen on prostate and breast cancer cells. The synthesis consisted in the initial synthesis of the trisaccharide building block from the one-pot reaction of the 3 suitable sugar intermediates under N-iodosuccinimide and triflic acid conditions in 67% yield (Figure 2.65).

2.1.10.5   Phosphate Glycosyl Donors

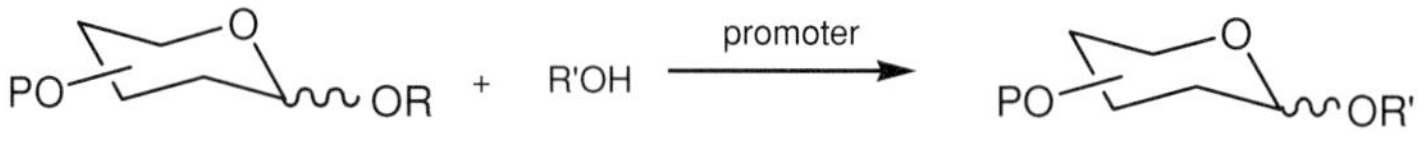

R	Promoter	Conditions
$P(=O)(OPh)_2$	TMSOTf	$CH_2Cl_2, -5°C$
$P(=S)(Me)_2$	TrClO$_4$	
$P(=O)(NMe_2)_2$	TMSOTf	$CH_3CN, -40°C$
$P(=NTs)(NMe_2)_2$	BF$_3$-Et$_2$O	CH_2Cl_2

Phosphorous glycosyl donors are another option for preparing oligosaccharides. These donors have been used for the preparation of sialyl oligosaccharides; however, the yield reported were moderate. This is the case of the preparation of sialyl tetrasaccharide derivative, which was carried out by condensation

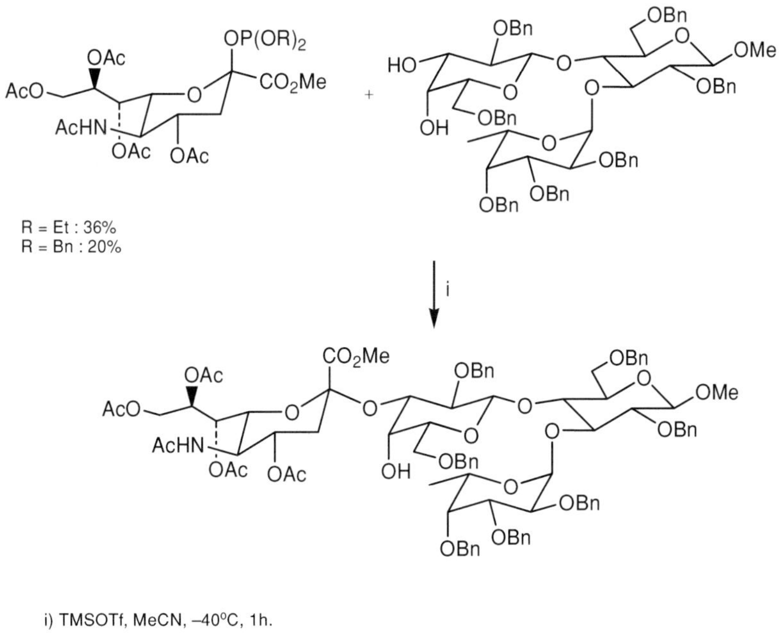

R = Et : 36%
R = Bn : 20%

i) TMSOTf, MeCN, −40°C, 1h.

FIGURE 2.66. Phosphorous glycosyl donors for oligosaccharide synthesis.

between sialyl phosphite with trisaccharide acceptor under TMSOTf as catalyst (Figure 2.66).[69]

2.1.11 Enzymatic Approach

Enzymes in organic chemistry has become an essential tool for the synthesis of important target molecules, and in many cases they are considered the first choice, especially for those key steps involving stereospecifically controlled reaction conditions. In general, enzymes are considered efficient catalysts that perform the desired transformation under mild conditions with high selectivity and specificity, usually avoiding epimerization, racemization, and rearrangements processes. Besides there is a current need of developing economical and environmentally friendly processes for synthesis. However, some aspects still need close attention in order to fulfill thoroughly the requirements especially for high-scale production. Thus, many enzymes are unstable, high cost, difficult to handle, and require expensive cofactors.

Glycosyltransferases are important enzymes involved in essential processes related to oligosaccharide biosynthesis, and they have been found also very useful as biocatalyst for the chemoenzymatic synthesis of interesting oligosaccharides

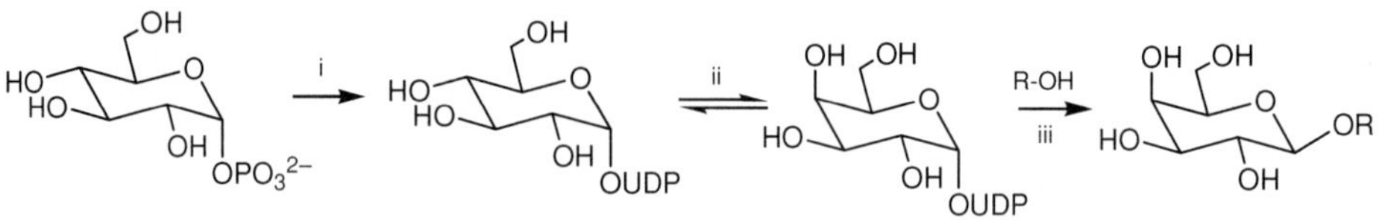

i) UTP, UDP-Glc pyrophosphorylase. ii) UDP-Glc 4-epimerase. iii) Gal transferase.

FIGURE 2.67. Glycosylation with galactosyltransferases.

and nucleotides.[70,71] They have been classified as Leloir if they are involved in the biosynthesis of most of N- and O-linked glycoproteins in mammalians, and requires mono- and diphosphates as glycosyl donors, and non-Leloir enzymes which utilize sugar phosphates as substrates.

Glycosylations with galactosyltransferases can be performed through the use of glucose-1-phosphate as donor. A general sequence consists in the conversion by using UDP-Glc pyrophosphorylase to give UDP-glucose. Epimerization with UDP-glucose epimerase affords UDP-galactose, which is used for glycosylation with galactosyltransferase (Figure 2.67).[72]

Several chemoenzymatic synthesis of $\alpha(2{\rightarrow}6)$ and $\alpha(2{\rightarrow}3)$-oligosaccharides have been reported through the use of sialyltransferases for glycosidic coupling reactions. One described approach involves the in situ regeneration of CMP-Neu5Ac, requiring catalytic amount of CMP-Neu5Ac (Figure 2.68).[73]

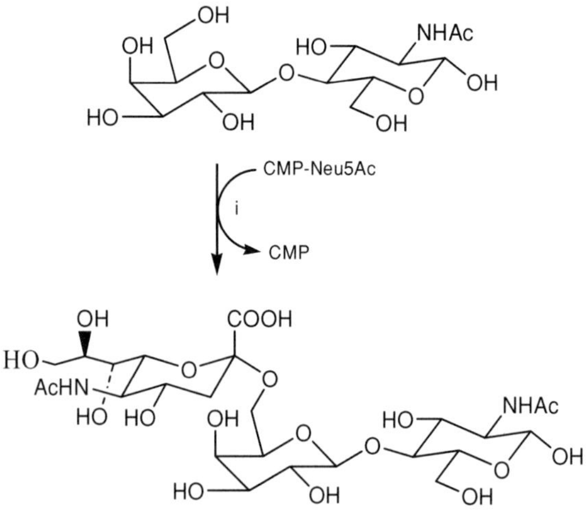

i) α-(2-6)-sialyltransferase

FIGURE 2.68. Synthesis of silayl trisaccharide mediated by silayl glycosiltransferase.

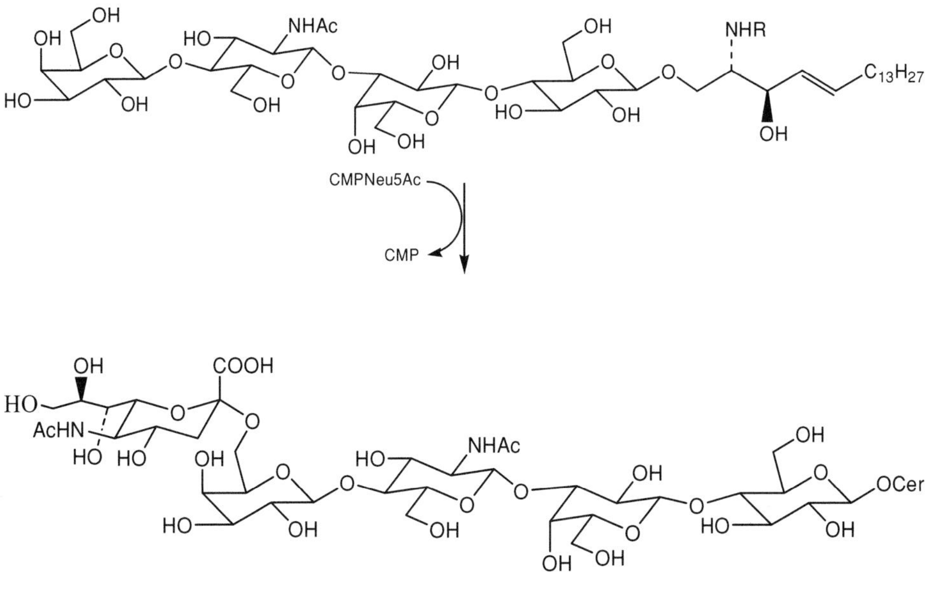

i) α-(2-6)-sialyltransferase

FIGURE 2.69. Enzymatic synthesis of ganglioside.

Sialyltransferases also proved to be efficient biocatalysts in the preparation of gangliosides, being involved in (2→6) linkage formation between the tetrasaccharide ceramide with CMP-Neu5Ac (Figure 2.69).[74]

Glucosamine may be enzymatically transformed to glucosamine 6-phosphate by treatment with hexokinase from yeast, and ultimately to glucosamine 1-phosphate by the action of phosphoglucomutase (Figure 2.70).[75]

UDP-glucuronic acid was prepared from UDP glucose by the action of UDP-Glc dehydrogenase along with NAD. This cofactor was regenerated with lactate dehydrogenase in the presence of piruvate (Figure 2.71).[76]

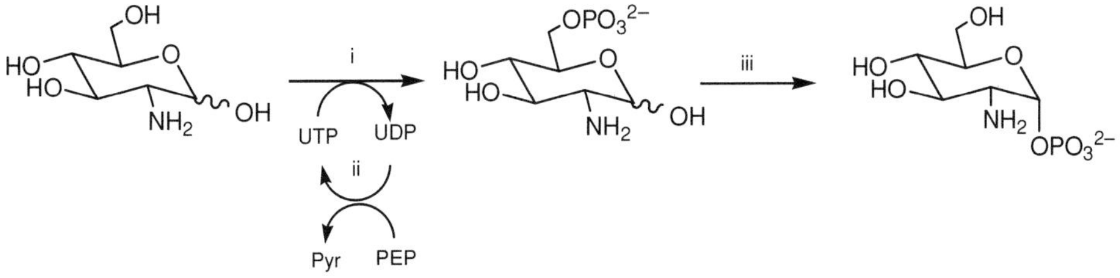

i) Hexokinase from yeast. ii) pyruvate kinase. iii) phosphoglucomutase.

FIGURE 2.70. Enzymatic preparation of glucosamine 6- and 1-phosphate.

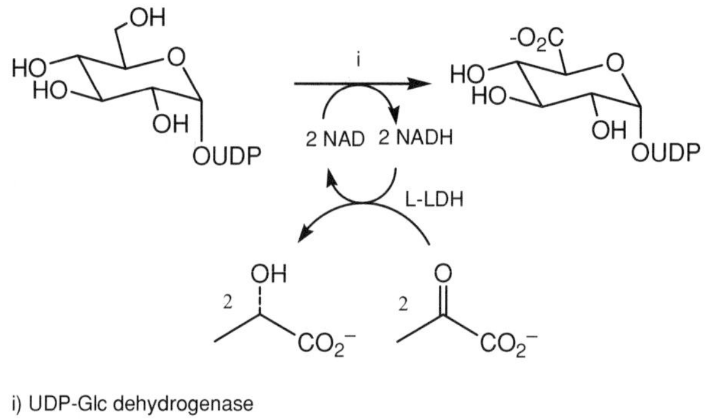

i) UDP-Glc dehydrogenase

FIGURE 2.71. Enzymatic preparation of UDP-glucuronide.

CMP-N-acetylneuraminic acid has been prepared from CTP and NeuAc under catalysis by CMP-NeuAc synthetase. In a cascade representation, it is observed that CTP is synthesized from CMP with adenylate kinase and pyruvate kinase (Figure 2.72).[77]

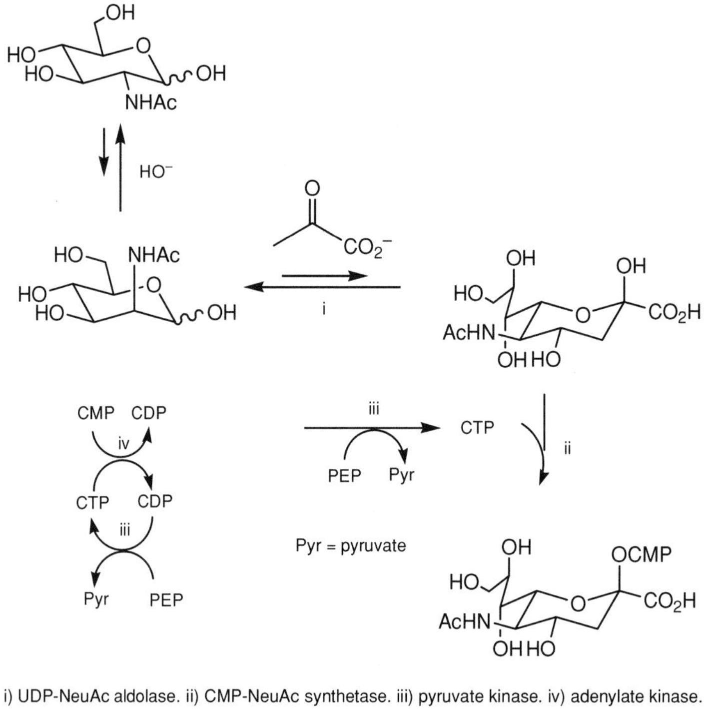

i) UDP-NeuAc aldolase. ii) CMP-NeuAc synthetase. iii) pyruvate kinase. iv) adenylate kinase.

FIGURE 2.72. Synthesis of CMP-N-acetylneuraminic acid.

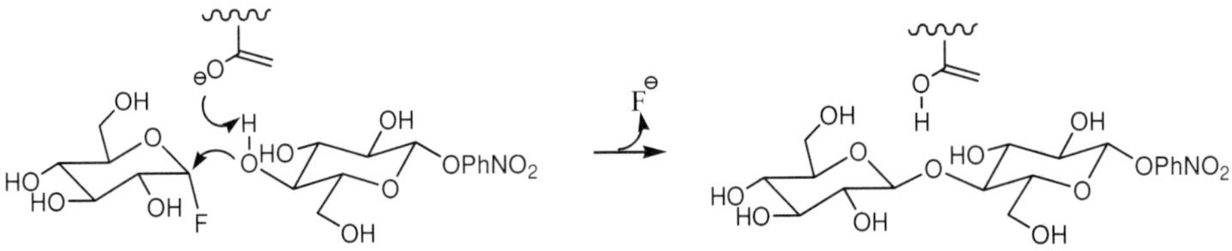

FIGURE 2.73. Glycosynthase-catalyzed oligosaccharide synthesis.

2.1.11.1   Enzymatic Synthesis of Oligosaccharides

Mutated glycosidase also known as glycosynthase AbgGlu358Ala in combination with activated glycosyl donors and suitable acceptors can generate synthetic oligosaccharides. Thus, for this transformation the conditions selected were α-glycosyl fluoride as glycosyl donor and p-nitrophenyl as glycosyl acceptor in the presence of ammonium bicarbonate buffer. The proposed mechanism of glycosynthase-catalyzed reaction is illustrated in Figure 2.73.[78]

The Regioselective preparation of α-1,3 and α-1,6 disaccharides by using α-glycosidase as biocatalyst has been described. Thus, by combining p-nitrophenyl-α-galactose functioning as glycosyl donor, with the glycosyl acceptor methoxygalactose, the expected 1,3- and 1,6-disaccharides were obtained in the form of α- and β- anomers (Figure 2.74).[79]

A transglycosylation reaction mediated by α-L-fucosidase from *Alcaligenes sp.* was performed by combination of p-nitrophenylglycosides donors, with different acceptors such as N-acetylglucosamine, lactose, D-GlcNAc, and D-Glc, providing the corresponding p-nitrophenyl glycosides of di- and trisaccharides containing a (1→2)-, (1→3)-, (1→4)-, or (1→6)-linked to the α-L-fucosyl group. In the general procedure illustrated in Figure 2.75 the p-nitrophenyl fucoside donor was combined with p-nitrophenyl lactosamine acceptor, being incubated with α-L-fucosidase at 50°C to produce the 2- and 3-linked trisaccharides.[80]

Sulfotransferases provides a versatile method for the preparation of glycoside sulfates. A recent report describes the use of 3'-phosphoadenosine -5'-phosphosulfate (PAPS), and GlcNAc-6-sulfotransferase as catalyst (Figure 2.76).[81]

A chemoenzymatic synthesis of rhodiooctanoside isolated from Chinese medicines was described. The synthesis was carried out by direct β-glucosidation between 1,8-octanediol and D-glucose using immobilized β-glucosidase from almonds with the synthetic propolymer ENTP-4000 to generate the glycoside in 58% yield (Figure 2.77).[82]

Lactosamine was prepared using an enzymatic approach consisting in the preparation of UDP glucose and condensation with N-acetyl glucosamine (GlcNAc) in the presence of galactosyl transferase (Figure 2.78).[83]

2.1.12   The Solid-Phase Methodology

Perhaps what remains the most challenging task for sugar chemistry is the synthesis of complex oligosaccharides such as those found in bacterial membranes or

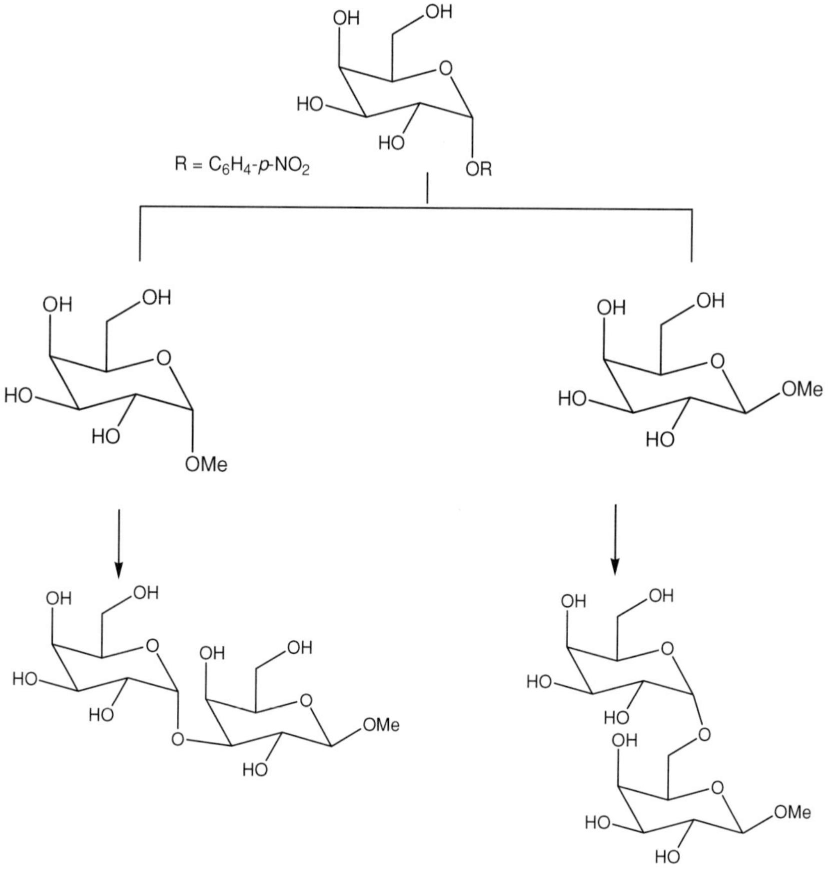

FIGURE 2.74. Example of microbial catalyzed coupling reaction.

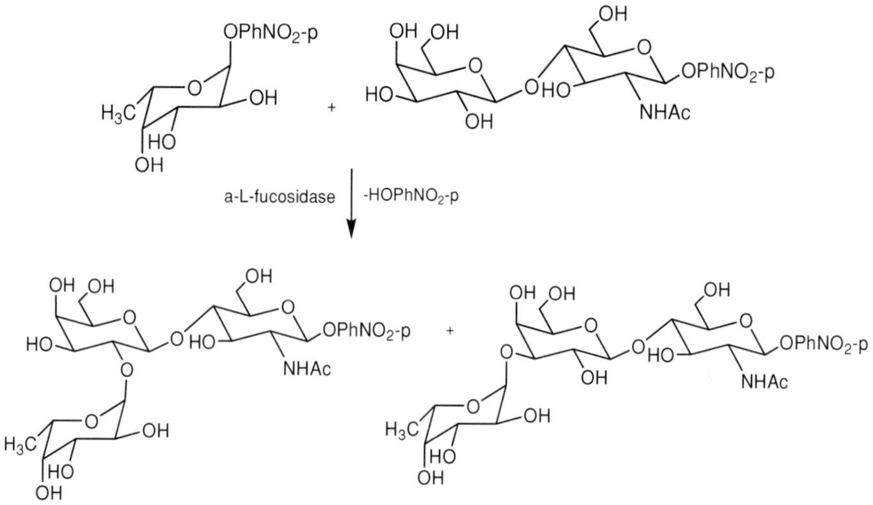

FIGURE 2.75. Transglycosylation reaction for the preparation of 2- and 3-linked trisaccharides.

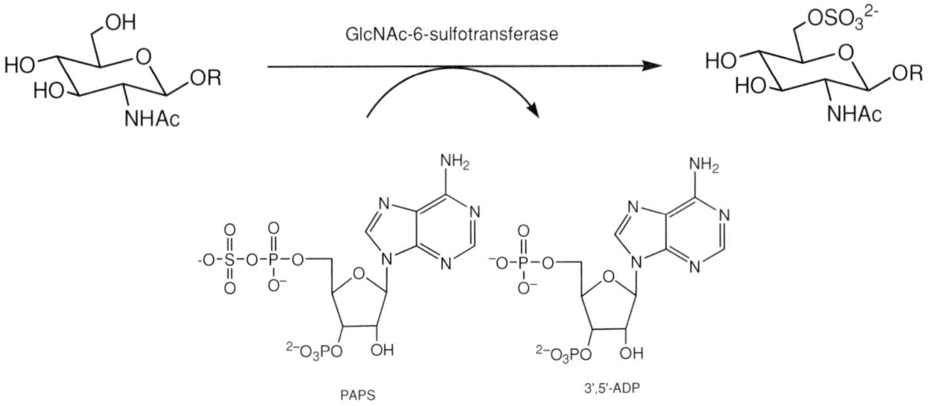
GlcNAc-6-sulfotransferase
PAPS
3',5'-ADP

FIGURE 2.76. Transfer of the sulfuryl group from PAPS to the glycoside.

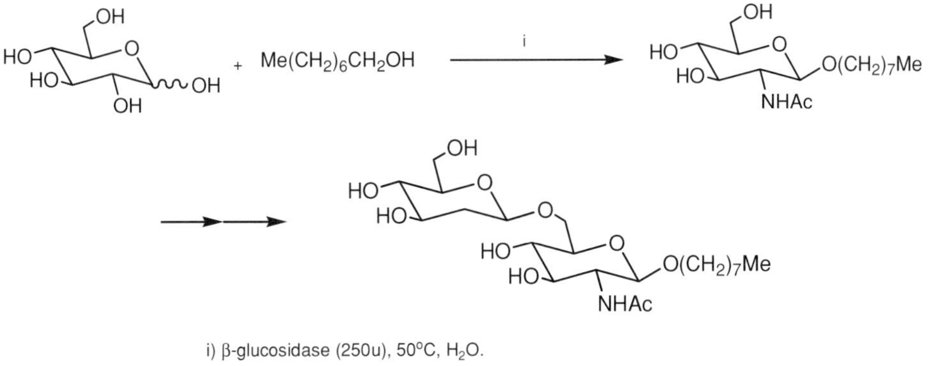
i) β-glucosidase (250u), 50°C, H2O.

FIGURE 2.77. Chemoenzymatic synthesis of rhodiooctanoside.

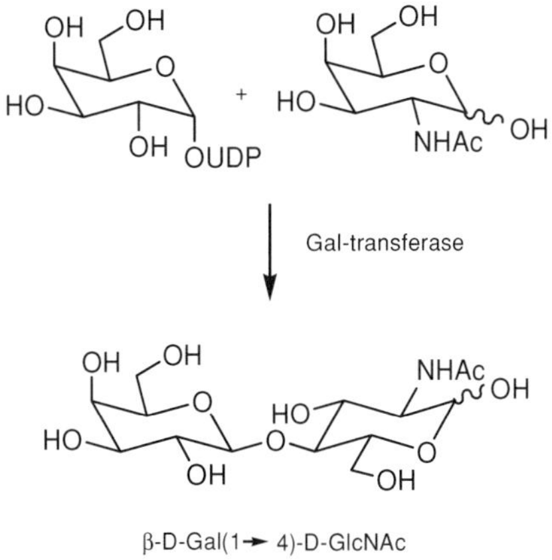
Gal-transferase
β-D-Gal(1→ 4)-D-GlcNAc

FIGURE 2.78. Enzymatic synthesis of lactosamine.

wall cells and are usually in the form of glycopeptides. Different types of monosaccharides can be present as constitutive parts such as glucose, galactose, mannose, N-acetylglucosamine, silaic acid, and L-fucose. Also, the order of linkage and stereoselectivity between them is rarely conserved.

The different nature, stereoselectivity, and linkage sequence have been a formidable obstacle for the development of general procedures of the type used for peptides and oligonucleotides which can be prepared on machine synthesizers with high efficiency.

The main advantage of the solid-phase methodology is the coupling of sugar units to the resin, which allows easy washing away of the nonreacted reagents, avoiding tedious purifications steps.

Nonetheless, despite the difficulties, interesting progress has been made for preparing oligosaccharides[84a] and glycopeptides,[84b] suggesting that in the solid phase technology for complex sugars will be affordable.

The solid-phase approach involves three elements, namely the glycosyl donor, glycosyl acceptor, and the resin, which is properly activated with a group susceptible for attachment either with the glycosyl donor or acceptor depending on the strategy of choice. Although it appears obvious, it is important to remain that the linkage between the resin and the sugar should be easily cleaved under compatible conditions for the glycoside bond.

According to a comprehensive review,[85] the synthetic strategies are classified by (a) Donor-bound, (b) Acceptor-bound, and (c) Bidirectional Strategies.

One general approach involves the initial attachment of a glycosyl donor (halides, trichloroacetimidate, sulfoxides, phosphates, phosphates, thio and pentenyl and glycols) to the resin (polystyrene-base). The attached sugar is selectively deprotected depending on the required position (1,2- 1,3- 1,4- 1,6), transforming the resin-sugar complex in a sugar acceptor which will be coupled to the next glycosyl donor to produce a second linkage. By repeating this sequence an elongated chain is obtained. The final release and full deprotection will produce the free oligosaccharide (Figure 2.79).[86]

An example of the donor-bound strategy is the bounding of sulfur glycoside to polistyrene resin to form a sulfur linkage between the donor and the resin (Figure 2.80). Suitable hydroxyl group from the donor will serve as linkage site with de next sugar unit for chain elongation.

It should be noted that the glycosyl donor also contains a position available for the linkage with the next sugar. In other words, the glycosyl donor once attached to the resin becomes a glycosyl acceptor, as can be seen for the next coupling sequence (Figure 2.81).[85]

The synthesis of β-(1→6) gentotetraose was accomplished by using a benzoyl propionate as resin linker. The glycosyl donor chosen was acetobromoglucose functionalized with trichloroacetate group as a temporary protecting group at position 5. Glycosylation reactions were effected under Helferich conditions and cleavage from resin was performed with hydrazinium acetate (Figure 2.82).

Polymer solid phase has been also exploited successfully by Crich et al,[87] for the synthesis of sensitive β-mannosides, using a variation of sulfhoxide method, consisting in the transformation of sulfoxide to triflic group as leaving group. The

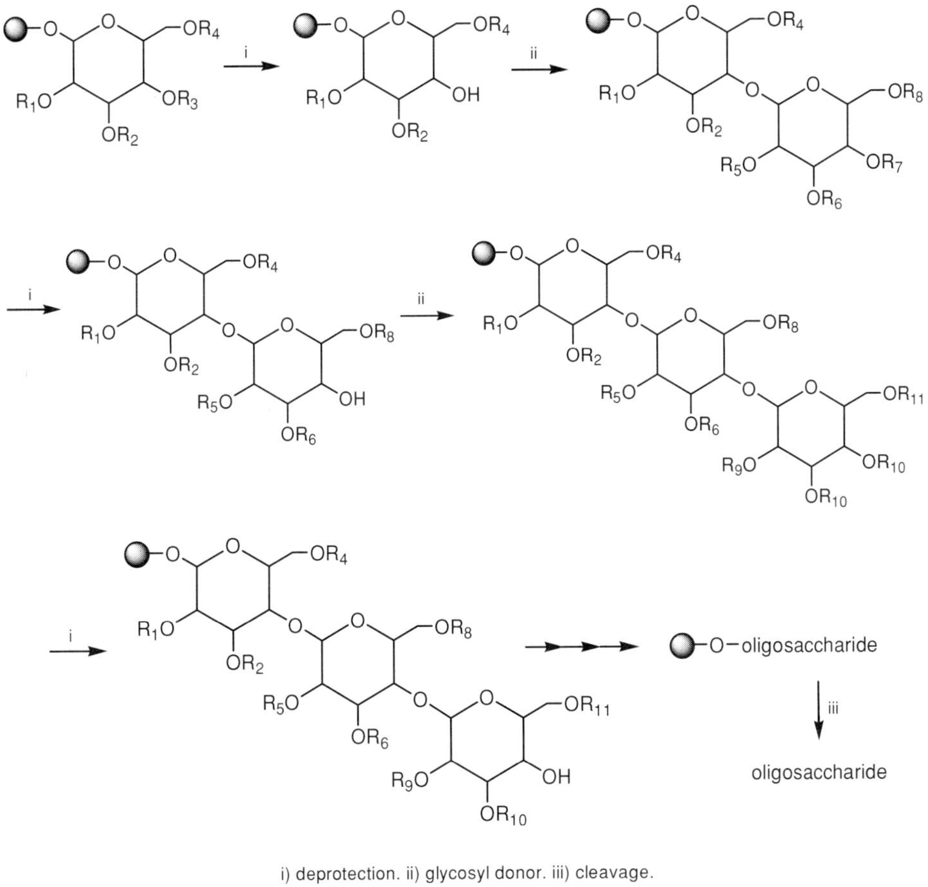

i) deprotection. ii) glycosyl donor. iii) cleavage.

FIGURE 2.79. General scheme for solid-phase oligosaccharide synthesis 1,4-linkage case.

subsequent addition of alcohol acceptor to the donor attached to the Wang resin will result in the glycoside β-mannoside formation (Figure 2.83).

The enzymatic solid-phase oligosaccharide synthesis relies mainly by the use of glycosyltransferases, glycosidases and glycosynthases. By taking advantage on their high stereo- and regioselectivity, various oligosaccharides and glycopeptides have been prepared usually under mild conditions without the need of using protecting groups. Unfortunately, the enzymatic approach is still in some cases

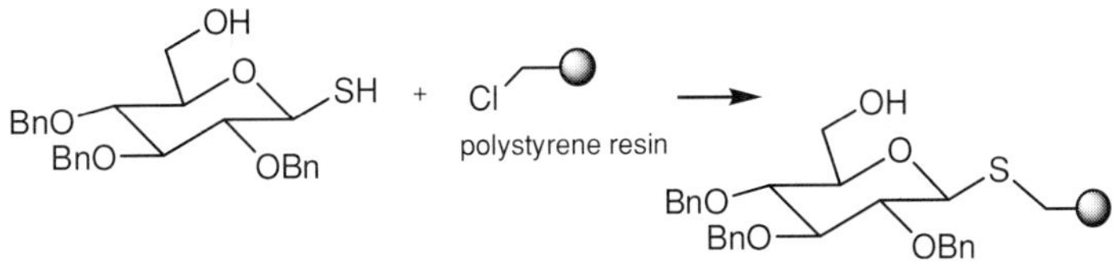

FIGURE 2.80. Example of donor-bound strategy for solid-phase glycosilation reactions.

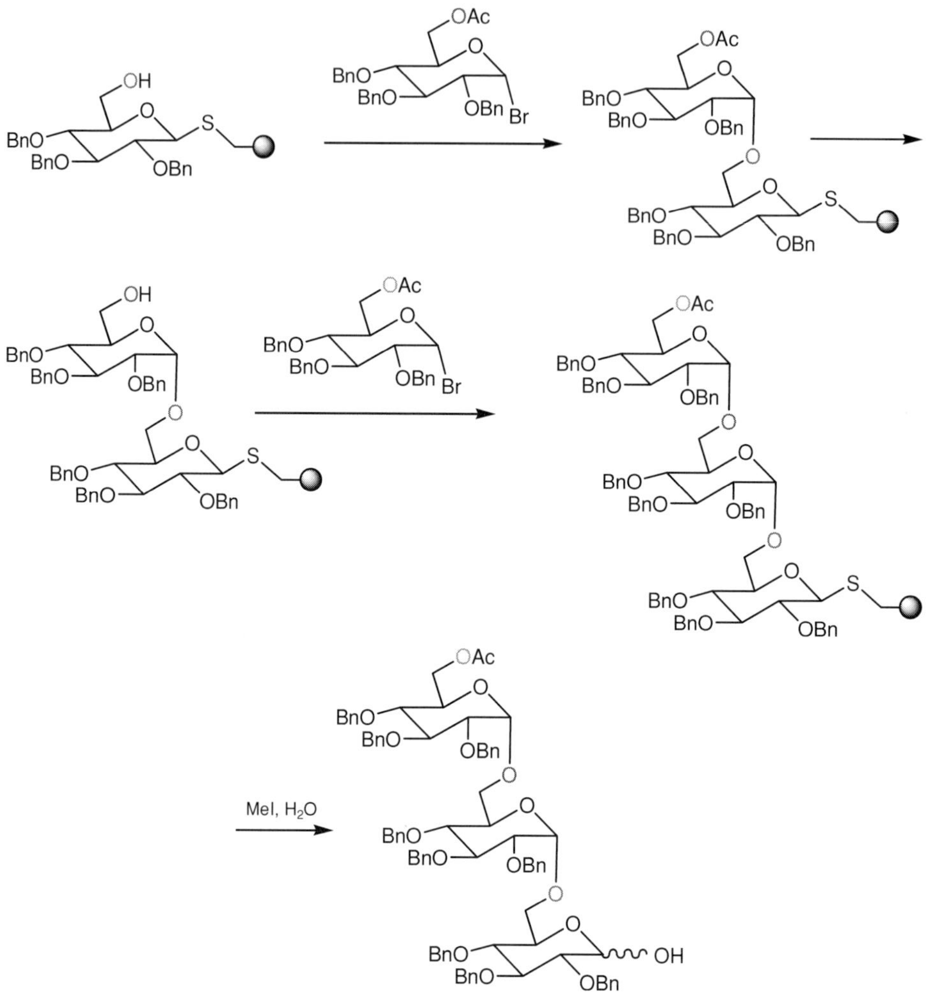

FIGURE 2.81. Sulfur mediated solid-phase coupling reaction.

unaffordable due its high cost for large-scale processes, lower yields provided, and their limited capability for recognizing a broad range of sugars, especially those not common. Two general approaches have been proposed for the preparation of oligosaccharides through the solid-phase approach (Figure 2.84).[88]

A solid-phase enzymatic approach for extending the oligosaccharide chain was described by Gijsen et al.[88] in which a disaccharide-linker fragment attached to a resin was coupled with the glycosyltransferases UDP-galactose and CMP-NeuAc in the presence of galacosyltransferases and Sialyltransferase as enzymatic catalyst. Final treatment with hydrazine was used to release the tetrasaccharide from the solid support (Figure 2.85).

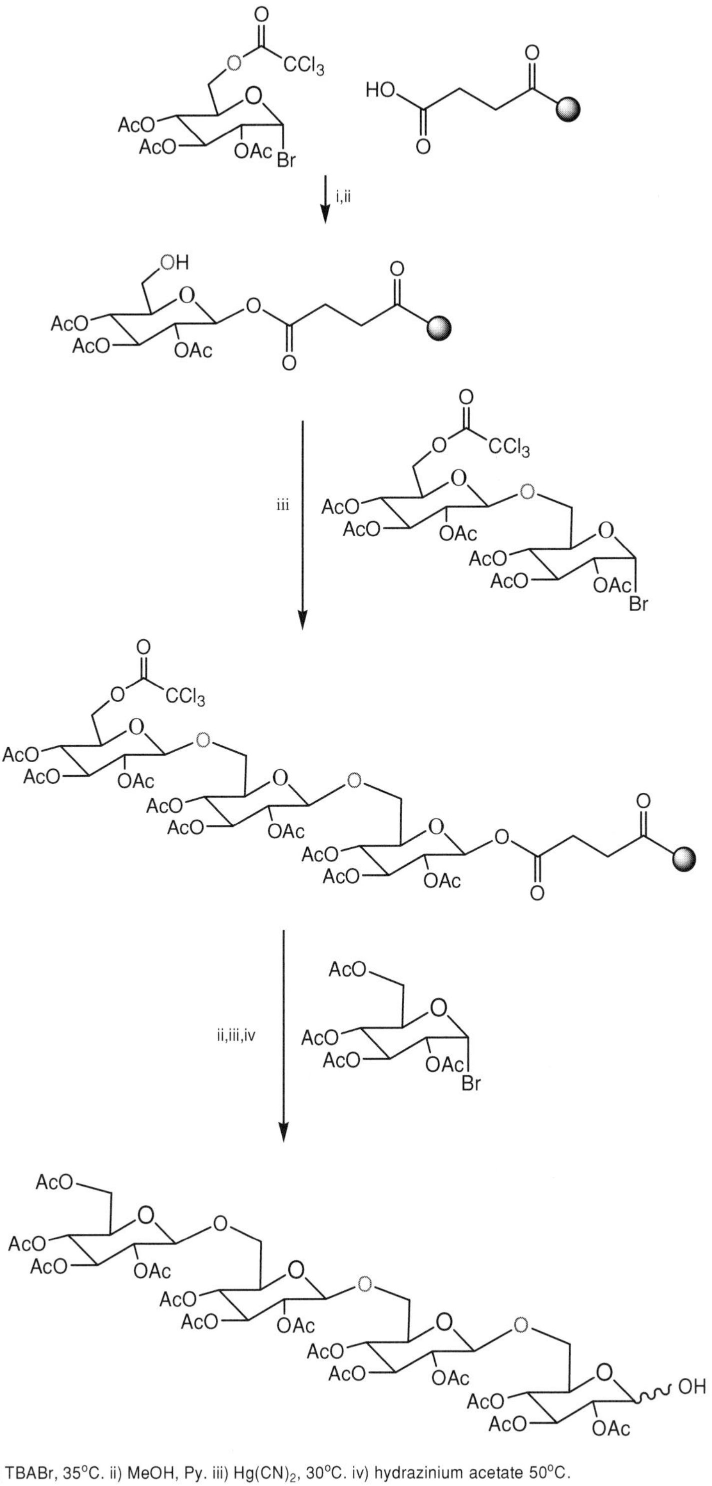

i) TBABr, 35°C. ii) MeOH, Py. iii) Hg(CN)$_2$, 30°C. iv) hydrazinium acetate 50°C.

FIGURE 2.82. Solid-phase coupling promoted by Helferich conditions.

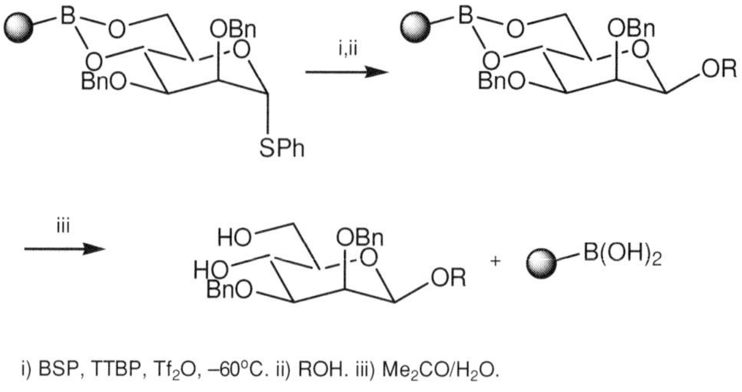

i) BSP, TTBP, Tf$_2$O, −60°C. ii) ROH. iii) Me$_2$CO/H$_2$O.

FIGURE 2.83. Solid-phase synthesis of β-mannoside glycoside.

2.2   Cyclic Oligosaccharides

The synthesis of cyclic oligosaccharides involves the preparation of linear saccharides, which ultimately are joined together to form a cyclic macromolecule. There are two main approaches proposed based on the cycloglycosylation step. The first involves the preparation of a long chain having at each end the donor and acceptor functionalities that will be interconnected through a glycosidic bond at a final step, and the second involving the polycondensation of smallest repeating unit called "saccharide monomers." It has been observed that the later strategy is considered less laborious however produce cyclic oligomers of different size since under these conditions the ring formation step is not controllable.

The chemical synthesis of cyclic oligosaccharides has been mainly driven to obtain cyclic (1→4)-linked oligopyranosides, however (1→3), and (1→6) linked cycloforms are also described. In the case of (1→2)-linked oligosaccharides, the ring closure requires about 17 or more glucopyranoside residues because (1→2)-linkage composed of pyranoside connected by one equatorial and one axial bond assumes rigid conformations and cannot cyclize.[89]

The pioneering total synthesis of cyclic oligosaccharide α-Cyclodextrin was carried out by Ogawa's group in 1985,[89] and since then alternative chemical or enzymatic methodologies appeared for preparing cyclic oligosaccharides. Nowadays the industrial production of cyclodextrins relies on the enzymatic conversion of prehydrolyzed starch into a mixture of cyclic and acyclic oligomers.

A full report about cyclic oligosaccharides[90] proposes four approaches to the synthesis of cyclic oligosaccharides developed during the last 10 years:

1. the stepwise preparation of a linear precursor that is subjected to cycloglycosylation;
2. the one-pot polycondensation /cycloglycosylation of a small "oligosaccharide monomer" typically, a di-, or trisaccharide that can yield a range of macrocycles of different sizes;
3. the enzyme-assisted synthesis of natural or unnatural cyclic oligosaccharides;

A) Immobilized substrate

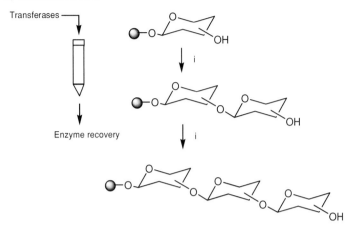

i) nucleotide sugar transferase

B) Immobilized enzymes

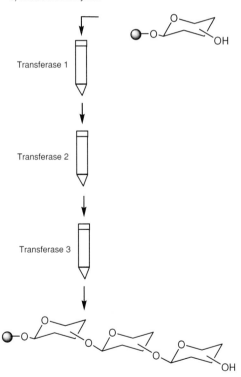

FIGURE 2.84. Two general approaches for immobilized solid-phase oligosaccharide synthesis.

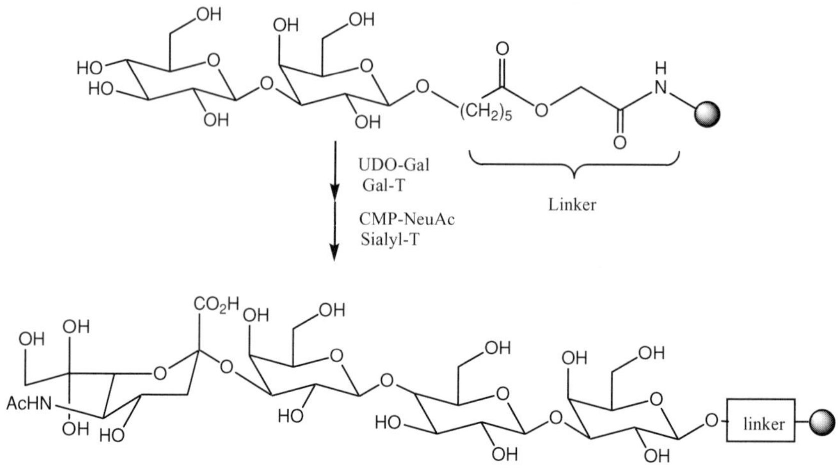

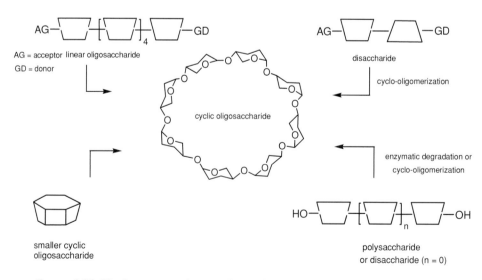

FIGURE 2.85. Enzymatic-solid phase glycosylation reaction.

4. the ring opening of cyclodextrins followed by oligosaccharide chain elongation and cycloglycosylation (Figure 2.86).

Despite the significant advances observed in cyclic oligosaccharide synthesis, their preparation is time-consuming, producing the target compounds with low regio- and stereoselective in low yields. The total synthesis of α-CD and γ-CD was described according to Figure 2.87.[91,92]

FIGURE 2.86. The four suggested approaches to the synthesis of cyclic oligosaccharides.

In 1990 it was reported the chemical synthesis of β-(1→3) linked hexasaccharide. The chemical approach involved the glycosidic reaction between benzylidene acceptor and protected glucosyl bromide as glycosyl donor, under silver triflate-promoter conditions. As can be seen in Figure 2.88, the construction of the linear oligosaccharide and its final cycloglycosylation was performed by using glycosyl bromides, which were prepared by photolytic brominolysis of 1,2-O-benzylidene glucose with $BrCCl_3$(Figure 2.88).[93]

The formation of (1→6)-glycopyranosidic linkages might produce cyclic di- tri- and tetrasaccharides. An early synthesis of β-(1→6)-glucopyranan under Helferich conditions, generated along with the linear oligomer, a cyclic di- and tetrasaccharide in 12 and 6%, respectively (Figure 2.89).[94]

An improved synthesis of cyclotetraoside was described by the same group 10 years later, consisting in the preparation from the peracetylated tetrasaccharide into the tetrasaccharide derivative having both the acceptor and the donor components. The final cyclization was performed under Helferich conditions providing a mixture of tri- and tetrasaccharide in 22% and 25% yield, respectively (Figure 2.90).[95]

2.2.1 Chemoenzymatic and Enzymatic Synthesis

The use of enzyme is as mentioned for many O- or N-glycosides the parallel possibility for preparing cyclic oligosaccharides. The limitation continues to be

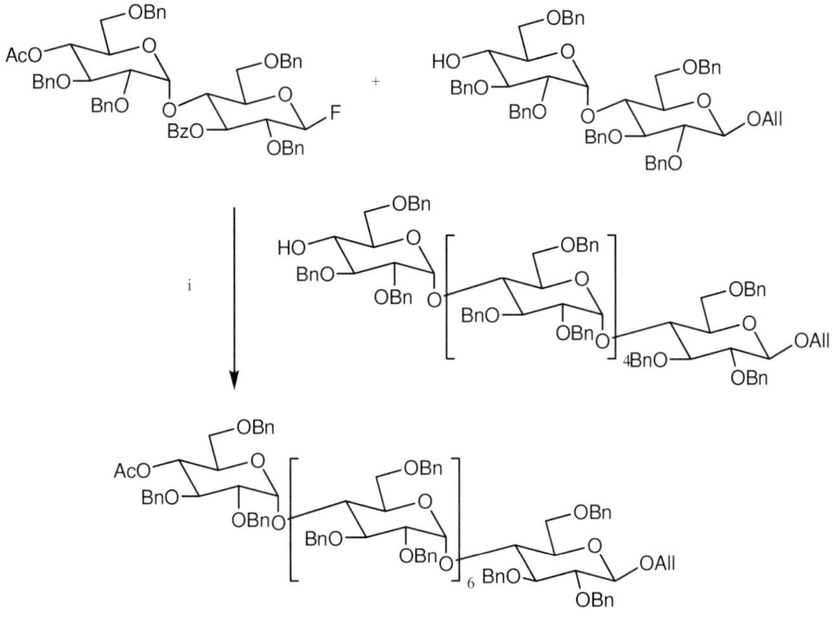

FIGURE 2.87. Chemical synthesis of cyclic α(1→4)-oligosaccharide γ-CD.

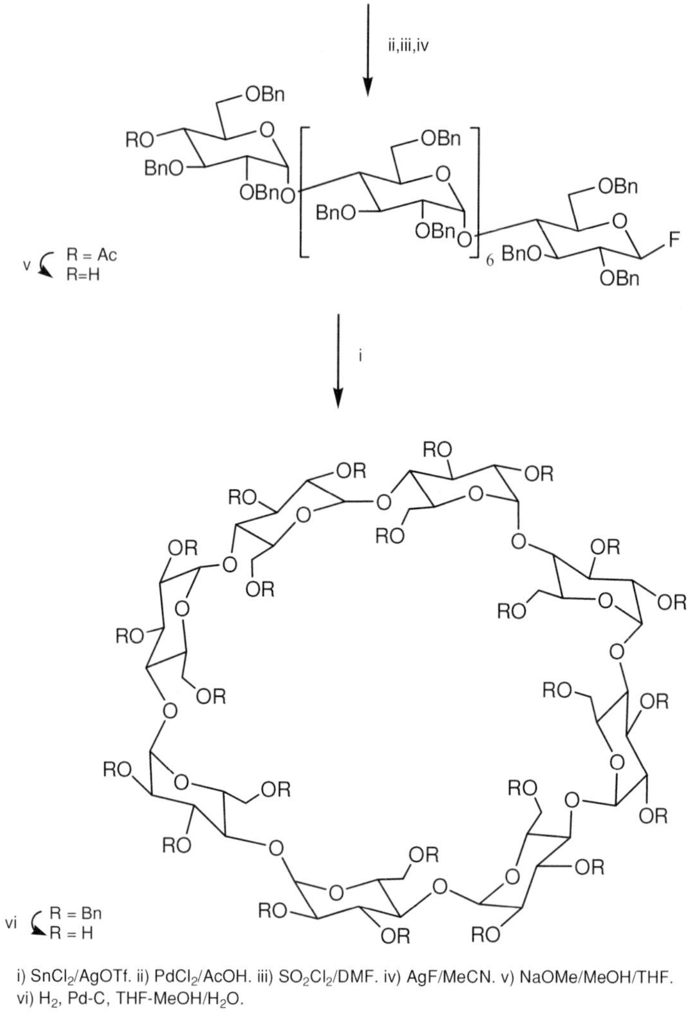

i) SnCl$_2$/AgOTf. ii) PdCl$_2$/AcOH. iii) SO$_2$Cl$_2$/DMF. iv) AgF/MeCN. v) NaOMe/MeOH/THF.
vi) H$_2$, Pd-C, THF-MeOH/H$_2$O.

FIGURE 2.87. (*continued*)

the availability and affordability, however. Some enzymes such as glycosidases and cycloglycosyltransferases (CGTases) that are involved in the preparation of cyclodextrins from starch and other α-(1→4)-glucans are accessible and more versatile.[95]

The feasibility of the chemoenzymatic approach was established in the preparation of cyclic β(1→4) hexa-, hepta- and octasaccharides, from 6-O-methylmaltosyl fluoride when incubated with CGTase. Thus, a mixture of 6^I, 6^{III}, 6^V-tri-O-methyl-α-CD (42%), 6^I, 6^{III}, 6^V-tetra-O-methyl-γ-CD (16 %) and in less proportion 6^I, 6^{III}, 6^V-tri-O-methyl-β-CD was obtained (Figure 2.91).[96]

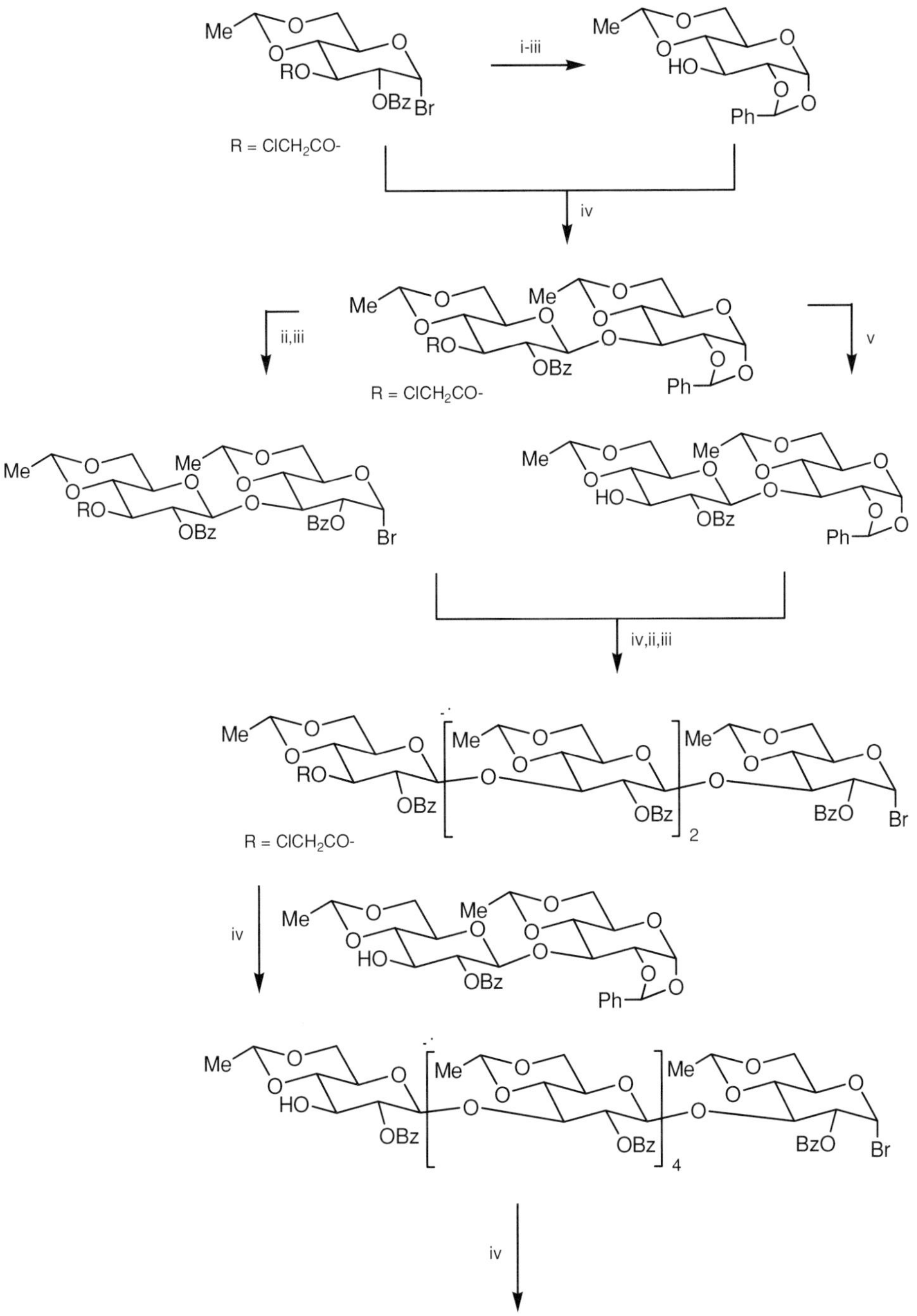

FIGURE 2.88. Synthesis of cyclic β-(1→3)-linked oligosaccharide.

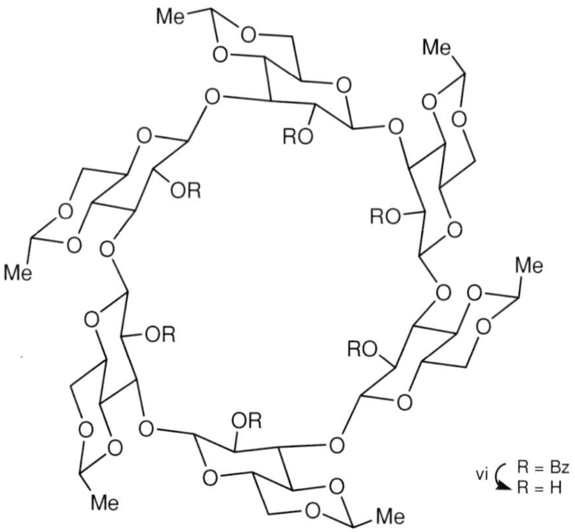

FIGURE 2.88. (*continued*)

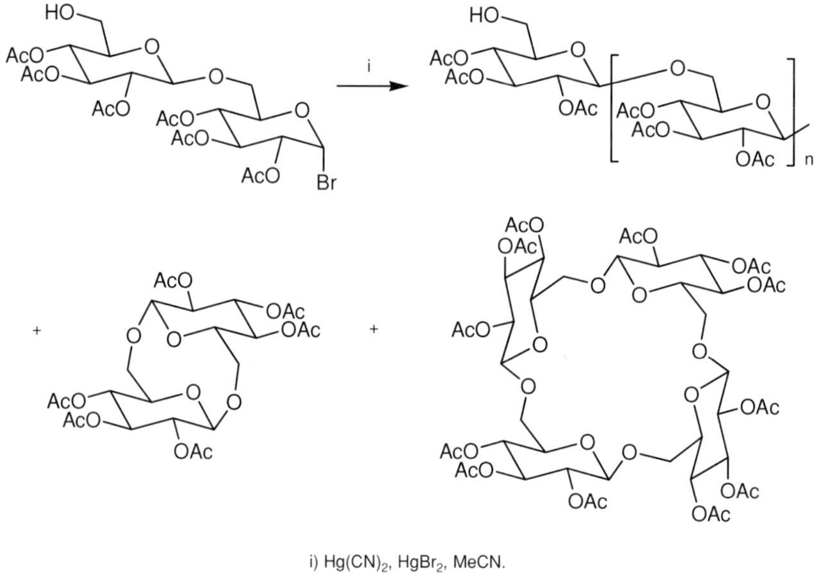

i) Hg(CN)$_2$, HgBr$_2$, MeCN.

FIGURE 2.89. Preparation of linear and cyclic $\beta(1\rightarrow6)$ di- and tetrasaccharides.

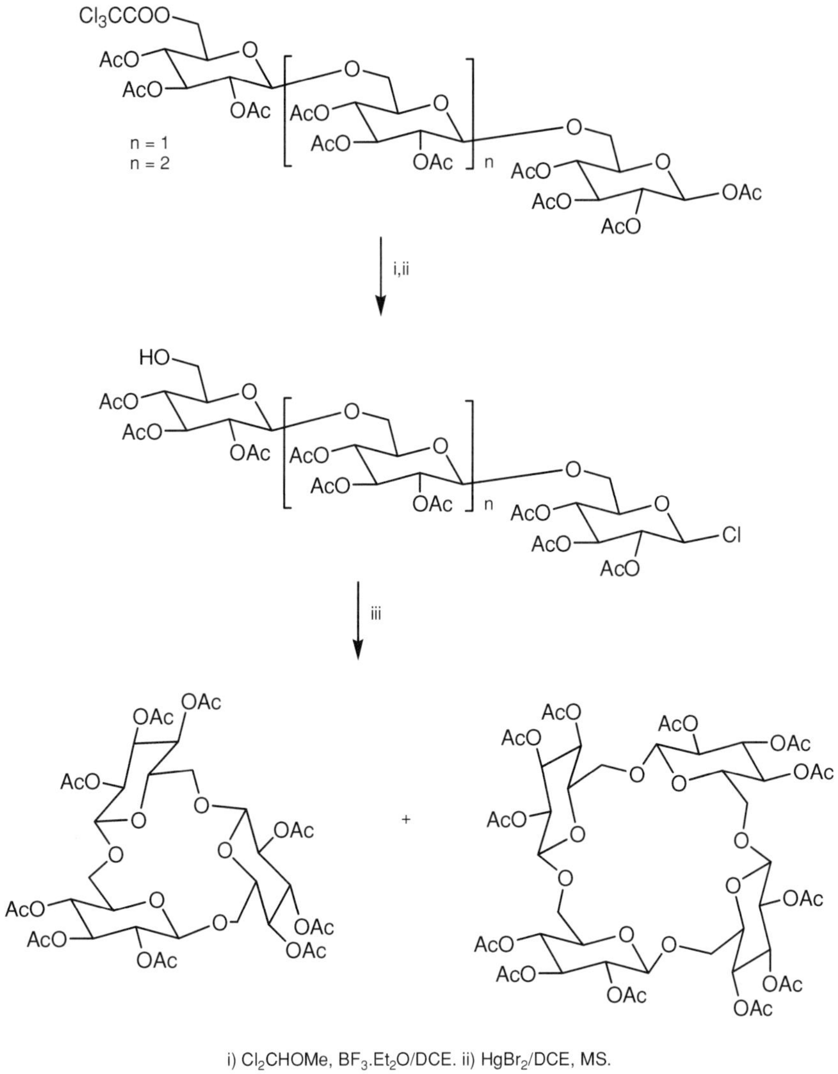
Cl3CCOO
AcO
AcO
OAc
n = 1
n = 2
OAc
OAc
i,ii
HO
Cl
iii
+
i) Cl2CHOMe, BF3.Et2O/DCE. ii) HgBr2/DCE, MS.

FIGURE 2.90. Improved synthesis of cyclic β(1→6) tri- and tetrasacharides.

Furthermore, under the same conditions it was possible to prepare from the maltotriosyl fluoride the cyclic α(1→4) hexasaccharide (6^{I}, 6^{II}-dideoxy-6^{I},6^{II}-diiodo-α-CD) in 38% (Figure 2.92).[97]

An alternative option for the enzymatic preparation of cyclic oligosaccharides besides CGTases are glycosidases, which exert their action on polysaccharides. This possibility was exploited in the preparation of cyclic fructins by conversion of β-(1→2)-fructofuranan by bacterial fructotransferases isolated from *Bacillus circulans* (Figure 2.93).[98]

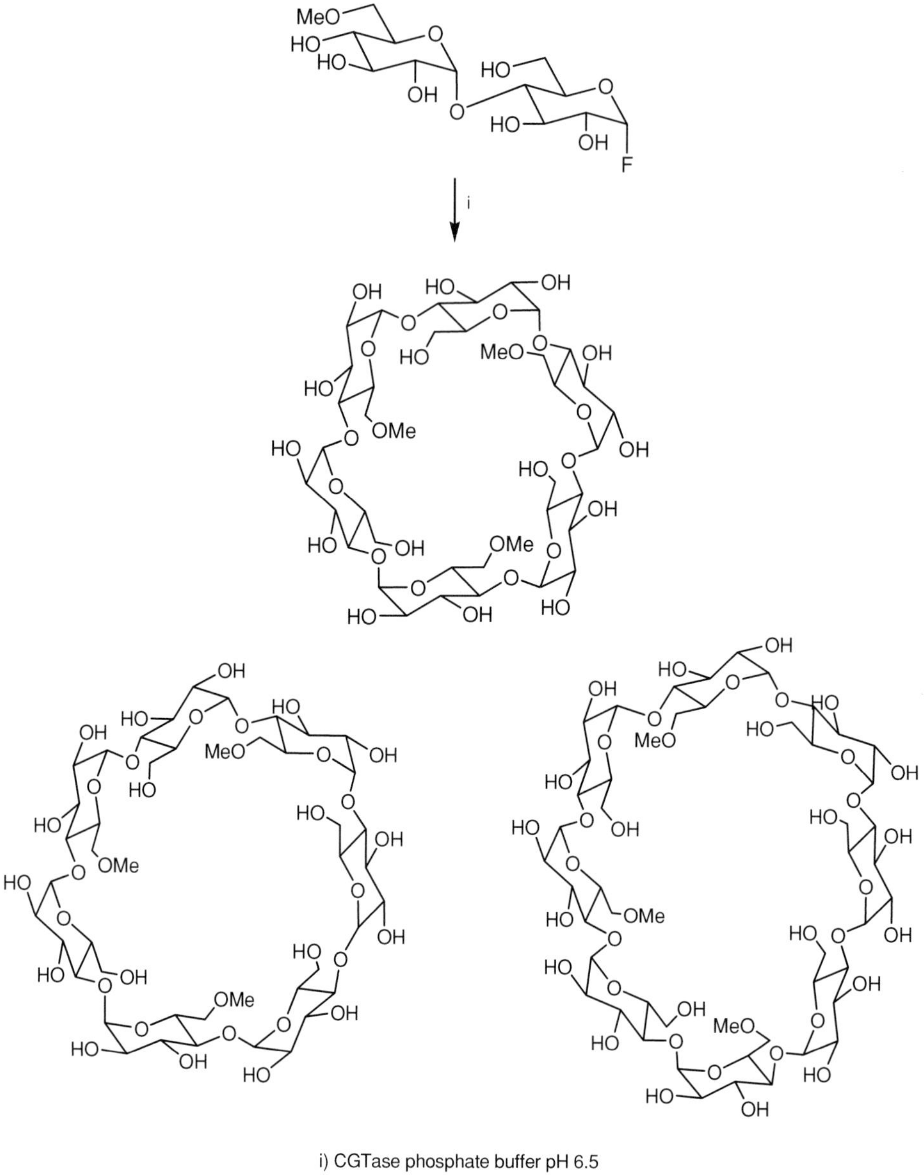

i) CGTase phosphate buffer pH 6.5

FIGURE 2.91. Synthesis of of 6^{I}, 6^{III}, 6^{V}-tri-O-methyl-α-CD, 6^{I}, 6^{III}, 6^{V}-tetra-O-methyl-γ-CD and 6^{I}, 6^{III}, 6^{V}-tri-O-methyl-β-CD.

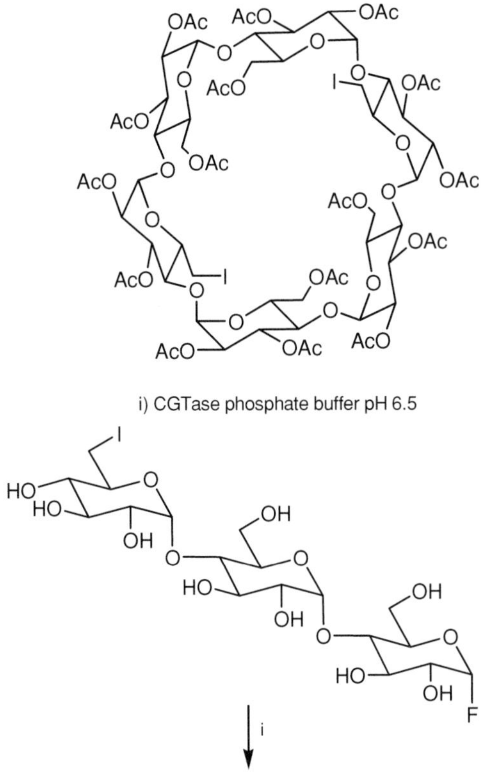

i) CGTase phosphate buffer pH 6.5

FIGURE 2.92. Enzymatic synthesis of 6^I, 6^{II}-dideoxy-6^I,6^{II}-diiodo-α-CD.

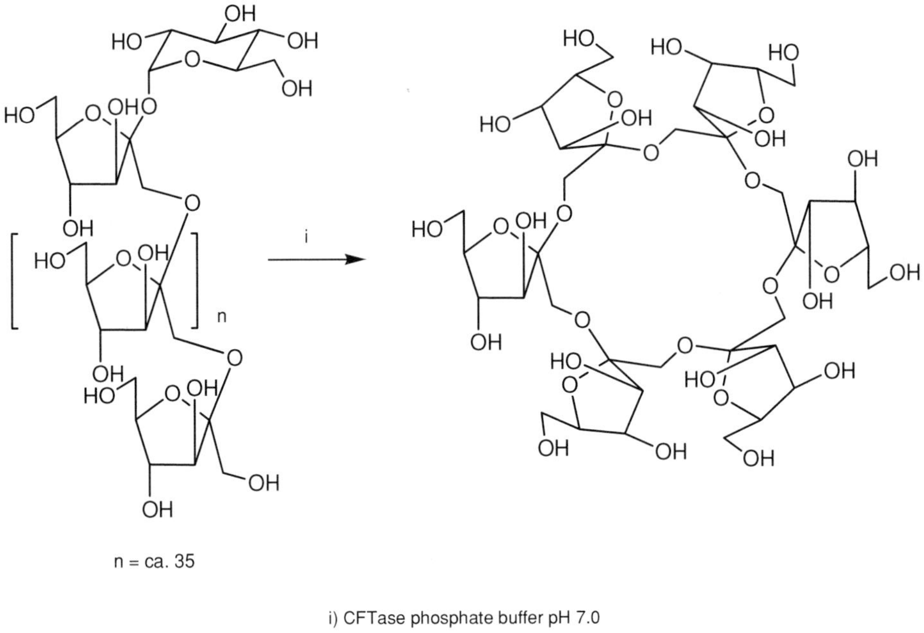

n = ca. 35

i) CFTase phosphate buffer pH 7.0

FIGURE 2.93. Enzymatic synthesis of cycloinulooligosaccharides.

Summary of Preparation of the Main Glycosyl Donors

References

1. K. Toshima, and K. Tatsuta, *Chem. Rev.* **93**, 1503 (1993).
2. F.B. Anderson, and D.H. Leaback, *Tetrahedron* **12**, 236 (1961).
3. H.P. Wessel, *Carbohydr. Chem.* **7**, 263 (1988).
4. Y.F. Shearly, C.A. O'Dell, and G. Amett, *J. Med. Chem.* **30**,1090 (1987).
5. W. Koenigs, and E. Knorr, *Chem. Ber.*, **34**, 957 (1901).
6. K. Igarashi, *Adv. Carbohydr. Chem.Biochem.*, **34**, 243 (1977).
7. J.-H. Kim, H. Yang, J. Park, and G-J. Boons, *J. Am. Chem. Soc.*, **127**, 12090 (2005).
8. M.P. De Ninno, P.A. McCarthy, K.C. Duplatiel, C. Eller, J.B. Etienne, M.P. Zawistowski, F.W. Bangerter, C.E. Chandler, L.A. Morehouse, E.D. Sugarman, R.W. Wilkins, H.A. Woody, and L.M. Zaccaro, *J. Med. Chem.* **40**, 2547 (1997).
9. R.B. Conrow, and S. Bernstein, *J. Org. Chem.* **36**, 836 (1971).
10. A.V. Stachulski, and G.N. Jenkins, *Natural Products Reports* 173 (1998).
11. A. Bredereck, A. Wagner, H. Kuhn, and H. Ott, *Chem. Ber.* **93**, 1201 (1960).

12. P. Bächli, E.G. Percival, *J. Chem. Soc.* 1243 (1952).

13. A.Y. Khorlin, I. M. Privalova, and I.B. Bystrova, *Carbohydr. Res.* **19**, 272 (1971).

14. H. Paulsen and H.Tietz, *Angew. Chem. Int. Ed. Engl.* **21**, 927 (1982).

15. H.P. Wessel, N. Iberg, M. Trumtel, and M.-C. Viaud, *BioMed. Chem. Lett.* **6**, 27 (1996).

16. K. Katano, H. An, Y. Aoyagi, M. Overhand, S.J. Sucheck, W.C. Stevens Jr., C.D. Hess, X. Zhou, and S.M. Hecht, *J. Am. Chem. Soc.* **120**, 11285 (1998).

17. (a) S. Umezawa, S. Koto, K. Tatsuta, H. Hineno, Y. Nishimura, and T. Tsumura, *Bull. Chem. Soc. Jpn.*, **42**, 529 (1969). (b) S. Hanessian, M. Tremblay, and E.E. Swayze, *Tetrahedron*, **59**, 983 (2003). (c) H. Tanaka, Y. Nishida, Y. Furuta, and K. Kobayashi, *Bioorg. Med. Chem. Lett.*, **12**, 1723 (2002).

18. H.J. Roth and A. Kleeman, *Pharmaceut.Chem.* **7**, 263 (1988).

19. T. Suami, T. Otake, T. Nishimura, and Y. Ikeda, *Bull. Chem. Soc. Jpn.* **46**, 1014 (1973).

20. N. Bagget, A.K. Samra, and A. Smithson, *Carbohydr. Res.* **124**. 63 (1983).

21. (a) R.R. Schmidt, *Angew. Chem. Int. Engl.*, **25**, 213 (1986). (b) R.R. Schmidt, and K.-H. Jung, *Carbohydr. Eur.*, **27**, 12 (1999). (c) R.R. Schmidt and W. Kinzy, *Adv. Carbohydr. Chem. Biochem.*, **50**, 21 (1994).

22. A. Fürstner and T. Müller, *J. Am. Chem. Soc.* **121**, 7814 (1999).

23. A. Hasegawa, K. Fushimi, H. Ishida, and M. Kiso, *J. Carbohydr. Chem.* **12**, 1203 (1993).

24. S. Danishefsky and M.D. Shair, *J. Org. Chem.* **16**, 61 (1996).

25. D.P. Larson, C.H. Heathcock, *J. Org. Chem.* **62**, 8406 (1997).

26. S.F. Lu, Q.O. O'Yang, Z.W. Guo, B. Yu, and Y.Z. Hui, *J. Org.Chem.* **62**, 8400 (1997).

27. M. Brito-Arias, R. Pereda-Miranda, and C.H. Heathcock, *J. Org. Chem.* **69**, 4567 (2004).

28. (a) G.J. Boons, S. Isles, *J. Org. Chem.* **61**, 4262 (1996). (b) G.-J. Boons, *Contemporary Organic Synthesis* **3**, 173 (1996).

29. (a) S. Komba, H. Galustian, H. Ishida, T. Feizi, R. Kannagi, and M. Kiso, *Angew. Chem. Int. Ed.* **38**, 1131. (1999). (b) Y. Zhang, A. Brodsky, P. Sinay, Tetrahedron: Asymmetry **9**, 2451 (1998). (c) A. Hasegawa, K. Ito, and H. Ishida, Kiso *J. Carbohydr. Chem.* **266**, 279 (1995).

30. (a) A. Koenig, R. Jain, R. Vig, K.E. Norgard,-Sumnicht, K.L. Matta, and A. Varki, *Glycobiology*, **7**, 79 (1997). W.J. Sanders, E. J. Gordon, O. Dwir, P. J. Beck, R. Alon, and L.L. Kiessling, *J. Biol. Chem.* **274**, 5271 (1999).

31. A. Lubineau, J. Alais, and R. Lemoine *J. Carbohydr. Chem.* **19**, 151 (2000).

32. K.C. Nicolaou, T. Ohshima, F.L. van Delft, D. Vourloumis, J.Y. Xu, J. Pfefferkorn, and S. Kim, *J. Am. Chem. Soc.* **120**, 8674 (1998).

33. H. Lonn, *Carbohydr. Res.* **115**, 139 (1985).

34. S.V. Ley, and H.W. Priepke, *Angew. Chem. Int. Engl.* **33**, 2292 (1994).

35. K. Hotta, H. Ishida, M. Kiso, and A. Hasegawa, *J. Carbohydr. Chem.* **14**, 491 (1995).

36. D. Crich, and H. Li, *J. Org. Chem.* **67**, 4640 (2002).

37. Y. Jing, ad X. Huang, *Tetrahedron Lett.* **45**, 4615 (2004).

38. D.R. Mootoo, P. Konradsson, U. Udodong, and B. Fraser-Reid, *J. Am. Chem. Soc.* **110**, 5583 (1988).

39. J.D.C. Codee, R.E.J.N. Litjens, R. den Heeten, H.S. Overkleeft, J.N. van Boom, and G.A. van der Marel, *Org. Lett.* **5**, 1519 (2003).

40. M. Yoshida, T. Kiyoi, T. Tsukida, and H. Kondo, *J. Carbohydr. Chem.* **17**, 673 (1998).

41. A.V. Demchenko, N.N. Malysheva, and C. De Meo, *Org. Lett.* **5**, 455 (2003).

42. E. Fischer, and K. Zach, *Sitz. ber. kgl. preuss. Akad. Wiss.*, **16**, 311 (1913).

43. B. Freiser-Reid, D.R. Kelly, D.B. Tulshian, and P.S. Ravi, *J. Carbohydrate Chem.* **2**, 105 (1983).

44. B.K. Shull, Z. Wu, and M. Koreeda, *J. Carbohydr. Chem.* **15**(8), 955 (1996).

45. R.W. Murray and R. Jeyaraman, *J. Org. Chem.* **50**, 2847 (1985).

46. W. Adam, J. Bialas, and L. Hadjiarapoglou, *Chem. Ber.* **124**, 2377 (1991).

47. C.H. Marzabadi, and C.D. Spilling, *J. Org. Chem.* **58**, 3761 (1993).

48. G. Belluci, G Catelani, G., Chiappe, C., D'Andrea, F., *Tetrahedron Lett.* **53**, 10471 (1997).

49. Y. Du and F. Kong, J. Carbohydr. Chem. **14**(3), 341 (1995).

50. R.L. Halcomb and S.J. Danishefsky, *J. Am. Chem. Soc.* **111**, 6661 (1989).

51. M. Upreti, D. Ruhela, and R.A. Vishwakarma, *Tetrahedron* **56**, 6577 (2000).

52. J. Broddefalk, K.-E. Bergquist, and J. Kihlberg, J. *Tetrahedron Lett.* **1996**, 37, 3011. Broddefalk, J.; Bäcklund, J.; Almqvist, F.; Johansson, M.; Holmdahl, R.; Kilhberg, J. *J. Am Chem. Soc.* **120**, 7676 (1998).

53. H. Kunz, *Angew. Chem. Int. Ed. Engl.* **26**, 294 (1987).

54. C-H. Wong, Y. Ichikawa, T. Krach, C. Gautheron,-Le Narvor, D.P. Dumas, and G.C Look, *J. Am. Chem. Soc.* **113**, 8137 (1991).

55. M. Shimizu, H. Togo, and M. Yokohama, *Synthesis* **6**, 779 (1998).

56. M. Morita, T. Natori, K. Akimoto, T. Osawa, H. Fukushima, and Y. Koezuka, *BioMed. Chem Lett* 5, 699 (1995).

57. T. Mukaiyama, Y. Hashimoto, S. Shoda, *Chem. Lett.* 1983, 935.

58. K.C. Nicolaou, and N.J. Bockovich,, and D.R. Carcanague, *J. Am. Chem. Soc.* **115**, 8843 (1993).

59. L.F. Tietze and R. Fischer, *Angew. Chem.* **5**, 902 (1983).

60. S. Metha, and B.M. Pinto, *J. Org. Chem.* **58**, 3269 (1993).

61. A. Sobti, K. Kim, and G.A. Solikowski, *J. Org. Chem.* **6**, 61 (1996).

62. K. Takeda, E. Kaji, H. Nakamura, A. Akiyama, A. Konda, and Y. Mizuno, and H. Takayanagi, and Y. Harigaya, *Synthesis* 341 (1996).

63. D.M. Garcia, H. Yamada, S. Hatakeyama, and M. Nishikawa, *Tetrahedron Lett.* **35**, 3325 (1994).

64. H. Nagai, S. Matsumura, and K. Toshima, *Tetrahedron Lett.* 43, 847 (2002).

65. K. Toshima, H. Nakai, and S. Matsumura, *Synlett* **9**, 1420 (1999).

66. K. Toshima, K. Kasumi, and S. Matsumura, *Synlett* **6**, 813 (1999).

67. M. Oikawa, T. Tanaka, N. Fukuda, S. Kusumoto, *Tetrahedron Lett.* **45**, 4039 (2004).

68. (a) F. Burkhart, Z. Zhang, S. Wacowich-Sgarbi, and C-H Wong, *Angew. Chem. Int. Ed.* **40**, 1274 (2001). (b) Lahmann, M., Oscarson, S. *Org Lett.* **2**, 3881 (2000).

69. J.M. Coteron, K. Singh, J.L. Asensio, M. Domingues-Dalda, A. Fernandez-Mayoralis, J. Jimenez-Barbero, and M. Martin-Lomas, *J. Org. Chem.* **60**, 1502 (1995).

70. K.G. Nilsson, *Carbohydr. Res.* **95**, 167 (1987).

71. G.A. Freeman, S.R. Shauer, J.L. Rideout, and S.A. Short, *Bioorg. Med. Chem.* **3**, 447 (1995).

72. E.S. Simon, S. Grabowski, G.M. Whitesides, *J. Org. Chem.* **55**, 1834 (1990).

73. Y. Ichikawa, G.J. Shen, C.-H. Wong, *J. Am. Chem. Soc.* **113**, 4698 (1991).

74. J.J. Gaudino, J.C. Paulson, *J. Am. Chem. Soc.* **116**, 1149 (1994).

75. J.E. Heidlas, W.J. Lees, P. Pale,and G.M. Whitesides, *J. Org. Chem.* **57**, 146 (1992).

76. E.J. Toone, E.S. Simon, and G.M. Whitesides, *J. Org. Chem.* **56**, 5603 (1991).

77. J.L.C. Liu, G.-J. Shen, Y. Ichikawa, J.F. Rutan, G. Zapata, W.F. Vann, and C.-H. Wong, *J. Am. Chem. Soc.* **114**, 3901 (1992).

78. L.F. Mackenzie, Q. Wang, Q., R.A.J. Warren, and S.G. Whiters, *J. Am. Chem. Soc.* **120**, 5583 (1998).

79. D.G. Drueckhammer. W.J. Hennen, R.L. Pederson, C.F. Barbas III, C.M. Gautheron, T. Krach, and C.-H. Wong, *Synthesis*, 499 (1991).

80. X. Zeng, T. Murata, and T. Usui, *J. Carbohydr. Chem.* **22**, 309 (2003).

81. B.N. Cook, S. Bhakta, T. Biegel, K.G. Bowman, J.I. Armstrong, S. Hemmerich, and C.R. Bertozzi, *J. Am. Chem. Soc.* **122**, 8612 (2000).

82. H. Akita, E. Kawahara, and K. Kato, *Tetrahedron Asymmetry* **15**, 1623 (2004).

83. C.-H. Wong, S.L. Haynie, G.M. Whitesides, *J. Org. Chem.* **47**, 5416 (1982).

84. (a) K.C. Nicolaou, N. Watanabe, J. Li, J. Pastor, and N. Winssinger, *Angew. Chem. Int. Ed.* **37**, 1559 (1998) K.C. Nicolaou, N. Winssinger, J. Pastor, and F. De Roose, *J. Am. Chem. Soc.* **119**, 449 (1997) Wong, C.-H., Ye, X.-S., Zhang, Z. *J. Am. Chem. Soc.* **120**, 7137 (1998). (b) S.A. Mitchell, M.R. Pratt, U.J. Hruby, and R. Polt, *J. Org. Chem.* **66**, 2327 (2001).

85. P.H. Seeberger and W.C. Haase, *Chem Rev.* **100**, 4349 (2000).

86. P. Sears. and C.-H. Wong, *Science* **291**, 2344 (2001).

87. D. Crich and M. Smith, *J. Am. Chem. Soc.* **124**, 8867 (2002).

88. H.J.M. Gijsen, L. Qiao, W. Fitz, and C.-H. Wong, *Chem Rev.* **96**, 443 (1996).

89. G. Gattuso, S.A. Nepogodiev, and J.F. Stoddart, *Chem. Rev.* **98**, 1919 (1998).

90. T. Ogawa and Y. Takahashi, *Carbohydr. Res.* **138**, C5 (1985).

91. Y. Takahashi and T. Ogawa, *Carbohydr. Res.* **169**, 277 (1987).

92. P.M. Collins and M.H. Ali, *Tetrahedron Lett.* **31**, 4517 (1990).

93. D. Bassieux, D. Gagnaire, and M. Vignon, *Carbohydr. Res.* **56**, 19 (1977).

94. G. Excoffier, M. Paillet, and M. Vignon, *Carbohydr. Res.* **135**, C10 (1985).

95. N. Nakamura, *Methods Carbohydr. Chem.* **10**, 269 (1994).

96. S. Cottaz, C. Apparu, and H. Driguez, *J. Chem. Soc., Perkin Trans.* 1 2235 (1991).

97. C. Apparu, S. Cottaz, C. Bosso, and H. Driguez, *Carbohydr. Lett.* **1**, 349 (1994).

98. M. Kamakura and T. Uchiyama, *Biosc. Biotechnol. Biochem.* **57**, 343 (1993).

3
N-Glycosides

These types of glycosides are generated when a sugar component is attached to an aglycon, through a nitrogen atom, establishing as a result a C-N-C linkage. Nucleosides are among the most relevant N-glycosides since they are essential components of DNA, RNA, cofactors, and a variety of antiviral and anti-neoplasic drugs.

Usually for nucleosides, a pyrimidine or purine base is linked to the anomeric carbon of a furanoside ring. The nucleosides responsible for the formation of the genetic material DNA and RNA are adenine, guanine, cytidine, and thymine, the latter exchanged by uracil in the case of RNA (Figure 3.1). Nucleosides can be classified in natural nucleosides such as those involved in the genetic storage of information, naturally modified nucleosides, and synthetic nucleosides.

Naturally modified nucleosides include a significant and diverse number of compounds, some of them with slight changes mostly at the base, or major structural modifications done by enzymes. So far most of them have unknown biochemical function;[1] nonetheless they have been strongly associated with antiviral, antitumoral, and growth regulation processes (Figure 3.2).

Representative examples of natural modified nucleosides includes queuosine (Q) and Wye base (W), which have been found in the tRNA of some plants and bacteria, and it plays an important rule in the inhibition of tumor processes. Derived from this relevant biological function the total synthesis of these unique nucleosides have been reported for Q[2–4] and W.[5]

Moreover, the synthesis of complex nucleoside antibiotics has been reviewed.[6] The analysis was focused on the challenging synthetic methods for carbohydrate and nucleoside chain elaboration, glycosidation, and methods for controlling stereochemistry and for joining subunits. As a result, the total synthesis of Polyoxin J,[7] sinefungin,[8] thuringiensin,[9] tunicamycin V,[10] nikkomycin B,[11] octosyl acid A,[12] hikizimycin,[13] and capuramycin[14] was completed (Figure 3.3).

Important cofactors playing a key rule as biological catalysts required by the enzymes for the optimal performance of biochemical transformations are nucleotides. Such is the case of adenosine triphosphate ATP and nicotinic acid adenine dinucleotide NAD that are constituted by an adenosine nucleoside combined with phosphate for the former, and phosphate and nicotinamide for the latter (Figure 3.4).

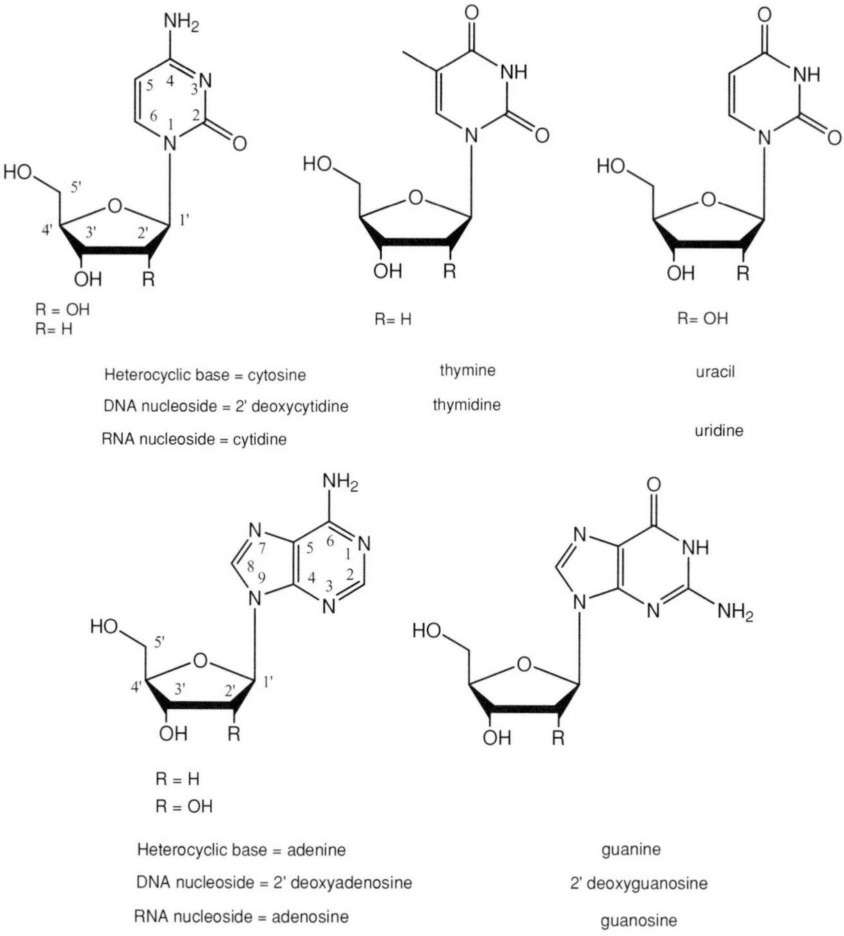

FIGURE 3.1. DNA and RNA nucleosides.

3.1   Nucleoside Formation

Considering a disconnection analysis, there are two major general routes for nucleoside syntheses.[15] The first is based on the attachment between the aglycon base and the protected sugar activated with a good leaving group at the anomeric position. Under these conditions, the stereoselectivity is conditioned by the protecting group attached at position 2. The second general procedure considers the coupling reaction between a base precursor and the sugar derivative, which contains the free amine linked to the anomeric carbon. The ring closure generally takes place after the glycosidation reaction and the configuration is predetermined by the nitrogen attached to the anomeric carbon. The latter approach has been most efficiently used for preparing carbocyclic nucleosides (Figure 3.5).

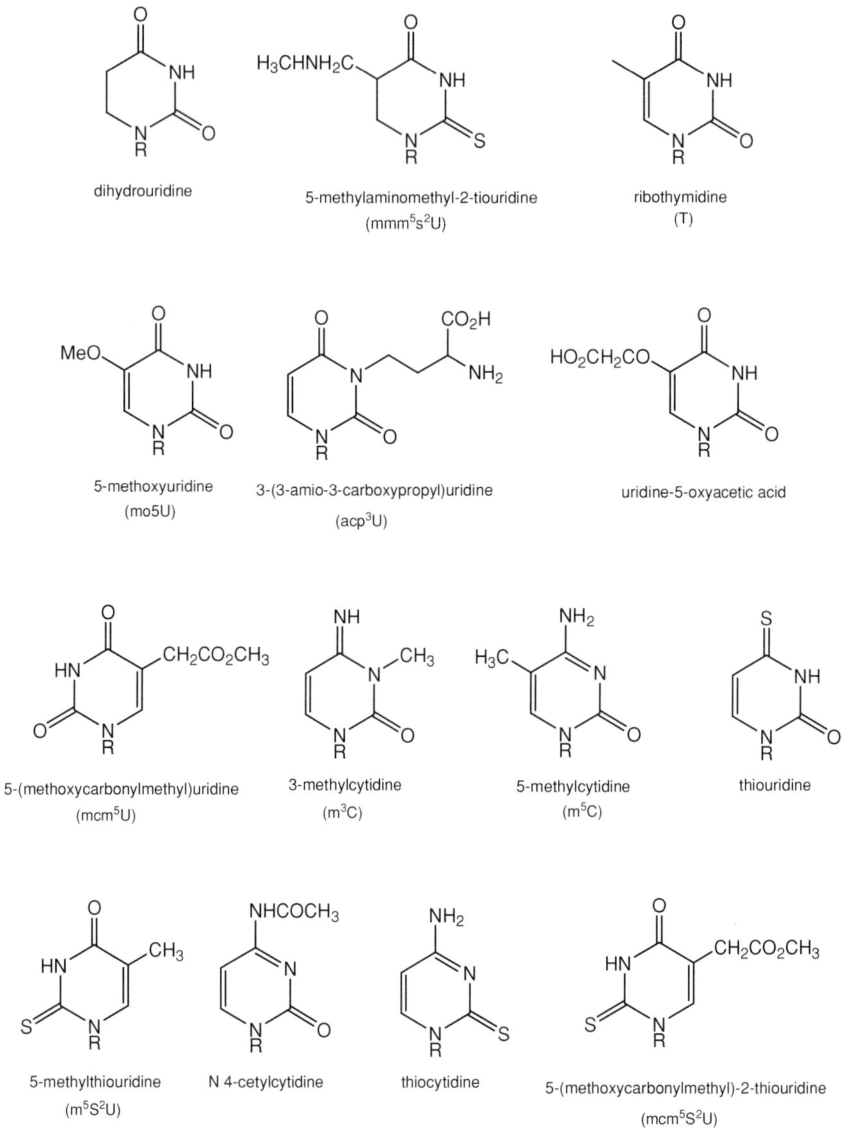

FIGURE 3.2. Naturally modified nucleosides.

3.2 Protecting Groups

It has been mentioned in the previous chapter that protecting groups are important components for most of the general methodologies designed for establishing glycosidic bonds. Usually the methods for glycoside formation require prior protection of those elements (usually heteroatoms) within the molecule that are needed

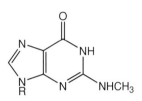

inosine
(I)

1-methylinosine
(m¹I)

N⁶-methyladenosine
(m⁶A)

1-methyladenosine
(m¹A)

1-methylguanosine
(m¹G)

2-methylguanosine
(m²G)

7-methylguanosine
(m⁷G)

base W

queuosine
(Q)

R₂ = mannose or galactose
R, R₂ = H queuine

FIGURE 3.2. (*continued*)

to remain unaltered. Also important is the fact that the cleavage of the protecting group should be carried out under preferentially mild conditions, and in the case of complex nucleosides the installation and removal of the protecting groups for nitrogen, oxygen, and sulfur should be accomplished under compatible conditions. The protection and deprotection of nucleosides can be done by chemical or enzymatic means. Some of the most commonly used protecting groups used in the preparation of O-glycosides are also useful in the synthesis of nucleosides (Figure 3.6).

3.2.1   Ribofuranoside Protecting Groups

Enzymes have been found to be interesting alternatives for installing protecting groups on nucleosides. Some of the enzymes used for this purpose are *subtilisin* mutant (8350)[16] and lipases mainly from *Pseudomona* and *Candida* strains.[17] Representative protections of purine and pyrimidine nucleosides are indicated in Figure 3.7.

By using the appropriate lipase it is possible to achieve regioselective acyl protections on nucleosides. For instance, the enzymatic transesterification reaction of 5′-fluorouridine with n-octanoic anhydride catalyzed with *Candida Antarctica* (CAL), *Pseudomona sp.* (PS), (KIWI-56), and *Mucor javanicus* (M) lipase was performed, producing 5′-, 3′- and 2′-acylnucleosides, respectively (Figure 3.8).[18]

Regioselective removal of certain protecting groups such as acetates attached to the ribosyl moiety of nucleosides might be carried out by enzymes. For instance, Subtilisin strain selectively hydrolyzes the 5′-position of purine and pyrimidine tri-O-acylated esters to produce 2′,3′-di-O-Acylribonucleosides in 40–92% (Figure 3.9).[19]

On the other hand, diastereoselective deacetylation of peracetylated 2′-deoxyribofuranosyl thymine was carried out using wheat germ lipase (WGL) and

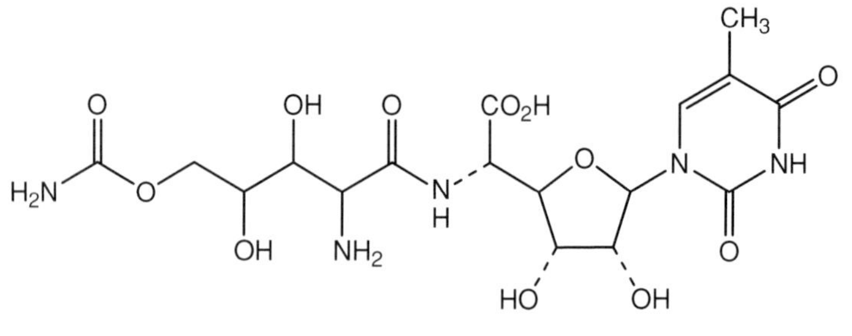

polyoxin J

FIGURE 3.3. Complex nucleoside antibiotics.

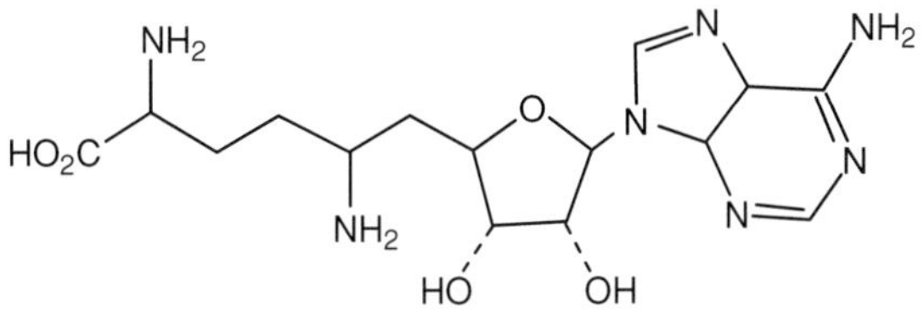

sinefungin

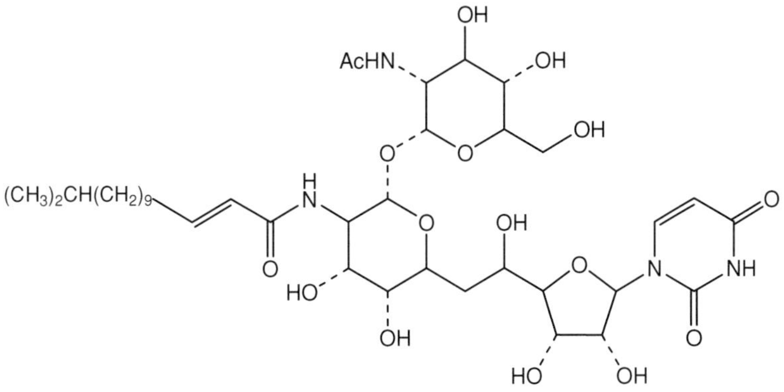

tunicamycin V

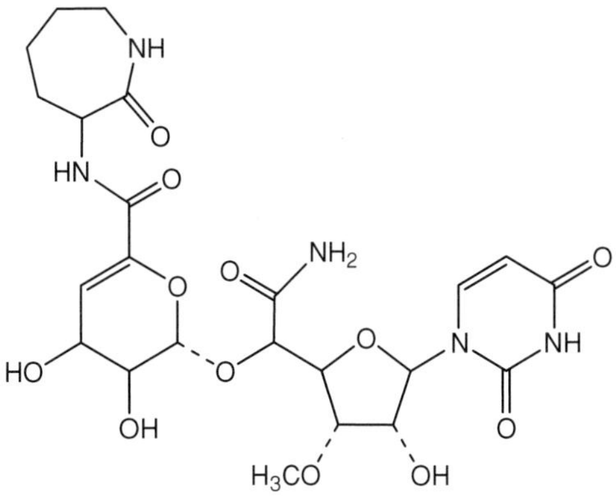

capuramycin

FIGURE 3.3. (*continued*)

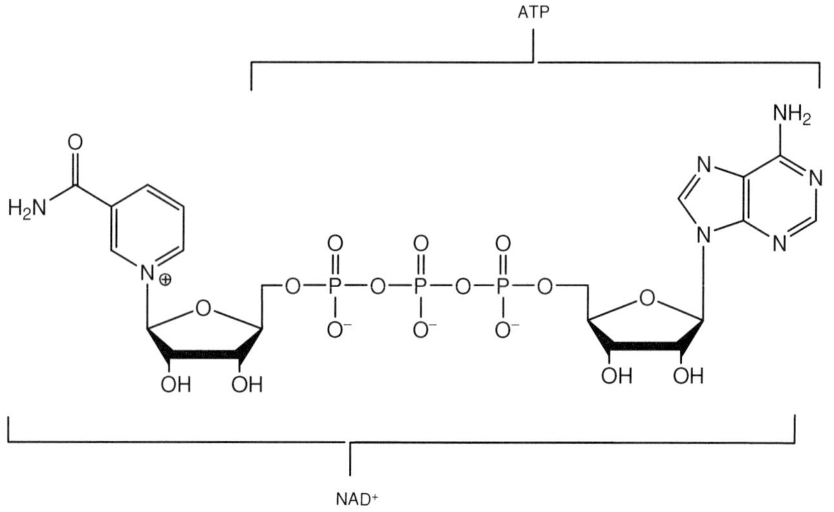

ATP
NAD+

FIGURE 3.4. Structure of nucleoside cofactors ATP and NAD.

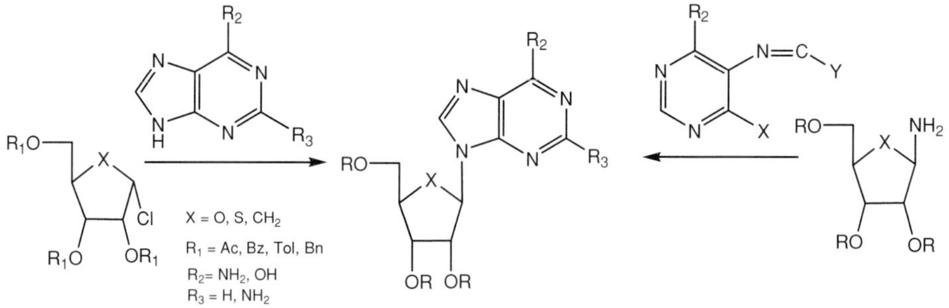

X = O, S, CH2
R1 = Ac, Bz, Tol, Bn
R2= NH2, OH
R3 = H, NH2

FIGURE 3.5. General procedures for N-glycoside formation.

Acetate (CH_3CO-)

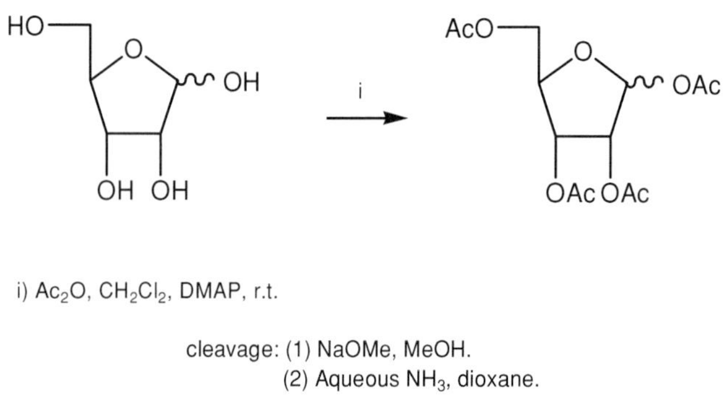

i) Ac2O, CH2Cl2, DMAP, r.t.
cleavage: (1) NaOMe, MeOH.
(2) Aqueous NH3, dioxane.

FIGURE 3.6. Common ribose protecting groups.

Benzoyl (PhCO-).

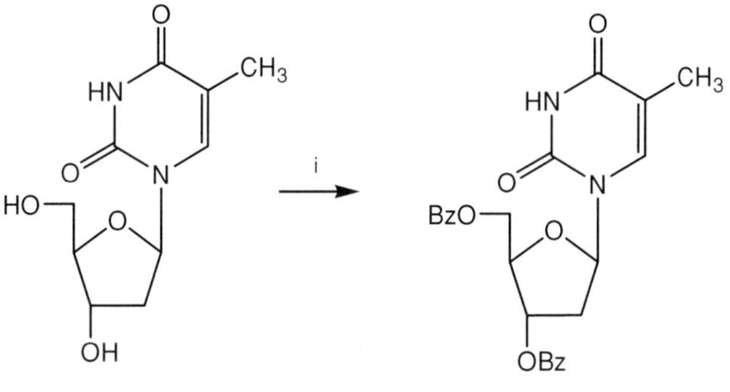

i) Bz-Cl, pyridine.

cleavage: (1) R-NH₂, EtOH, 100°C.
          (2) EtOH, KOH, reflux, 3 h.
          (3) NH₃, MeOH

Toluyl (p-MePhCO-)

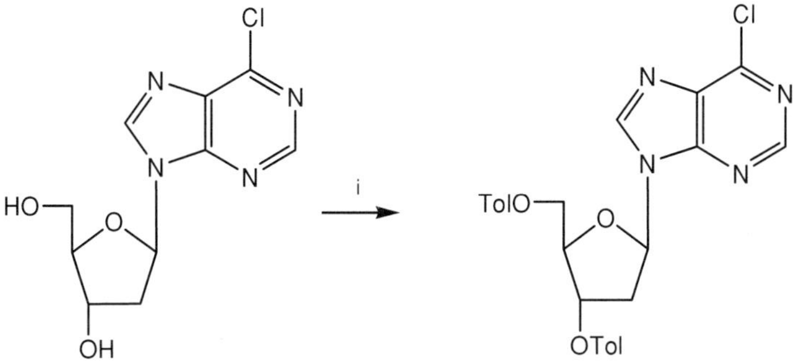

i) Tol-Cl, pyridine.

cleavage: NH₃, MeOH, 100°C, 78%.

Pivaloyl (Me₃CCOCl)

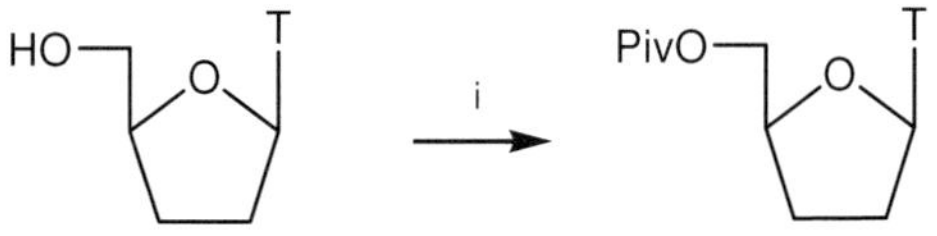

i) Piv-Cl, pyridine.

cleavage: NaOMe, MeOH.

FIGURE 3.6. (*continued*)

Trityl (Ph₃C-)

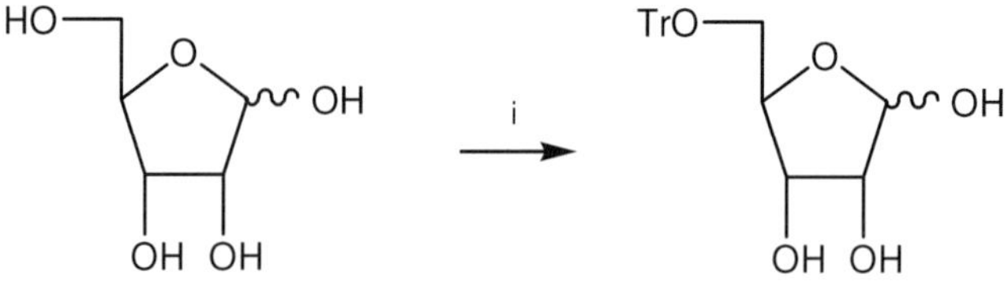

i) Tr-Cl, pyridine, r.t.

cleavage: (1) 80%, AcOH, 60°C.
(2) HCO_2H, Et_2O.

Benzyl (PhCH₂-)

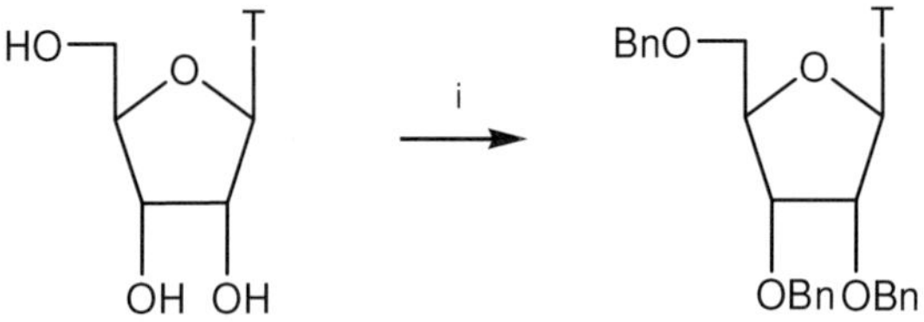

i) BnBr, NaH, DMF.

cleavage: $H_2/Pd(OH)_2$, EtOH.

Tertbutyldimethylsilyl (^{t}BuMe₂Si-)

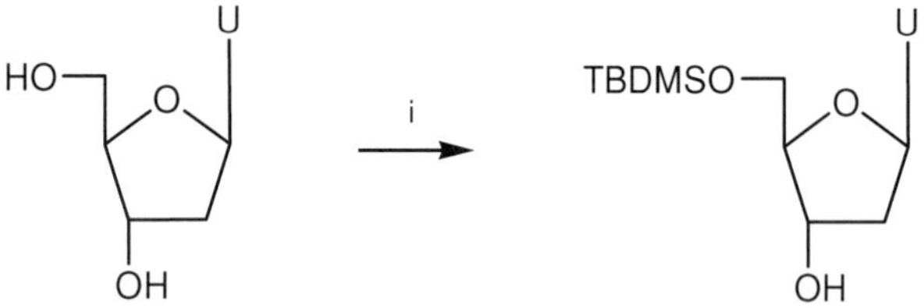

i) TBDMS-Cl, pyridine, r.t.

cleavage: (1) tetrabutylammonium fluoride (TBAF).
(2) pTsOH, MeOH, H_2O, 7h.

FIGURE 3.6. (*continued*)

Tetraisopropylsilyl ([(^{i}Pr)$_2$Si]$_2$O-)

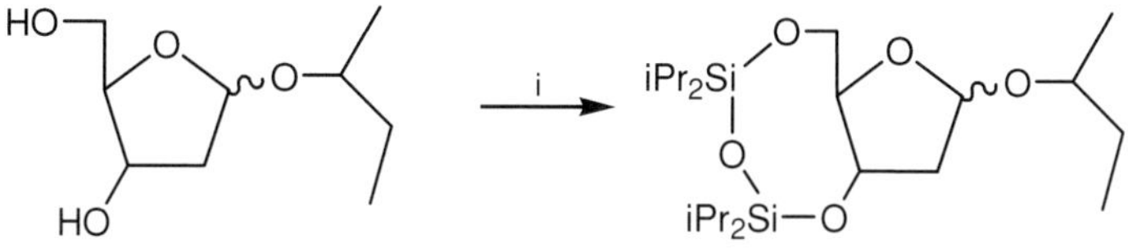

i) Pr$_2$iSi(Cl)OSi(Cl)Pr$_2$i, imidazole, THF, rt, 90 min.

cleavage: Bu$_4$NF, THF.

FIGURE 3.6. (*continued*)

porcine liver esterase (PLE), affording pure β-anomer thymidine in 29 and 31%, respectively (Figure 3.10).[20]

When porcine pancreas lipase (PPL) in phosphate buffer is used for deacetylation of 3′,5′-di-O-acetylthymine, the removal of the acetyl group at the 5′-position is achieved, leading to the 3'-O-acetylthymidine (Figure 3.11).[20]

Other suitable selective protections and deprotections useful for chemical manipulations that might occur at the ribosyl moiety are illustrated in Figure 3.12.

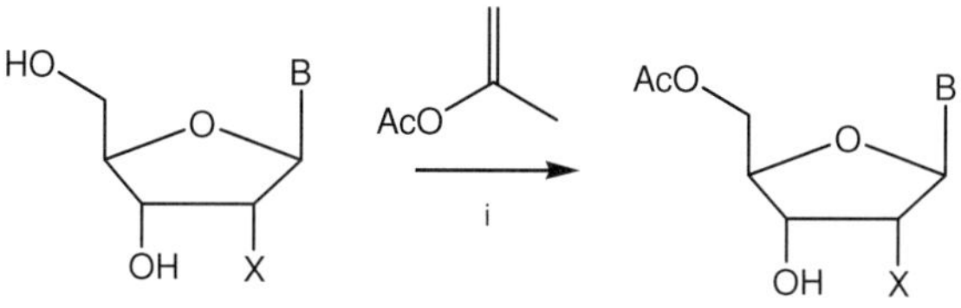

B = T, U, C, A
X = H, OH

i) Subtilisin 8350, DMF. 65-100%

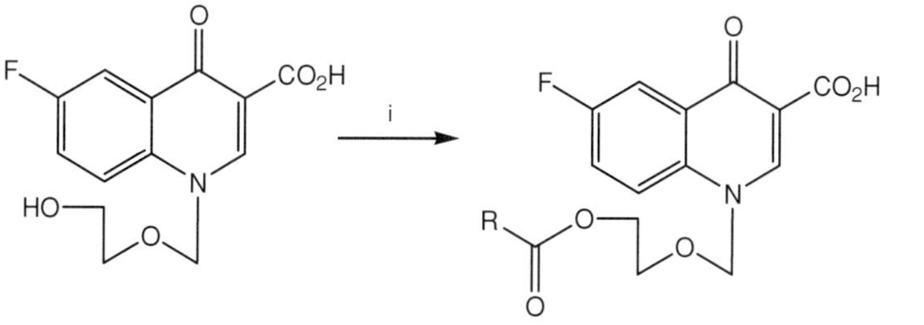

i) Pseudomona cepacea lipase, RCO$_2$Et, AcOEt, rt, 72h.

FIGURE 3.7. Enzymatic regioselective acylation by oximeacetates and lipases.

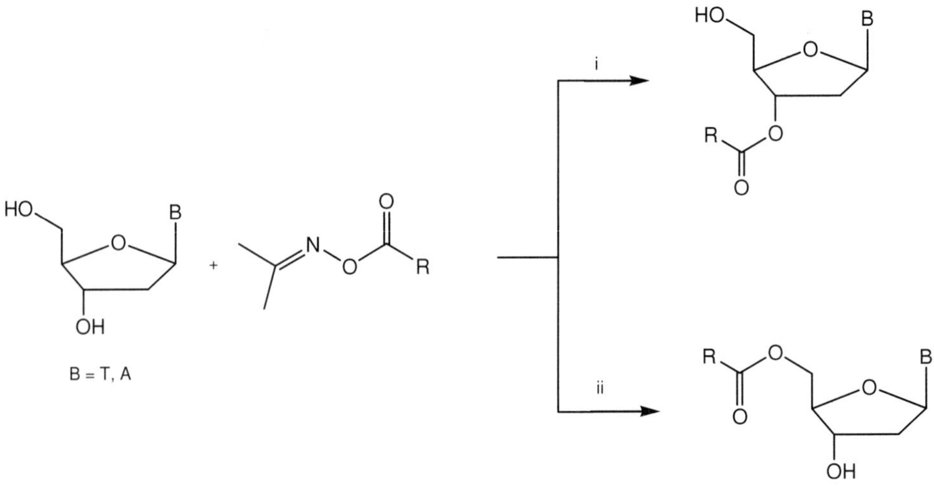

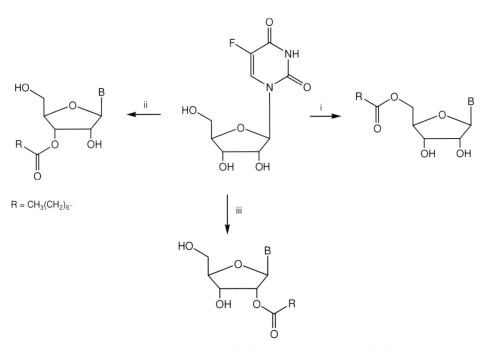

i) Pseudomona cepacea llipase (PSL), Pyridine. ii) Candida antartica lipase (CAL), THF.

FIGURE 3.7. (*continued*)

R = CH₃(CH₂)₆⁻

i) Candida antarctica lipase (CAL), 90%. ii) Pseudomona sp. lipase (PS), 92%.
iii) Mucor javanicus lipase (M), 42%.

FIGURE 3.8. Regioselective acyl protection by lipase.

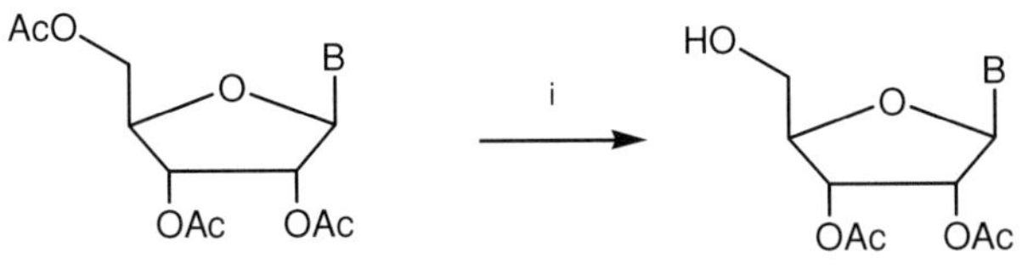

B = U, C, A, G, N-2AcG, H

i = Subtillisin or PPL, organic solvent, phosphate buffer, pH 7.

FIGURE 3.9. Selective enzymatic 5'-acetyl deprotection.

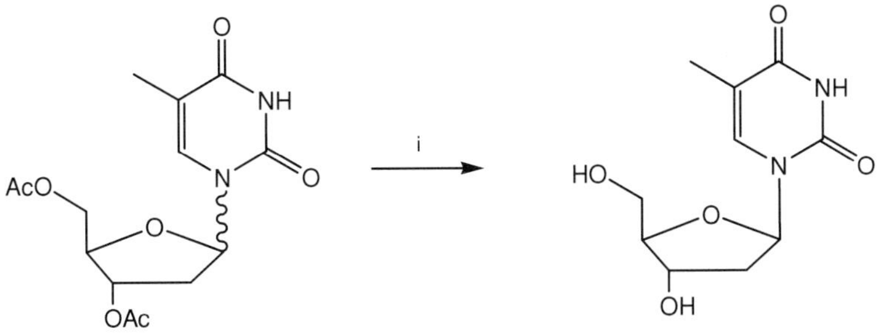

i) WGL, phosphate buffer, 29%. or PLE, phosphate buffer, 31%.

FIGURE 3.10. Lipase-catalyzed deacetylation of anomeric nucleoside.

Regioselective protections and deprotections is often a critical step especially for the preparation of complex nucleosides. Some suitable deprotections of complex nucleosides that do not alter the original composition of the structure have been described (Figure 3.13).[6]

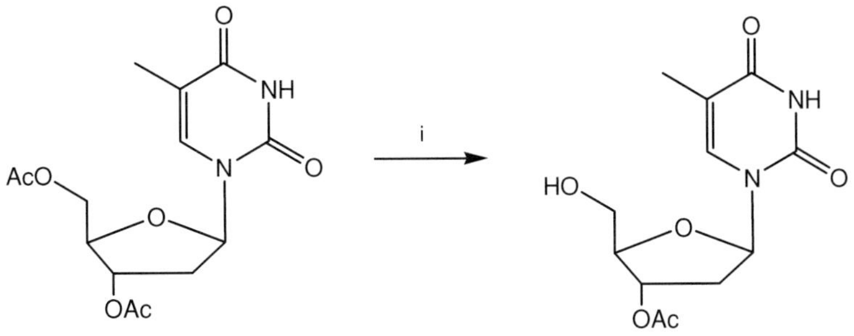

i) PPL, phosphate buffer, 98%.

FIGURE 3.11. Selective enzymatic.5'-deacetylation of 3',5'-di-O-acetyl thymidine.

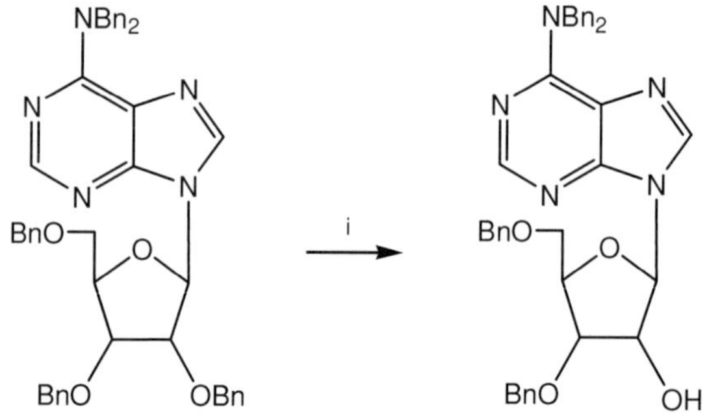

i) t-BuMgBr, PhH, 80°C, 69%

Ref.[21]

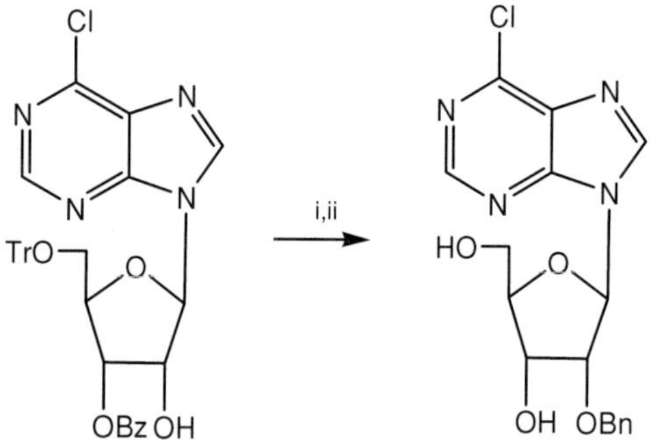

i) BnOH, Me₂NCON=NCONMe₂. ii) NH₃/MeOH

Ref.[22]

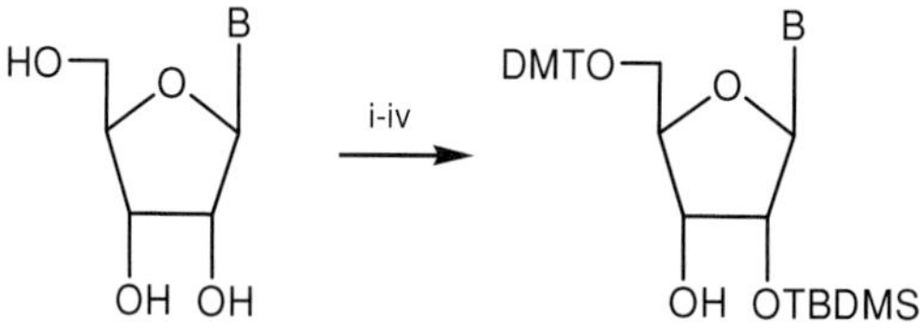

B = G, A, C, U

i) t-Bu₂Si(OTf)₂, Im, DMF, 0°C. ii) t-BuMe₂SiCl, Im, DMF, 60°C, 80-87%. iii) HF-Py, CH₂Cl₂, 0°C, 90%
iv) DMT-Cl, Py, 0°C, 90%.

Ref.[23]

FIGURE 3.12. Miscellaneous chemical protection and deprotection.

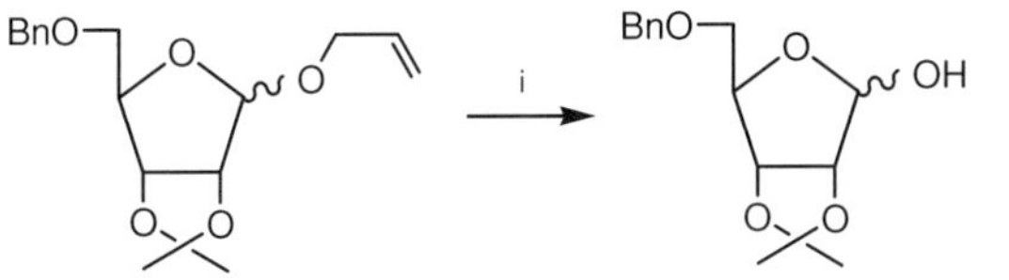

i) DIBAL, NiCl$_2$, Et$_2$O, 0°C, 55%

Ref.[24]

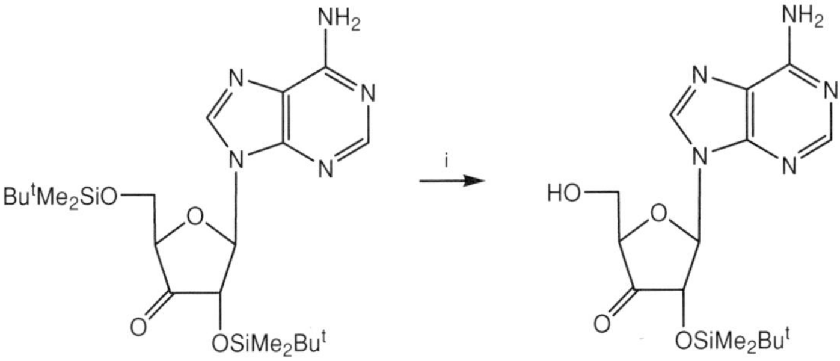

i) CF$_3$COOH-H$_2$O (9:1), 0°C, 95 %.

Ref.[25]

FIGURE 3.12. (*continued*)

3.3   General Methods

The following are general methods that we will explain in this section:

- Michael reaction
- Fischer-Helferich reaction

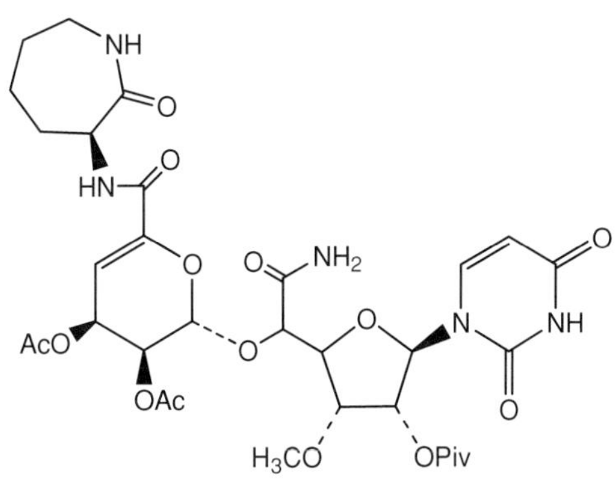

FIGURE 3.13. Suitable deprotection
of complex nucleosides.

1. NaOH, aq MeOH, 2.5 h (cleaves O-Ac and O-Piv)

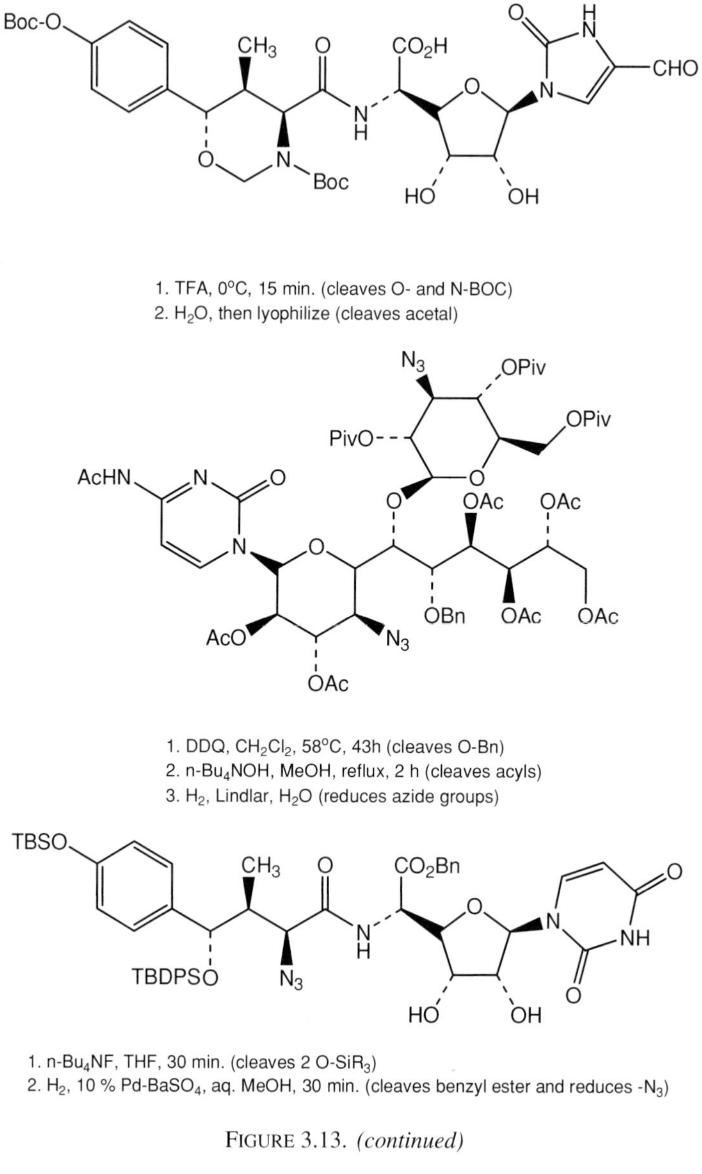

1. TFA, 0°C, 15 min. (cleaves O- and N-BOC)
2. H₂O, then lyophilize (cleaves acetal)

1. DDQ, CH₂Cl₂, 58°C, 43h (cleaves O-Bn)
2. n-Bu₄NOH, MeOH, reflux, 2 h (cleaves acyls)
3. H₂, Lindlar, H₂O (reduces azide groups)

1. n-Bu₄NF, THF, 30 min. (cleaves 2 O-SiR₃)
2. H₂, 10 % Pd-BaSO₄, aq. MeOH, 30 min. (cleaves benzyl ester and reduces -N₃)

FIGURE 3.13. *(continued)*

- Davol-Lowy reaction
- Silyl-mediated reaction
- Sulfur-mediated reaction
- Mitsunobu reaction
- Palladium-mediated reaction
- Microbial/enzymatic approach

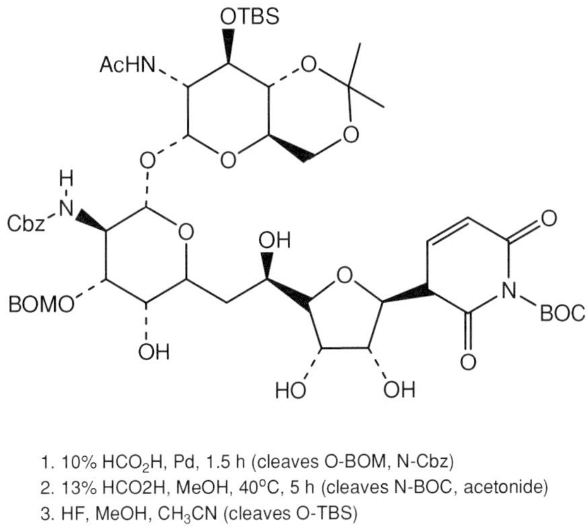
OTBS
AcHN
O
O
O
O
H
N
Cbz
O
OH
BOMO
OH
O
O
N
BOC
O
HO
OH
1. 10% HCO2H, Pd, 1.5 h (cleaves O-BOM, N-Cbz)
2. 13% HCO2H, MeOH, 40°C, 5 h (cleaves N-BOC, acetonide)
3. HF, MeOH, CH3CN (cleaves O-TBS)

FIGURE 3.13. *(continued)*

3.3.1 The Michael Reaction

3.3.1.1 General Figure and Conditions

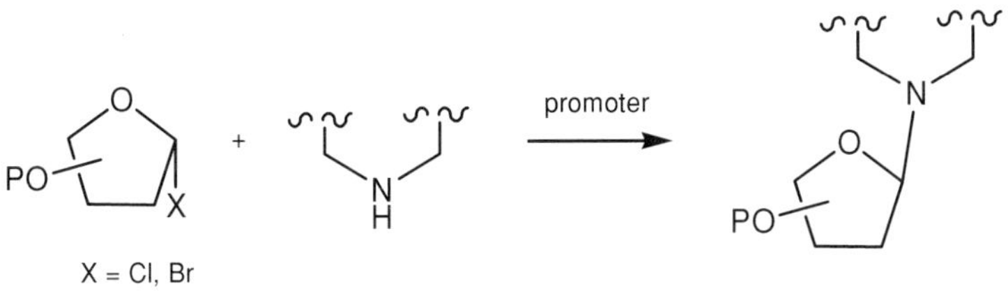
O
PO
X
+
N
H
promoter
N
O
PO
X = Cl, Br

Promoter	Conditions
NaH	DMF
K_2CO_3	DMF
KOH-TBA	CH_2Cl_2

It is a classical procedure for preparing nucleosides, and it can be considered a modified *O*-glycoside approach. In this way, the sugar derivative is an R-*O*-furanosyl halide, where R can be acyl-, benzoyl, benzyl, tosyl, or silyl, and the halogen commonly chlorine instead of bromine, since it has probed to be more stable for furanose derivatives than its counterpart. The nitrogen base (purine or pyrimidine) is reacted under basic conditions, usually NaH or K_2CO_3 in DMF (Figure 3.14).

A variety of antibiotics have been prepared according to this method, as in the case of the nucleoside known as methyltubercidine. For achieving this goal, the 7-deazaguanine was used as the nitrogen base, which was condensed to

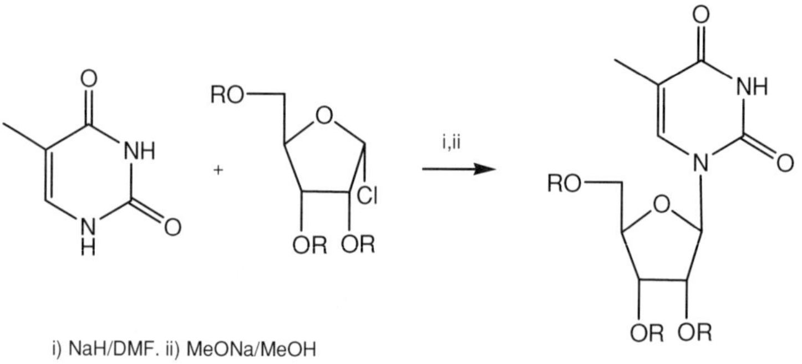

i) NaH/DMF. ii) MeONa/MeOH

FIGURE 3.14. The Michael modification.

2,3,5-tri-O-benzylribofuranosyl bromide under NaH/DMF conditions to afford a 1:1 anomeric mixture of the N-glycoside (Figure 3.15).[26]

More recently, Battaharya[27] reported the synthesis of fluoroarabinotubercidine, toyocamicine, and sangivamicine, under the current N-glycoside formation

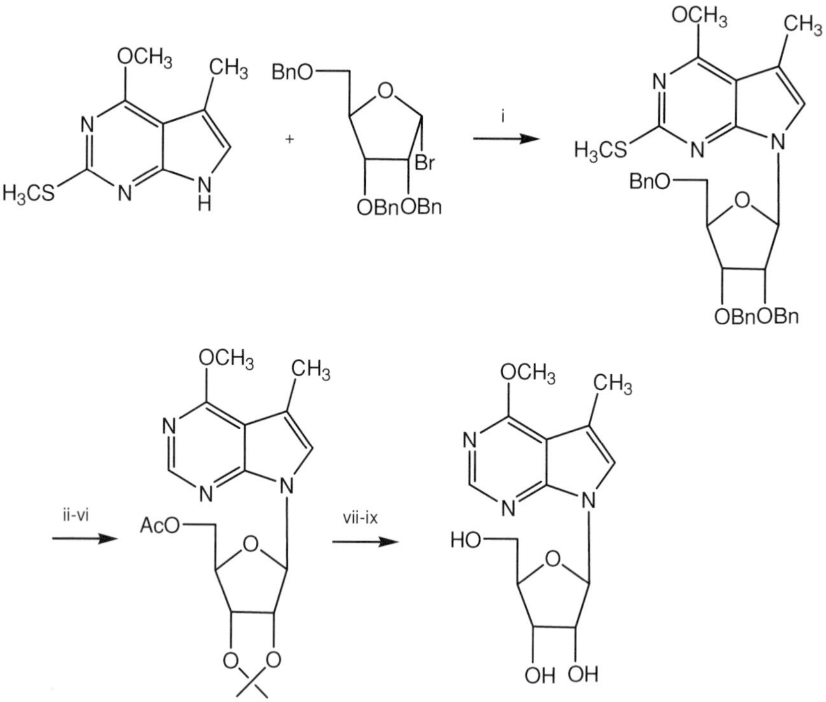

i) NaH/DMF. ii) Ni/EtOH-PhH. iii) HCl/dioxane. iv) H$_2$, Pd-C. v) a) acetone/p-TsOH. b) Ac$_2$O/Py. vii) POCl$_3$. viii) NH$_3$/MeOH. ix) F$_3$CCOOH/H$_2$O.

FIGURE 3.15. Synthesis of methyltubercidine.

procedure. Other deazapurines have been described by Seela et al.[28] involving the condensation between the purine base, with protected ribosyl halides under basic conditions. According to Seela[29] and Kazimierczuk,[30] the stereoselective glycosylation of the sodium salts of halopurines, with 2-deoxy-3,5-di-O-p-tolouyl-α-D-$erytro$-pentofuranosyl chloride, gave β-nucleosides via Walden inversion. This was demonstrated in the preparation of 2-amino 2'-desoxytubercidine and 2-aminotubercidine by condensation of 3,5-di-O-(p-tolyl)-α-D-pentafuranosylchloride and 5-O-[(1,1-dimethylethyl)dimethylsilyl]-2,3-O-(1-methylethyliden)-α-D-ribofuranosylchloride with the halopurine under Michael conditions. Final ammonia treatment provided the target deazanucleoside (Figure 3.16).

The 7-deazapurine nucleoside Cadeguomycin isolated from strain of the actinomycete culture filtrate *Streptomyces hygroscopicus* was also synthesized under this approach. Thus, coupling reaction between protected 7-deazapurine derivative with 1-chloro-2-deoxy-3,5-ditoluyl-α-D-erythro-pentofuranose was effected with preference for the β-isomer. Subsequent transformations provided the target molecule 2'-deoxycadeguomycin (Figure 3.17).[31]

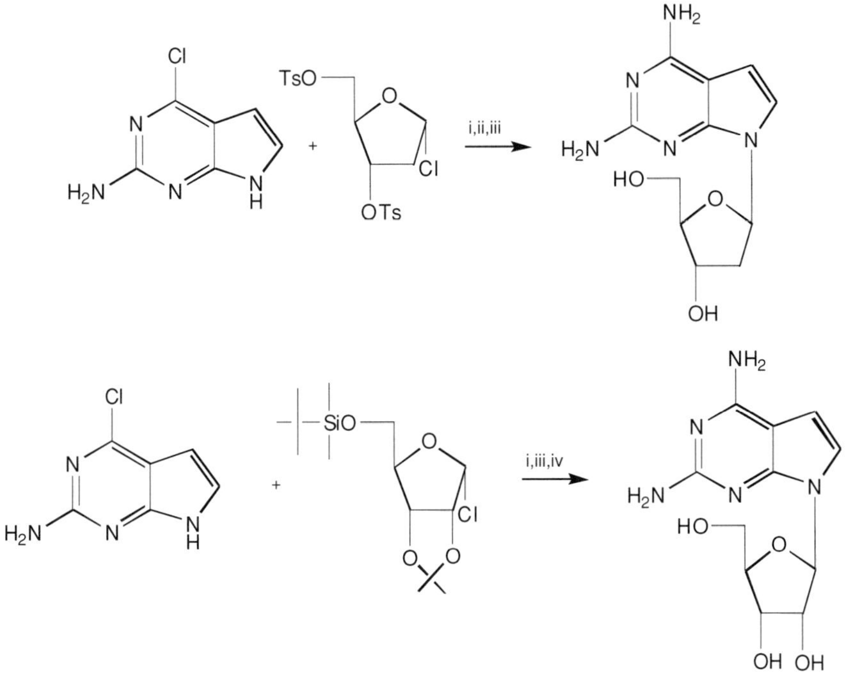

i) KOH, TBA/CH$_2$Cl$_2$. ii) MeONa/MeOH. iii) NH$_3$/MeOH. iv) CF$_3$COOH/H$_2$O.

FIGURE 3.16. Synthesis of 2-aminotubercidine and 2-amino-2'-deoxytubercidine.

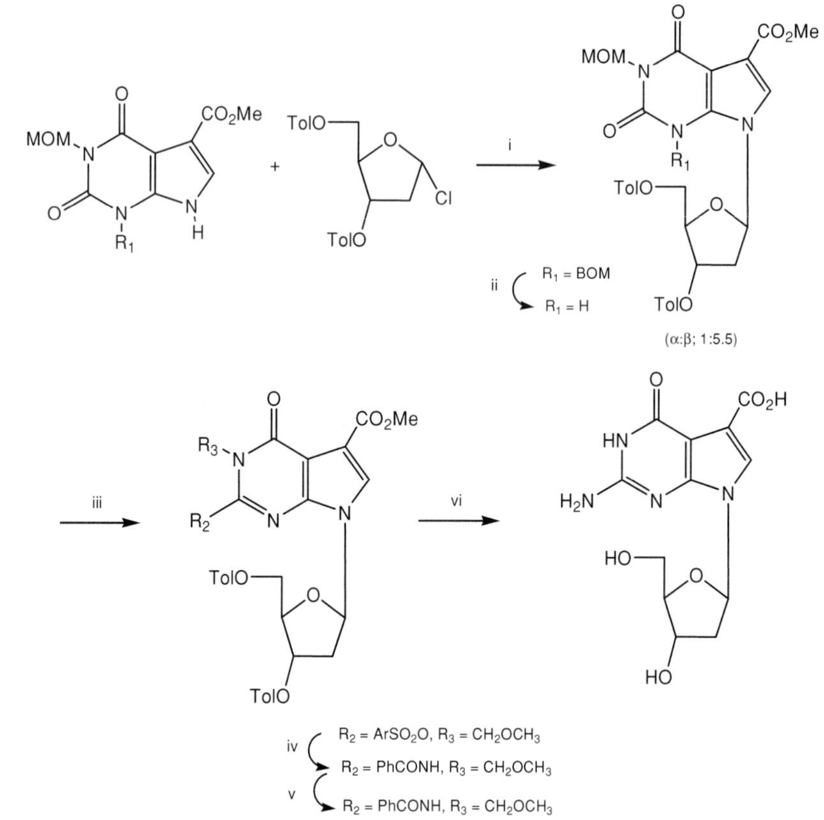

i) NaH, CH_3CN, 25°C. ii) $H_2/Ph(OH)_2$, MeOH, 25°C. iii) NaH, THF, TIPBS-Cl. iv) $PhCONH_2$, NaH, THF. v) TFA, CH_2Cl_2.

FIGURE 3.17. Synthesis of 7-deazapurine nucleoside 2-deoxycadeguomycin.

3.3.2   The Fischer-Helferich Reaction

3.3.2.1   General Figure and Conditions

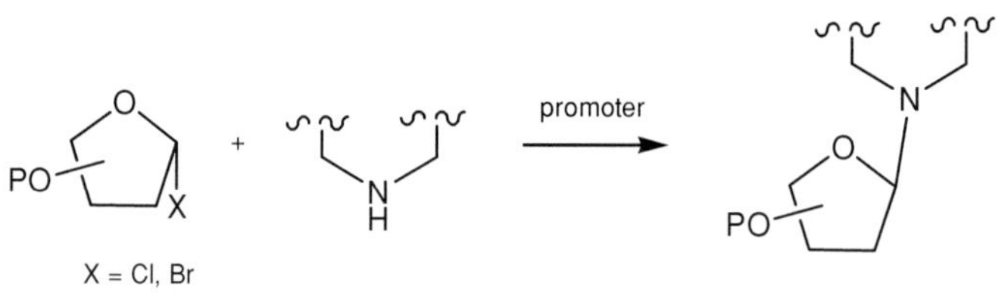

Promoter	Conditions
Silver salts	xilene

This general procedure consists in the use of an acylfuranoside or acylpyranoside, which is reacted with the silver or mercury salts of a nitrogen base. The

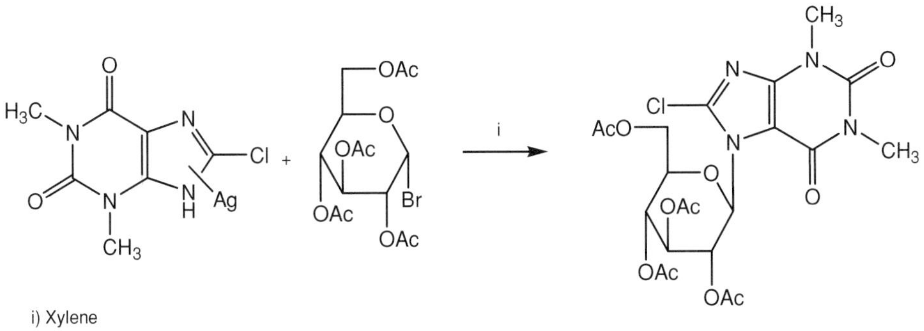

i) Xylene

FIGURE 3.18. The Fischer-Helferich method.

original reaction involves the condensation between silver salt of theophylline with acetobromoglucose in hot xylene, giving preferentially the N-7 regioisomer (Figure 3.18).

The feasibility of this method is observed in the synthesis of adenosine and guanosine by condensation of tri-O-acetyl-α-D-ribofuranosyl chloride with the silver salt of 2,8-dichloroadenine to generate an intermediate which under the conditions described below can generate either adenosine or guanosine (Figure 3.19).[32]

The stereochemistry of this reaction can be predicted by applying the "trans rule" proposed by Tipson[33] and extended by Baker. The rule establishes that the condensation between the purine or pyrimidine sal with the acyl-O-glycosyl halide will generate a nucleoside with C1-C2 trans configuration regardless of the initial configuration of C1-C2 of the sugar.

The trans rule is demonstrated in the preparation of thymidine acetoglucopyranose and mannopyranose, where -OH at position 2 for the former is equatorial, and axial for the latter. By following the rule, the coupling reaction generates the β- and α-anomers, respectively, having both of them a trans disposition between substituents at positions 1 and 2 (Figure 3.20).

3.3.3   The Davol-Lowy Reaction

3.3.3.1   General Figure and Conditions

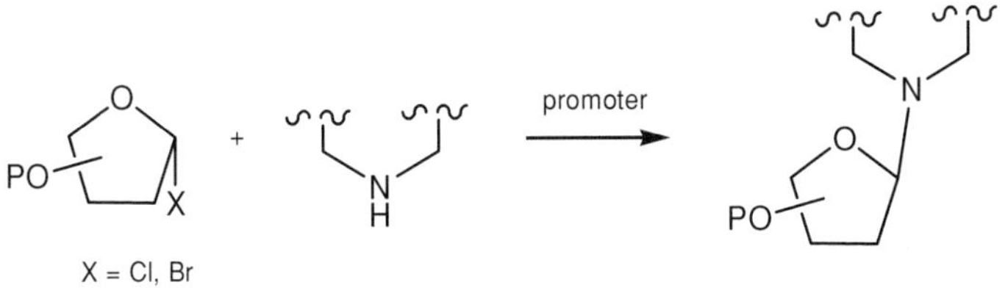

X = Cl, Br

Promoter	Conditions
Hg(CN)$_2$	CH$_3$NO$_2$, reflux
Hg(CN)$_2$	xilene

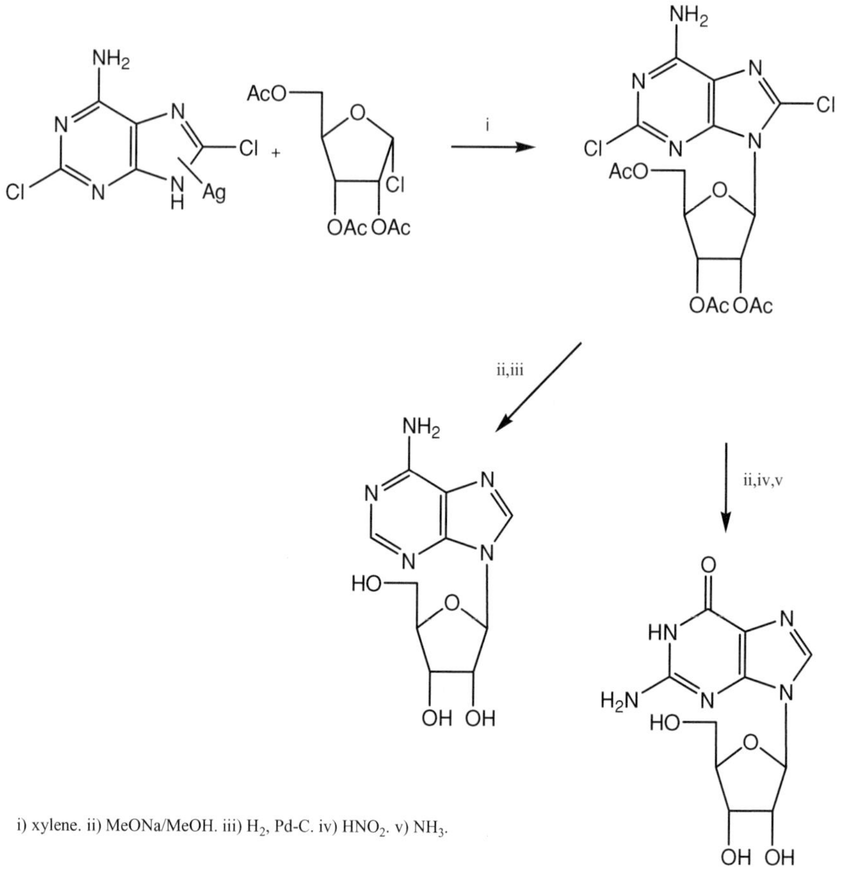

i) xylene. ii) MeONa/MeOH. iii) H$_2$, Pd-C. iv) HNO$_2$. v) NH$_3$.

FIGURE 3.19. Synthesis of adenosine and guanosine.

This method has been also considered a modified Fischer-Helferich procedure and involves the use of mercury chloride instead of silver salts. Under these conditions the useful intermediate chloropurine nucleoside has been prepared under mild conditions (Figure 3.21).

The nature of the glycosyl halide is important for determine the regioselectivity of the glycosidic linkage. If the condensation reaction occurs between purines with acetobromoglucose the N-7 regioisomer is obtained preferentially. On the other hand, if acetoribosyl chloride is condensed with the same purine, the N-9 regioisomer is the major product observed (Figure 3.22).

Another purine nucleoside prepared under these conditions is shown in Figure 3.23, consisting of the coupling reaction between the protected guanine with protected furanosyl chloride in nitromethane under refluxing conditions produced by the corresponding N-glycoside in 50% yield.[34]

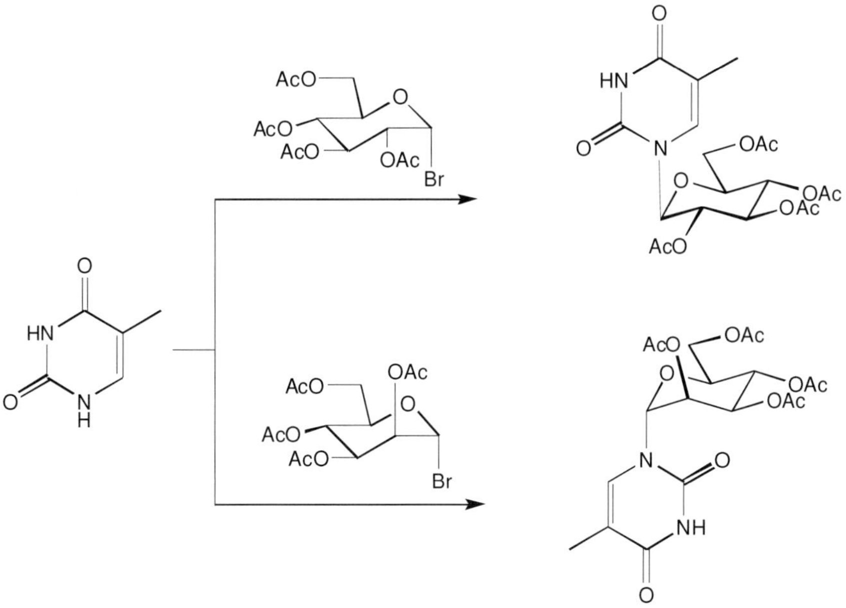

FIGURE 3.20. Tipson's trans rule.

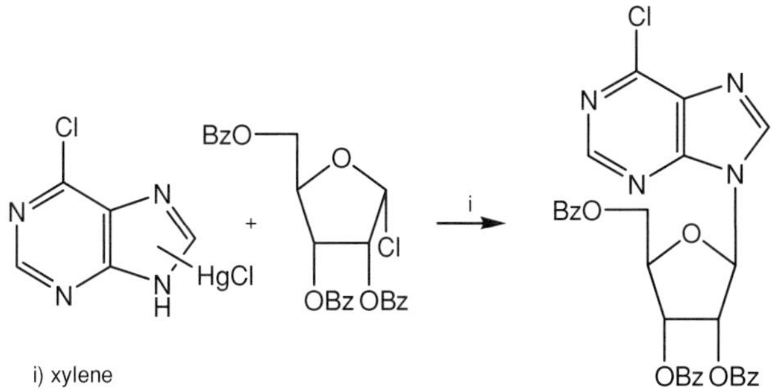

i) xylene

FIGURE 3.21. The Davol-Lowy method.

3.3.4 Silyl Coupling Reaction

3.3.4.1 General Figure and Conditions

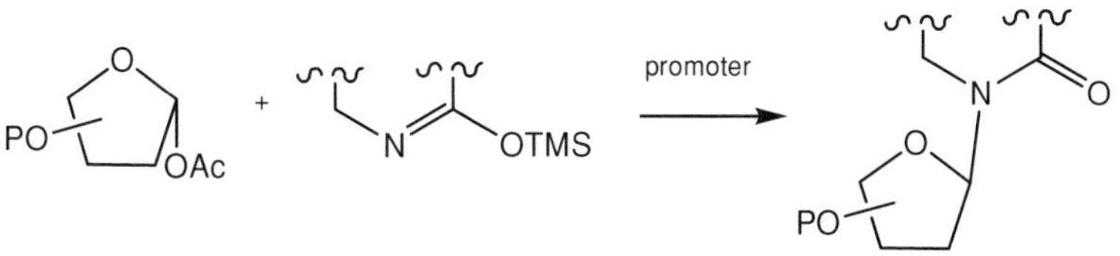

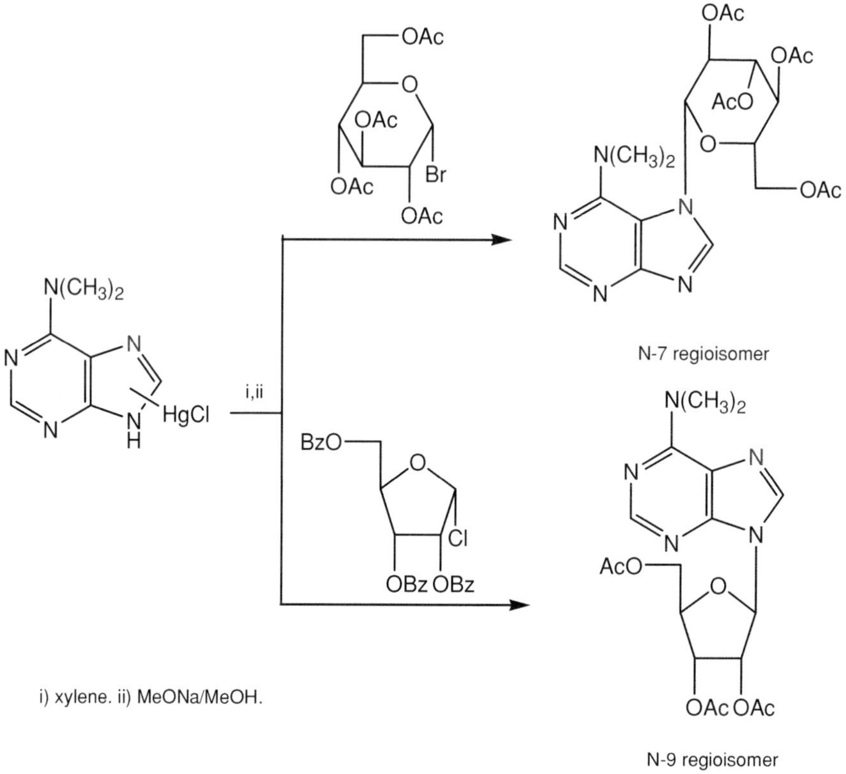

FIGURE 3.22. Preparation of N-7 and N-9 regioisomers.

Promoter	Conditions
TMS-OTf	CH$_3$CN, 0°C→r.t
TMS-OTf	PhNO$_2$,127°C
SnCl$_4$	CH$_3$CN
HMDS-TMDS	
(MeSi)$_2$NAc	
CF$_3$(CF$_2$)$_3$SO$_3$K/HMDS-TMSCl	
HMDS/(NH$_4$)$_2$SO$_4$	

Various types of silyl agents have been tested as either protecting groups and/or N-glycoside promoters. Among them trimethylsilyl chloride (TMS-Cl), bis(trimethylsilyl) acetamide, trimethylsilyltriflate, and hexamethyldisilasane are representative examples.

De Clercq et al.[35] prepared purine and pyrimidine α-D-lixofuranosylnucleosides employing HMDS, TMS, and TMSF as silyl coupling agents. Nucleoside α-D-lyxofuranosyl thymine was prepared by condensation between 1,2,3,5-

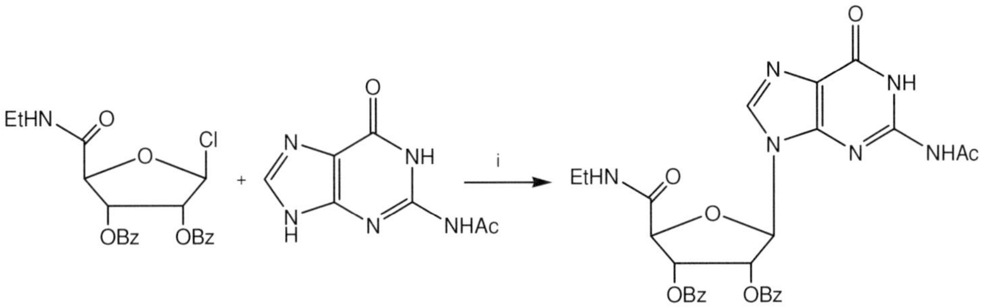

i) Hg(CN)₂, CH₃NO₂, reflux, 16 h.

FIGURE 3.23. Glycosidation reaction for preparation of guanine derivative.

tetra-O-acetyl-α-D-lyxose with thymine in the presence of HMDS-TMSCl mixture (Figure 3.24).

Likewise cytidine has been synthesized in 95% through condensation of silyl cytidine obtained from cytosine with bis [trimethylsilyl] acetamide, and sugar derivative 2,3,5-tri-O-benzoylribose, as represented in Figure 3.25.

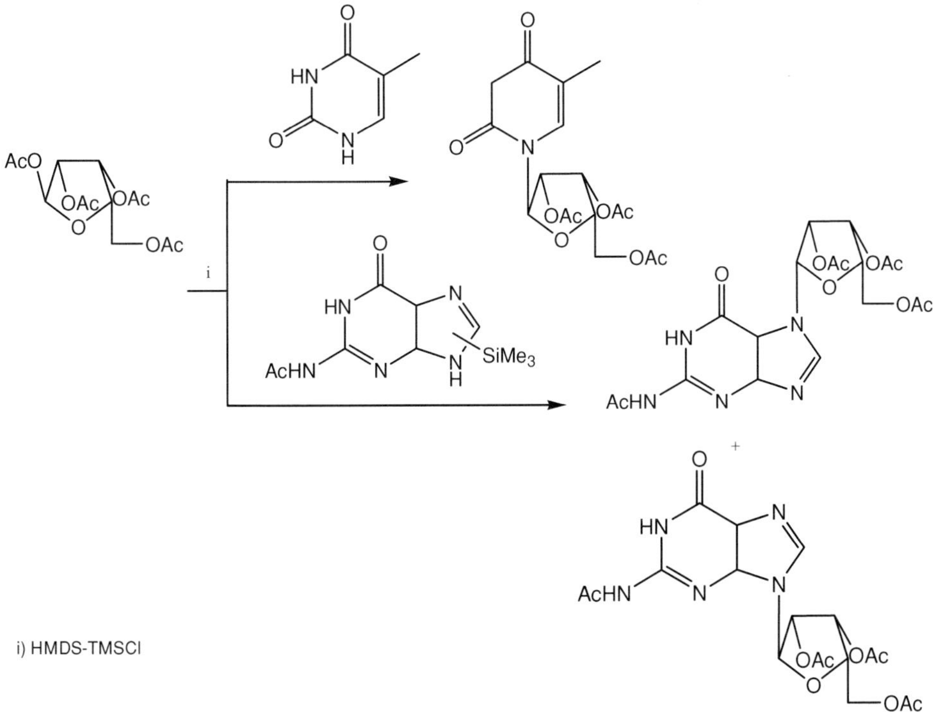

i) HMDS-TMSCl

FIGURE 3.24. Preparation of α-D-lyxofuranosyl thymine and guanine protected nucleosides.

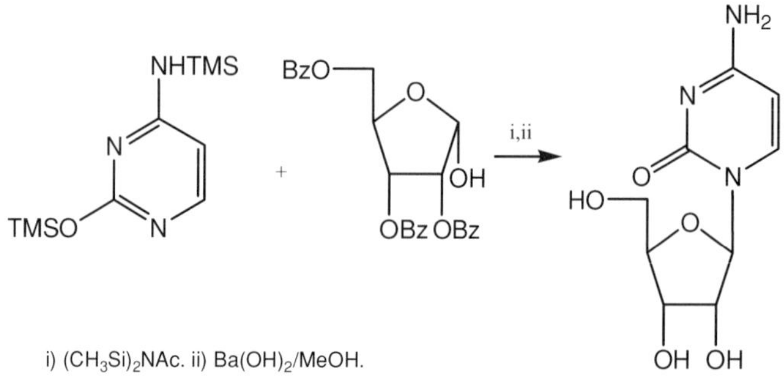

FIGURE 3.25. Silyl mediated coupling reaction.

Hilbert and Johnson[36] developed a procedure for preparing nucleosides employing a mixture of hexamethyldisilane (HMDS), trimethylsilane chloride, and potassium nonaflate. According to this procedure, 5-methoxyuridine was prepared by condensing 5-methoxyuracil, with 1-O-acetyl-2,3,5-tri-O-benzoyl-β-D-ribofuranose (Figure 3.26).

A widespread silyl-based methodology was developed by Vorbrüggen[37] based on the use of persilylated purines or pyrimidines, which are condensed with peracylated sugars in the presence of Lewis acid catalysis. Usually silylation of the base is achieved with hexamethyldisilazane (HMDS) or N,O-bis(trimethylsilyl)acetamide, the latter less difficult to remove during the work-up process. Among the Lewis acids employed as catalysts, trimethylsilyl triflate (TMSOTf) has been the most suitable condensing agent for this reaction.

AZT alkylthioanalogs have been synthesized under the method reported by Vorbrüggen. This condition requires hexamethyldisilane for activation of the anomeric center, and trimethylsilyltriflate as condensing agent (Figure 3.27).

Vörbruggen-type coupling reaction has been method of choice in the N-glycoside bond formation of various complex nucleosides such as octosyl acid A, tunicaminyl-uracil, sinefungin, and hikizimycin. Some of the general conditions

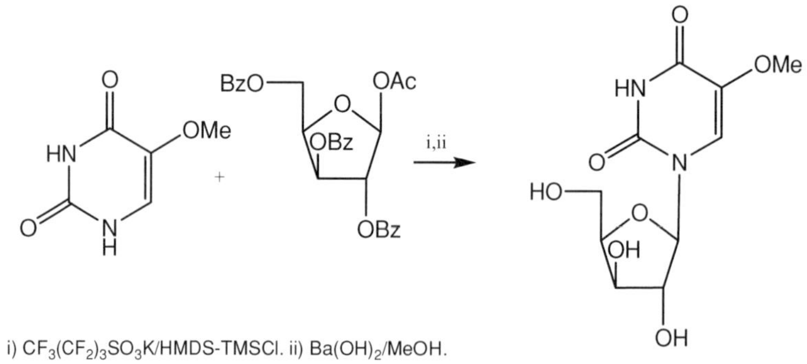

i) CF$_3$(CF$_2$)$_3$SO$_3$K/HMDS-TMSCl. ii) Ba(OH)$_2$/MeOH.

FIGURE 3.26. Hilbert and Johnson approach.

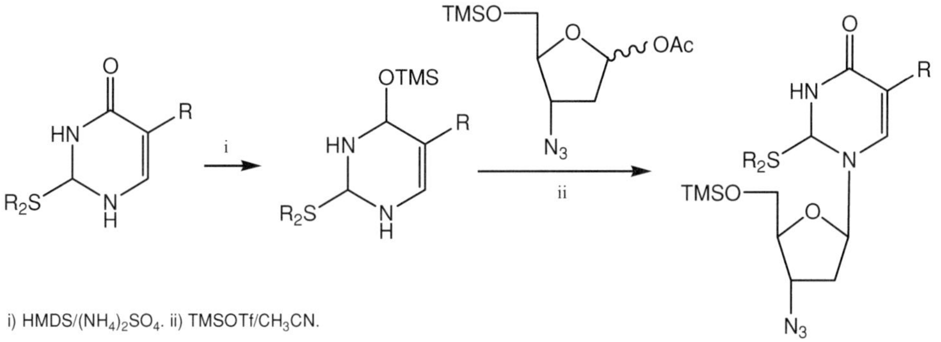

i) HMDS/(NH₄)₂SO₄. ii) TMSOTf/CH₃CN.

FIGURE 3.27. Vorbrüggen's synthesis of AZT thioderivatives.

reported for the accomplishment of the mentioned synthesis are described in Figure 3.28.[6a−b]

3.3.5 *Sulfur-mediated Reaction*

3.3.5.1 General Figure and Conditions

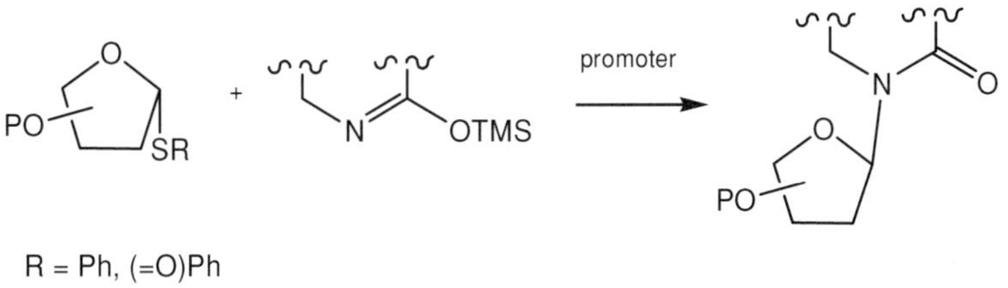

R = Ph, (=O)Ph

Promoter	Conditions
NIS-OTf	CH₂Cl₂
TMS-OTf	DCE r.t.
Br₂	DMF

Derived from their extensive use in the preparation of *O*-glycosides, the sulfur glycosyl donors have become another standard procedure for N-glycosylations. The conditions reported for the coupling reactions involve the sulfur glycosyl donor, the silyl protected heterocycle acceptor, and usually N-iodosuccinimide, triflic acid as catalyst (Figure 3.29).[38]

3.3.6 *Mitsunobu Reaction*

This reaction has been selected as another strategy for preparing N- and carboxyclic nucleosides. The mechanism involves a nucleophilic substitution displacement with inversion of the configuration between species bearing poor leaving groups with nucleophiles. The reaction mechanism involves the initial reaction

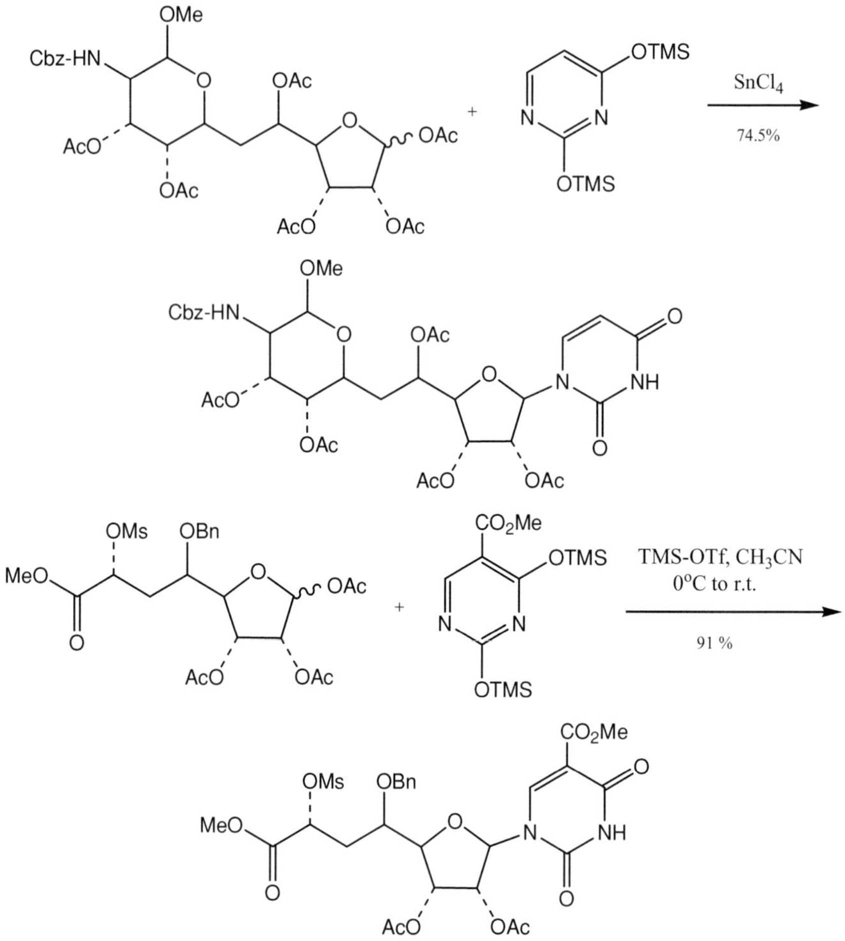

FIGURE 3.28. Vörbruggen-type coupling reactions.

of triphenylphosphine (Ph$_3$P) with diethylazodicarboxylate (DEAD) to produce a dipolar intermediate that will react with an alcohol to form an alcoxyphosphonium salt and diimide. Then the nucleophile will displace triphenylphospine oxide to give the substitution product. (Figure 3.30).[39]

This procedure was used successfully for preparing the N-glycoside shown in Figure 3.31 by reacting 2,3,4,6-tetraacetyl glucose with the heterocyclic base under the Mitsunobu conditions.[40]

3.3.7   Palladium-mediated Reaction

Palladium catalysis is a well-established and versatile methodology for the preparation of nucleosides. Also known as the Heck reaction, it was developed initially

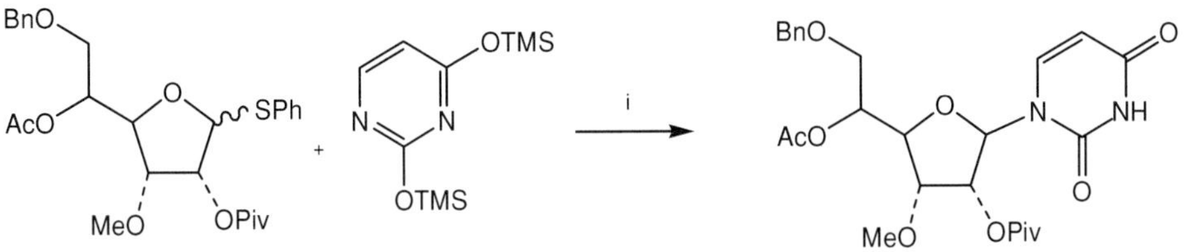

i) NIS, TfOH, CH₂Cl₂, –20°C to 0°C, 10 min, 85%

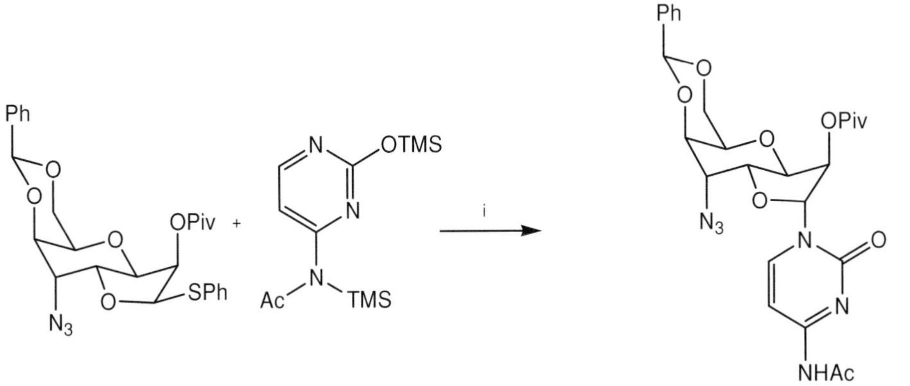

i) NIS, TfOH, CH₂Cl₂, 1h, 95%

FIGURE 3.29. N-glycoside formation via sulfur glycosyl donor.

for C-C bond formation and consists in the coupling of an aryl halide with activated olefin in the presence of palladium (0) as catalyst (Figure 3.32).[41]

More recently, other palladium-mediated reaction have been developed with great potential for heterocycle coupling reaction with furanosides, to produce an interesting variety of nucleosides. The group of reactions includes the Suzuki

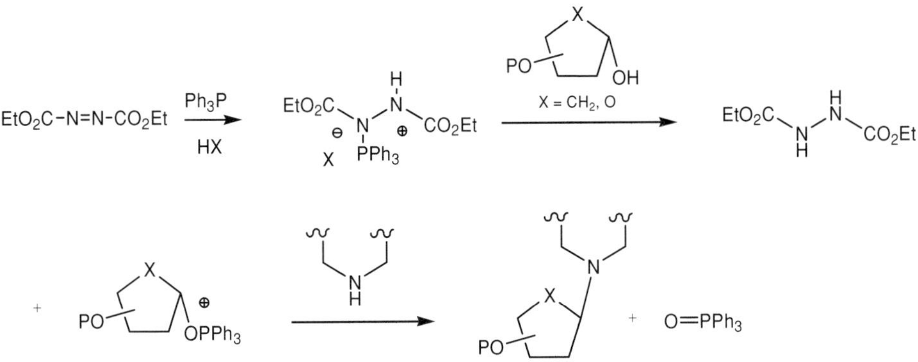

FIGURE 3.30. The Mitsunobu reaction for the construction of a glycosidic bond.

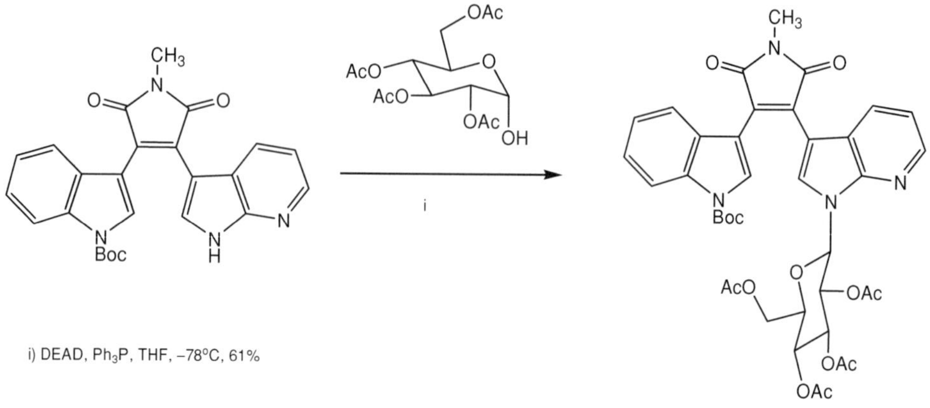

FIGURE 3.31. The Mitsunobu reaction for preparation of N-glycosides.

(organoboranes),[42] Stille (organostannanes),[43] Negishi (zincated),[44] Sonogashira (alkyne-CuI),[45] Hiyama (organosilicon),[46] and Tsuji-Trost[47] (Figure 3.33).

Early reports in the use of Heck-type reactions for the preparation of nucleosides were described by Bergstrom.[48] More recently a comprehensive overview about palladium-mediated reactions for N-glycoside bond formation or modifications at the base or the sugar moieties were described. A general figure summarizing such possibilities is shown in Figure 3.34.[49]

3.3.8   Microbial/Enzymatic Approach

The synthesis of nucleosides by enzymatic methods is another extended possibility, and for this purpose the enzyme nucleoside phosphorylase has been selected as one of the most appropriate one. Usually the conversion proceeds by the reversible formation of a purine or pyrimidine nucleoside and inorganic phosphate from ribose-1-phosphate (R-1-P) and a purine or pyrimidine base. The general approach consists of the reaction of R-1-P as the glycosyl donor, which is condensed with purine or pyrimidine analogs. Following this method any heterocycle recognized by this enzyme can be glycosilated (Figure 3.35).

The enzyme synthetase phosphoribosyl pyrophosphate PRPP was used for nucleotide synthesis of UMP. The sequence involves the conversion of ribose-6-phosphate with PRPP synthetase to produce phosphoribosyl pyrophosphate which was condensed with orotate in the presence of O5P-pyrophosphorylase

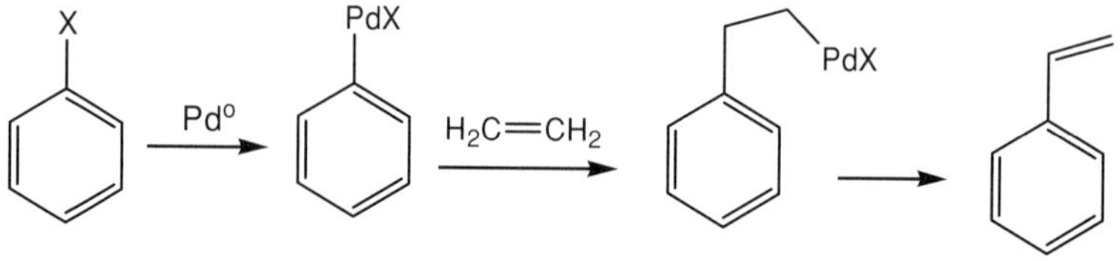

FIGURE 3.32. The Heck reaction.

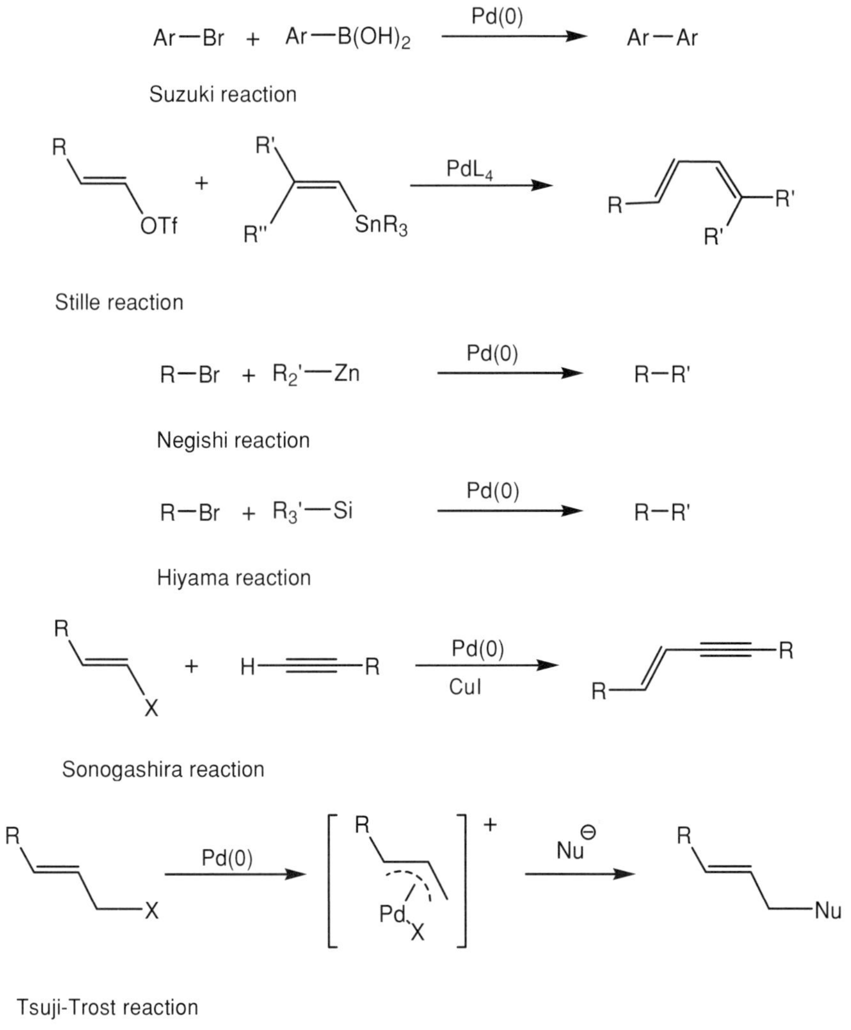
Ar—Br + Ar—B(OH)2
Pd(0)
Ar—Ar
Suzuki reaction
R
OTf
+
R'
R''
SnR3
PdL4
R
R'
R'
Stille reaction
R—Br + R2'—Zn
Pd(0)
R—R'
Negishi reaction
R—Br + R3'—Si
Pd(0)
R—R'
Hiyama reaction
R
X
+
H—≡—R
Pd(0)
CuI
R
R
Sonogashira reaction
R
X
Pd(0)
R
Pd
X
+
Nu
R
Nu
Tsuji-Trost reaction

FIGURE 3.33. Palladium-mediated coupling reactions.

to yield the nucleotide intermediate orotidine 5′-phosphate that, after descarboxylation produced by the action of O 5P-decarboxylase the nucleotide Uridine monophosphate (Figure 3.36).[50]

3.4 Oligonucleotide Synthesis

Deoxyribonucleic acid (DNA) and ribonucleic acid (RNA) are very important natural polymers responsible for the processing of the genetic information of all organisms.

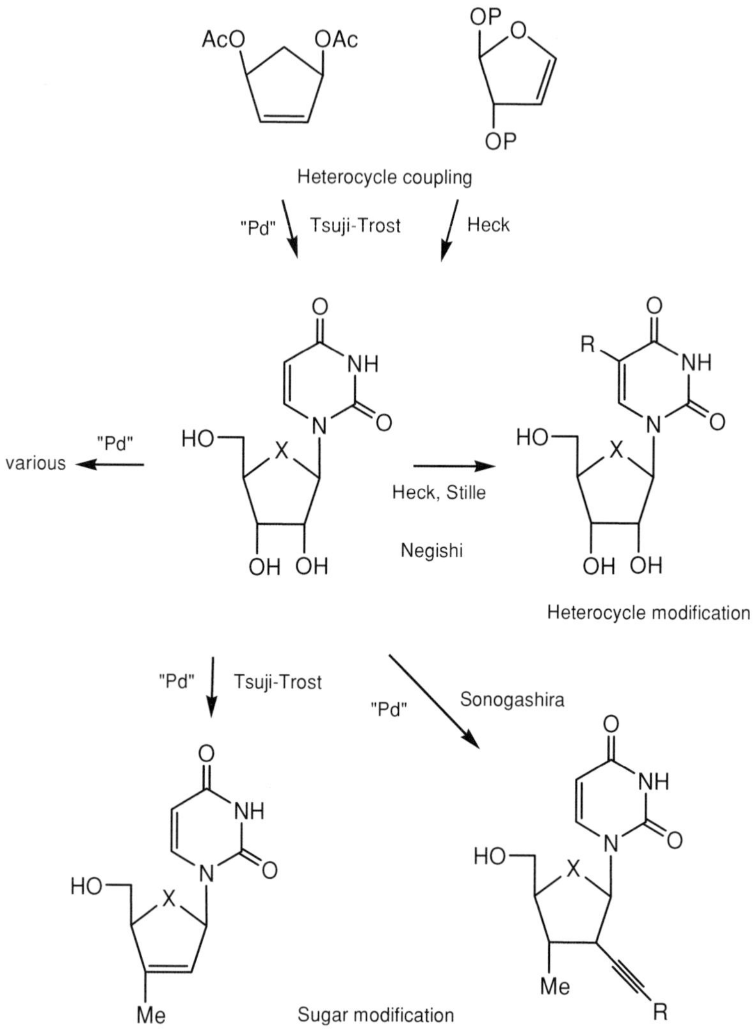

FIGURE 3.34. Palladium-assisted modifications.

The basic repetitive unit known as nucleotide is composed by a nucleotide base, a sugar moiety, and a phosphate. The combinatorial pattern of the four different nucleosides constituted by the heterocyclic bases cytosine, thymine, guanine, and adenine are the base of the DNA structure. In RNA strands uracil replaces thymine and the furanoside is ribose instead of 2-deoxyribose. The phosphate group is attached at position 3′ of one sugar unit and the 5′ position of the next one, forming a 3′-5′ elongation chain (Figure 3.37).

Oligonucleotide synthesis does not involve N-glycoside bond formation, but it requires the design of nucleoside donors and nucleoside acceptors, following

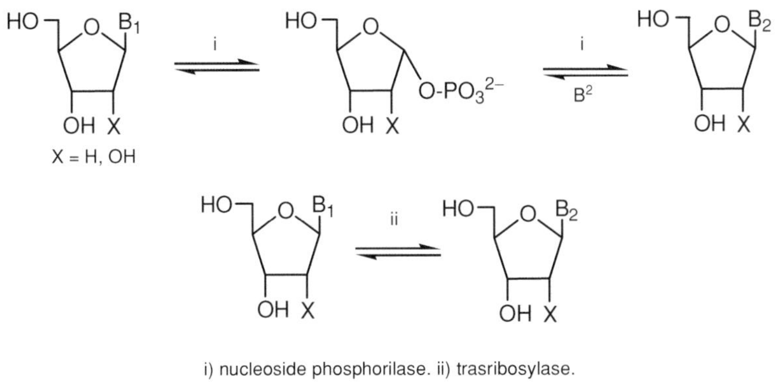

i) nucleoside phosphorilase. ii) trasribosylase.

FIGURE 3.35. General figure for enzyme-mediated nucleoside synthesis.

the same principle that applies for glycoside coupling reactions where suitable protecting groups, glycosyl donors and acceptors are requested.

Solid-phase procedures appear to be of great advantage for the coupling of nucleosides, and unlike for oligosaccharide solid-phase chemistry, the attachment positions are always the same (3′ and 5′). The sequence of reactions that occurs in oligonucleotide synthesis starts on the attachment of 3′-OH position of 5′-protected nucleoside to a resin. Next is deprotection of 5′-OH and subsequent attachment to a nucleoside donor, which contains a phosphate precursor, which in turn will be converted to a phosphate group.

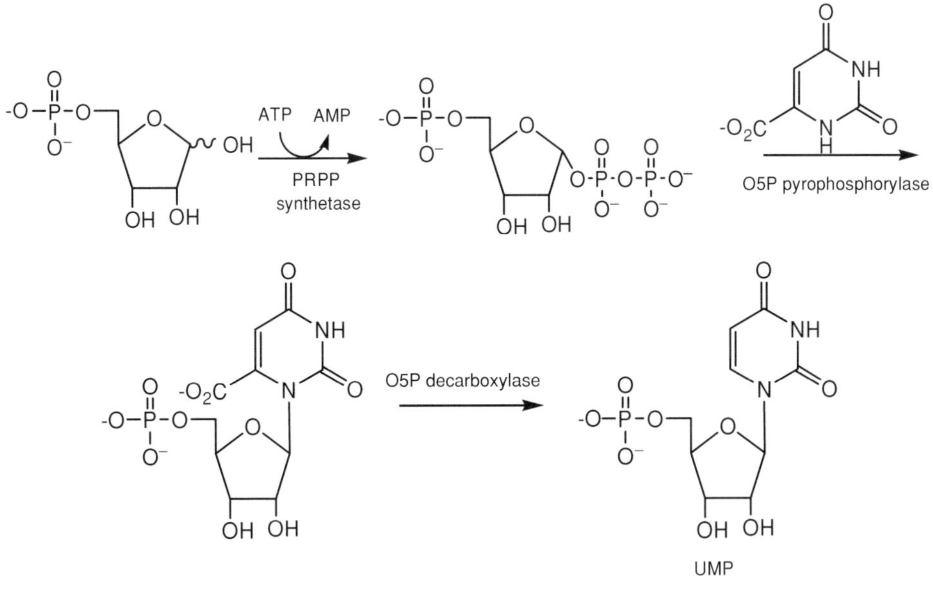

FIGURE 3.36. Enzyme-catalyzed synthesis of nucleotide.

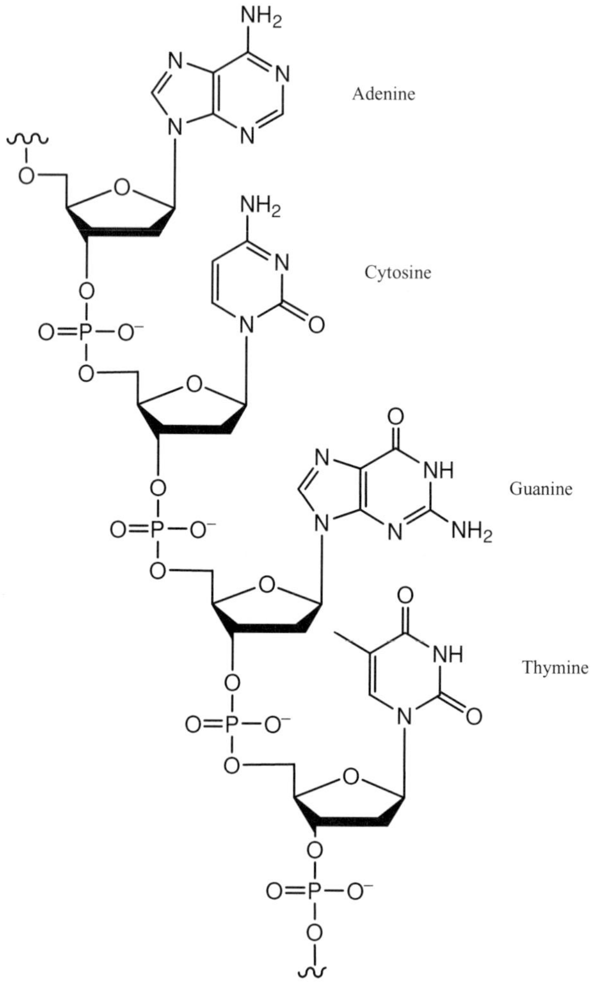

FIGURE 3.37. Fragment of a single strand of DNA structure.

There are mainly two procedures for oligonucleotide synthesis: the phosphoramidite and the phosphonate methods.[15,51]

3.4.1 Phosphoramidite Method

This methodology involves the use of the air-sensitive reagent 2-cyanoethyl tetraisopropylphosphorodiamidite $\{[(CH_3)_2CH]_2N\}POCH_2CH_2CN$ or   2-cyanoethyl N,N-diisopropylchlorophosphoramidite $(iPr)_2NP(Cl)OCH_2CH_2CN$ for activation of nucleoside donor.[52] This intermediate can be obtained by treatment of PCl_3 with 2 eq of diisopropylamine, and 1 eq of cianoethylethanol. The general phosphoramidite approach is outlined in Figure 3.38 and begins with

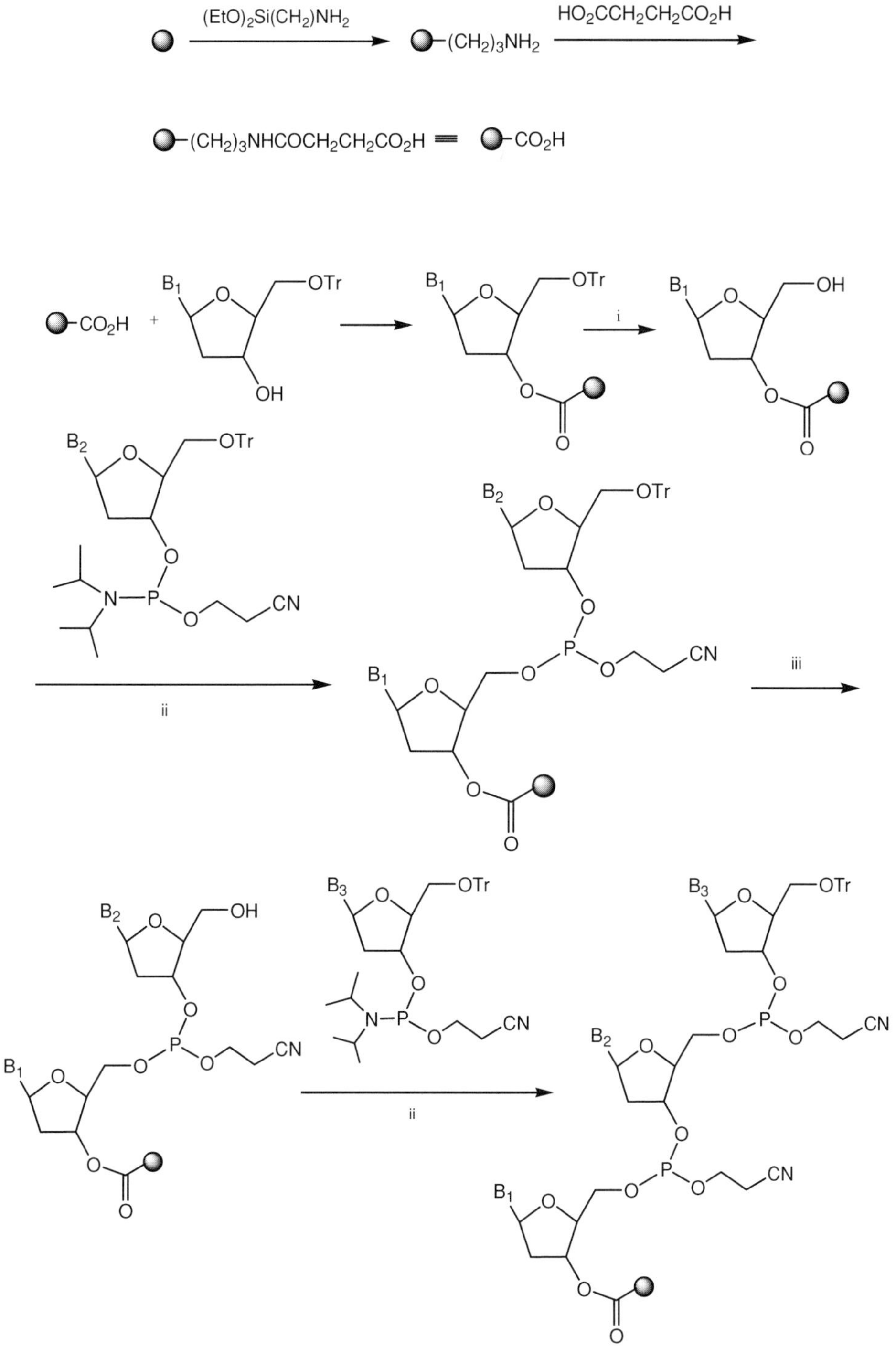

FIGURE 3.38. The phosphoramidite oligonucleotide strategy.

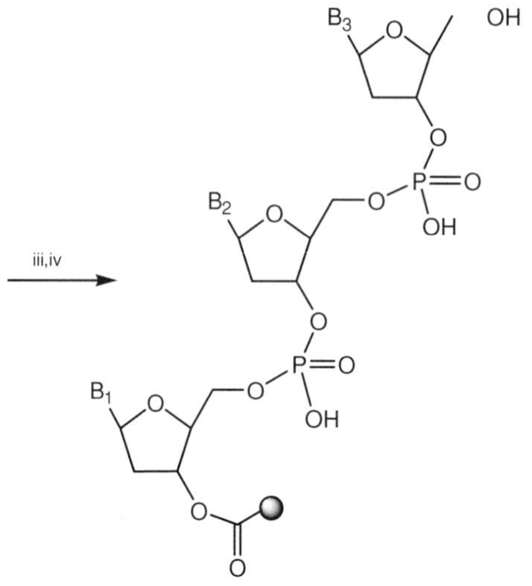

FIGURE 3.38. (*continued*)

i) Cl₃CCOOH. ii) tetrazol. iii) Cl₃CCOOH. iv) a) I₂/H₂O. b) NH₄OH

a nucleoside previously protected at the 5'-OH position with 4,4'-dimethoxytrityl group (Tr-), also attached to a silica support. The trityl group is then removed from the 5-OH position and allowed to react with a nucleoside donor protected at position 5-OH with Trityl group and activated at position 3' with 2-cyanoethyl diisopropylphophoroamidite. The coupling reaction being the critical step is catalyzed by tetrazol, and the process is repeated for the installation of subsequent nucleoside unit. Once the oligonucleotide chain is formed, the phosphoramidite group is transformed to phosphate with I_2-H_2O and released from resin with ammonia.

3.4.2   Phosphonate Method

In this method the nucleoside donor functions as a phosphotriester sugar derivative that reacts with the nucleoside acceptor at the 5-OH position, which is available for linkage. An advantage of this method is the possibility of introducing substituents to the phosphate position giving place to the preparation of modified oligonucleotides (Figure 3.39).

3.4.3   Modified Oligonucleotides

Modified oligonucleotides are another important application of solid-phase oligonucleotide synthesis. It is known that natural oligonucleotides used as therapeutic strategy against viral infections as *antisense* for targeting RNA sequences may undergo enzymatic hydrolysis by endonucleases. Series of modified oligonucleotides carrying the modification either on the base, sugar, or phosphate moiety

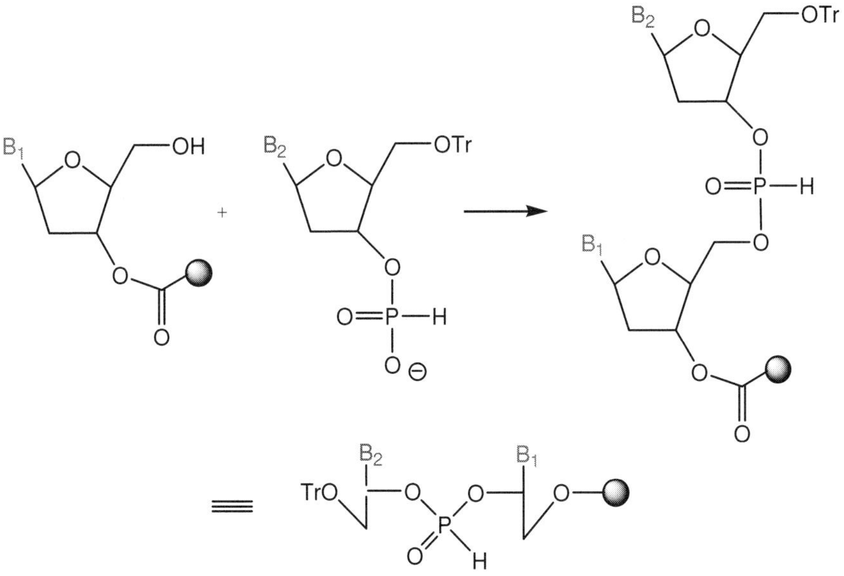

FIGURE 3.39. The phosphonate method.

provides ideally endonuclease resistance as well as high affinity for complementary RNA sequences.

Phosphodiester bond is the primary target for endonuclease breakage. Therefore, the effort has been focused mainly on the modification of this segment of the chain. As a result of this, a first generation of modified phosphorous oligonucleotides such as phosphothioates, methylphosphonates, phosphoramidates, phosphotriesters, and phosphodithioates was synthesized. Although these phosphorous derivatives showed increased resistance to endonuclease activity, the affinity for complementary sequences was decreased.[53-55] For instance, the

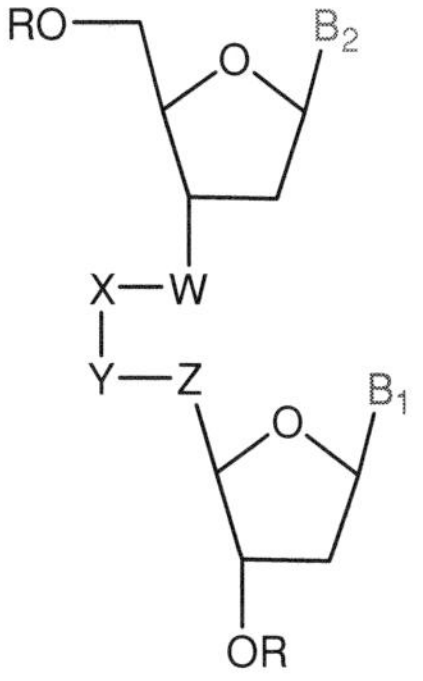

W	X	Y	Z
O	-P(O)S-	O	CH_2
O	-P(O)CH$_3$-	O	CH_2
O	-P(O)NHR-	O	CH_2
O	-P(O)OR-	O	CH_2
NH	-P(O)O-	O	CH_2

FIGURE 3.40. Modified oligonucleotides.

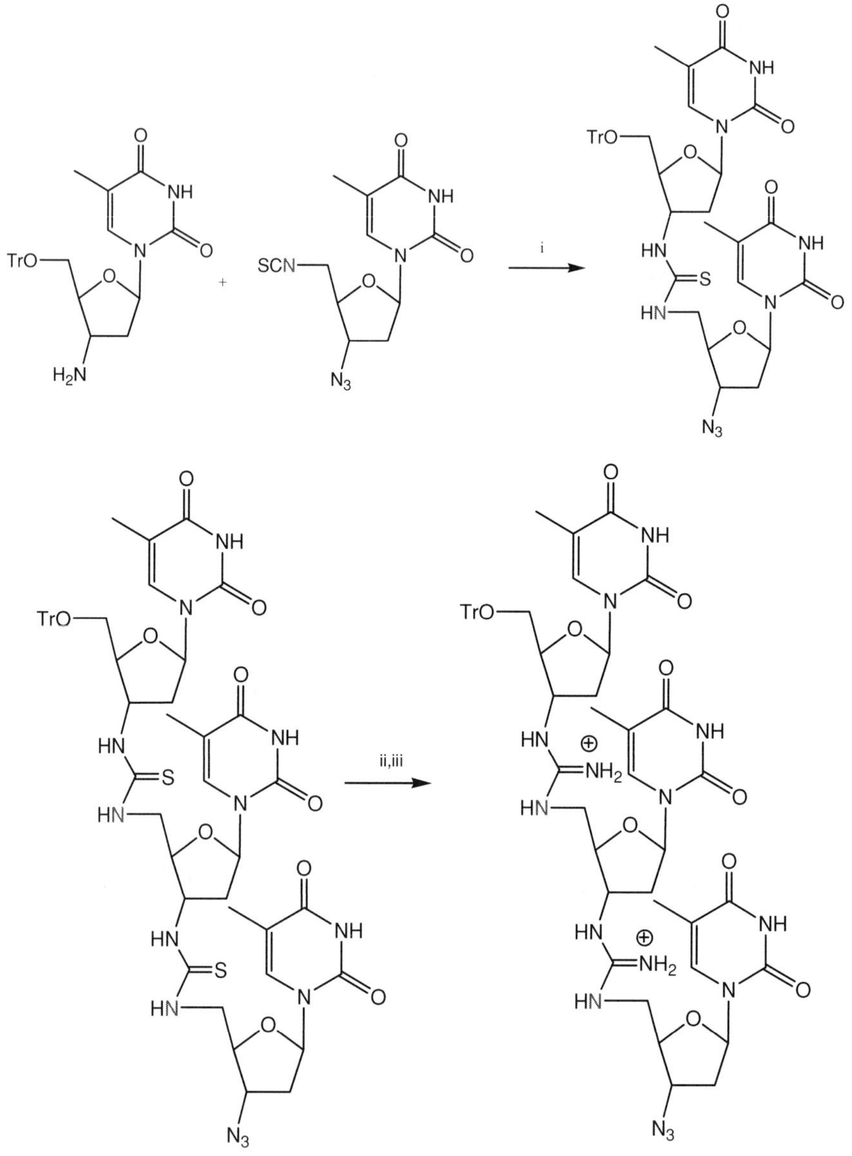

i) DMF. ii) H₂S. iii) 3′-azido-5′-isothiocyano-3′,5′-deoxythymidine. iv) a) TFA.
b) H₂NC(=NH)SO₂H. c) NH₄OH.

FIGURE 3.41. Preparation of guanidinium oligonucleotides.

synthesis of the antisense oligomer phosphorothioate analog of a 28-nucleotide
homo-oligodeoxycytidine (S-dC$_{28}$) was achieved, and tested as a potent inhibitor
of HIV in vitro, showing significant inhibition of reverse transcriptase activity and
syncytium formation between HIV-1 producing cells and CD4$^+$.[56]

A second generation proposed the replacement of phosphodiester group by a bioisoster such as amides, urea, and carbamate (Figure 3.40). In general the observations reveal better enzymatic hydrolysis resistances, but again poor affinity toward RNA complementary sequences.

Alternatively, Dempcy et al.[57] reported the synthesis of modified guanidine-timidine oligonucleotide following the procedure depicted in Figure 3.41. The reactions involved are the condensation between 3'-amino-5'-O-trityl-3'-deoxythymidine and 3'-azido-5'-isothiocyano-3',5'-deoxythymidine, to generate 5'→3' thiourea-nucleoside dimer. Reduction followed by coupling reaction of dimer with the later nucleoside produced chain elongation reaction. Guanidine conversion was done with aminoiminosulfonic acid and ammonium hydroxide, affording guanidinium thymidyl pentamer.

The unit assemble for oligoribonucleotide synthesis is to some extent similar to deoxyribonucleotides synthesis. However, an additional consideration should be taken into account, which is the suitable protection of position 2-OH of ribose. The use of the silyl protecting group is one of the best choices so far reported, in particular the hindered *tert*-butyldimethyl silyl (TBDS) group. The protection of tritylribonucleoside produced a mixture of isomers, being the 2-OH silyl derivative

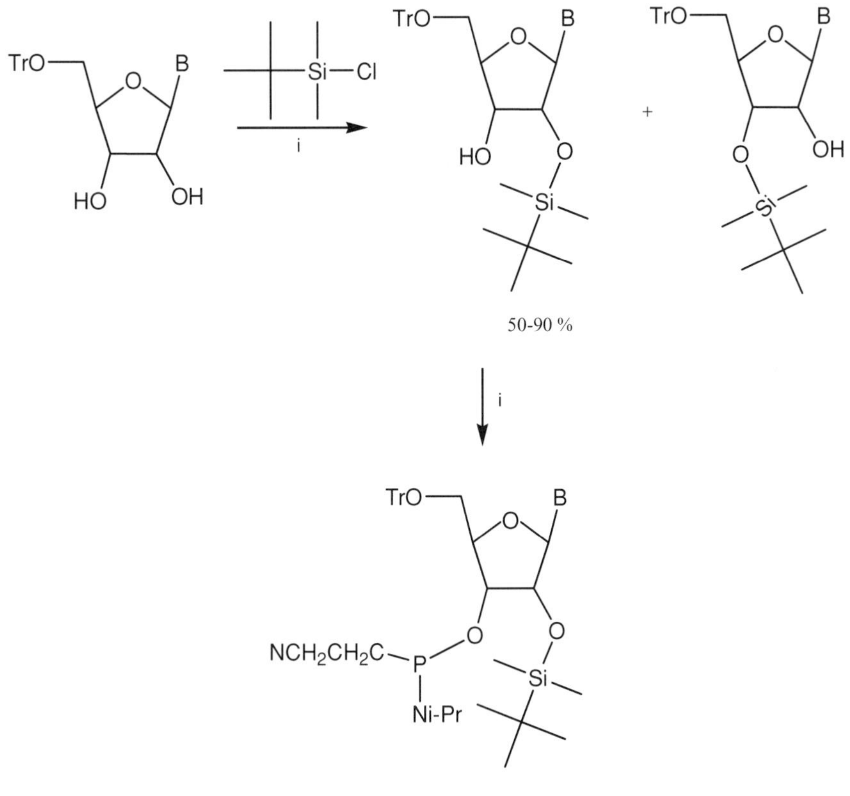

FIGURE 3.42. Ribose protecting groups for oligoribonucleotide synthesis.

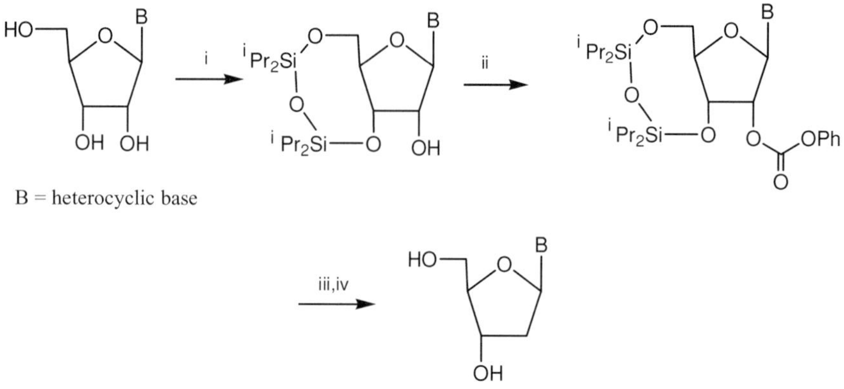

B = heterocyclic base

i) $[(iPr)_2SiCl]_2O$, Py. ii) ClC(S)OPh, DMAP, MeCN. iii) Bu_3SnH, AIBN, $PhCH_3$.
iv) Bu_4NF, THF.

FIGURE 3.43. The Barton-McCombie procedure for the preparation of 2′deoxynucleosides.

generated in between 50–90% yield. Final removal of this protecting group is usually achieved with 1 M tetrabutylammonium fluoride in THF (Figure 3.42).

Some other choices for 2-OH protection are tertahydropyran-1-il, 4-methoxytetrahydropyran-4-il, and modified ketal of 1-(2-fluorophenyl)-4-methoxypiperidin-4-il (Fpmp). However, it has been found that acid conditions for removal of these protecting groups are not compatible with the trityl protecting group.

Simultaneous protection of positions 3′ and 5′ can be achieved by using the silyl protecting group tetraisopropyldisiloxychloride (TIPS-Cl) in pyridine. This type of protection has been useful in the conversion of adenosine to 2′-deoxyadenosine under the conditions reported by Barton and McCombie[58] (Figure 3.43).

References

1. S. Nishimura, *Progress Research Molecular Biology* **12**, 49 (1972).
2. T. Kondo, T. Ohgi, and T. Goto, *Agric. Biol. Chem.* **41**, 1501 (1977).
3. H. Akimoto, E. Imayima, T. Hitaka, H. Nomura, and S. Nishimura, *J. Chem Soc Perkin Trans. I* 1637 (1988).
4. C.J. Barnett and L.M. Grubb, *Tetrahedron* **56**, 9221 (2000).
5. T. Itaya, *Chem. Pharm. Bull.* **38**, 2656 (1990).
6. (a) S. Knapp, *Chem. Rev.* **95**, 1859 (1995). (b) M.T. Migawa, L.M. Risen, R.H. Griffey, and E.E. Swayze Org. Lett. **7**, 3429 (2005).
7. N. Chida, K. Koizumi, Y. Kitada, C. Yokohama, and S. Ogawa, *J. Chem. Soc. Chem. Comm.* 11 (1994).
8. M.P. Maguire, P.L. Feldman, and H. Rapoport, *J. Org. Chem.* **55**, 948 (1990).
9. L. Kalvoda, M. Prystas, and F. Sorm, *Collect. Czech. Chem. Commun.* **41**, 788 (1976).
10. A.G. Myers, D.Y. Gin, and D.H. Rogers, *J. Am. Chem. Soc.* **116**, 4697 (1994).

11. H. Hahn, H. Heitsch, R. Rathmann, G. Zimmermann, C. Bormann, H. Zahner, and W. Konig, *Liebigs Ann. Chem.* 803 (1987).

12. S. Hanessian, J. Kloss, and T. Sugawara, *J. Am. Chem. Soc.* **108**, 2758 (1986).

13. N. Ikemoto and S.L. Schreiber, *J. Am. Chem. Soc.* **114**, 2524 (1992).

14. S. Knapp and S.R. Nandan, *J. Org. Chem.* **59**, 281 (1994).

15. G.M. Blackburn and M. Gait, *Nucleic Acids in Chemistry and Biology*, IRL 77 (1990).

16. C.-H. Wong, S.T. Chen, W.J. Hennen, J.A. Bibbs, Y-F. Wang, J.L-C. Liu, M.W. Pantoliano, M. Whitlo, and P.N. Bryan, *J. Am. Chem. Soc.* **112**, 945 (1990).

17. A. De la Cruz, J. Elguero, V. Gotor, P. Goya, A. Martínez, and F. Moris, *Synth. Commun.* **21**, 1477 (1991). V. Gotor, and F. Morís, *Synthesis* 626 (1992).

18. S. Ozaki, K. Yamashita, T. Konishi, T. Maekawa, M. Eshima, A. Uemura, and L. Ling, *Nucleosides Nucleotides* **14**, 401 (1995).

19. H.K. Singh, G.L. Cote, and R.S. Sikirski, *Tetrahedron Lett.* **34**, 5201 (1993).

20. D.L. Damkjaer, M. Petersen, and J. Wengel, *Nucleosides Nucleotides* **13**, 1801 (1994).

21. Kawana et al., *Chem Lett.* 1541 (1981).

22. S. Kozai, T. Fuzikawa, K. Harumoto, T. Maruyama *Nucleosides, Nucleotides & Nucleic Acids* **22**, 779 (2003).

23. V. Serebryany and L. Beigelman, *Tetrahedron Lett.* **43**, 1983 (2002).

24. Taniguchi et al., *Angew. Chem. Int. Ed.* **37**, 1136 (1988).

25. M.J. Robins, V. Samano, and M.D. Johnson, *J. Org. Chem.* **55**, 410 (1990).

26. T. Kondo, T. Ohgi, and T. Goto, *Chemistry Lett.* 419 (1983).

27. B.K. Battacharya, T.S. Rao, and G.R. Revankar, *J. Chem. Soc.* 1543 (1995).

28. F. Seela, H. Steker,. H. Driller, and U. Bindig, *Liebigs Ann. Chem.*15 (1987).

29. F. Seela and A. Kehne, *Liebigs Ann. Chem.* 876 (1983).

30. Z. Kazimierczuk, H.B. Cottam, G.R. Revankar, and R.K. Robins, *J. Am. Chem. Soc.* **106**, 6379 (1984).

31. E. Edstrom and Y. Wei, *J. Org. Chem.* **60**, 5069 (1995).

32. J. Davoll, B. Lythgoe, and A.R. Todd, *J. Chem. Soc.* 967 (1948).

33. R. S. Tipson, *J. Biol. Chem.* **55**, 130 (1939).

34. K.H. Jung, and R.R. Schmidt, *Liebigs, Ann., Chem.* 1013 (1988).

35. E. De Clercq, G. Gosselin, M.C. Bergogne, J. De Ruddes, and J.L. Imbach, *J. Med. Chem.* **30**, 982 (1987).

36. G.E. Hilbert and T.B. Johnson, *J. Am. Chem. Soc* 52, 4489 (1930).

37. (a) H. Vorbrüggen, K. Krolikiewicz and B. Bennua, *Chem. Ber.* **114**, 1234 (1981). (b) H. Vorbrüggen and G. Höfle, *Chem. Ber.* **114**, 1256 (1981).

38. S. Maier, R. Preuss, and R.R. Schmidt, *Liebigs Ann. Chem.* 483 (1990).

39. O. Mitsunobu, *Synthesis* 1 (1981).

40. Marminon et al., *J. Med. Chem.* **46**, 609 (2003).

41. R.F. Heck, *Organic Rect.* **27**, 345 (1982).

42. N. Miyaura and A. Suzuki, *Chem. Rev.* **95**, 2457 (1995).

43. W.J. Scott, G.T. Crisp, and J.K. Stille, *J. Am. Chem. Soc.* **106**, 4630 (1984).

44. E.-I. Negishi, *J. Organomet. Chem* **653**, 1 (2002).

45. K. Sonogashira, *In Comprehensive Organic Chemistry*, Trost, B. M., Fleming, I., Eds.; Pergamon Press: N.Y. **3**, 521 (1991).

46. Y. Hatanaka and T. Hiyama, *Synlett* 845 (1991).

47. (a) B.M. Trost, and D.L. Van Vranken, *Chem Rev.* **96**, 395 (1996). (b) Tsuji, J., Yamakawa, T., *Tetrahedron Lett.* **20**, 613 (1979).

48. J.L. Ruth and D.E. Bergstrom, *J. Org. Chem.* **43**, 2870 (1978). D.E. Bergstrom and J.L. Ruth, *J. Am. Chem. Soc.* **98**, 1587 (1976). D.E. Bergstrom, J.L. Ruth, and P. Warwick, *J. Org. Chem.* **46**, 1432 (1981).
49. L.A. Agrofolio, I. Gillaizeau, and Y. Saito, *Chem. Rev.* **103**, 1877 (2003).
50. A. Gross, O. Abril, J.M. Lewis, S. Geresh, and G.M. Whitesides, *J. Am. Chem. Soc.* **105**, 7428 (1983).
51. F. Eckstein, *Oligonucleotides and Analog:- a Practical Approach,* IRL (1991).
52. N.-S Li and J.A. Piccirilli, *J. Org. Chem.* **69**, 4751 (2004).
53. A. De Mesmaeker, R. Haner, P. Martin, and H.E. Moser, *Acc. Chem. Res.* **28**, 366 (1995).
54. M.J. Damha, P.A. Giannaris, P.A. Marfey, and L.S. Reid, *Tetrahedon Lett.* **32**, 2573 (1991).
55. M. Sekine, H. Tsuruoka, S. Iimura, H. Kusuoku, and T. Wada,. *J. Org. Chem.* **61**, 4087 (1996).
56. H. Mitsuya, R. Yarchoan, and S. Broder, *Science* **249**, 1533 (1990).
57. R. Dempcy, K.A. Browne, and T.C. Bruice, *J. Am. Chem. Soc.* **117**, 6140 (1995).
58. D.H.R. Barton and S.W. McCombie, *J. Chem. Soc. Perkin Trans 1* 1574 (1975).

4
Nucleoside Mimetics

Modified nucleosides are useful therapeutic agents being currently used as antitumor, antiviral, and antibiotic agents. Despite the fact that a significant variety of modified nucleosides displays potent and selective action against the mentioned diseases, the challenge still attracts full attention since most of them do not discriminate between normal and tumor cell and in viral infection resistant strains usually appear during the course of the treatment.

Synthetic acyclic, carbocyclic, C-nucleosides, and modified N-nucleosides have shown remarkable action against AIDS, hepatitis, and Herpes infections, among others. Some of the nucleosides used as approved drugs are acyclovir, carbovir being the treatment of choice against Herpes, AZT, ddI, ddC, ddG, Abacavir, which in combination with protease inhibitors is indicated in the treatment against HIV, and C-nucleoside Ribavirin in the treatment against hepatitis.[1,2]

Representative examples of chemotherapeutic agents modified at the heterocyclic base, the sugar fragment, L and Cnucleosides, carbocyclic and acyclic nucleosides are depicted in Figure 4.1.

A significant number of synthetically modified nucleosides were designed as antiretroviral drugs in the therapy of human immunodeficiency virus (HIV) infection. During retroviral infection, the viral RNA is used as a template for proviral DNA synthesis, a process mediated by the viral DNA polymerase better known as reverse transcriptase. Thus, the process involves the initial formation of an RNA-DNA hybrid, which is then degraded by an RNAse to release the DNA strand that will be the template for the synthesis of the doublestranded viral DNA, a process also catalyzed by the reverse transcriptase.[3]

The proposed mechanism of action of modified agents such as AZT during viral infection involves the interruption of the viral replication process that occurs between the virus and host, particularly the replication inhibition inside T cells, monocytes, and macrophages.

When the modified nucleoside is introduced into the cell, a sequential 5′phosphorylation process mediated by kinases occurs on the furanoside ring, which is subsequently incorporated into the DNA as triphosphate (Figure 4.2).

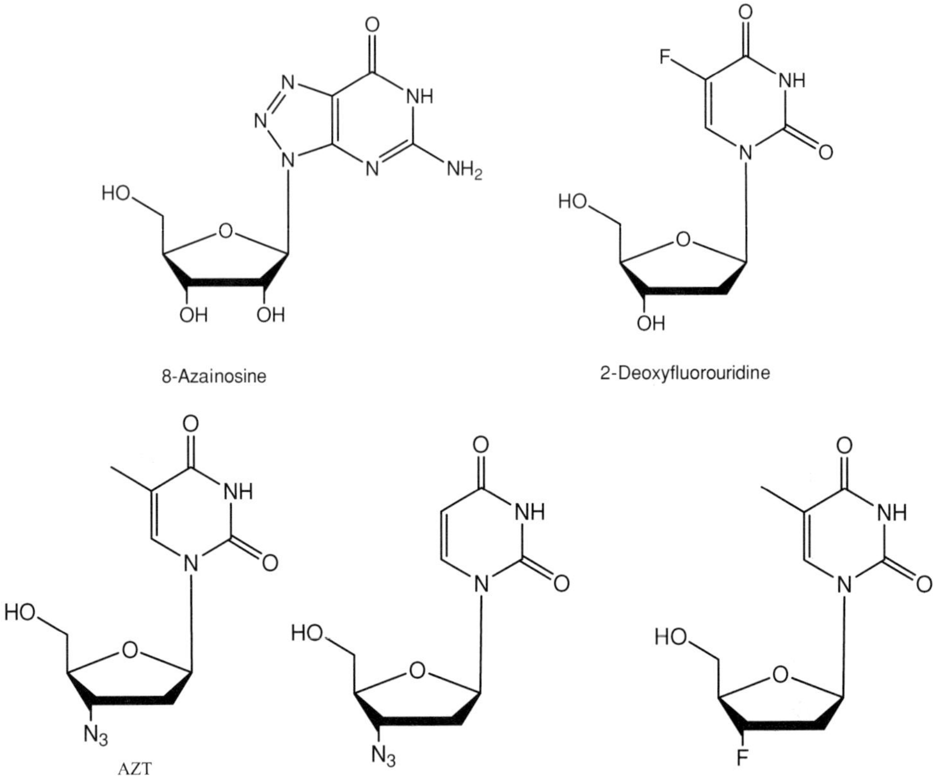

FIGURE 4.1. Representative synthetically modified nucleosides.

An important collection of active nucleosides mimetics has been synthesized and classified for better understanding as follows:[4]

Modified Nnucleosides
Lnucleosides (Disomers)
Cnucleosides
Carbocyclic nucleosides
Acyclic nucleosides
Thionucleosides

4.1 Modified Nucleosides

A broad number of modified nucleosides have been developed and tested on clinical trials, some being highly promising. The chemical manipulations have been made at the heterocyclic base, the sugar of both. Some representative examples of chemical modifications leading to key intermediates or active nucleosides are

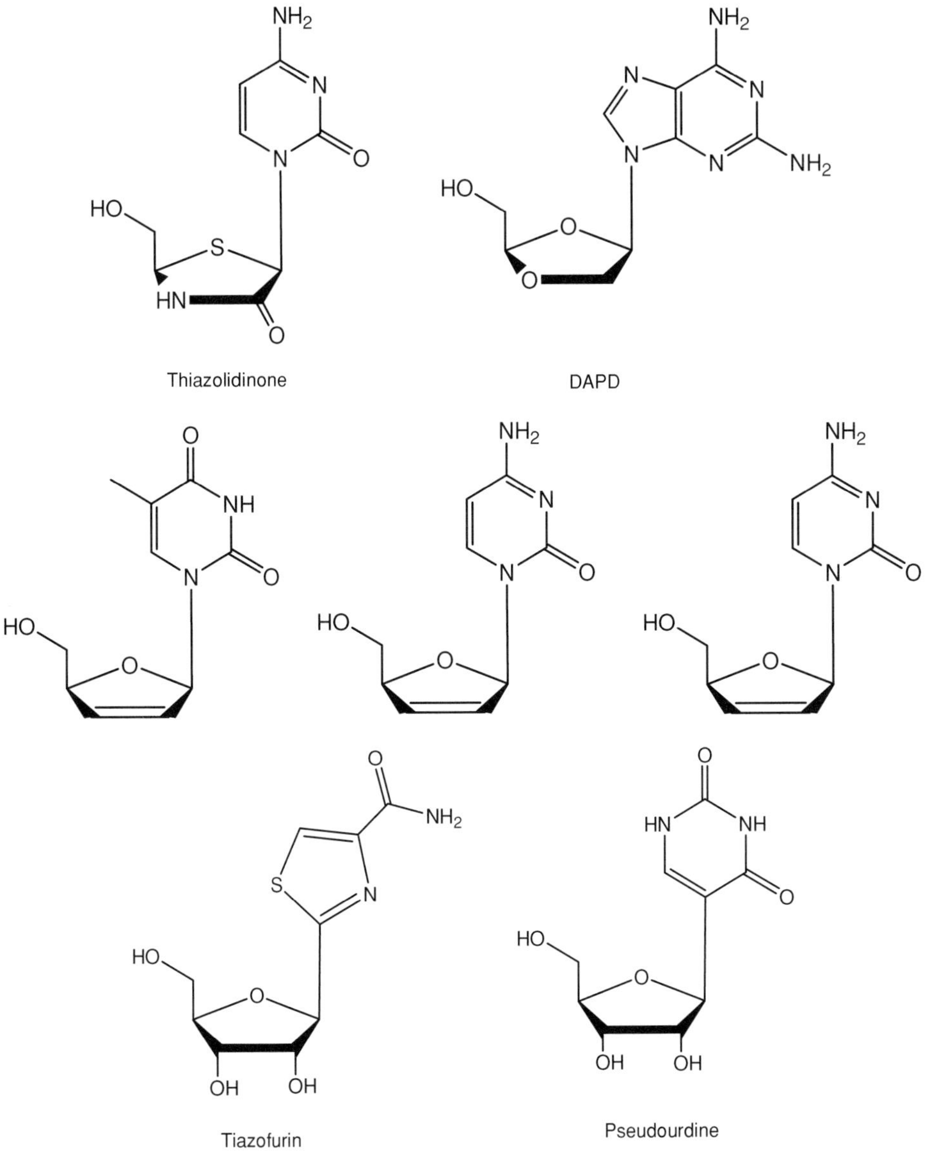

Thiazolidinone

DAPD

Tiazofurin

Pseudourdine

FIGURE 4.1. (*Continued*)

4.1.1   *Heterocycle Modifications*

4.1.1.1   C-5 Substituted Pyrimidines

Several nucleoside analogs bearing modifications at the 5-position have been found to be active as antiviral and anticancer drugs. Examples of this are BVDU, IDU, and FIAU (Figure 4.3) [5]

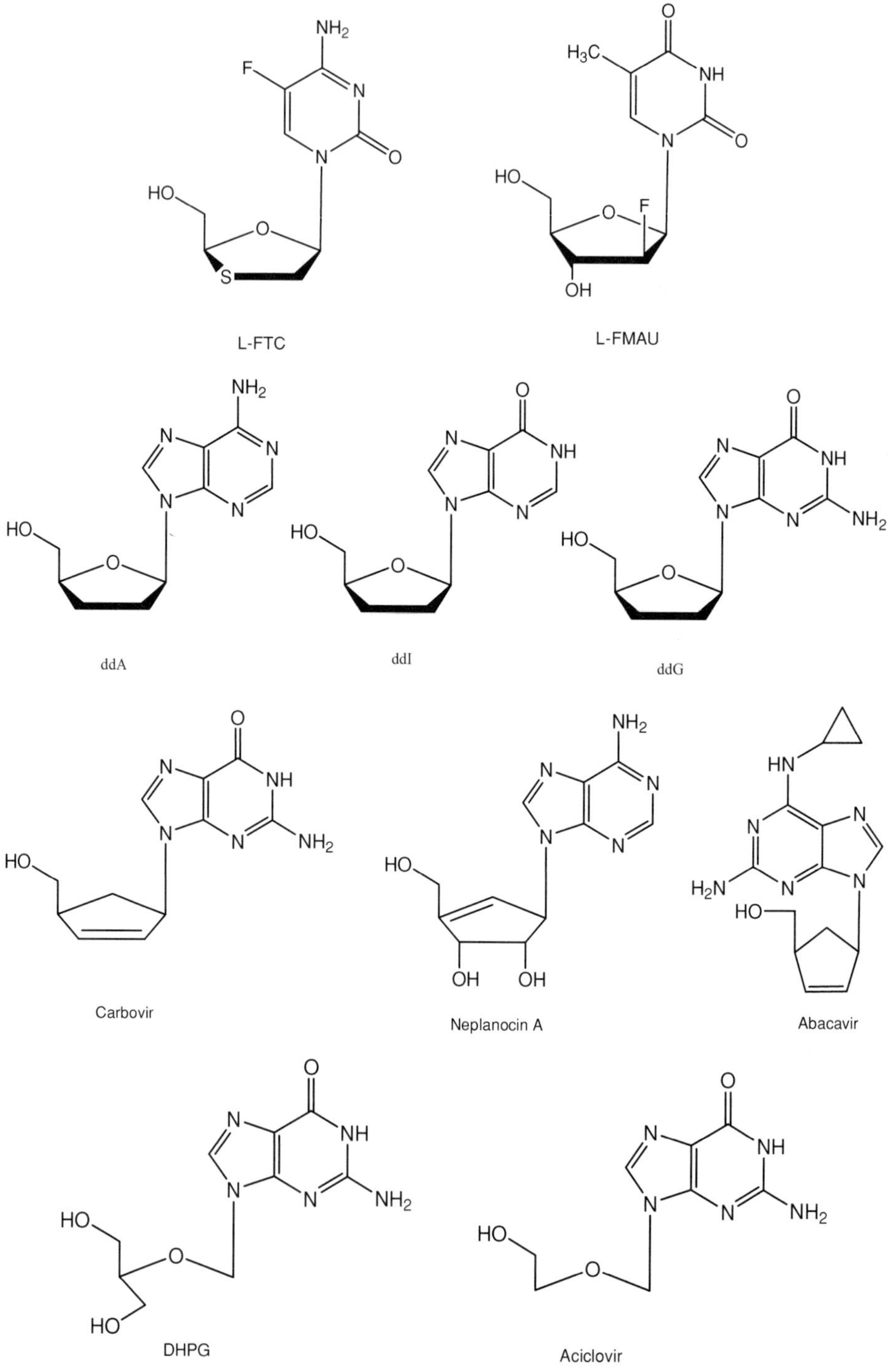

FIGURE 4.1. (*Continued*)

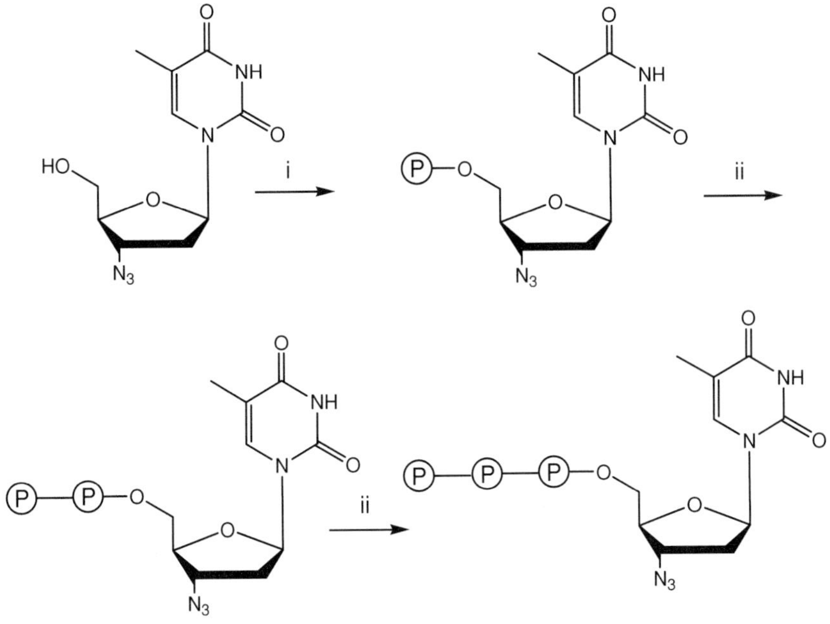

i) Timidinkinase. ii) Timidilatokinase. iii) Nucleosidediphosphatekinase.

FIGURE 4.2. Phosphorylation of AZT.

Palladium-mediated transformations is a suitable strategy for introducing substituents at C-5. Some of the reactions implemented for this purposes are the Sonogashira,[6,7] Stille,[8,9] Heck,[10] and Hiyama[11] (Figure 4.4).

4.1.1.2   C-6 Substituted Pyrimidines

By following palladium-mediated substitutions, a more limited number of C-6 substituted pyrimidines have been described in comparison with C-5. For instance, applying the Stille reaction allows one to prepare C-6 substituted aryl, vinyl, alkynyl derivatives (Figure 4.5). [12]

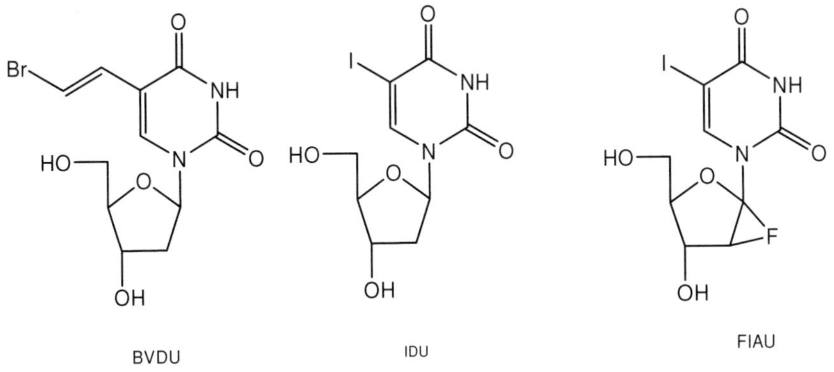

BVDU   IDU   FIAU

FIGURE 4.3. Active C5-substituted pyrimidines.

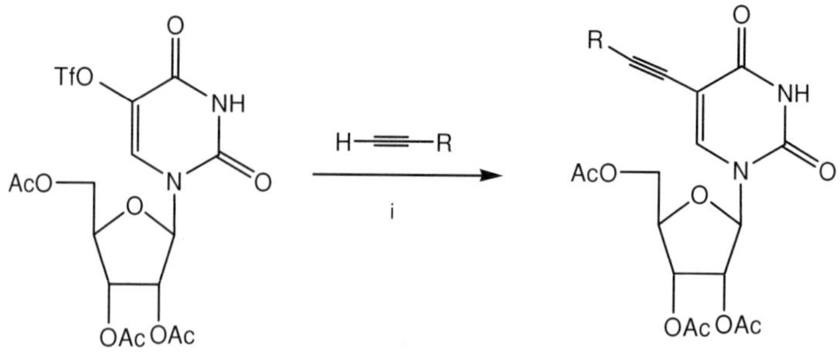

i) Pd(PPh$_3$)$_4$, 10%, CuI, 20%, Et$_3$N 1.2 eq./DMF.

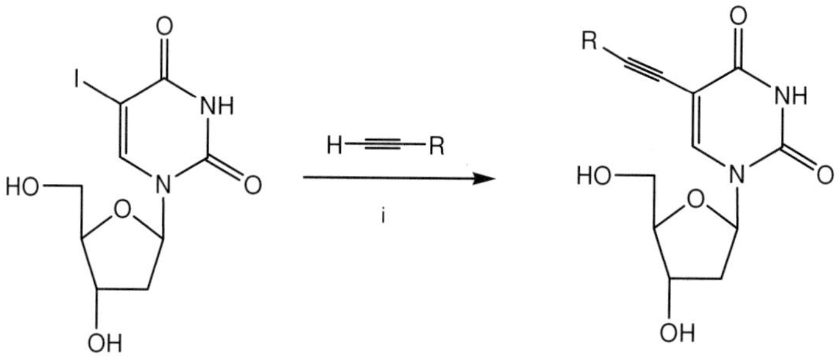

i) Pd(PPh$_3$)$_4$, CuI, 20%, iPrEtN 40-60%.

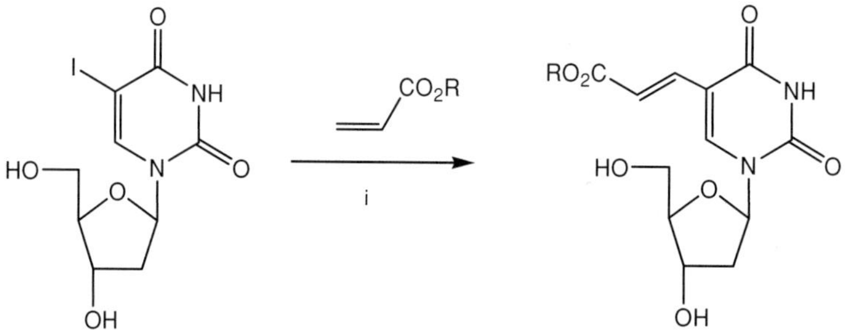

i) Pd(OAc)$_2$, PPh$_3$, Et$_3$N, dioxan, 40%.

FIGURE 4.4. Palladium-mediated substitutions at C-5 pyrimidine position.

4.1.1.3   Purine Formation

The conventional methods of preparation of C-C purines are based on heterocyclization.[13,14] The classical procedures involve.

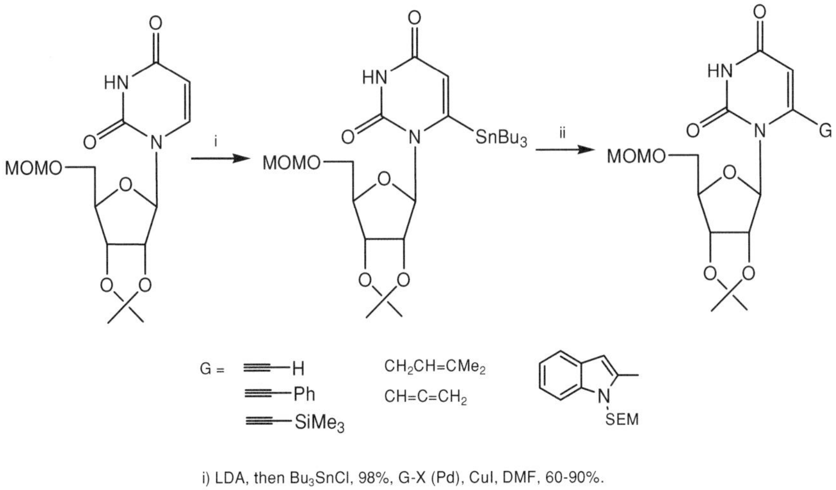

G =   ≡—H $CH_2CH=CMe_2$

≡—Ph $CH=C=CH_2$

≡—SiMe$_3$

i) LDA, then Bu$_3$SnCl, 98%, G-X (Pd), CuI, DMF, 60-90%.

FIGURE 4.5. Palladium-mediated substitution of 6-C substituted pyrimidines.

1. 2-C-Cpurines cyclization of 4-aminoimidazole-5-carboxamides or nitriles with carboxylic acid equivalents.
2. 8-C-Cpurines from 5,6-diaminopyrimidines and carboxylic acid derivatives; and for 6-C-C-purines from 4-alkyl or 4-aryl-substituted 5,6-diaminopyrimidines (Figure 4.6).[15]

Other explored methods involves radical[16,17] or nucleophilic substitution,[18] sulfur extrusion,[19] and Wittig-type reactions.[20,21] Despite their usefulness, other methods based on the use of organometallic complex are getting particular significance especially in the synthesis of substituted purines (Figure 4.7).[15]

Usually the cross-coupling reactions involving organometallic compounds include organolithium,[22] magnesium,[23] aluminum,[24] cuprates,[25] zinc,[26] stannanes,[27] and boron[28] reagents, in the presence of palladium catalyst, and the purine base bearing a good leaving group is usually halides or tosyl (Figure 4.8).

Deazapurines are pyrrolo[2,3]pyrimidines of natural or synthetic source with significant antitumor, antiviral, and antibacterial activities. Some compounds included in this class are tubercidin, toyocamycin, sangivamycin, and the hypermodified nucleoside queuosine. A flexible route for the preparation of pyrrolo[2,3]pyrimidines (7-deazapurines) has been developed, consisting of the condensation of protected uracil with ethyl N-(p-nitrophenethyl)glycinate and subsequent treatment with acetic anhydride and amine base with heating to afford 5(-acetyloxy)pyrrolo[2,3-d]pyrimidine 2,4-dione in 74% yield (Figure 4.9).[29]

4.1.2   Sugar Modifications

4.1.2.1   2'3'-Dideoxysugars

A significant number of saturated and unsaturated dideoxysugars have been synthesized and tested as antiviral or anticancer drugs. Remarkably, ddI and ddC are

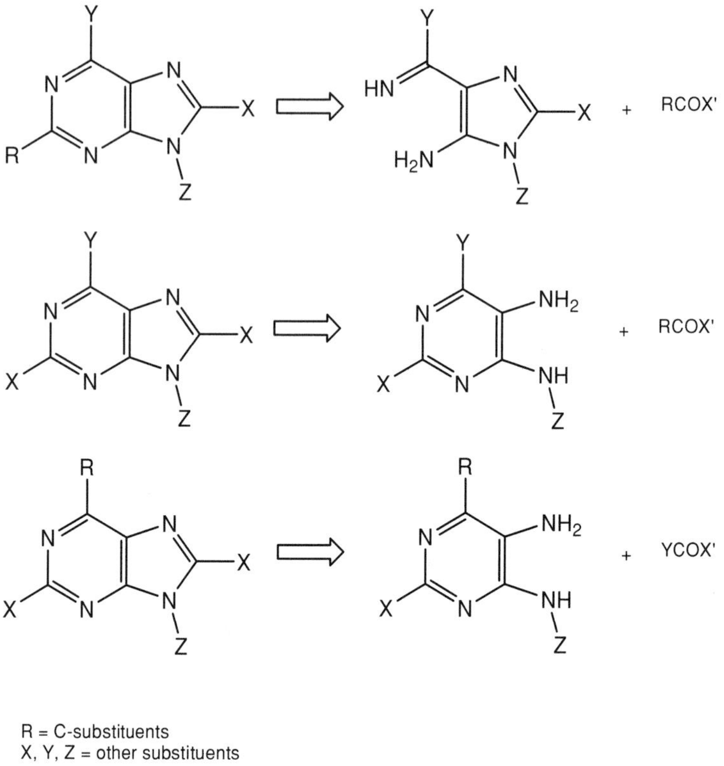
Y
N
N
X
R
N
N
Z
HN
H2N
+ RCOX'
X
NH2
NH
+ RCOX'
R
+ YCOX'
R = C-substituents
X, Y, Z = other substituents

FIGURE 4.6. Conventional methods of preparation of C-C purines.

approved drugs for the treatment of AIDS,[3] and others such as d4T being currently under clinical studies (Figure 4.10).[30,31]

A method for preparing ddC was described involving bromoacetylation with HBr in acetic acid of N4-acetylcytidine followed by reductive elimination with zinc-cooper couple in acetic acid to provide the corresponding 2′3′-unsaturated derivative. Final hydrogenation over 10% palladium on charcoal gave ddC in 95% accompanied by some N-C cleavage in 5% (Figure 4.11).[32] Similar reaction conditions were used for preparing 2′3′-dideoxyadenosine in 81% yield from adenosine.[33]

The design and synthesis of potent inhibitors for Human Hepatitis B Virus (HBV) 2′,3′-dideoxy-2′3′-didehydro-β-L-cytidine (β-L-d4C) and 2′,3′-dideoxy-2′3′-didehydro-β-L-5-fluorocytidine (β-L-Fd4C) nucleosides were carried out according to the pathway shown in Figure 4.12.[34] The key starting material 3′,5′-dibenzoyl-2′-deoxy-β-L-uridine was submitted to transglycosilation reaction with silylated 5-fluorouracil using TMSOTf as catalyst, affording an anomeric mixture separated by chromatography. After benzoyl deprotection, the anomeric nucleosides were treated with mesyl chloride followed by base to form cyclic ethers.

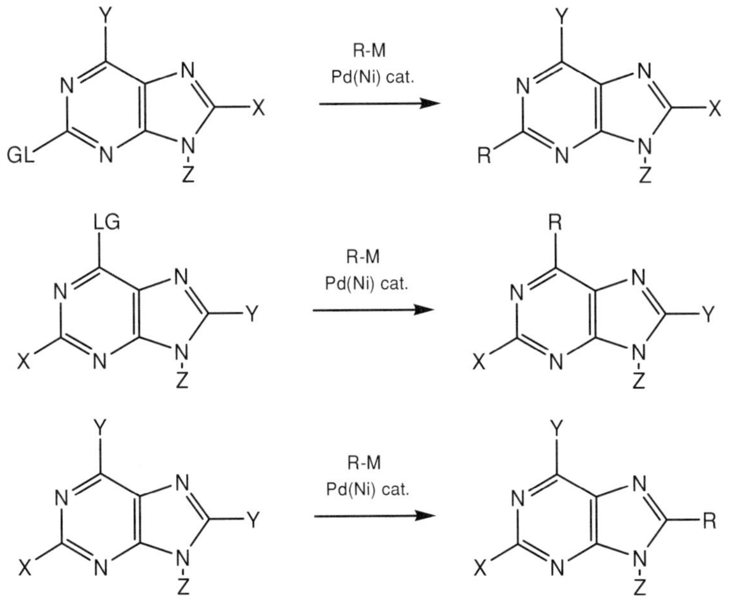

LG = Cl, Br, I, OTs

R = alkyl, alkenyl, alkynyl, aryl, hetaryl

M = MgBr, AlR$_2$, (Cu), ZnCl, SnR'$_3$, B(OH)$_2$

X, Y, Z = diverse substituents

FIGURE 4.7. General figure between purines and organometallic compounds.

Further transformation at the pyrimidine ring was followed by potassium *ter*-*t*butoxide treatment to furnish β-L-d4C and β-L-Fd4C.

Other methods designed for the preparation of 2'3'-unsaturated and saturated deoxyfuranosides are based on (a) the Corey-Winter reaction involving cyclic thionocarbonate,,[35,37] (b) the Eastwood olefination process in which a five-membered cyclic orthoformate suffer a fragmentation to give in the presence of acetic anhydride the desired olefin (successfully applied in the preparation of ddU),[38,39] and (c) the Barton deoxygenation involving the cyclic thionocarbonate or the bisxantate, and then treated with tributyltin hydride,[40,41] or alternatively diphenylsilane[42] (Figure 4.13).

The synthesis of modified nucleosides from natural nucleosides is another useful alternative for preparing pharmaceutically active dideoxy nucleosides. The potent antiviral inhibitors ddC, ddG, d4C, and d4G have been obtained from the corresponding protected natural nucleosides, as shown in Figure 4.14.[43]

The chemoenzymatic approach has been also explored for the synthesis of 2',3'dideoxynucleosides. Such is the case of the antiviral 2',3'-dideoxyguanosine, which was synthesized from guanosine in 40% overall yield using as a key step the commercially available mammalian adenosine deaminase (ADA) (Figure 4.15).[44]

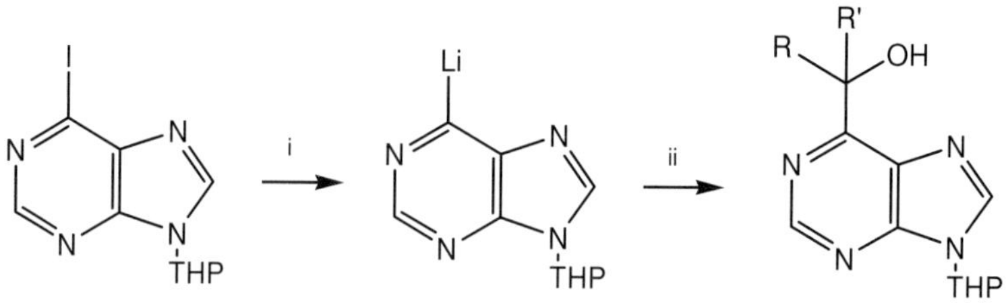

i) BuLi, –130°C. ii) RCOR'.

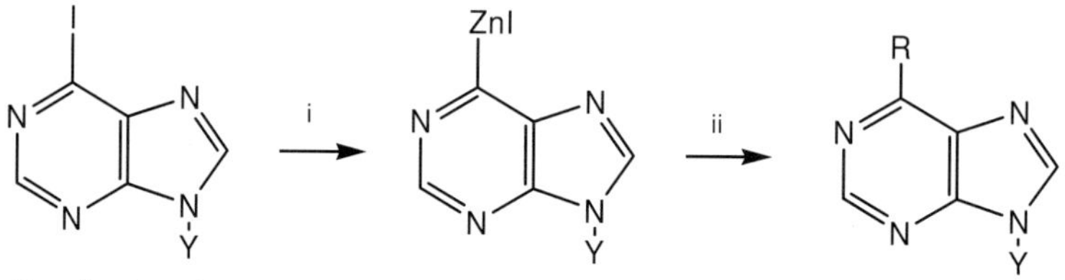

R = alkenyl, aryl

Y = protected sugar

i) Zn, THF. ii) R-X, Pd. cat.

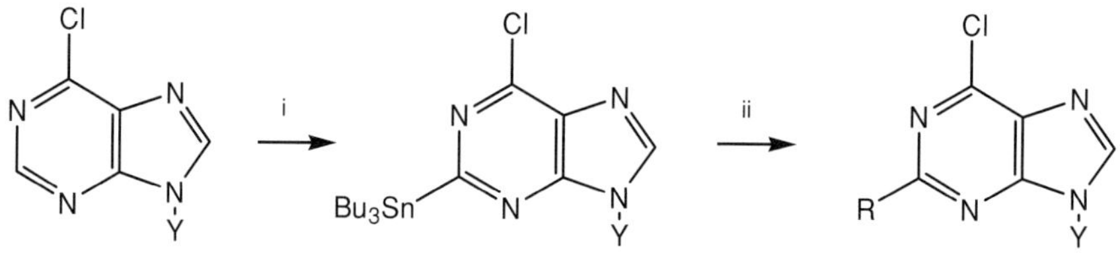

i) a) LMPT. b) Bu₃SnCl. ii) R-X, Pd, cat..

FIGURE 4.8. Cross-coupling reactions for purine modification.

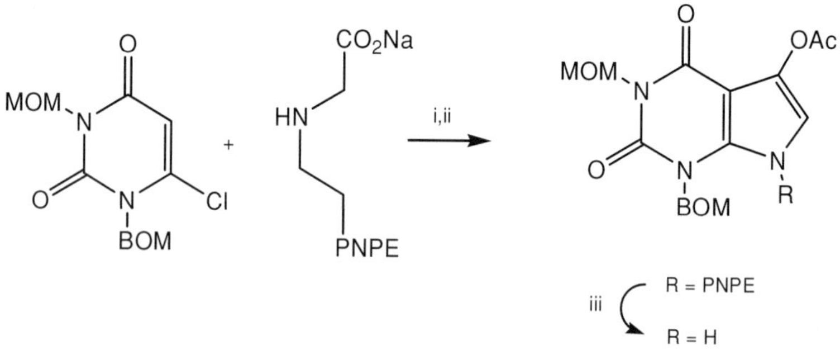

i) EtOH/H₂O, Δ. ii) Ac₂O, n-Pr₃N, 100°C. iii) DBU, CH₃CN, 25°C 81%

FIGURE 4.9. Synthesis of 7-deazapurine analogs.

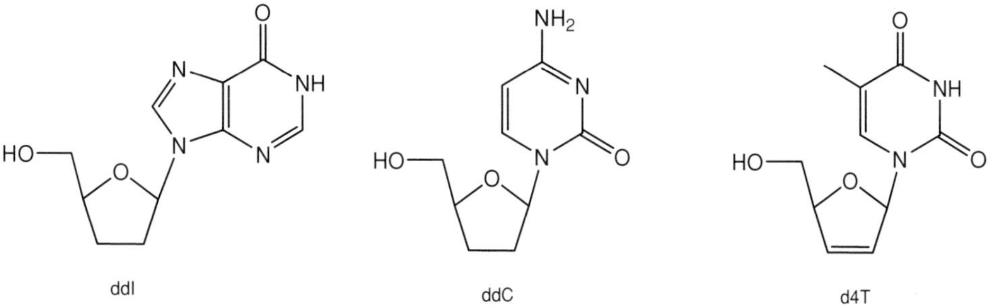

FIGURE 4.10. Anti-AIDS 2′3′-dideoxy nucleosides.

A strategy for preparing D- and L-2′-fluoro-2′3′-unsaturated nucleosides has been described and their anti-HIV activity evaluated. This approach requires 1-acetyl-5-*O*-benzoyl-2,3-dideoxy-3,3-difluoro–D-ribofuranose as key starting material, which was condensed under Vörbruggen's conditions with purines and pyrimidines to provide the corresponding nucleosides. The resulting nucleosides were subjected to β-elimination to generate the fluoro unsaturated nucleosides (Figure 4.16).[45]

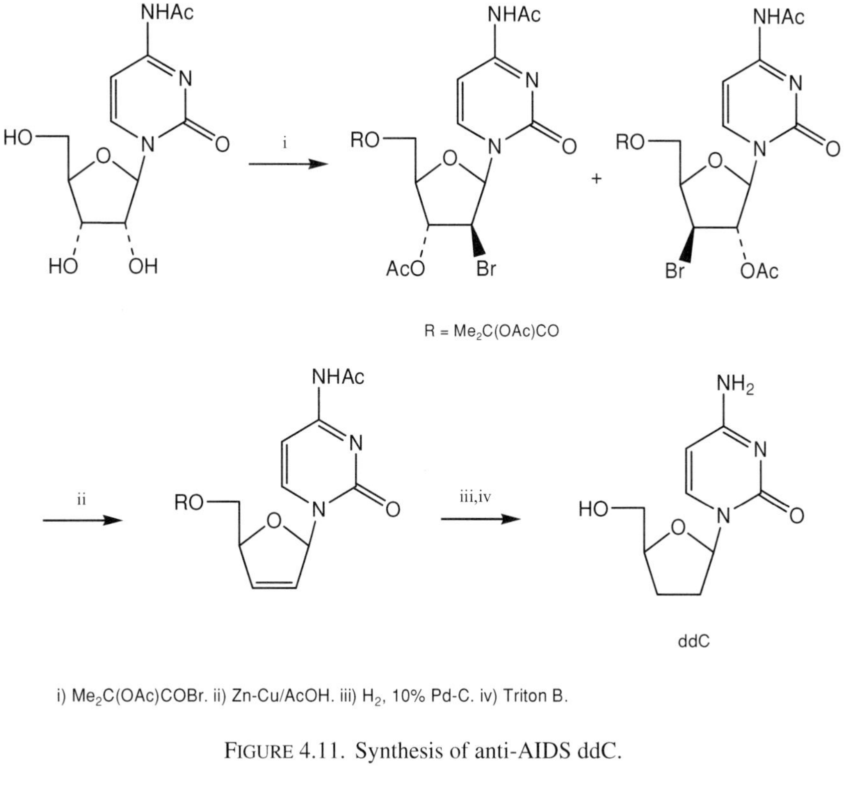

R = Me₂C(OAc)CO

i) Me₂C(OAc)COBr. ii) Zn-Cu/AcOH. iii) H₂, 10% Pd-C. iv) Triton B.

FIGURE 4.11. Synthesis of anti-AIDS ddC.

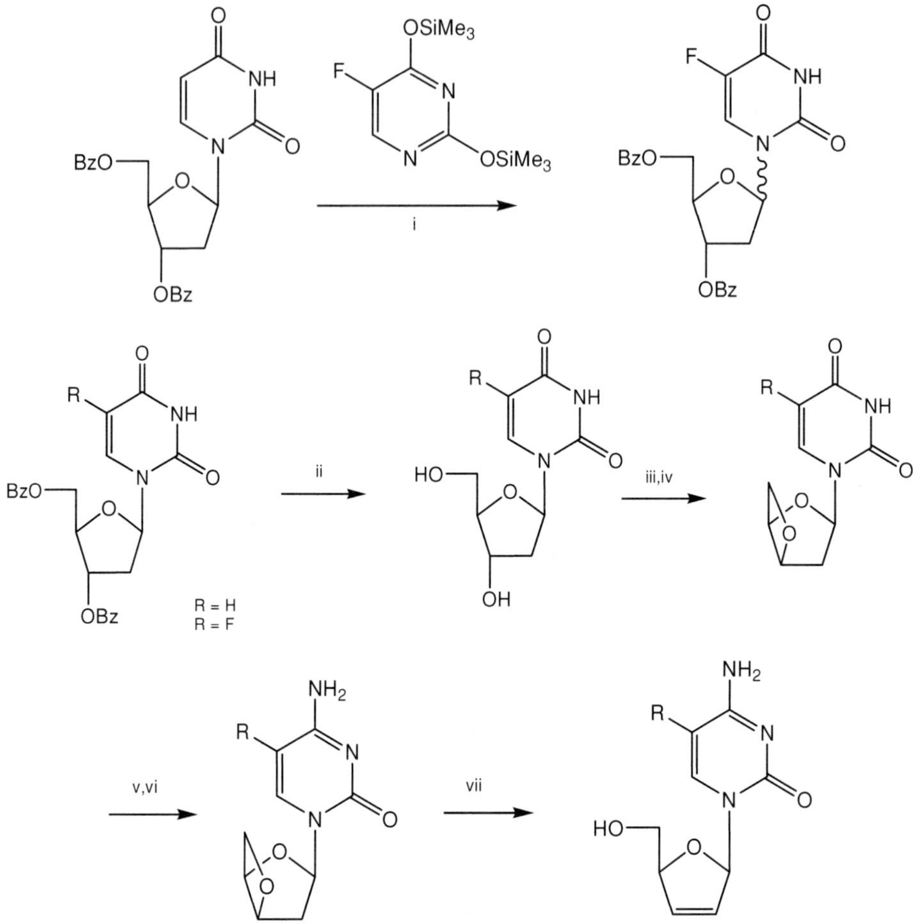

i) $CF_3SO_3SiMe_3$, CH_3CN. ii) $NH_3/MeOH$. iii) MsCl, Py. iv) 1N NaOH, $EtOH/H_2O$. v) 1,2,4-triazole, p-ClC$_6$H$_4$OPOCl$_2$, Py. vi) NH_4OH, dioxane. vii) t-BuOK, DMSO.

FIGURE 4.12. Synthesis of anti-hepatitis B virus β-L-d4C and β-L-Fd4C.

4.1.2.2   2′Deoxynucleosides

The Barton deoxygenation provides another useful method for preparing 2′- and 3′-deoxynucleosides (obtained as a mixture) and involves as a key step the hydride reduction of the cyclic thionocarbonate with tributyltin hydride.[42] On the other hand, 2′-monotosylate nucleoside, when treated with excess of lithium triethyl-borohydride, produces the 2′-deoxy-3′β-hydroxy nucleoside in high yield (Figure 4.17).[46]

2′-deoxynucleosides have been obtained from starting materials of different composition such as α,β-unsaturated aldehydes[47] chiral epoxy alcohols,[48] butenolides,[49,50] and polyfunctionalized acetals, among others.[51]

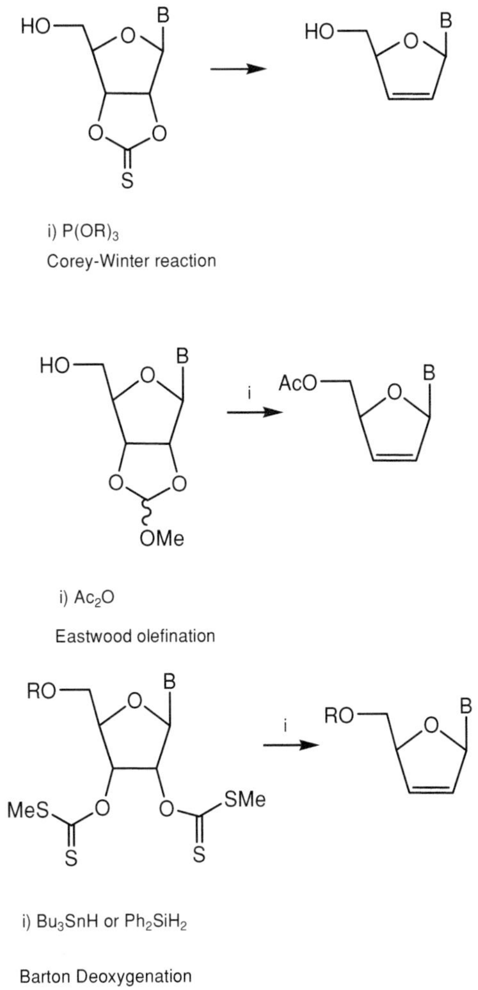
HO
B
O
O
O
S
HO
B
O
i) P(OR)3
Corey-Winter reaction
HO
B
O
i
AcO
B
O
O
O
OMe
i) Ac2O
Eastwood olefination
RO
B
O
i
RO
B
O
MeS
O
O
SMe
S
S
i) Bu3SnH or Ph2SiH2
Barton Deoxygenation

FIGURE 4.13. Alternative procedures for preparing 2′3′-unsaturated nucleosides.

The remarkable 2′-deoxynucleoside AZT widely prescribed as an anti-AIDS drug was originally prepared from thymidine by Horwitz and coworkers,[52] and since then, several other syntheses have been developed, some of them starting with either a nucleoside or a sugar derivative,[5356] and others relaying on the use of noncarbohydrate starting materials.[56,57]

The procedure developed by Chu et al.[50] consisted of the use of mannitol as staring material, which was subsequently transformed to provide the protected key intermediate 3′azide-2′deoxyribofuranose. The next step involved the coupling reaction with silylated thymine under Vörbruggen's conditions to produce an anomeric mixture of nucleosides in 66%. Final desilylation and separation by chromatography column provided AZT in overall yield of 25% from the furanoside intermediate (Figure 4.18).

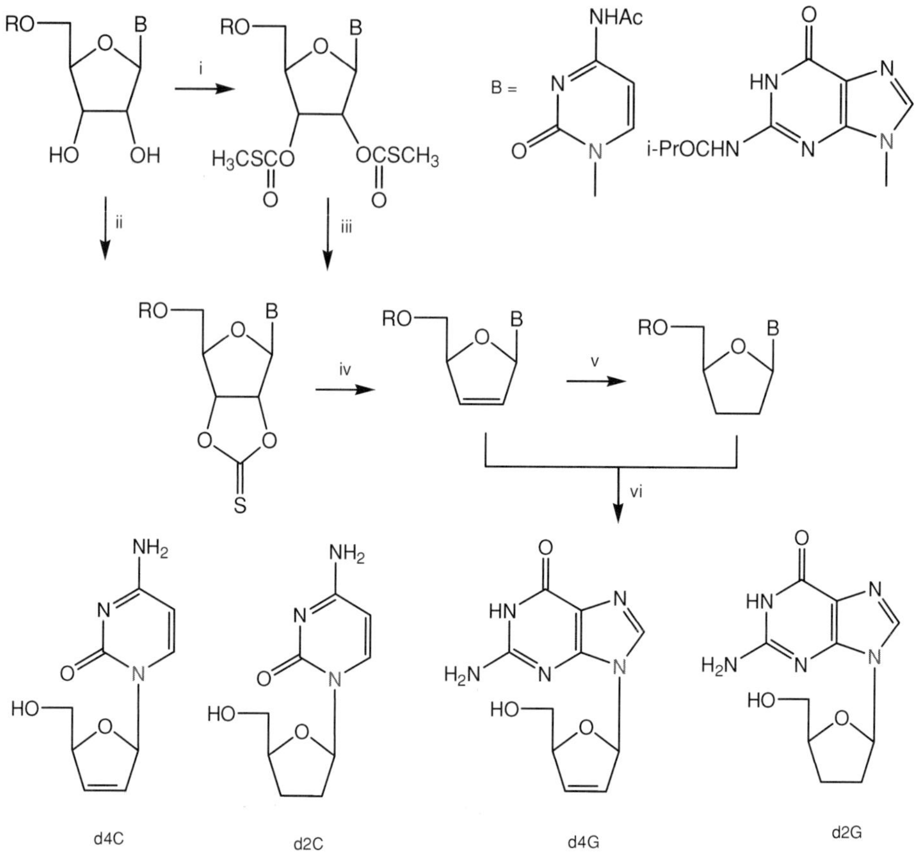

i) NaOH, CS$_2$/CH$_3$I. ii) (Im)$_2$C=S. iii) Bu$_3$SnH. iv) (EtO)$_3$P. v) H$_2$, Pd-C. vi) NH$_3$/MeOH.

FIGURE 4.14. Antiviral modified nucleosides from natural sources.

Another possibility was described by Hager and Liotta involving the coupling re-action between the azido diol intermediate and silylated thymine under Vörbruggen conditions to yield a diastereomeric mixture of azido diol nucleoside. Finally, when exposed to concentrated acidic conditions, the open form is converted into the βanomer of AZT in 67% yield (Figure 4.19).[57]

Transglycosidic reaction mediated by a deoxyribosyl transferase obtained from *E.coli* has been used in the synthesis of 3′-azido-2′,3′-dideoxyguanosine. The en-zymatic reaction occurs between AZT, which acts as glycosyl donor, with substi-tuted 2-amino-6-purines to generate the desired purine nucleoside and thymine as byproduct (Figure 4.20).[58]

4.1.2.3   3′-Deoxynucleosides

These deoxynucleosides may be readily prepared from 3′-O-tosylate via a [1,2]hy-dride shift from C3′ to C2′ position with accompanying inversion of the C2′ center

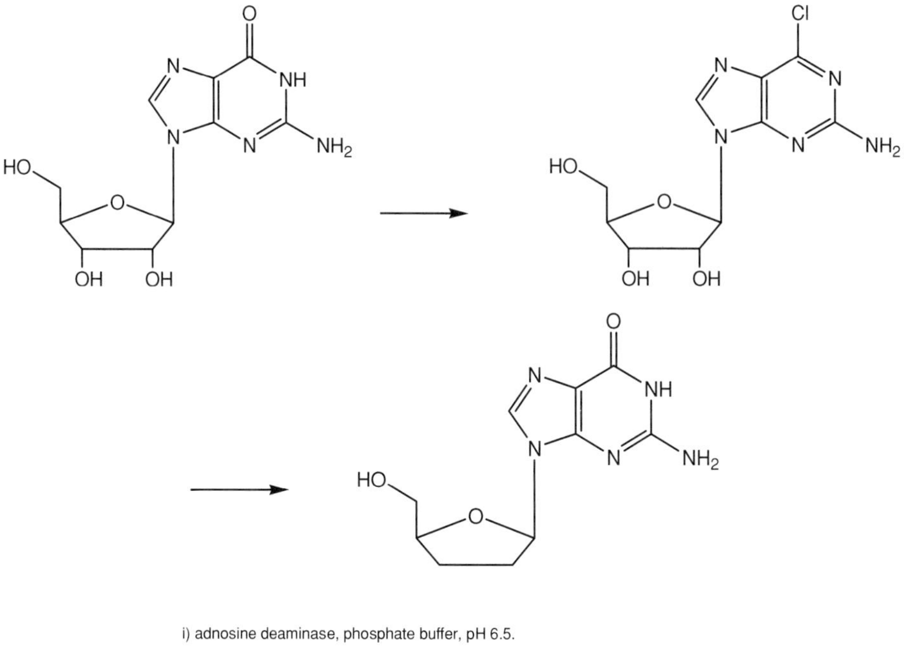

i) adnosine deaminase, phosphate buffer, pH 6.5.

FIGURE 4.15. Chemoenzymatic synthesis of 2′,3′-dideoxyguanosine.

affording a 3′-ketone that was stereoselectively reduced by the hydride to produce 3′-deoxynucleoside (Figure 4.21).[46,2]

Also, 3′-deoxyguanosine was synthesized by an enzymatic transglycosylation of 2,6-diaminopurine using 3′-deoxycytidine as a donor of the sugar moiety. The diaminopurine nucleoside was transformed to 3′-deoxyguanosine by the action of adenosine deaminase (Figure 4.22).[59]

Lodenosine   [9-(2,3-dideoxy-2-fluoro-β-D-threo-pentofuranosyl)]   adenine (FddA) is a reverse transcriptase inhibitor with activity against HIV. This purine analog was evaluated as one of the most selective inhibitors in a series of 2′3′-dideoxyadenosines, although less active than ddA. An efficient method was developed starting from chloropurine riboside, which was tritylated and selectively benzoylated at 3′-position. Before fluorination the 2′-hydroxyl group was converted to imidazolesulfonate or trifluoromethanesulfonate. Fluorination proceeds smoothly with 6 equiv. of $Et_3N._3HF$ at reflux in 88% yield. Simultaneous 6-amination and 3′-debenzoylation was done with ammonia in high yield. Elimination of the 3′-hydroxy group was carried out under the Barton-McCombie procedure involving the formation of the 3′-O-thiocarbonyl followed by silane treatment. Final removal of trityl group afforded FddA (Figure 4.23).[60]

4.1.2.4   4′-Substituted Nucleosides

4′-Substituted nucleosides have attracted much attention because of the discovery of potent anti-HIV agents 4′-azido- and 4′-cyano thymidine (Figure 4.24).

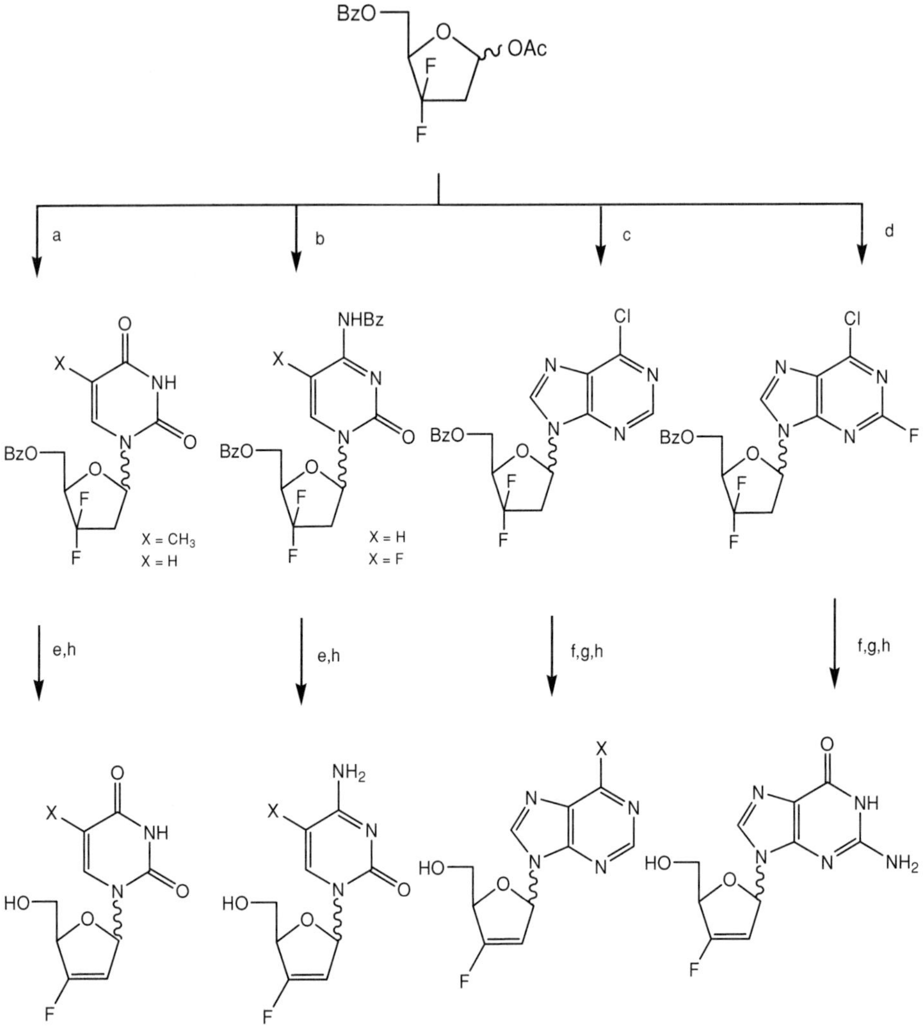

i) silylated thymine, TMSOTf, MeCN. b) silylated N^4-Bz-cytosine derivatives, TMSOTf, MeCN. c) silylated 6-chloropurine, TMSOTf, MeCN. d) silylated 6-Cl-2-F-purine, TMSOTf, MeCN. e) NH$_3$/MeOH, r.t. f) NH$_3$/MeOH, 90°C. g) HSCH$_2$CH$_2$OH, MeONa, MeOH, reflux. h) t-BuOK, THF.

FIGURE 4.16. Preparation of D- and L-2′-fluoro-2′3′-unsaturated nucleosides.

One procedure involves the epoxidation of the exoglycal with dimethyldioxirane and ring opening of the resulting 4′,5′-epoxynucleosides to produce with high stereoselectivity the 4′-C-branched nucleosides (Figure 4.25).[61]

Likewise, other 4′-substituted nucleosides such as 4′-C-Ethynyl-β-D-arabino and 4′-C-Ethynyl-2′-deoxy-β-D-ribopentofuranosyl pyrimidines have been reported by a different approach outlined in Figure 4.26.[62]

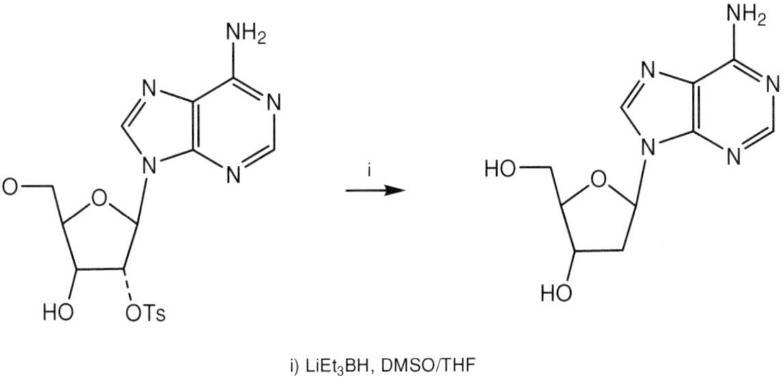

i) LiEt$_3$BH, DMSO/THF

FIGURE 4.17. The Barton deoxygenation for preparing 2'-deoxynucleosides.

4.1.3   Complex Nucleosides

The hypermodified Q base Queuine found in tRNA of plants and animals has been strongly associated with tumor growth inhibition. Three different approaches for preparing queuine have been described,[63,65] the more recent in 11 steps from ribose. Completion of the synthesis involved the condensation of bromo aldehyde intermediate with 2,3-diamino-6-hydroxypyrimidine to give the desired heterocyclic product in 45%. Final removal of protecting groups provided Q base (Figure 4.27).

Capuramycin is a complex nucleoside antibiotic isolated from *Streptomyces griseus* 446-S3, which exhibits antibacterial activity against *Streptococcus pneumoniae* and *Mycobacterium smegmatis* ATCC 607. The total synthesis was reported by Knapp and Nandan[66] consisting of the glycosylation reaction between the key thioglycoside donor and silylated pyrimidine to produce the corresponding L-*talo*-uridine. The next glycosidic coupling reaction was carried out with L-*talo*-uridine and imidate glycosyl donor under TMSOTf conditions to afford

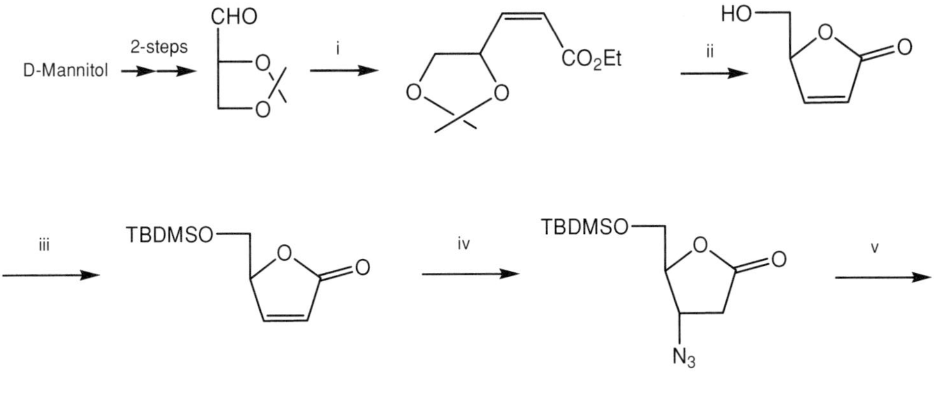

FIGURE 4.18. Synthesis of AZT from mannitol.

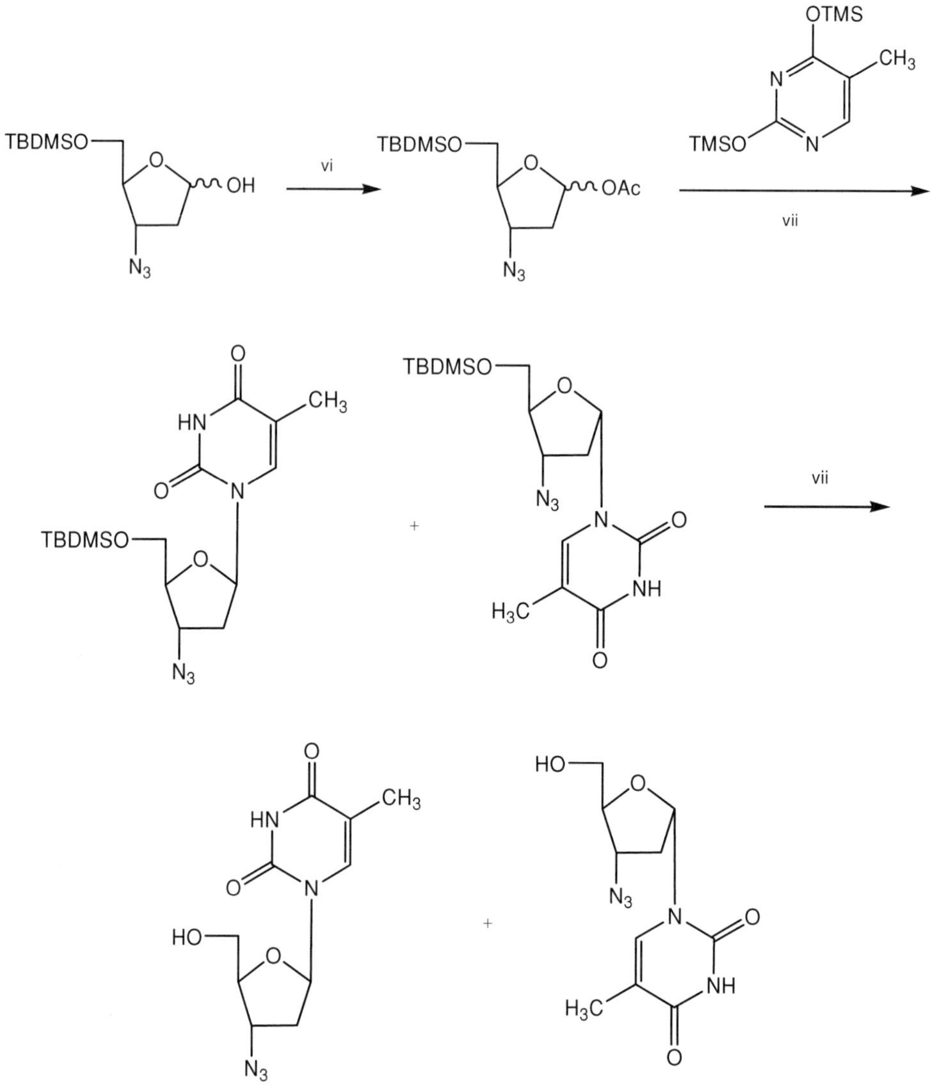

i) Ph$_3$P=CHCO$_2$Et, MeOH, 0°C. ii) HCl dil. iii) t-Bu(Me)$_2$SiCl. imidazole, DMF. iv) LiN$_3$, THF, AcOH, H$_2$O. v) DIBAL, CH$_2$Cl$_2$, -78°C. vi) Ac$_2$O, Py. vii) TMS-triflate, ClCH$_2$CH$_2$Cl. viii) n-Bu$_4$NF, THF.

FIGURE 4.18. (*Continued*).

the disaccharide nucleoside. Further transformations led to the target molecule (Figure 4.28).

Capuramycin has been also chemically transformed in an attempt to extend the antibacterial spectrum. Thus, radical oxygenation gave unexpected lactone

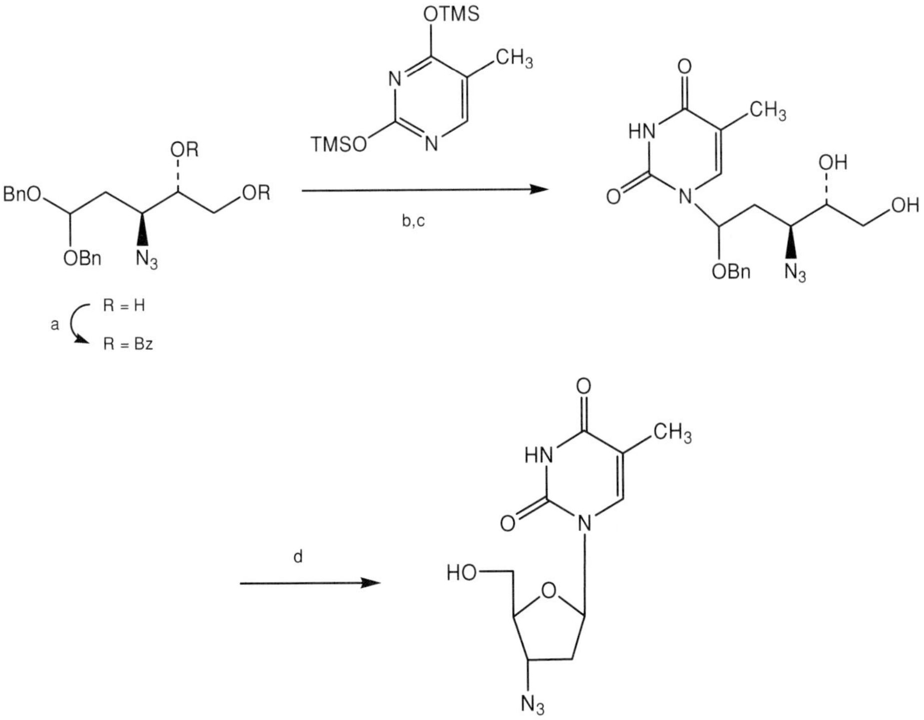

i) PhCOCl (2.2 equiv.), NEt$_3$, DMAP, CH$_2$Cl$_2$. b) (CH$_3$)$_3$SiOTf, ClCH$_2$CH$_2$Cl. c) NaOH (2 equiv.), MeOH. b) 4.7 N H$_2$SO$_4$ in MeOH.

FIGURE 4.19. Synthesis of AZT from azido diol intermediate.

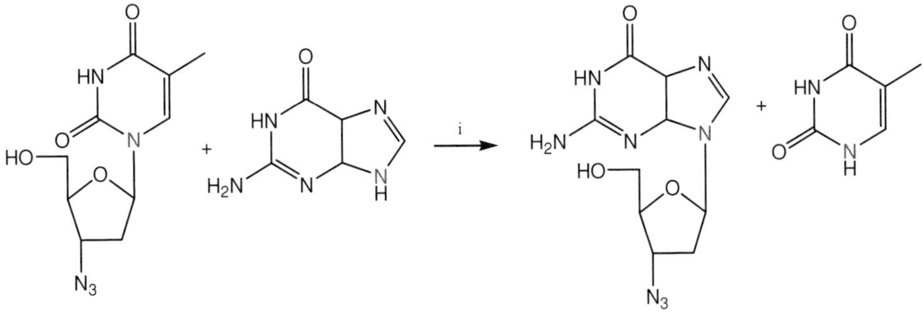

i) glycosyltransferases, pH 6.0, 50°C.

FIGURE 4.20. Enzymatic synthesis of 3′-azido-2′,3′-dideoxyguanosine.

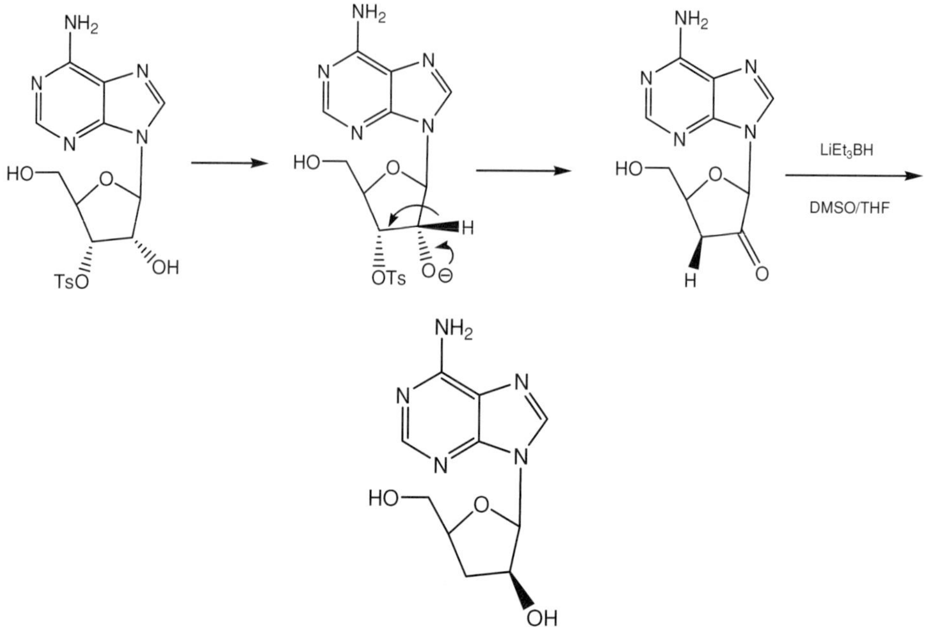

FIGURE 4.21. Method for preparation of -3′-deoxynucleoside.

in moderate yield via an intramolecular radical ArC glycosylationlactonization reaction (Figure 4.29).[67]

Synthestic studies of unique class Tunicamycin antibiotics leading to the preparation of (+)Tunicaminyluracil, (+)Tunicamycin-V, and 5′-*epi*Tunicamacyn-V were described by Myers et al.[68] The key features are the development and application of a silicon-mediated reductive coupling of aldehydes, the allylic alcohols to construct the undecose core of the natural product, and the development of an

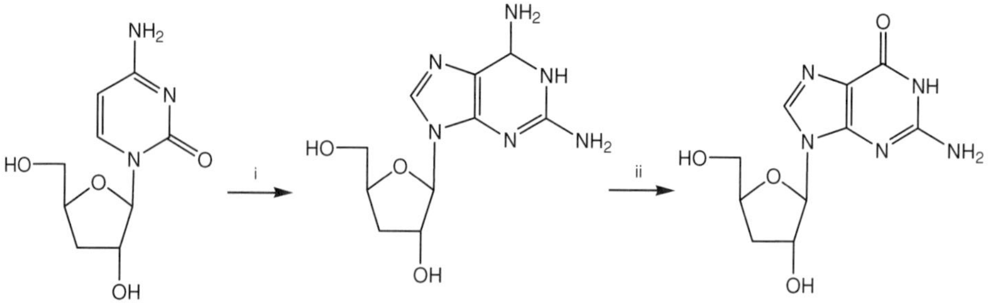

i) 2,6-diaminopurine, *E. coli* BM-11 and BMT-4D/1A, K-phosphate buffer, 52°C, 26 h, 64%. ii) Adenosine deaminase (ADase), r.t., 16 h, 68%.

FIGURE 4.22. Enzymatic synthesis of 3′-deoxyguanoside.

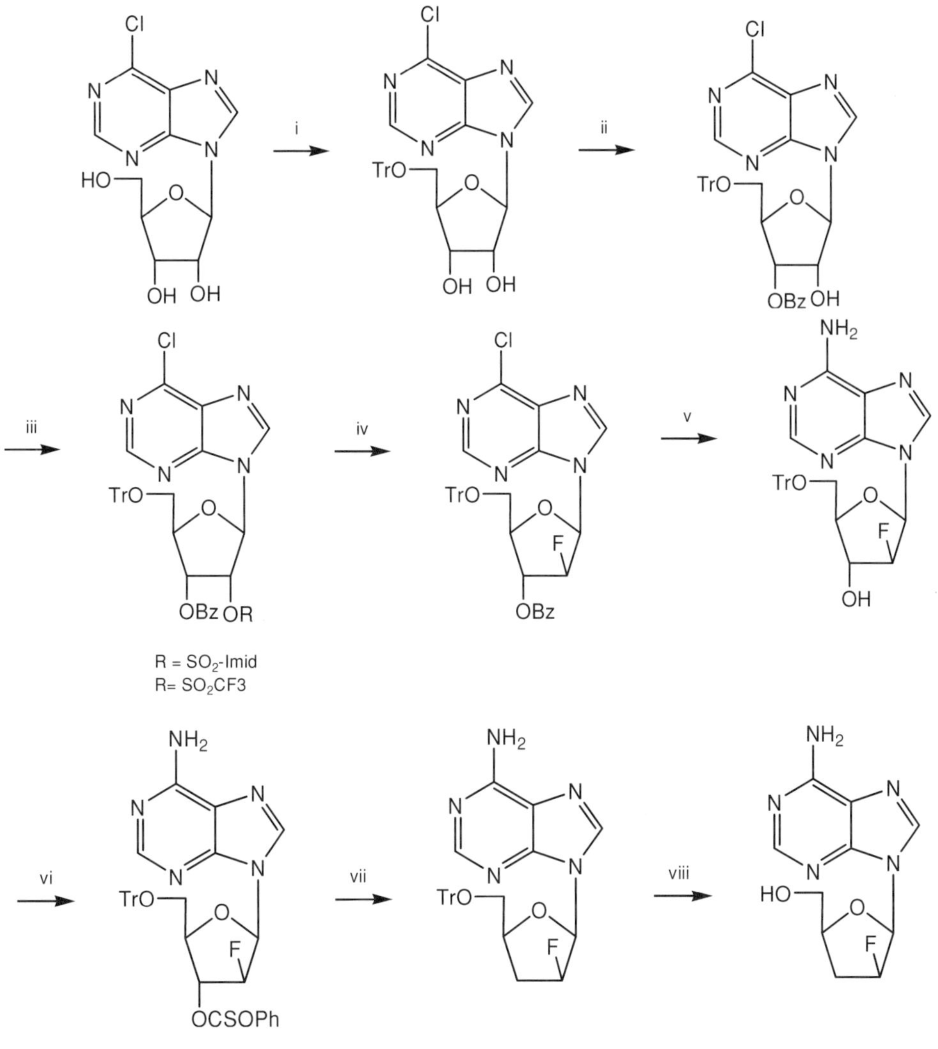

i) TrCl-iPrNH, DMF, 79%. ii) a) BzCl-Py, toluene. b) cat. Et₃N, toluene, 70%. iii) a) SO₂Cl₂-Py, CH₂Cl₂, b) imidazole. or CF₃SO₂Cl, DMAP, toluene. iv) Et₃.3HF, Et₃N, 70 and 78%. v) NH₃-MeOH, toluene 98%. vi) ClC(S)(OPh), DMAP, CH₃CN, 92%. vii) Ph₂SiH₂, AIBN, dioxane, 81%. viii) 80% AcOH, 100°C, 85%.

FIGURE 4.23. Preparation of antiviral 2′3′-fluoro dideoxyadenosine FddA.

efficient procedure for the synthesis of the tetrahalose glycosidic bond within the antibiotic (Figure 4.30).

4.1.3.1   Fused Heterocyclic Nucleosides

Selective and potent antiVaricella Zoster Virus (VZV) bicyclic furanopyrimidine deoxynucleosides were synthesized. The bicyclic formation was performed by

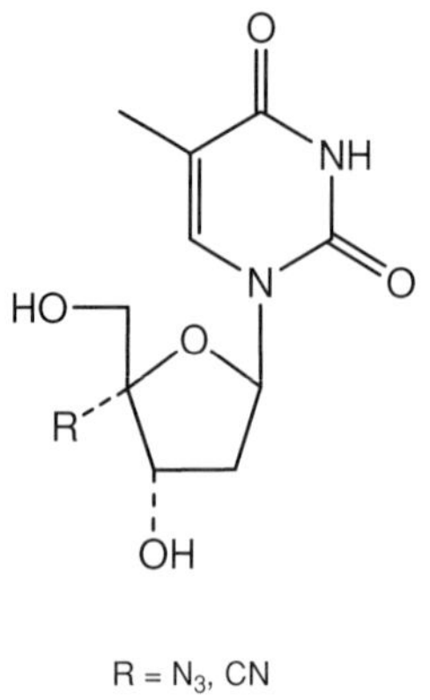

FIGURE 4.24. Structure of potent anti-HIV -4'-substituted nucleosides.

R = N$_3$, CN

palladium-catalyzed coupling of aryl acetylenes with 5-iodo-2'-deoxyridine affording the desired fused furanenucleoside (Figure 4.31).[69]

Triciribine is a tricyclic nucleoside with antineoplasic and antiviral properties, synthesized in an improved fashion from 6-Bromo-5-cyanopyrrolo [2,3-d] pyrimidin-4-one intermediate. Series of transformations including N-glycoside coupling reaction afforded 4-amino-5-cyano-7-[2,3,5-tri-O-benzoyl-β-D-ribofuranosyl] pyrrolo [2,3-d] pyrimidine that was then converted to the desired tricyclic nucleoside (Figure 4.32).[70]

4.2   C-Nucleosides

These modified nucleosides are structurally distinct to their counterparts N-nucleosides because of the presence of a C-C linkage instead of C-N between

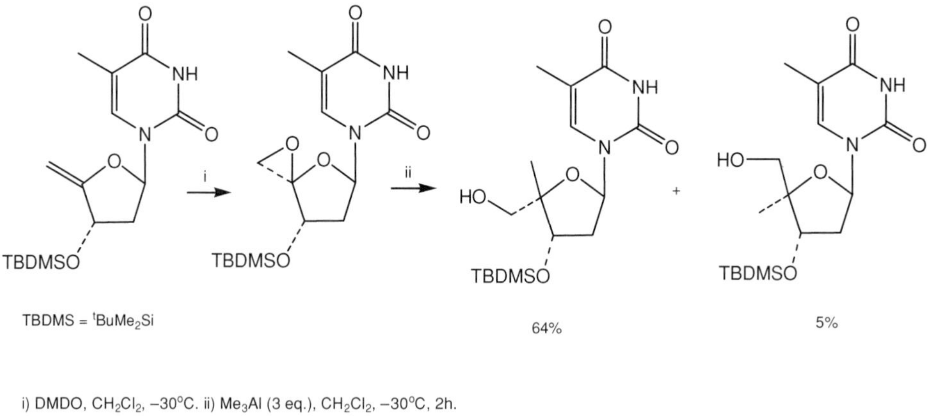

TBDMS = ^{t}BuMe$_2$Si

64%   5%

i) DMDO, CH$_2$Cl$_2$, −30°C. ii) Me$_3$Al (3 eq.), CH$_2$Cl$_2$, −30°C, 2h.

FIGURE 4.25. Ring opening of 4',5'-epoxynucleosides.

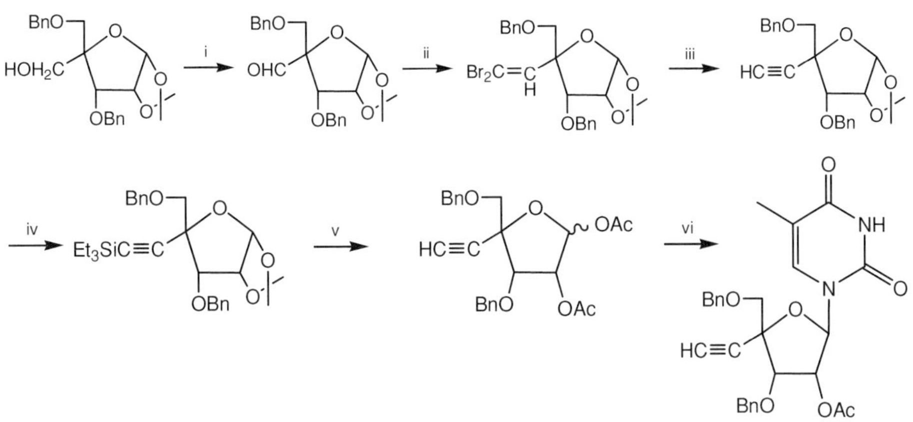

i) (COCl)$_2$, DMSO, Et$_3$N, CH$_2$Cl$_2$. ii) CBr$_4$, PPh$_3$, CH$_2$Cl$_2$. iii) n-BuLi, THF.
iv) n-BuLi, THF, then Et$_3$SiCl. v) a) 70% AcOH, TFA. b) Ac$_2$O, Py. vi) N,O-bis
(trimethylsilyl)acetamide, thymine, CH$_2$Cl$_2$, reflux, 1h, 96%.

FIGURE 4.26. Synthesis of 4'-C-Ethynyl-β-Darabino and 4'-C-Ethynyl-2'-deoxy-β-
Dribopentofuranosyl pyrimidines.

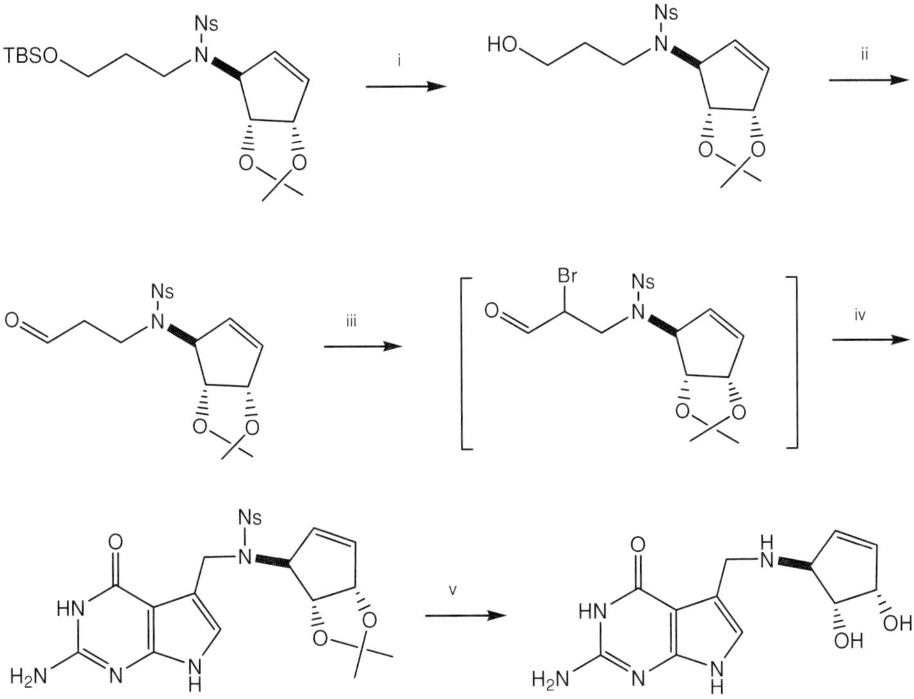

i) TBAF, THF, 87%. ii) TEMPO, NaOCl, KBr, CH$_2$Cl$_2$, 88%. iii) TMSBr, DMSO, MeCN. iv) NaOAc, H$_2$O/MeCN, 45%.
v) a) HSCH$_2$CH$_2$OH, DBU, DMF, 46%. b) HCl, MeOH, 84%.

FIGURE 4.27. Synthesis of hypermodified base Queuine.

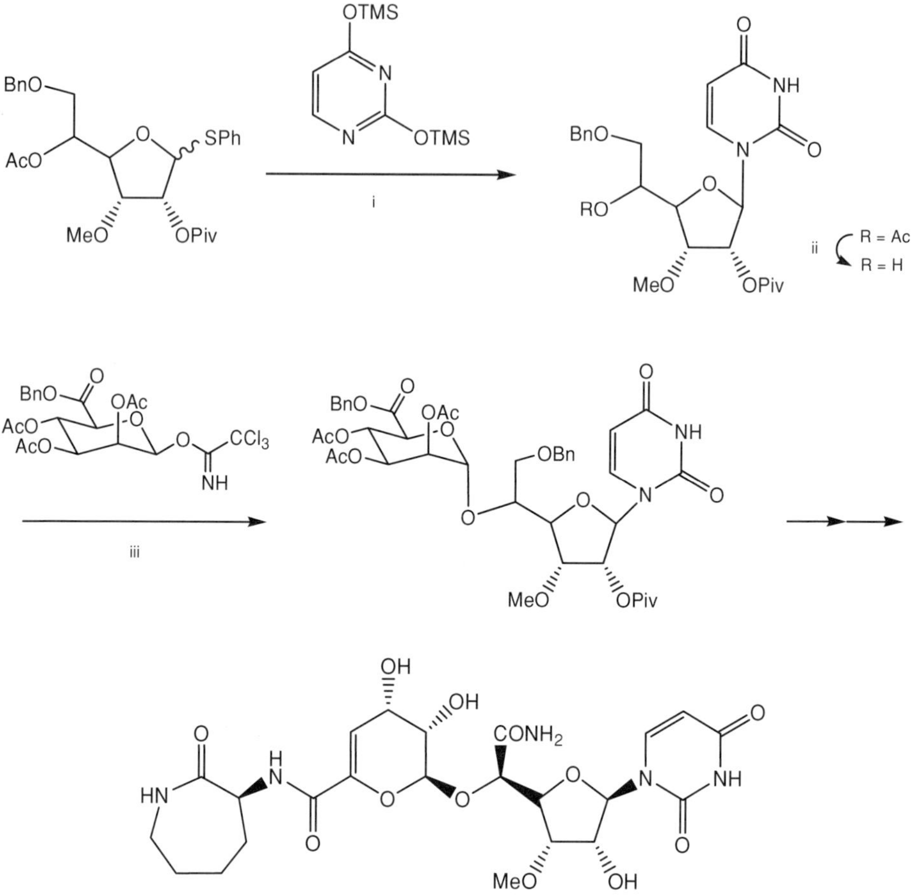

i) NIS, TfOH, CH$_2$Cl$_2$, −20°C. ii) NaOMe, MeOH, 77%. iii) TMS-OTf, CH$_2$Cl$_2$, −25°C, 16 h, 85%.

FIGURE 4.28. Synthesis of Capuramycin.

the furanoside and the heterocyclic aglycon. Their source could be either naturally occurring (pyrazomycin, showdomycin, formycin) or synthetic (thiazofurin), having in either case significant antitumor and antiviral activity. Also, some of them have been found in tRNA codons (pseudouridine) and others (tiazofurine and oxazofurine) designed as competitive inhibitors of cofactor nicotin adenin dinucleotide (Figure 4.33).

An early approximation for the preparation of C-nucleosides proposed two basic possibilities depending on the nature of the atoms surrounding the C-C bond (Figure 4.34).[71]

1. If there is one heteroatom adjacent to the C-glycosidic bond, for example, tiazofurine, formicine, pyramicine.

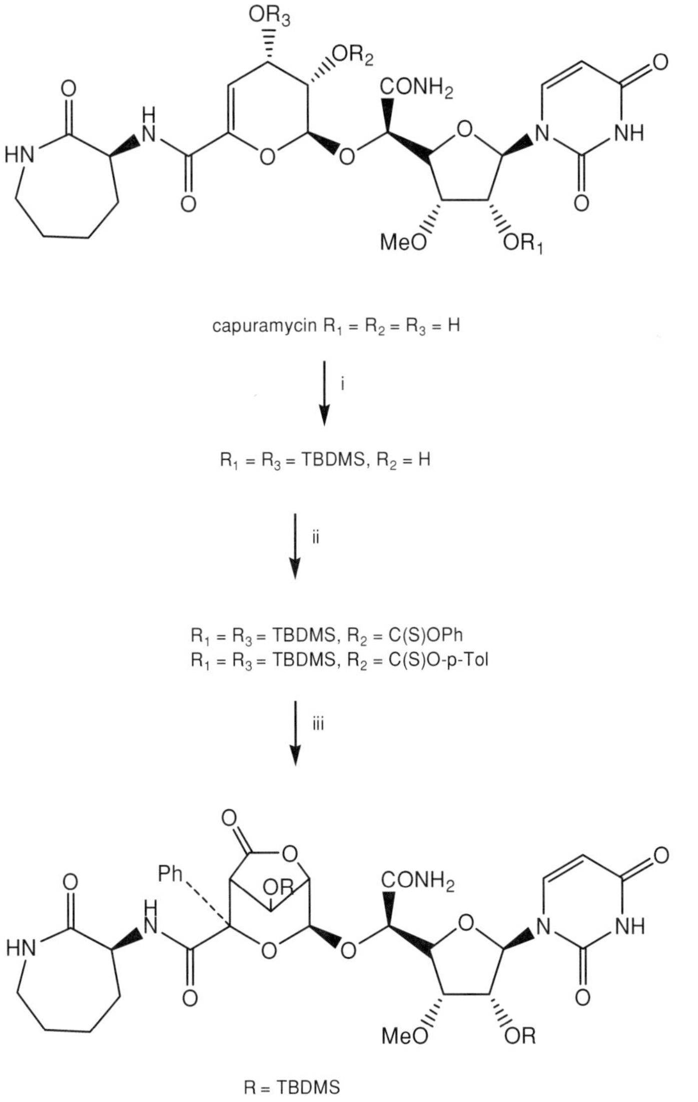

capuramycin $R_1 = R_2 = R_3 = H$

i

$R_1 = R_3 = TBDMS, R_2 = H$

ii

$R_1 = R_3 = TBDMS, R_2 = C(S)OPh$
$R_1 = R_3 = TBDMS, R_2 = C(S)O-p-Tol$

iii

R = TBDMS

i) TBDMS-Cl, pyridine. ii) $C_6H_5OC(S)Cl$, DMAP, CH_2Cl_2. iii) Bu_3SnH, AIBN, PhMe, reflux

FIGURE 4.29. Chemical transformations of capuramycin

2. If there is no heteroatom adjacent to the C-glycosidic bond.

Alternatively, other authors considers three general pathways for preparing C-nucleosides depending on the precursor employed as starting material.[72]

An early synthesis of modified C-nucleoside from naturally occurring pseu-douridine was carried out via ring opening with ozone to generate an intermediate,

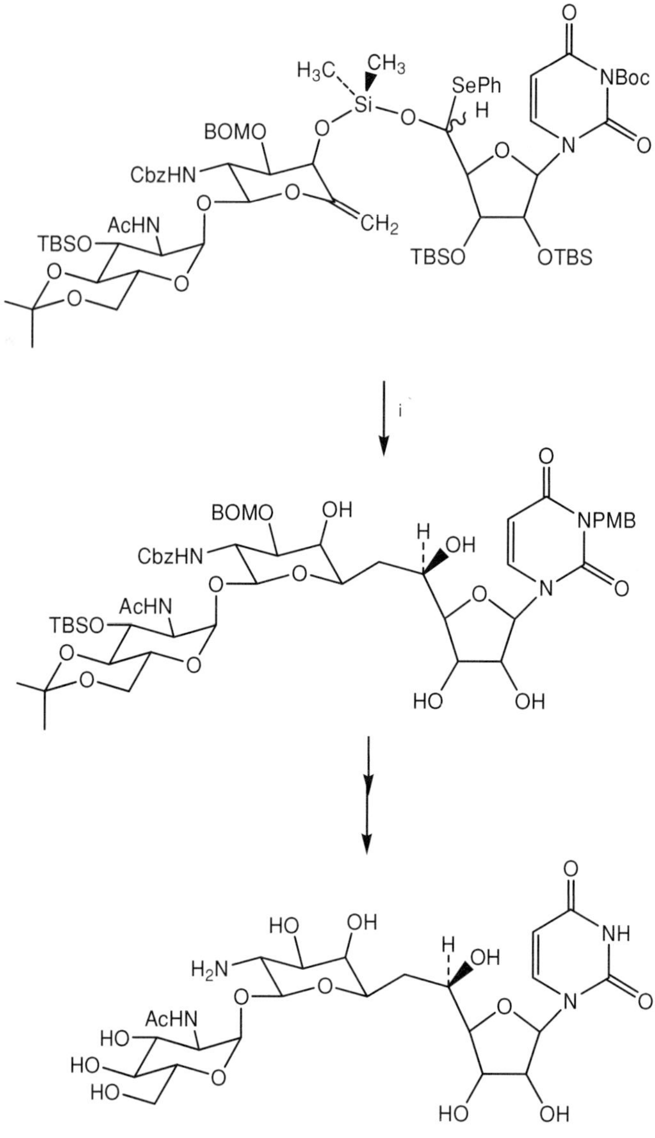

i) triethyborate, Bu$_3$SnH, toluene, 0°C. 2 h. b) KF.H$_2$O, MeOH. 60%.

FIGURE 4.30. Key step for the synthesis of Tunicamycin antibiotic.

which was treated with thiosemicarbazone to afford 6-azathiopseudouridine. Treatment with iodomethane in acid medium produced the desired C-nucleoside as shown in Figure 4.35.[73]

The synthesis of the C-nucleoside pseudouridine was reported by Asburn and Binkley,[74] involving the condensation between 5-O-acety-2,3-O-

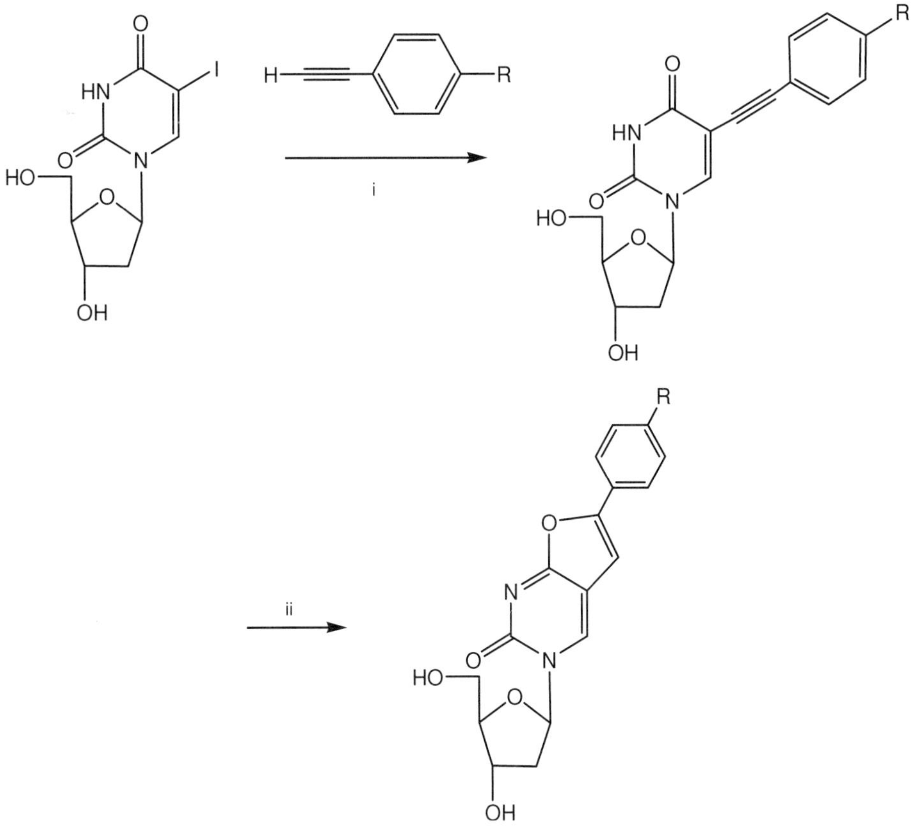

i) Pd(PPh$_3$)$_4$, iPr$_2$EtN, CuI, DMF, r.t, 19 h. ii) Et$_3$N/MeOH. CuI, Δ , 4h.

FIGURE 4.31. Synthesis of bicyclic furano pyrimidine.

isopropylidene-D-ribonolactone with 2,4-dibenzyloxypirimidin-5-il lithium to provide the condensation product, which was subjected to hydride reduction and hydrogenolysis to yield pseudouridine (Figure 4.36).

Antitumor C-nucleoside Tiazofurine was synthesized by Robins et al.[75] from 2,3,5-tri-O-benzoyl-β–Dribofuranosyl cyanide, which undergoes ring closure under conditions described in Figure 4.37.

A new report for the synthesis of Tiazofurine is described, avoiding the use of H$_2$S gas, which is unsafe on large-scale production. The synthesis initiates with the preparation of 1-cyano-2,3-O-isopropylidene-5-O-benzoyl-β-D-ribofuranose, which was reacted with cysteine ethyl ester hydrochloride to give thiazoline derivative in 90%. Further steps including oxidative aromatization under MnO$_2$ in benzene and acetonide deprotection with iodide in methanol produced the desired C-nucleoside (Figure 4.38).[76]

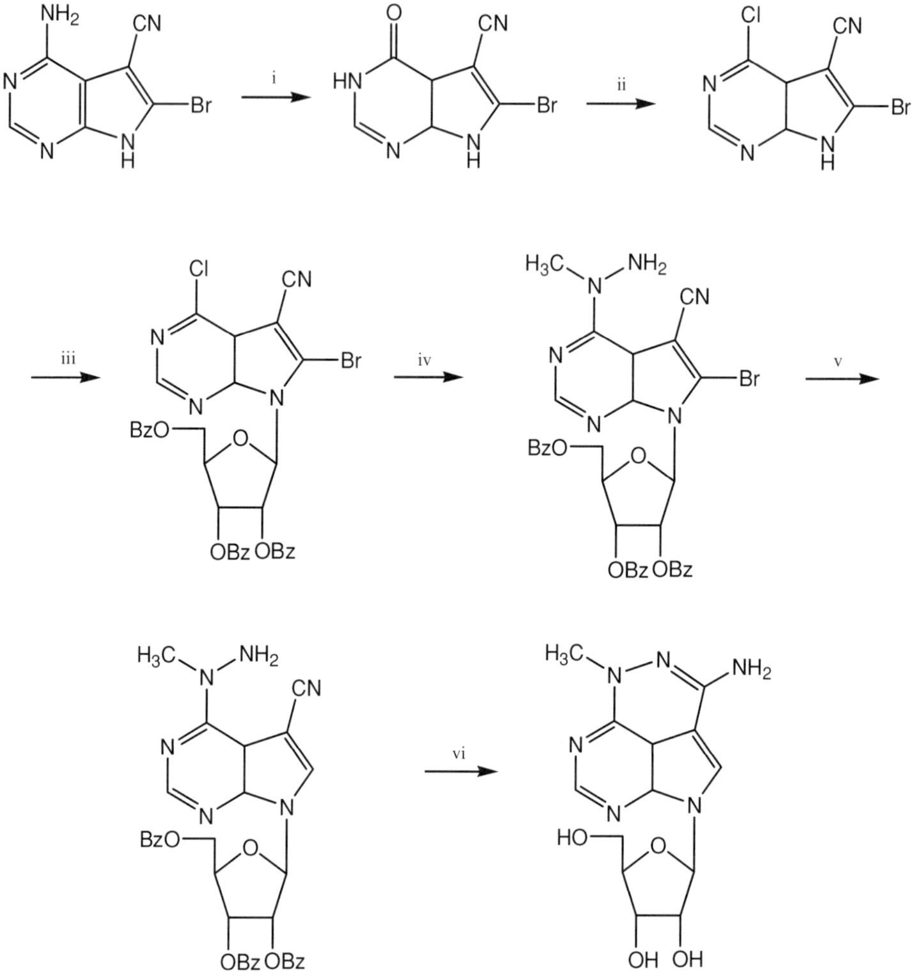

i) NaNO$_2$, AcOH, H$_2$O, ii) POCl$_3$. iii) BSA, CH$_3$CN then 1-O-acetyl-2,3,5-tri-O-benzoyl-b-D-ribofuranoside, TMSOTf. iv) NH$_2$NHCH$_3$, EtOH, CHCl$_3$. v) HCO$_2$NH$_4$, 10% Pd-C, EtOH, reflux. vi) NaOMe, MeOH, reflux.

FIGURE 4.32. Synthesis of tricyclic nucleoside Triciribine.

Another biologically important *C*-nucleoside known as showdomicine was prepared by Trumnlitz and Moffat.[77] The aldehyde used as starting material was converted first to an α–hidroxyacid and then to α–ketoacid. Wittig reaction on this intermediate and Lewis acid catalysis produced ring closure (Figure 4.39).

Pyrazine riboside derivative was synthesized by treatment of glycine riboside with formaldehyde and cyanide (Strecker conditions) to generate cyanide intermediate as a mixture of isomers. Sulfenylation and sodium methoxide treatment produce the *C*-nucleoside (Figure 4.40).[78]

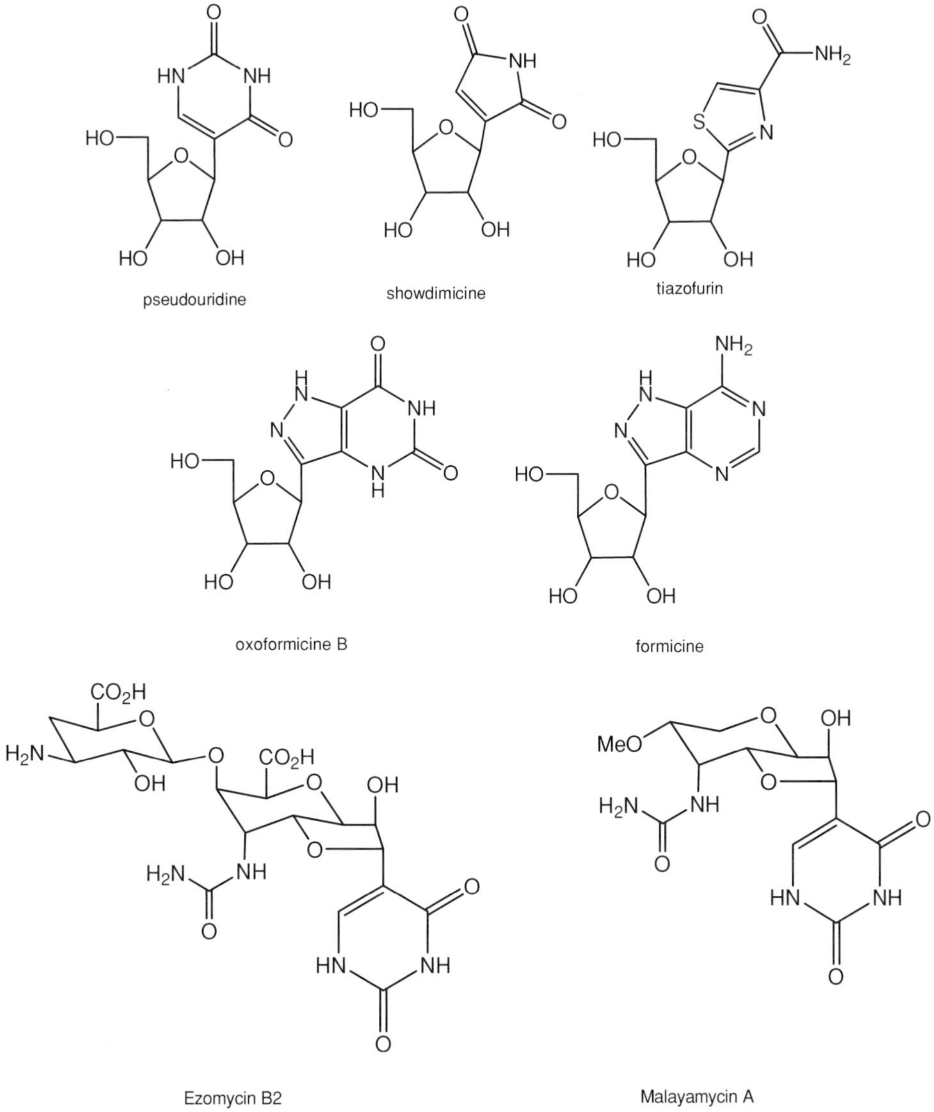

FIGURE 4.33. Biologically active *C*-nucleosides.

Analogs of antiviral C-nucleoside Formicine have been synthesized by using the palladium-mediated glycosidic reaction between the furanoid glycal and the iodinated heterocycle. Similar conditions were used for preparing the pyrimidine analogs (Figure 4.41).[79]

Radical cyclization of ribo-phenylselenoglycoside tethered with propargyl moieties on C-5 hydroxyl group afforded cyclic intermediates potentially useful for the

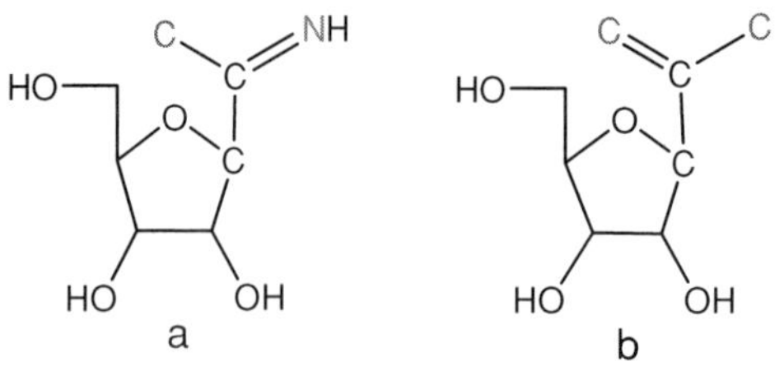

FIGURE 4.34. C-nucleosides partial representations, with and without heteroatom attached to the C-glycosidic bond.

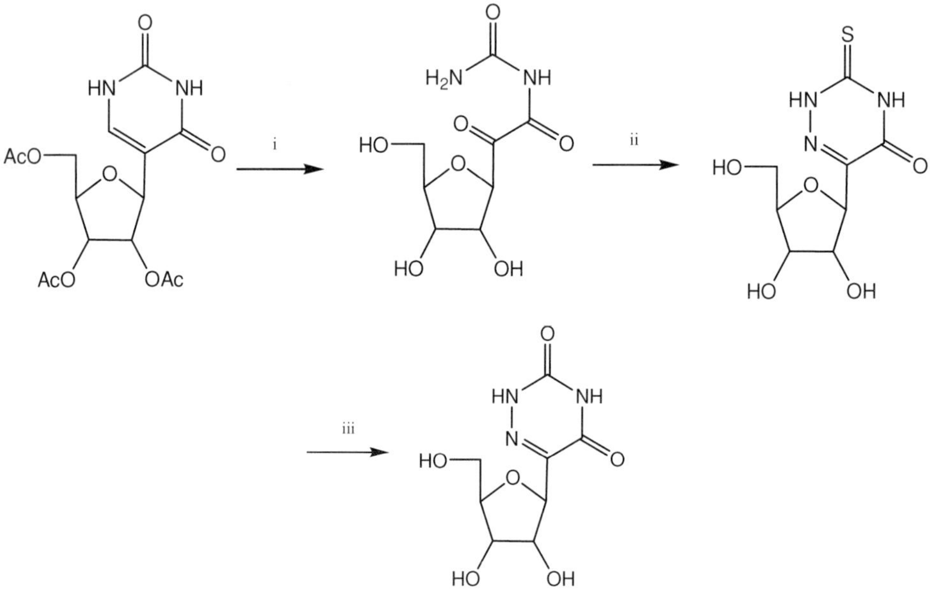

i) O$_3$. ii) NH$_2$NHCNNH$_2$=S. iii) MeI/H$_3$O$^+$

FIGURE 4.35. Preparation of 6azapseudouridine.

synthesis of C-nucleoside derivatives. Propargyl intermediate was prepared from ribo-phenylselenoglycoside via two-step sequence and then under radical reaction conditions (Bu$_3$SnH/AIBN) transformed to the cyclic intermediates in high yields. Further ring opening produced an aldehyde intermediate, which was subjected to a coupling reaction with 1,2-phenylenediamine to produce the pyrazine C-glycoside (Figure 4.42).[80]

Polyhalogenated quinoline C-nucleosides were synthesized as potential antiviral agents. The key step reaction for quinolin-2-one ring formation consisted in the condensation between 2-aminophenoneallose derivative with keteneylidene(triphenyl)-phosphorane in benzene under reflux to provide the desired 6,7-dichloroquinolin-2-one nucleoside in 50% yield (Figure 4.43).[81]

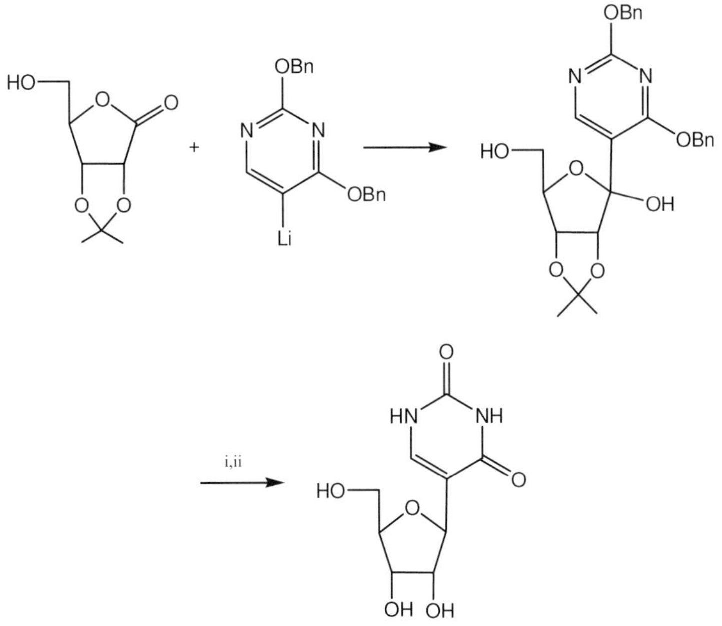

i) NaBH$_4$. ii) a) H$_2$, Pd-C. b) H$_3$O$^+$

FIGURE 4.36. Preparation of pseudouridine.

The Novel bicyclic C-nucleoside Malayamycin A from *Streptomyces malaysiensis* was elegantly synthesized from D-Ribonolactone, which was transformed to the target molecule according to the pathway indicated in Figure 4.44.[82]

4.3 Carbocyclic Nucleosides

This class of modified nucleosides in which the furanose ring has been replaced by a cycloalcane ring (mainly cyclopentane) has been prepared by chemical or

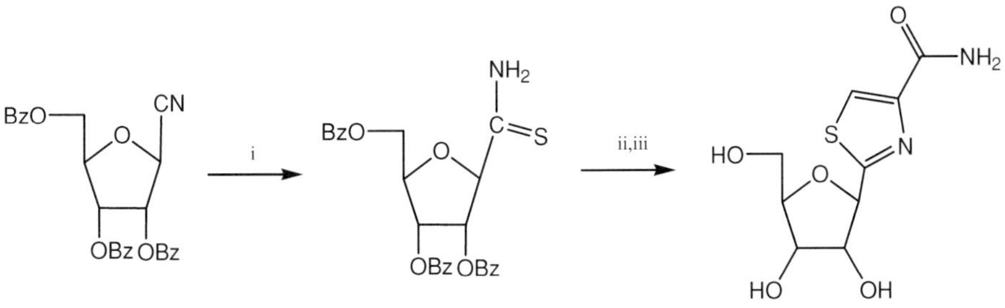

i) H$_2$S, 4-DMAP. ii) ethylbromopyruvate. iii) NH$_3$/MeOH.

FIGURE 4.37. Synthesis of tiazofurine.

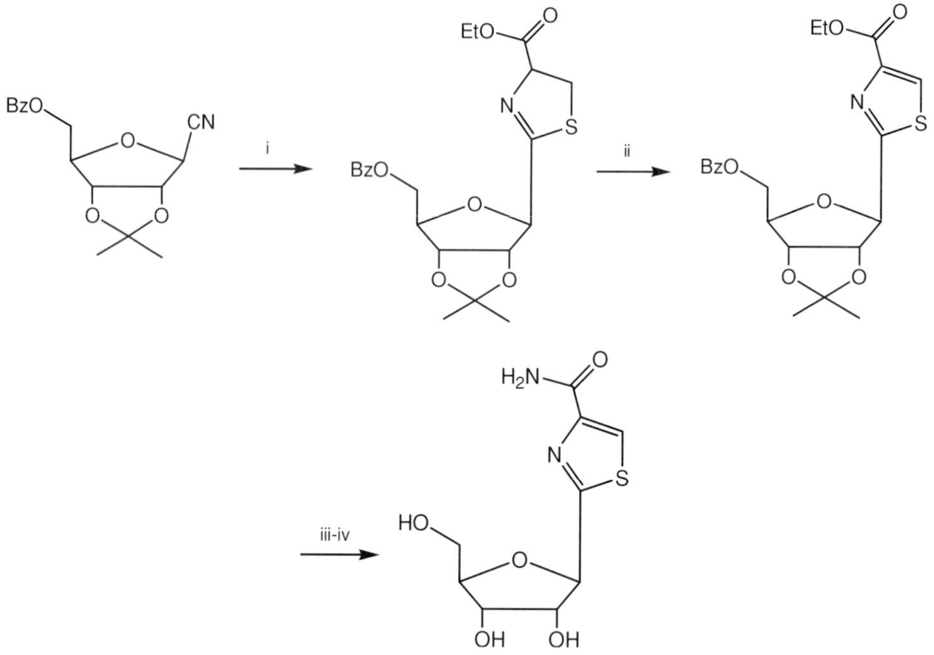

i) Cysteine ether ester hydrochloride/TEA. ii) MnO_2/Ph, reflux. iii) 90% TFA. iv) MeOH/NH_3

FIGURE 4.38. A new synthetic methodology for tiazofurine.

enzymatic methods. Besides their potent antitumor and antiviral activity for some of them, they have also shown high resistance to phosphorylases.

The use of enzymes, particularly lipases, for protections and deprotections is an important strategy for preparing carbocyclic nucleosides. This approach has been advantageous especially for the resolution of enantiomeric forms, leading to high enantiomeric purity. Constrained three[83] and four[84] member ring carbocyclic nucleosides have been obtained by applying chemoenzymatic methodologies involving lipase for enantiomeric resolution and stereoselective deprotections. In the case of more abundant five-member rings, the use of lipases for enzymatic resolution and regioselective deprotections has been under intense study. Special attention has been paid to cyclopentenyl diacetates, which have been used as building blocks for the preparation of important carbocyclic nucleosides such as Neplanocin and Aristeromycin. To achieve this goal, the hydrolase enzyme acetylcholinesterase (EEAC)[85] showed high efficiency for obtaining the desired enantiomer (1R,4S)-4-hydroxy-2-cyclopentenyl derivative in enantiomeric excess (ee) up to 96% (Figure 4.45).[86,87]

Racemic cyclopentenyl derivatives have been used as starting material in the preparation of the antiviral carboxyclic nucleoside (-)-5'-Deoxyaristeromycin. The

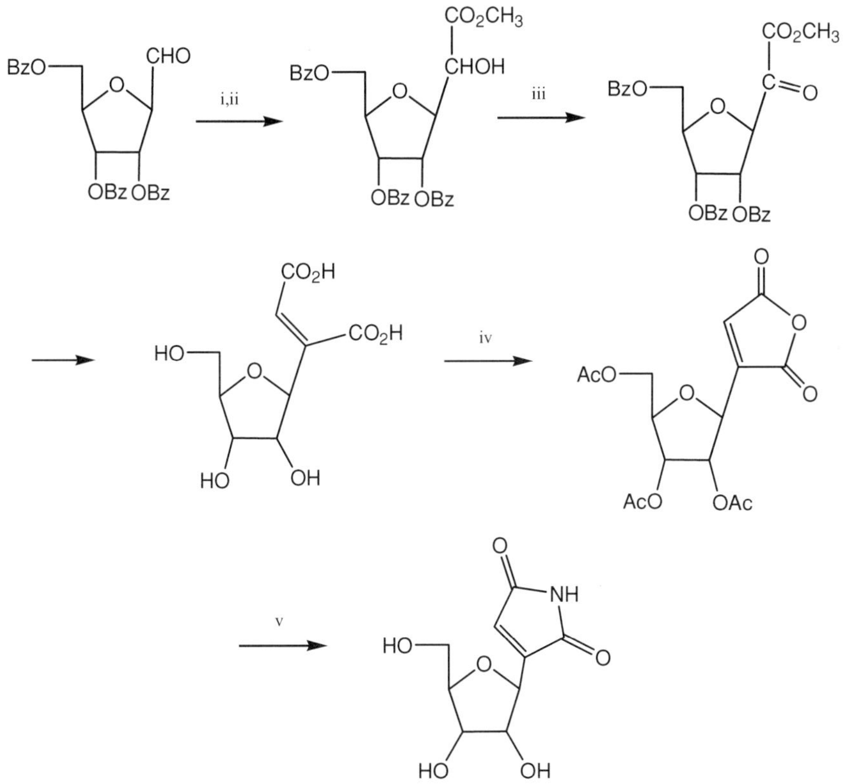

i) NaCN. ii) a) MeOH-H$_3$O$^+$. b) Me$_2$SO-DCC. iii) Ph$_3$P=CHCONH$_2$. iv) Ac$_2$O. v) a) NH$_3$. b) EPP. v) H$^+$.

FIGURE 4.39. Preparation of showdomicine.

key step reaction was the enzymatic resolution with *Pseudomona sp* lipase (PSL) of the racemic mixture affording the (+)-enantiomer, which was transformed chemically to the desired carbocyclic nucleoside (Figure 4.46).

The separation of racemic carbocyclic nucleosides by enzymatic means has been reported as an alternative approach. Thus, racemic aristeromycin was treated with adenosine deaminase (ADA) to give (−)-carbocyclic inosine and pure destrorotatory enantiomer (Figure 4.47).[88]

4.3.1 Cyclopropane Carbocyclic Nucleosides

Conformationally constrained cyclopropane nucleosides have been prepared following a chemoenzymatic approach.[83] Thus, the racemic resolution of *trans*-1-(diethoxyphosphyl)difluoromethyl-2-hydroxymethylcyclopropane followed by acetate hydrolysis was carried out with porcine pancreas lipase (PPL) to yield (+)- and (−)-cyclopropanes in high enantiomeric excess. Further

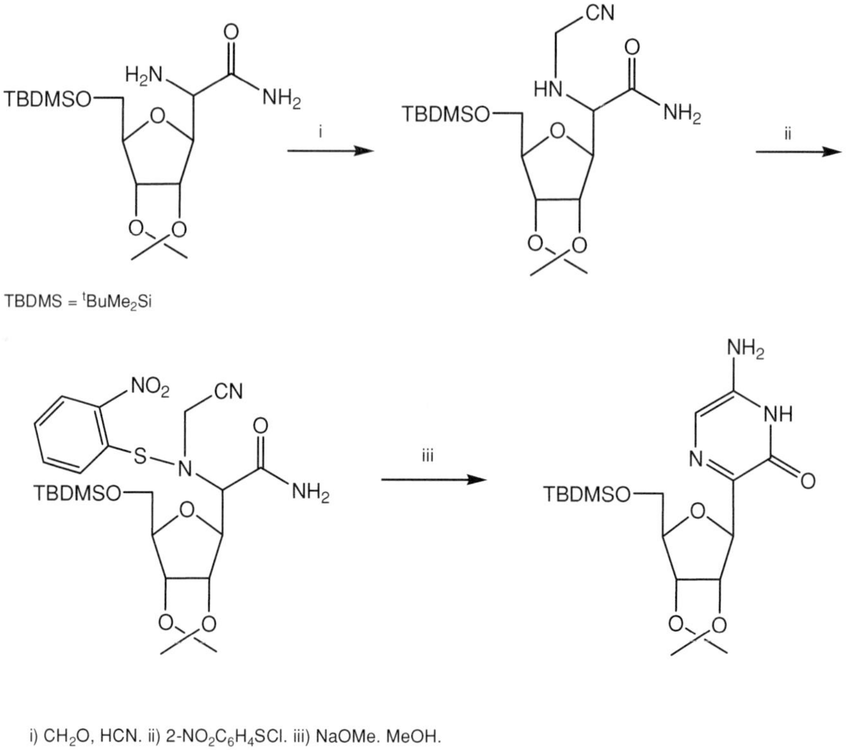
TBDMSO
H2N
O
NH2
i
HN
CN
ii
TBDMS = tBuMe2Si
NO2
S
N
iii
NH
N
i) CH2O, HCN. ii) 2-NO2C6H4SCl. iii) NaOMe. MeOH.

FIGURE 4.40. Synthesis of C-nucleoside by pyrazine ring formation.

transformation led to the preparation of the target cyclopropane nucleoside (Figure 4.48).

4.3.2 Cyclobutane Carbocyclic Nucleosides

Lubocavir is a synthetic potent inhibitor of DNA polymerase, active against cytomegalovirus[89] (Figure 4.49).

The carboxyclic four-membered *C*-nucleoside Cyclobut-A was prepared following the Barton decarboxylation method. The method is based on the reaction between carboxylic acids with heteroaromatic compounds (Figure 4.50).[90]

Other carbocyclic oxetanocin analogs have been prepared from oxetanocin A[91] 3,3-diethoxy-1,2-cyclobutanedicarboxylate-[92] and enantiomeric cyclobutanone intermediates[93] as starting materials.

4.3.3 Cyclopentane Carbocyclic Nucleosides

The Mitsunobu reaction has become a valuable alternative approach for preparing cyclopentane carbocyclic nucleosides. This has been demonstrated in the preparation of conformationally locked carbocyclic AZT triphosphate analogs under this

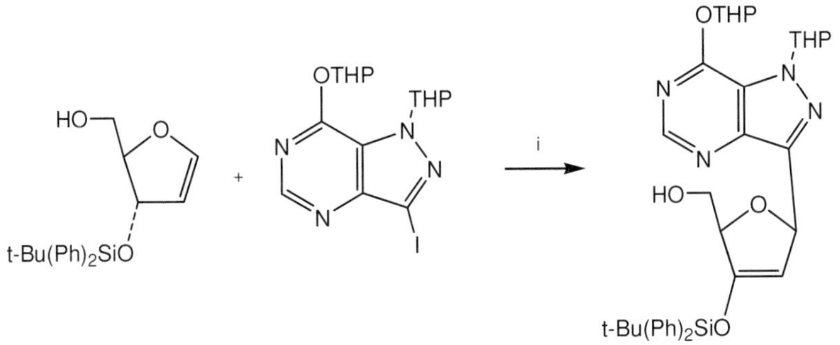

i) Pd(dba)₂, AsPh₃, MeCN.

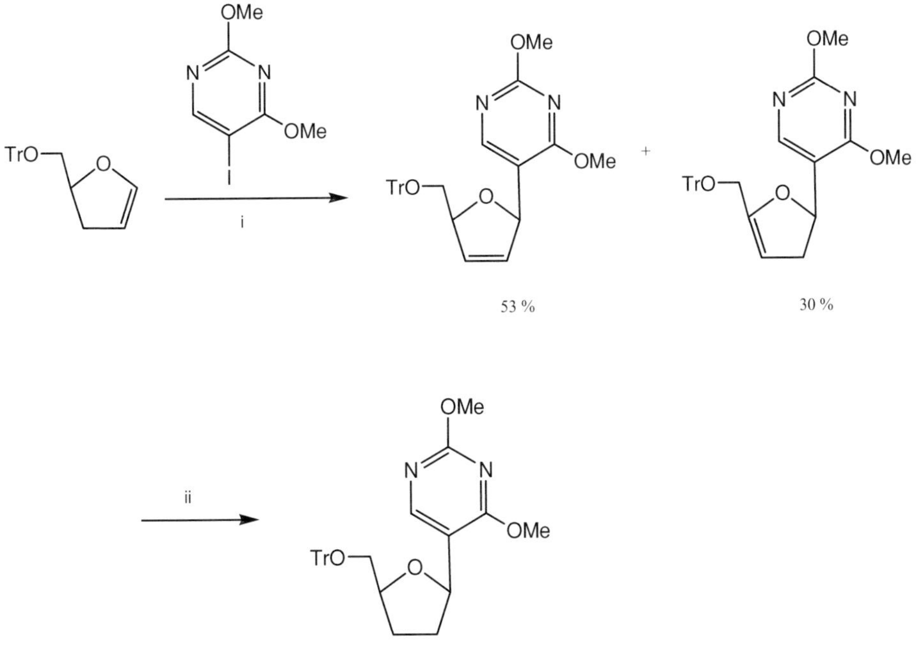

53 %   30 %

i) Pd(OAc)₂, NaOAc, n-Bu₄NCl, Et₃N, DMF. ii) H₂, Pd/C, ammonium formate, ETOH.

FIGURE 4.41. Palladium-mediated synthesis of *C*-nucleoside formycin analogs.

versatile condition.[94] The standard procedure usually takes place with diethyl or diisipropylazocarboxylate (DEAD or DIAD) with triphenylphosphine (Ph)₃P in THF to yield carbocyclic purines or pyrimidines nucleosides in high yield (Figure 4.51).[95]

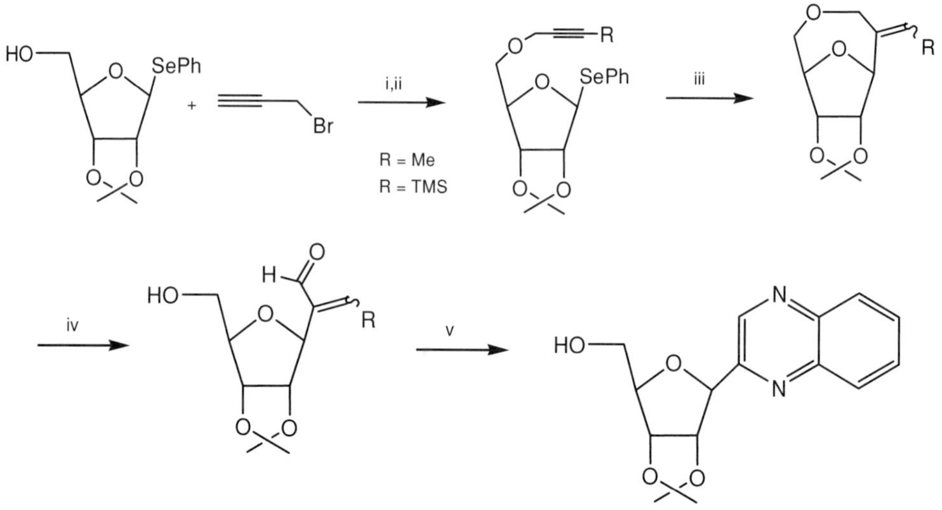

i) NaH, ii) n-BuLi, TMSI or MeI. iii) n-Bu₃SnH. iv) SeO₂-AcOH, 1,4-Dioxane. v) a) O₃. b) DMS. 3) 1,2-phenylenediamine.

FIGURE 4.42. C-nucleoside derivative formation via radical cyclization.

Another example on the applicability of this method was observed in the preparation of the carbocyclic thymidine nucleoside. It is worth mentioning that the desired stereochemistry of the hydroxyl group is obtained also through the Mitsunobu reaction (Figure 4.52).[96]

4.3.4 Palladium Mediated

Based on the widespread palladium-coupling methodologies, several dideoxy, carbocyclic and C-nucleosides have been efficiently prepared. For instance, the

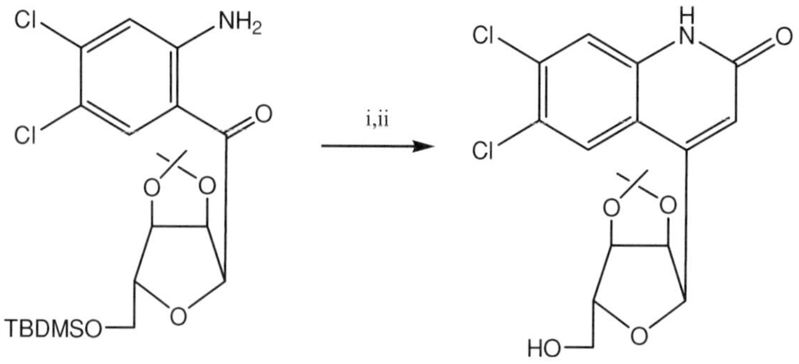

i) Ph₃P=C=C=O, PhH, reflux. ii) TBAF, THF, rt.

FIGURE 4.43. Quinolin-2-one C-nucleoside formation via Wittig reaction.

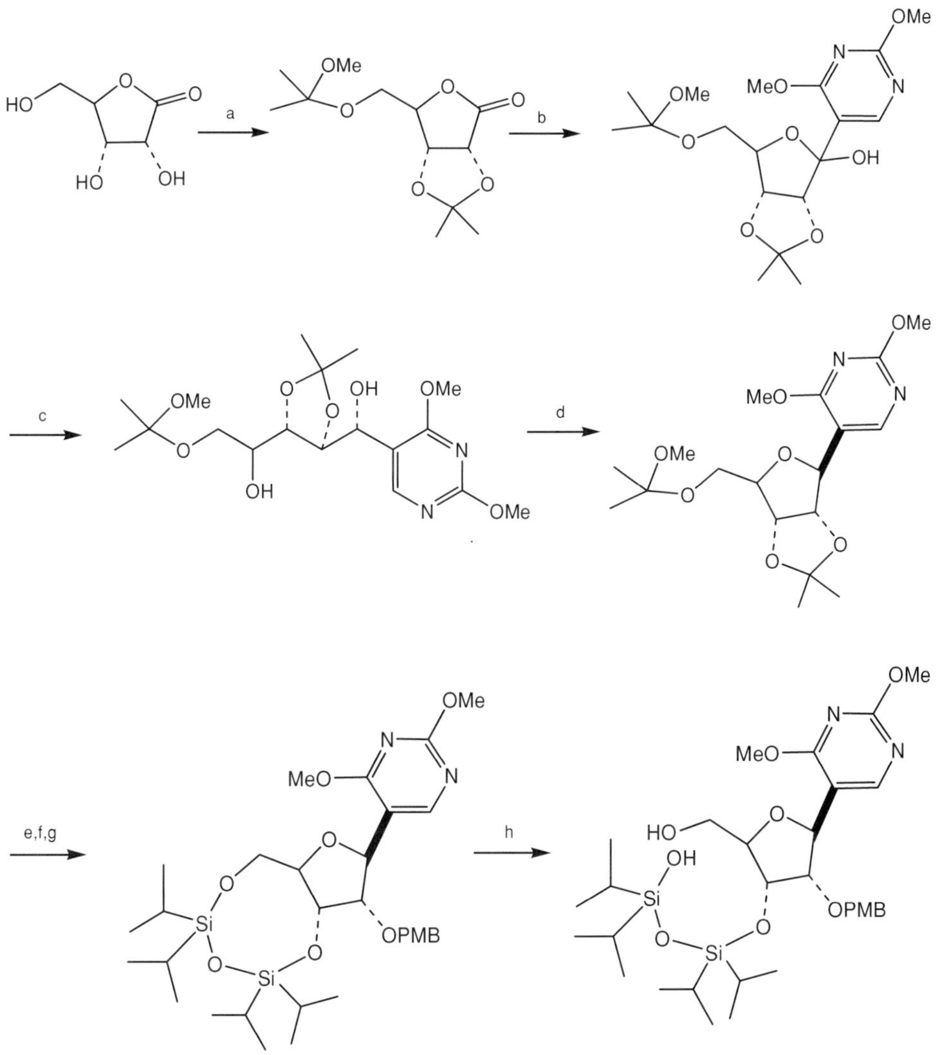

FIGURE 4.44. Total synthesis of C-nucleoside Malayamycin A.

antiviral C-nucleosides 2′-deoxyformycin B was prepared by condensation reaction between the heterocycle iodide intermediate with the glycal, under Pd(dba)$_2$ as palladium catalyst in 62% yield (Figure 4.53).[97]

Solid-phase synthesis of carbovir analogs under palladium catalysis was recently reported.[98] The carbocyclic derivative was linked to the Wang resin and then coupled with chloropurines under Pd(0) catalyst (Figure 4.54).

The Tsuji-Trost approach was used to prepare (−)-neplanocin A and its analog.[99] This synthesis proceeds via an allylic rearrangement of the hydroxyl group from the (+)-allylic alcohol to the (−)-allylic acetate (Figure 4.55).

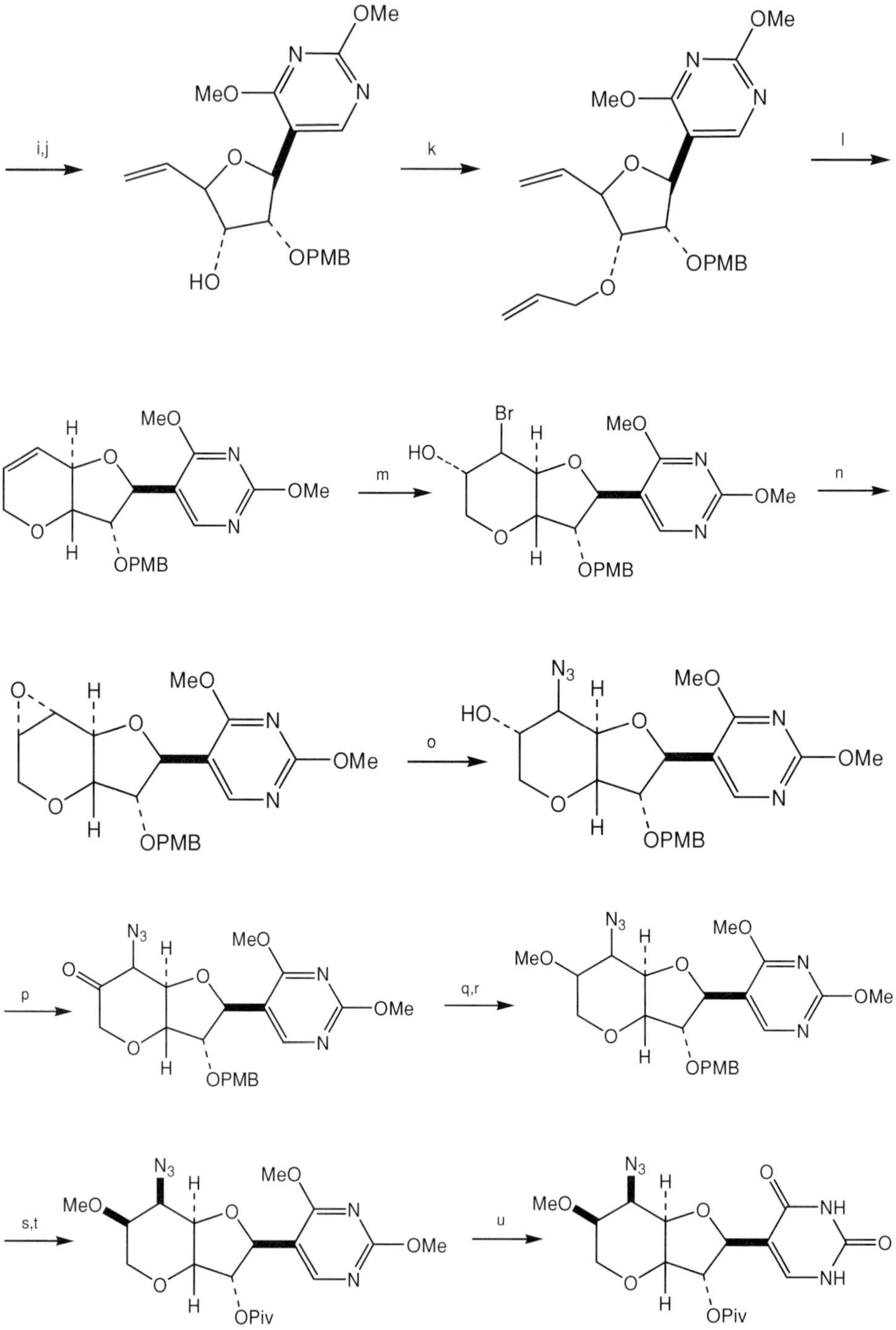

FIGURE 4.44. (*Continued*)

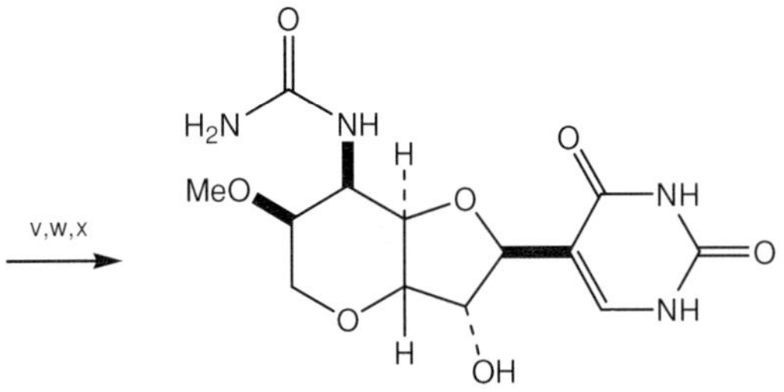

a) 2,2-dimethoxypropane, Na$_2$SO$_4$, PPTS, 94%. b) 2,4-dimethoxy-5-iodopyrimidine, t-BuLi, 75%. c) L-Selectride, ZnCl$_2$, DCM, 86%. d) DIAD, Ph$_3$P, THF, 91%. e) 70% AcOH, 85%. f) 1,3-dichloro-1,1,3,3-tetraisopropyldisiloxane, pyridine, 89%. g) NaH, PMBBr, DMF/THF, 84%. h) 1 N HCl, Dioxane, 88%. i) DMSO, (COCl)$_2$, i-Pr$_2$NEt, DCM. j) Ph$_3$PCH$_3$Br, NaHMDS, THF (36%, 2 steps). k) allyl bromide, NaH, DMF, 93%. l) Cl$_2$RuCHPh(PCy$_3$)$_2$, DCM, reflux (89%). m) NBS, H$_2$O, THF. n) NaOH, THF. o) NaN$_3$, methoxyethanol (5:1, 41%, 3 steps for 14). p) Dess-Martin periodinate, DCM. q) NaBH$_4$, MeOH. r) NaH, MeI, DMF (93%, 3 steps). s) DDQ, H$_2$O, DCM (84%). t) Piv-Cl, DMAP, NEt$_3$, pyridine. u) TMS-Cl, NaI, actonitrile (42%, 2 steps). v) PMe$_3$, H$_2$O, THF. w) trichloroacetylisocyanate, DCM. x) MeNH$_2$, MeOH, H$_2$O (60%, 3 steps).

FIGURE 4.44. (*Continued*)

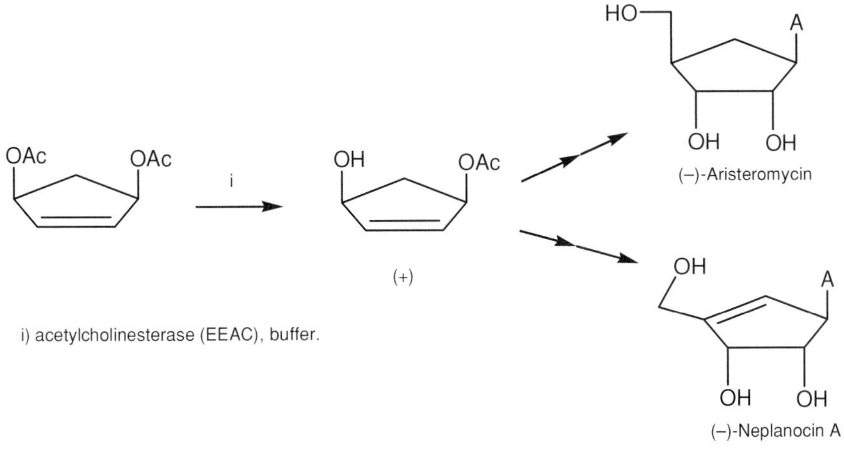

i) acetylcholinesterase (EEAC), buffer.

FIGURE 4.45. Enantiomeric resolution of prochiral cyclopentene diacetate.

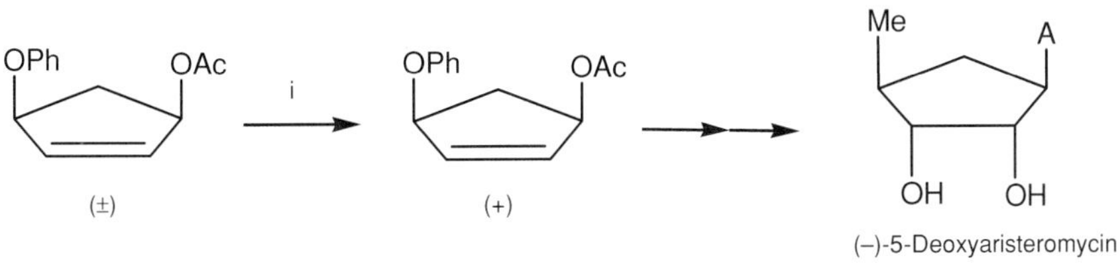

i) PSL, buffer.

FIGURE 4.46. Enzymatic resolution of racemic cyclopentene building blocks.

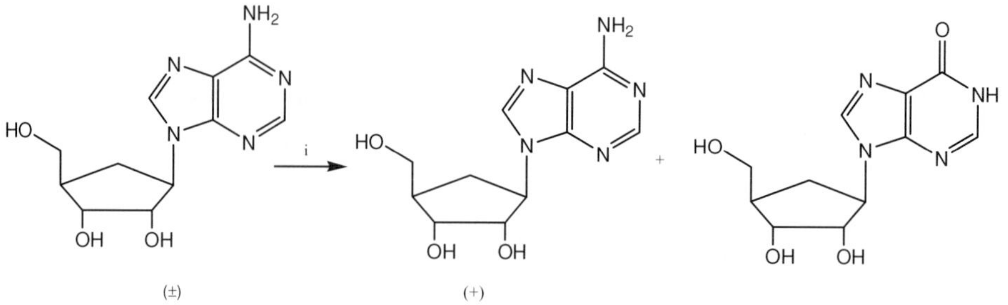

FIGURE 4.47. Enzymatic resolution of carbocyclic nucleoside.

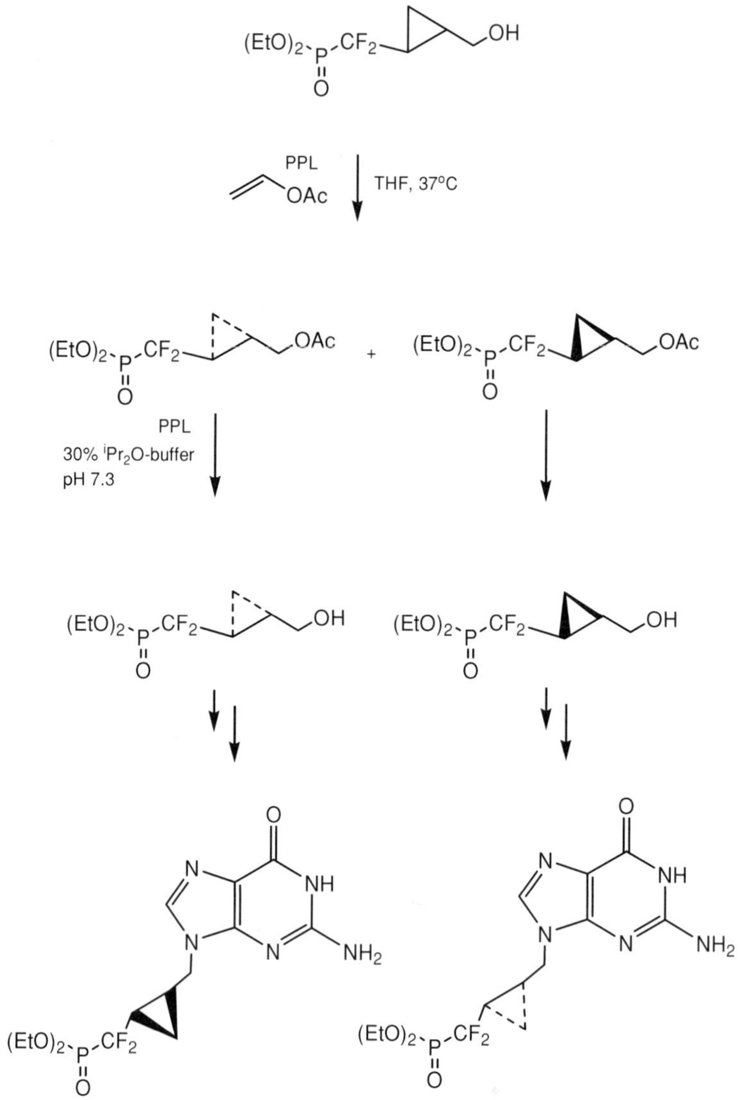

FIGURE 4.48. Chemoenzymatic syntheses of cyclopropane nucleosides.

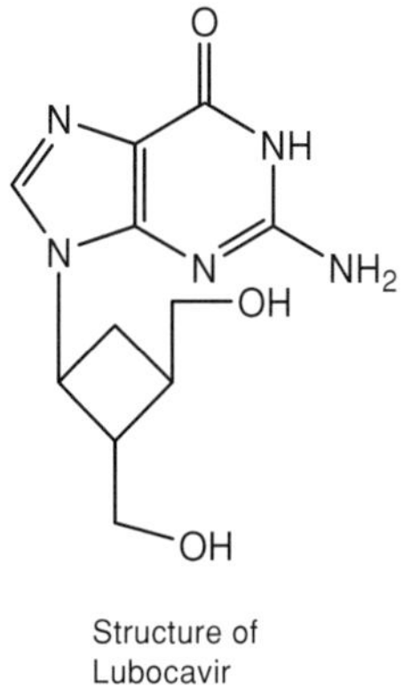

FIGURE 4.49. Structures of carboxetan carbocyclic nucleoside.

Structure of
Lubocavir

Carbocyclic nucleoside Aristeromycin with antitumor and antiviral activity was prepared by condensation of the carbocyclic diacetate intermediate with the sodium salt of adenosine base under Pd(0) in 75% yield (Figure 4.56).[100]

Palladium-mediated coupling of purine base with carbocyclic acetates, carbonates, or benzoates led to a mixture of N-7 and N-9 isomers. The regioselectivity of purine alkylations depends on the size and nature of the ligands, being the most typical Ph_3P, BINAP, $P(OMe)_3$, $P(OiPr)_3$, $P(OPh)_3$. (Figure 4.57).[101]

Another straightforward methodology for preparing carbocyclic nucleosides involves the direct condensation of mesylated carboxyclic intermediate with the

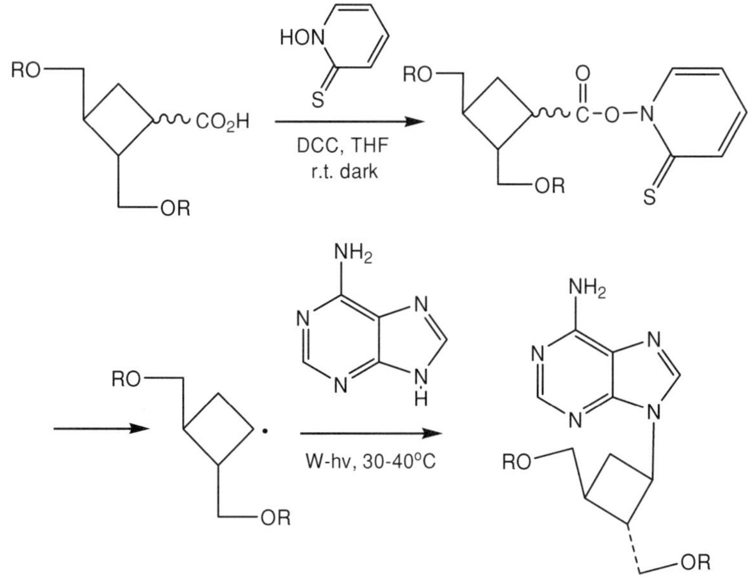

FIGURE 4.50. The Barton decarboxylation method for the preparation of carbocyclic C-nucleosides.

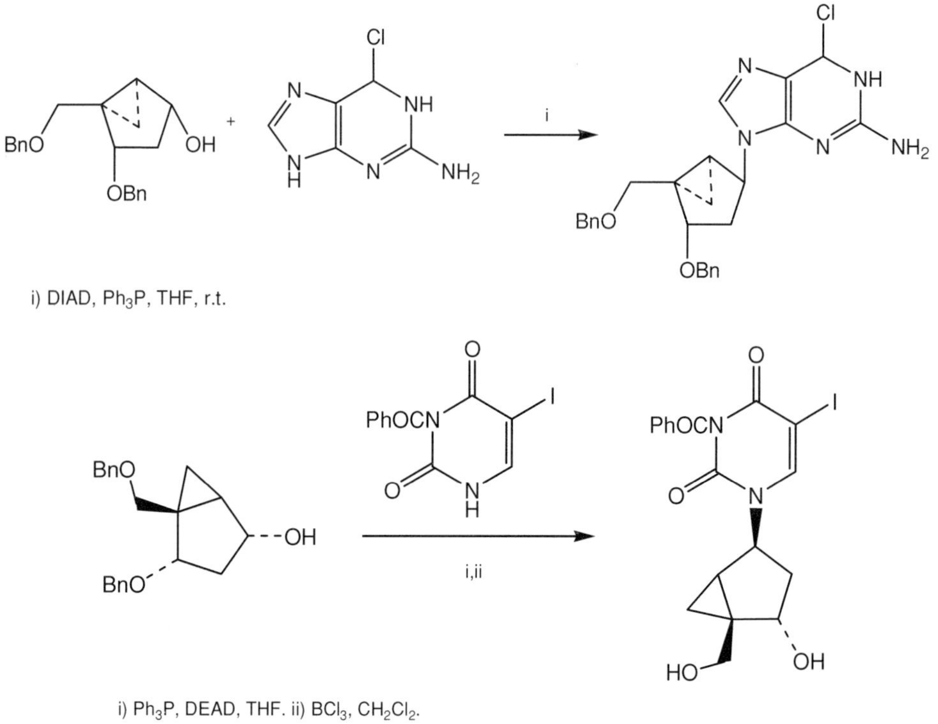
BnO
OH
OBn
+
Cl
N
NH
N
H
N
NH2
i
i) DIAD, Ph3P, THF, r.t.
BnO
OH
PhOCN
O
I
NH
i,ii
HO
i) Ph3P, DEAD, THF. ii) BCl3, CH2Cl2.

FIGURE 4.51. Synthesis of conformationally locked carbocyclic purine and pyrimidines under the Mitsunobu approach.

heterocyclic base in the presence of potassium carbonate and crown ethers as coupling conditions (Figure 4.58).[102]

4.3.5 Enzymatic Synthesis

Likewise, carbocyclic nucleosides aristeromycin and neplanocin A can be biosynthetically prepared by using a mutant strain of *S. citricolor* as is observed in Figure 4.59.

The cyclopropylamino carbocyclic nucleosides (−)-Abacavir is a potent anti-HIV drug with promising results on clinical trials.[103] An improved synthesis has been described by Crimmins et al.[104] involving the treatment of key carbocyclic 2-amino-6-cloropurine intermediate with cyclopropilamine producing Abacavir along its parent anti-HIV carbocyclic nucleoside (−)-Carbovir (Figure 4.60).

4.3.5.1 Base Ring Formation

Another useful strategy used for preparing carbocyclic nucleosides involves the use of intermediates in which the amino group is already attached to the sugar

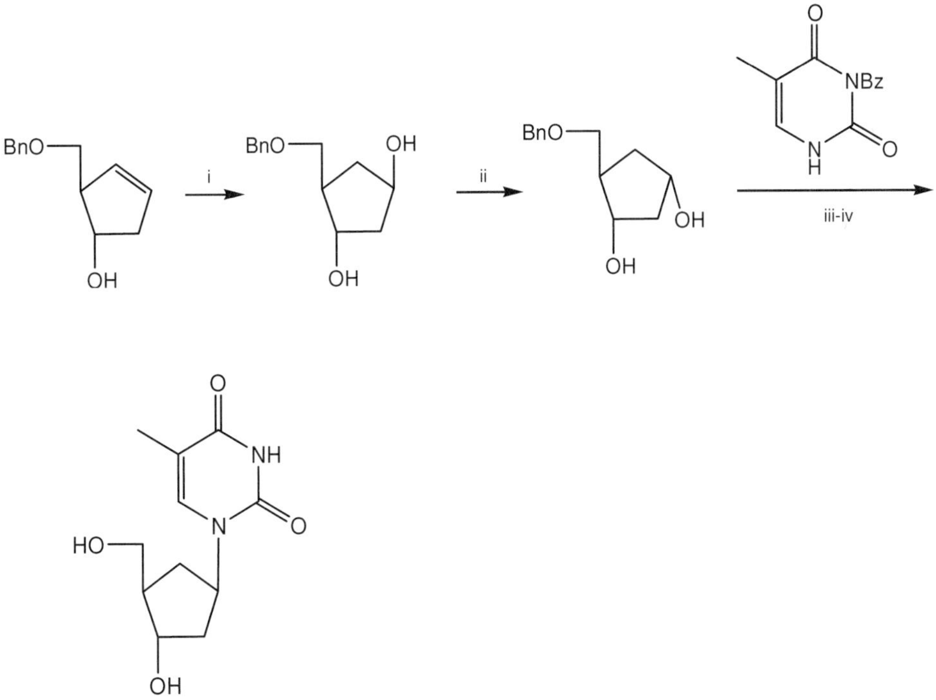

i) NaH, BnBr. ii) 9-BBN, H$_2$O$_2$/NaOH, 87 %. iii) PPh$_3$, DIAD. iv) a) NaOH/MeOH. b) H$_2$, Pd/C, 90%

FIGURE 4.52. The Mitsunobu reaction for preparation of the carbocyclic thymidine nucleoside.

moiety and once the coupling reaction is achieved, a ring closure process takes place to generate the expected nucleoside. According to this procedure Roberts et al.[105] prepared the potent antiviral inhibitor (−)-carbovir, which possesses similar activity than AZT against HIV in MT-4 cells. Thus, the starting material (±)-2-azabiciclo [2.2.1] hept-5-en-3-one was submitted to microbial treatment with *Pseudomona solanacearum* to provide enantiomerically pure (−) isomer. The enantiomerically pure carbocyclic amine was then conjugated to 2-amino-4,6-dichloropirimidine to produce the carbocyclic precursor which was ultimately cyclized to afford the desired (−)-carbovir (Figure 4.61).

Anti-leukemia carboxylic nucleoside Neplanocin A has been synthesized by Marquez et al., under the ring closure approach mentioned above. Thus, condensation of pyrimidine intermediate with isopropylideneaminocyclopentene-diol furnished intermediate, which was further cyclized to the purine base with

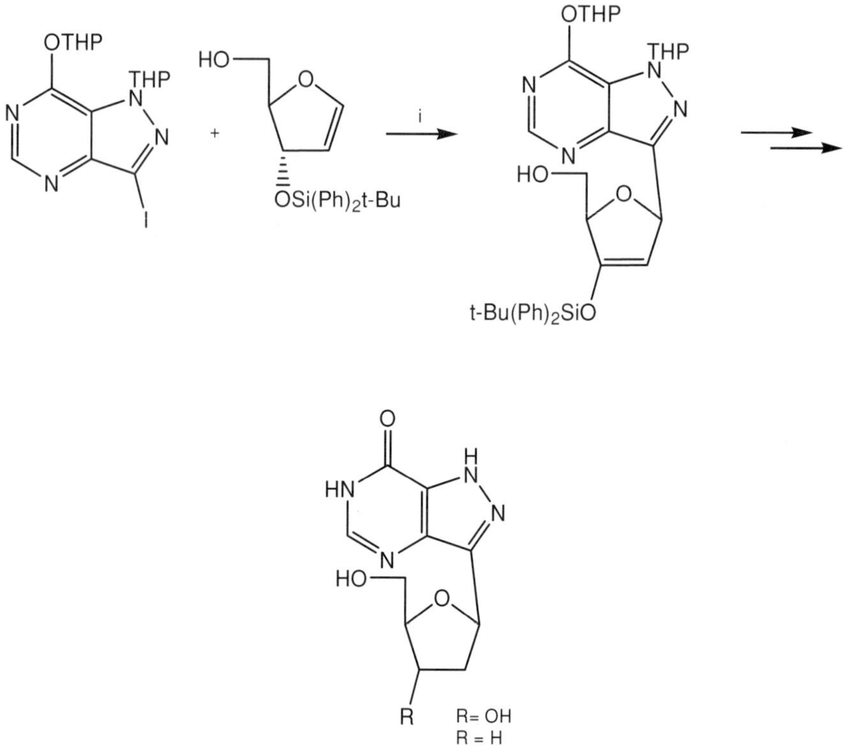

i) Pd(dba)$_2$, AsPh$_3$, CH$_3$CN.

FIGURE 4.53. Palladium-mediated 2′-deoxyformycin B and 2′,3′-dideoxyformycin B.

triethylortoformiate. Final conversion to adenine ring with ammonia and protecting group removal gave place to Neplanocin A (Figure 4.62).[106]

Likewise, this procedure was applied for the preparation of the close related pyrimidine analog by condensation of the previous carbocyclic amine with the unsaturated ether to produce the pyrimidine precursor who was transformed to thiopyrimidine and then to carbocyclic cytosine, as Figure 4.63 shows. This compound has been found to be active against leukemia type L1210 *in vivo*.[107]

An antiviral carbocyclic purine nucleoside was also reported[108] by following a ring closure step for purine formation. Condensation between pyrimidine intermediate with carbocyclic amine afforded condensation product, which is activated with diazonium salt for amino introduction. Ring closure was achieved with triethyl orthoformate in acid medium (Figure 4.64).

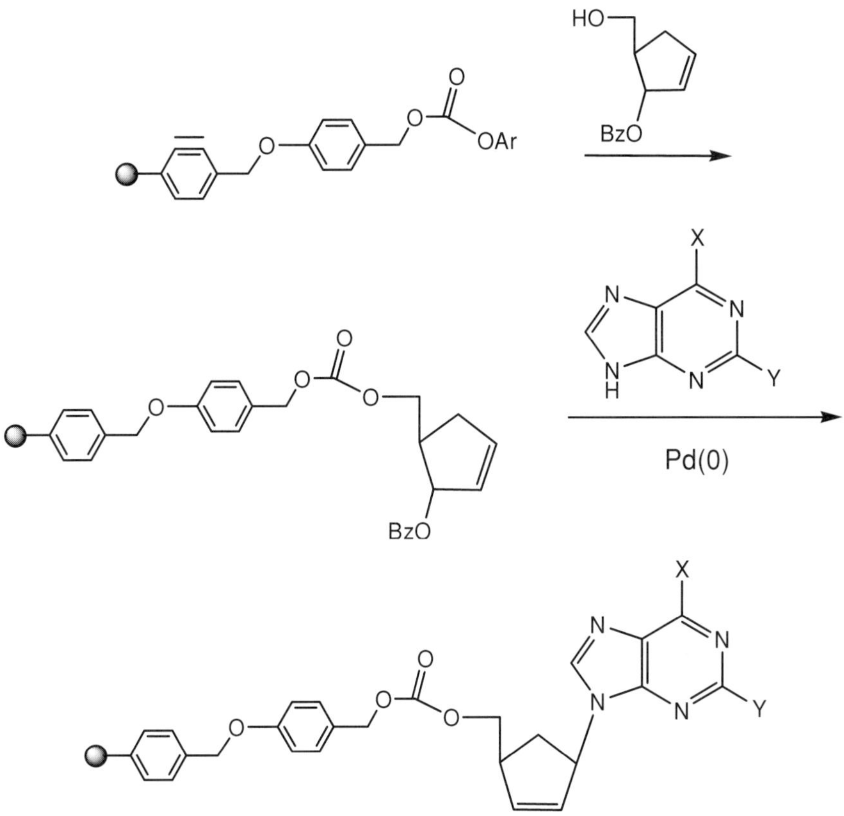

FIGURE 4.54. Solid-phase synthesis of carbocyclic nucleosides under palladium catalysis.

4.3.6   *Carbocyclic C-Nucleosides*

This class of C-nucleosides in which a methylene group replaces the furan oxygen ring does not show significant biological activity so far; however, there is an interest to synthesize C-nucleoside with natural heterocycle moieties in a stereocontrolled fashion. A recent stereocontrolled synthesis of carbocyclic C-nucleosides has been proposed involving as key starting material the cyano carboxylic intermediate which was condensed to 9-deazapurine to produce saturated and unsaturated carbocyclic 9-deazapurine nucleosides (Figure 4.65).[109]

4.4   Acyclic Nucleosides

Since the discovery of Acyclovir as an anti-herpes drug, important efforts have been made toward the synthesis of analogs of acyclovir and other acyclic nucleosides. A comprehensive review made by Chu and Cutler[110] summarizes the major

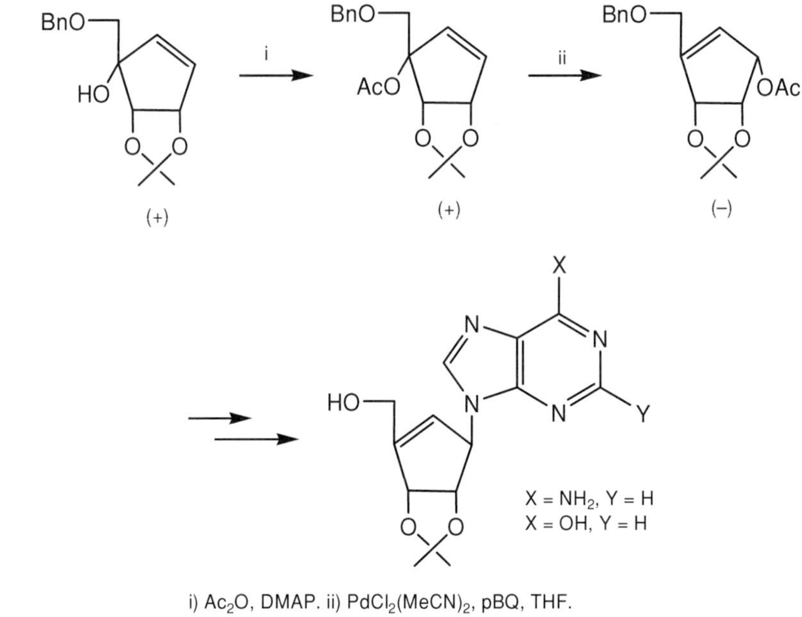
BnO
HO
(+)
i
BnO
AcO
(+)
ii
BnO
OAc
(−)
X
N
N
HO
N
N
Y
X = NH2, Y = H
X = OH, Y = H
i) Ac2O, DMAP. ii) PdCl2(MeCN)2, pBQ, THF.

FIGURE 4.55. Tsuji-Trost reaction for the preparation of neplanocin A.

achievements carried out for preparing acyclonucleosides defined as those heterocyclic compounds containing one or more hydroxyl groups on the alkyl side chain.

At least three representative synthesis of acyclovir have been made, the first by Schaeffer et al.[111] involving a condensation reaction of dichloropurine with ether chloride intermediate, and further purine transformation to generate 9-(2-hydroxyethoxymethyl)guanine (acyclovir) (Figure 4.66).

An improved version introduced by Barrio et al.[112,113] consists of the initial reaction of 1,3-dioxolane with trimethylsilyl iodide to produce the side chain,

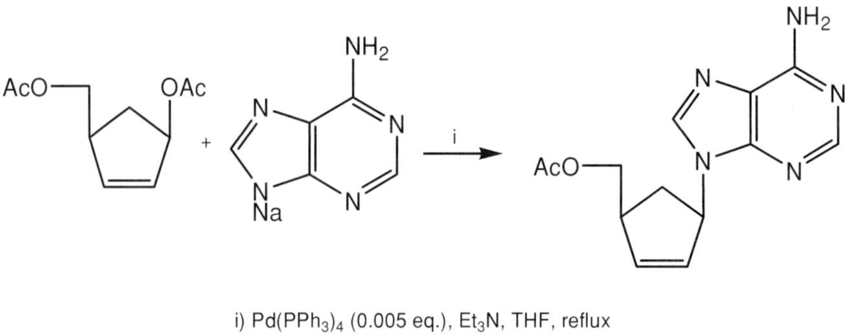
AcO
OAc
+
NH2
N
N
N
Na
N
i
AcO
NH2
N
N
N
N
i) Pd(PPh3)4 (0.005 eq.), Et3N, THF, reflux

FIGURE 4.56. Palladium catalyzed synthesis of aristeromycin.

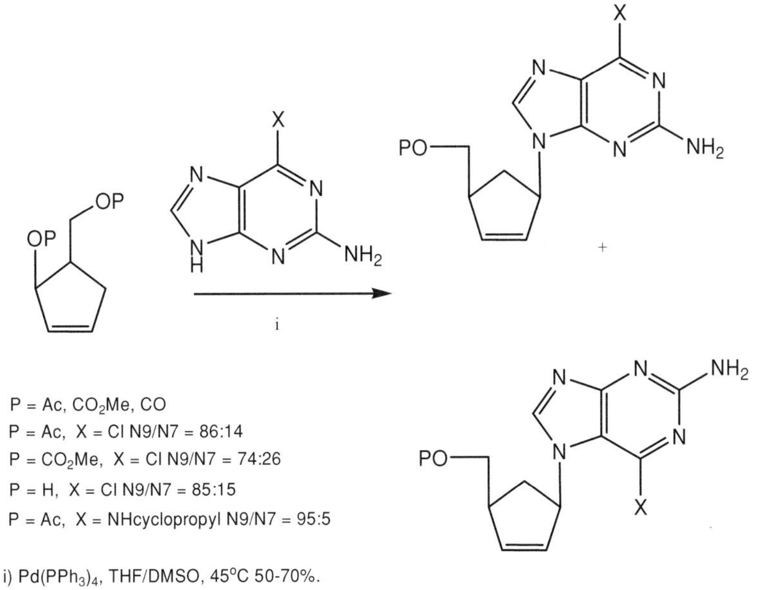

P = Ac, CO₂Me, CO
P = Ac, X = Cl N9/N7 = 86:14
P = CO₂Me, X = Cl N9/N7 = 74:26
P = H, X = Cl N9/N7 = 85:15
P = Ac, X = NHcyclopropyl N9/N7 = 95:5

i) Pd(PPh₃)₄, THF/DMSO, 45°C 50-70%.

FIGURE 4.57. Palladium-catalyzed coupling with purine base.

which was then condensed with the halogenated purine to yield after hydrolysis and ammonolysis the target acyclovir (Figure 4.67).

Robins and Hatfield[114] employed a chemoenzymatic approach for preparing acyclovir consisting initially in the use of mercury salts and hexamethyldisilane (HMDS) and in the final step an enzymatic conversion. Thus, the procedure

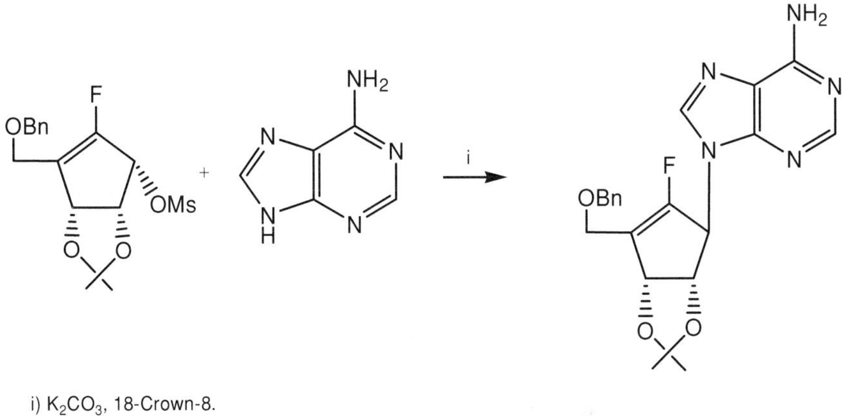

i) K₂CO₃, 18-Crown-8.

FIGURE 4.58. Preparation of carbocyclic nucleosides with mesylated carboxyclic intermediates.

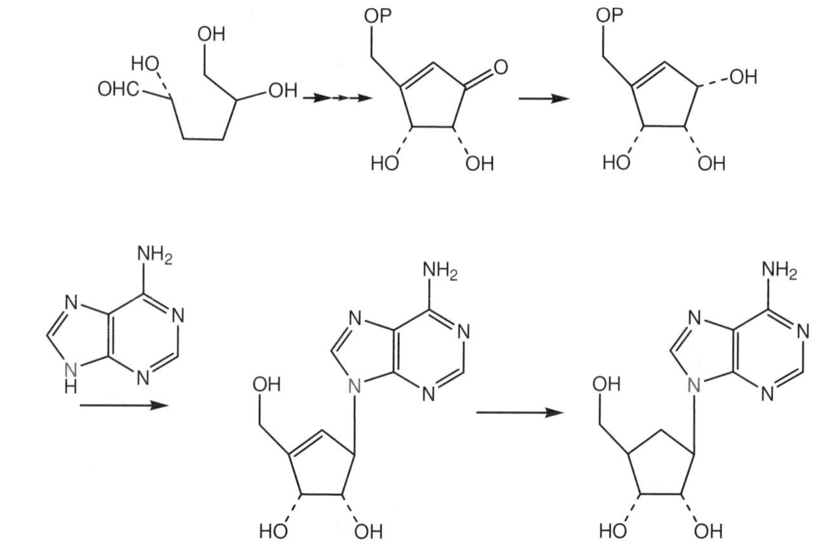

Neplanocin A Aristeromicin

FIGURE 4.59. Biosynthetic pathway of neplanocin A and aristeromicin.

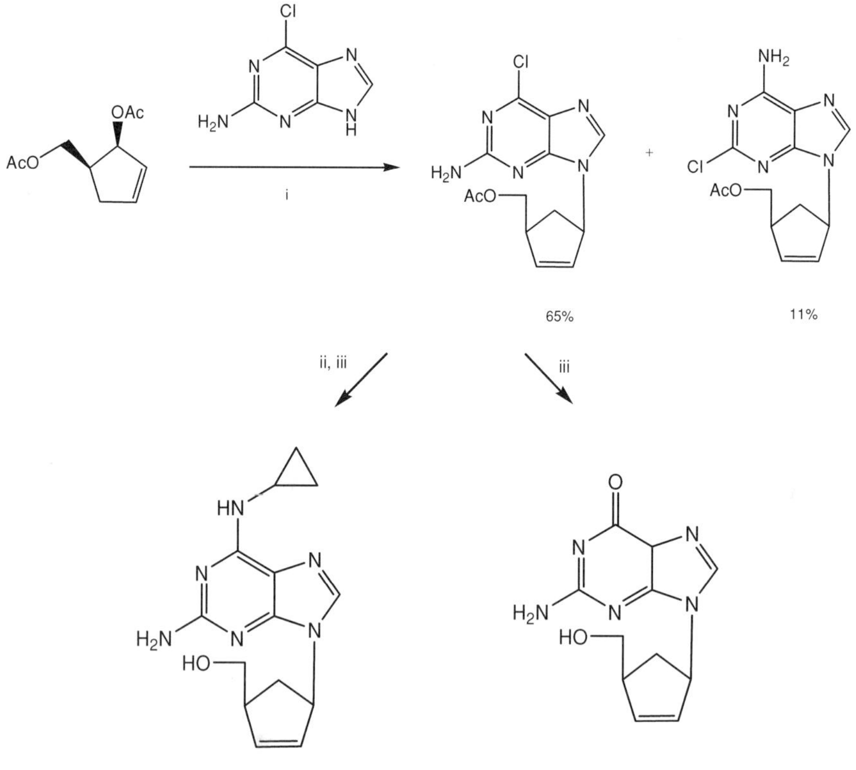

i) NaH, $Pd(PPh_3)_4$, 1:1 THF:DMSO. ii) cyclopropylamine, EtOH. iii) NaOH, H_2O.

FIGURE 4.60. Synthesis of anti-HIV (-)-Abacavir and (-)-Carbovir.

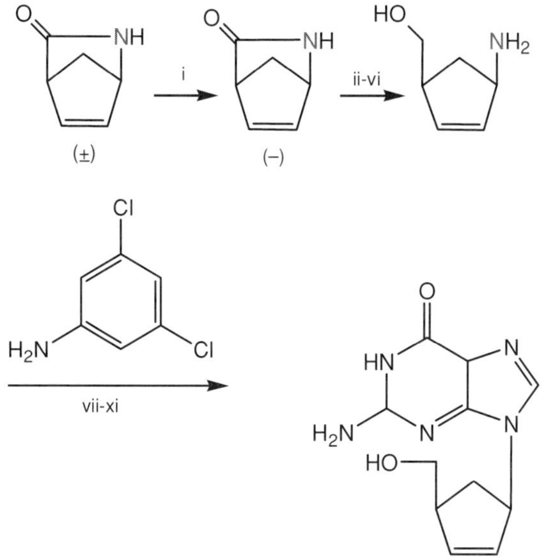

(±)   (−)

(−)-carabovir

i) P. solanacearum NCIB 40249. ii) HCl-H$_2$O. iii) (MeO)$_2$CMe$_2$. iv) Ac$_2$O/Py.
v) Ca(BH$_4$)$_2$/THF. vi) HCl-H$_2$O/EtOH. vii) PrNEt, nBuOH. viii) 4-Cl-C$_6$H$_4$N$_2$+ Cl-
-AcOH, AcONa/H$_2$O. ix) Zn, AcOH/EtOH-H$_2$O. x) (EtO)$_3$CH/HCl. xi) NaOH/H$_2$O.

FIGURE 4.61. Synthesis of (−)-Carbovir.

involves the condensation between 2,6-dichloropurine with the bromoether, affording regioisomer N-7 shown in Figure 4.68. Further amination and final transformation to guanine with the enzyme adenosine-deaminase produces the desired
antiviral compound.

The phosphonate acyclic nucleoside 9(2-phosphonomethoxyethyl)adenine
(PMEA) was founded as a good antiviral analog with prolongated action.[115] A
regio-defined synthesis base on the purine ring formation was described involving the initial attachment of the phosphonate amine intermediate by nucleophilic
substitution to the 5-amino-4,6-dichoropyrimidine base, and then ring formation
followed by amination to produce the desired phosphonate acyclic adenine PMEA
(Figure 4.69).[116]

The effectiveness of acyclovir as an antiviral drug encouraged different groups
to synthesize more potent acyclic analogs. As a result of these efforts, the acyclic
nucleoside 9-[(1,3-Dihydroxy-2-prpoxy)methyl]guanine (DHPG)[117] was prepared
and tested as antiviral nucleoside, showing similar potency as acyclovir against
simple herpes but stronger against encephalitis and vaginitis herpes.

Various report of DHPG were described, one of them involving the use of
hexamethyldisilylasane (HMDS) as condensing agent (Figure 4.70).[110]

An alternative route for preparing DHPG involved the condensation reaction
of acetylguanine base and triacetate derivative in the presence of ethanesulfonic

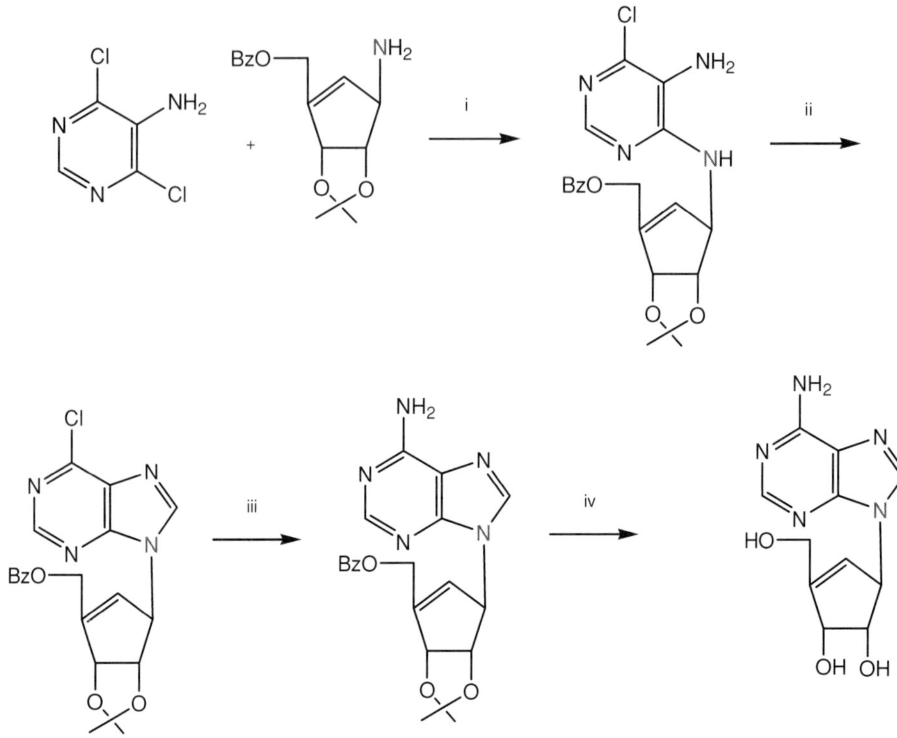

i) EtN$_3$/EtOH. ii) HC(OEt)$_3$, Ac$_2$O. ii) NH$_3$/MeOH. iii) BCl$_3$/CH$_2$Cl$_2$-MeOH

FIGURE 4.62. Synthesis of neplanocin A.

acid, at temperatures ranging from 155 to 160°C. As result two regioisomers were obtained from which one of those was converted to the desired antiviral compound Figure 4.71.[110]

4.5 Thionucleosides

Nucleosides having the sugar ring oxygen replaced by sulfur are known as thionucleosides. The synthesis and therapeutic evaluation mainly as antiviral and anticancer drugs of these nucleoside mimics have been reviewed.[118] A comparative analysis of thionucleosides and nucleosides showed that sulfur replacement in some cases produced equivalent or higher potency[119,120] and do not undergo enzymatic cleavage of the glycosidic bond, although it has been also observed increased toxicity as in the case of β-4′-thiothymidine.[121] Some thionucleosides displaying antiviral and/or anticancer activity are shown in Figure 4.72.

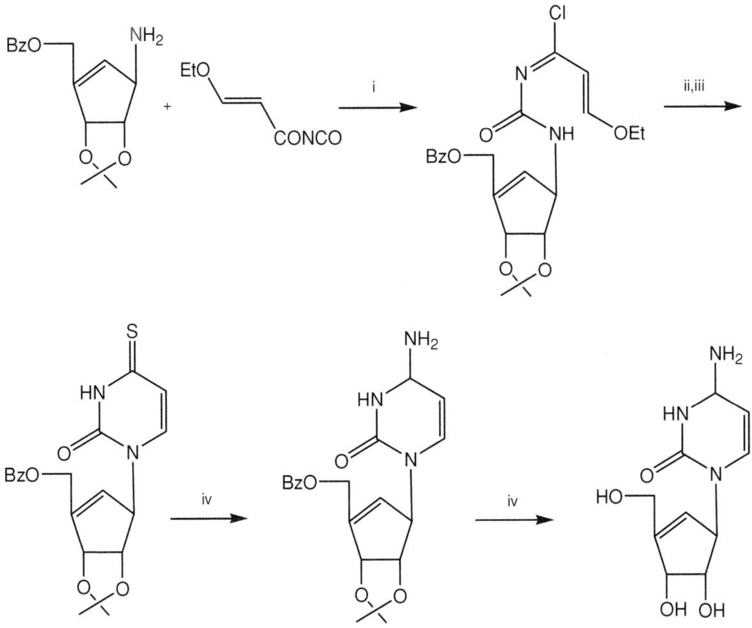

i) PhH. ii) DMF, NH₄OH. iii) Lawesson. iv) NH₃liq. v) a) BCl₃/CH₂Cl₂-MeOH. b) Dowex H⁺

FIGURE 4.63. Synthesis of carbocyclic pyrimidine nucleoside.

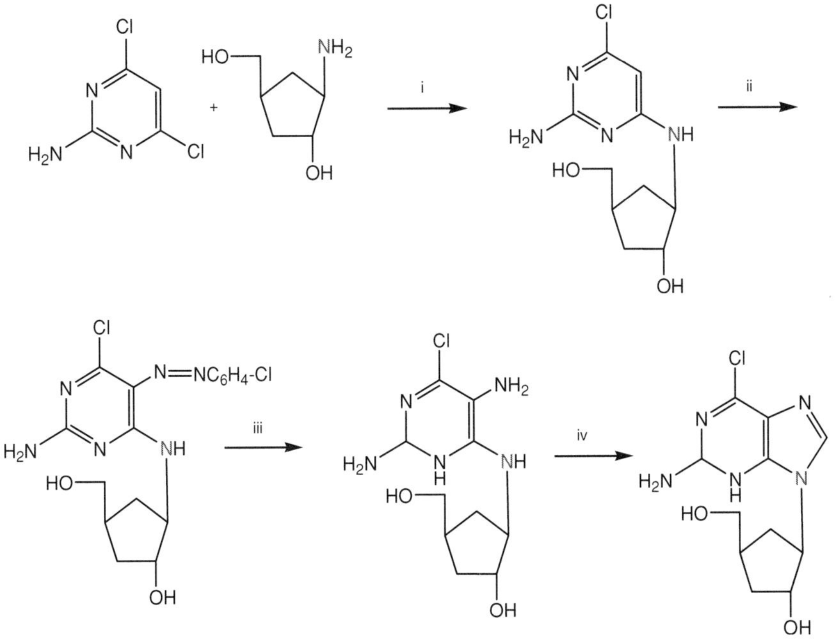

i) Et₃N/EtOH. ii) 4-Cl-C₆H₄N₂⁺Cl⁻, Na₂CO₃, AcOH/H₂O. iii) Zn/AcOH. iv) CH(OEt)₃-HCl/DMF.

FIGURE 4.64. Ring closure approach for preparation of carbocyclic purine.

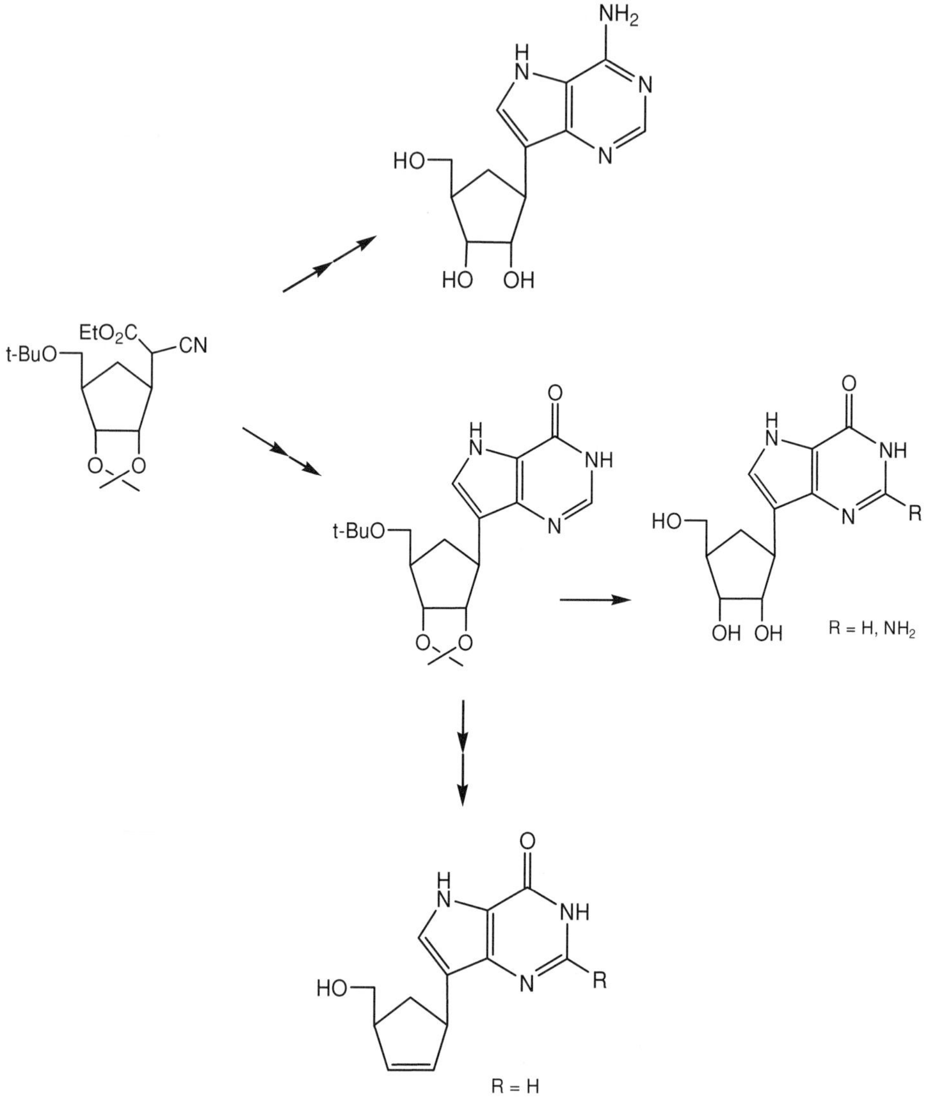

FIGURE 4.65. Stereocontrolled syntheses of carbocyclic 9-deazapurine nucleosides.

Based on their structural features N-thionucleosides defined also as thioribosyl sugars are classified into four groups (Figure 4.73):

4.5.1 Preparation of Thioribofuranosyl Intermediates

A number of approaches oriented to replace or insert a sulfur atom instead or besides the cyclic oxygen into the ribose ring have been described. One of the

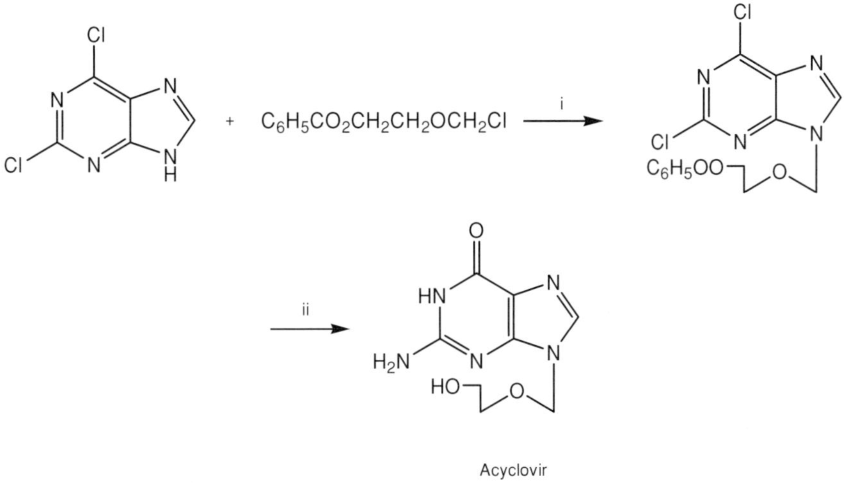

i) Et₃N. ii) NH₃.

FIGURE 4.66. First synthesis of Acyclovir.

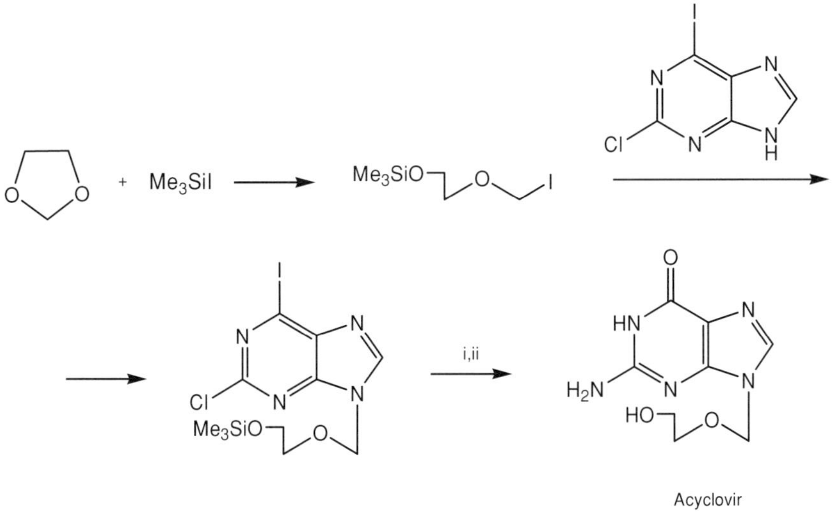

i) K₂CO₃. ii) NH₃.

FIGURE 4.67. Improved synthesis of acyclovir.

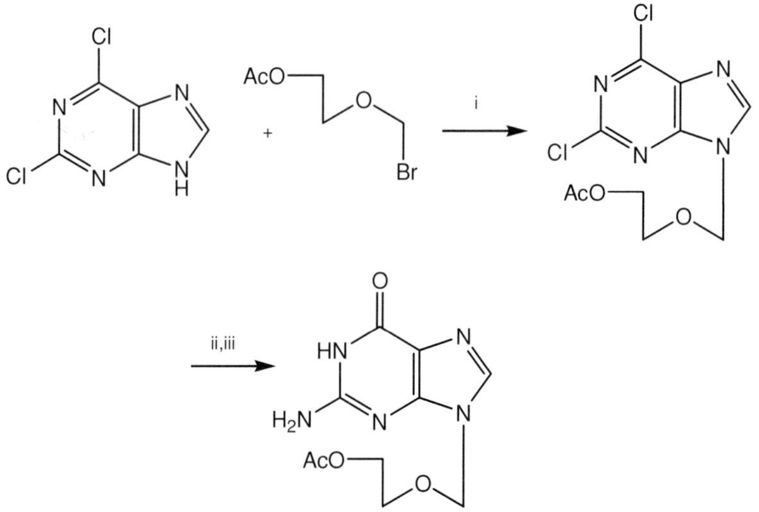

i) Hg(CN)$_2$-HMDS. ii) NH$_3$. iii) adenosine-deaminase.

FIGURE 4.68. Acyclovir synthesis.

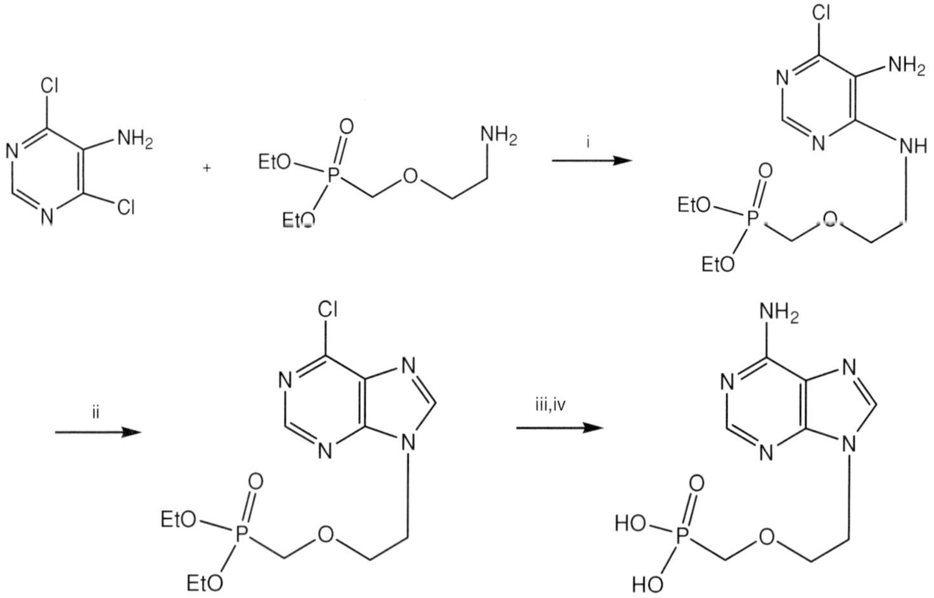

i) Et$_3$N, EtOH, reflux. ii) diethoxymethyl acetate, 120°C. iii) NH$_3$. iv) TMSBr, MeCN.

FIGURE 4.69. Synthesis of phosphonate acyclic adenine PMEA.

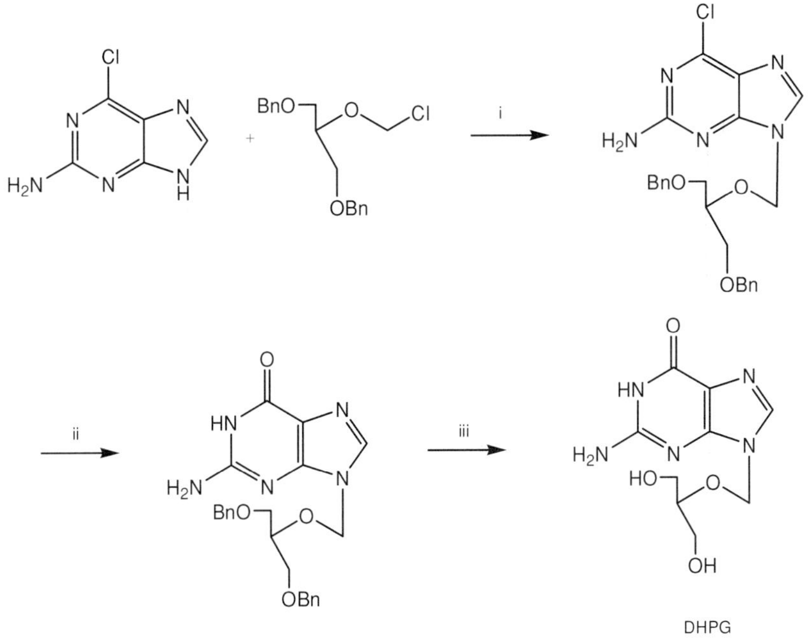

i) HMDS. ii) NaOMe, HSCH$_2$CH$_2$OH. iii) H$_2$, Pd-C

FIGURE 4.70. Synthesis of antiviral acyclic nucleoside DHPG.

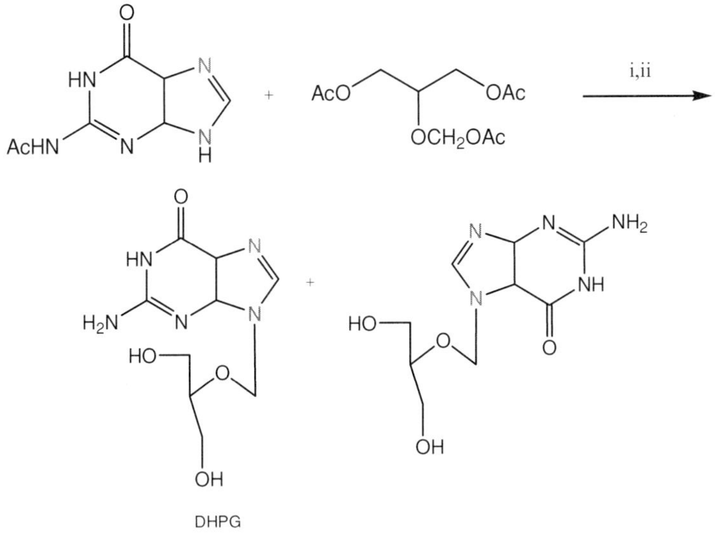

i) EtSO$_3$H, 155-160°C. ii) MeONa/MeOH

FIGURE 4.71. Alternative synthesis of DHPG.

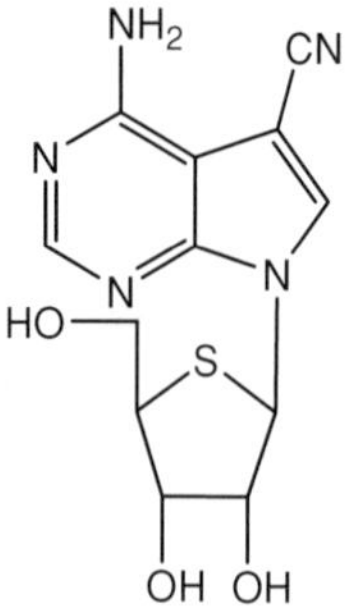

Thiotoyocamycin

Leukemia growth inhibitor

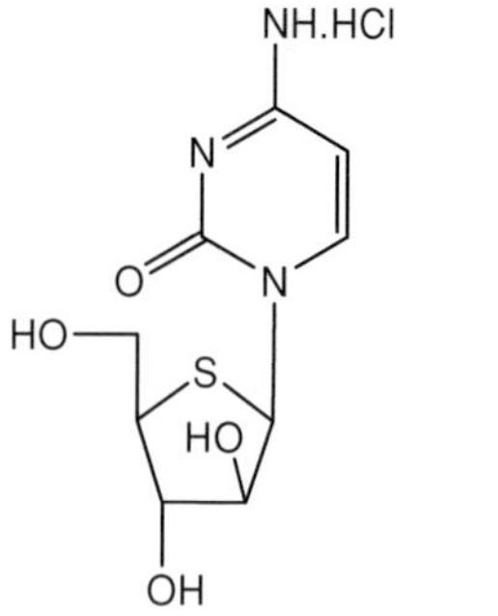

Thioarabinofuranosylcytosine

KB cell growth inhibitor

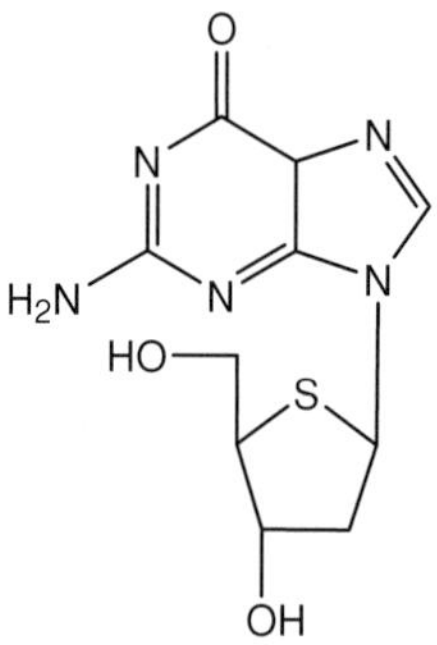

2'-Deoxy thioguanosine

Antiviral against HBV and HCMV

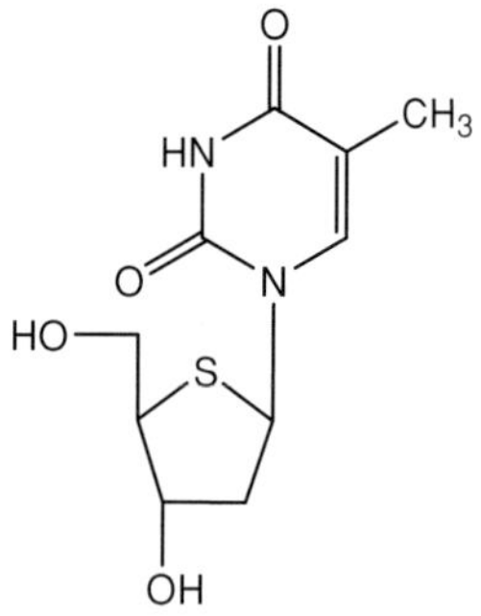

Thiothymidine

Carcinoma growth inhibitor

FIGURE 4.72. Some active N-thionucleosides.

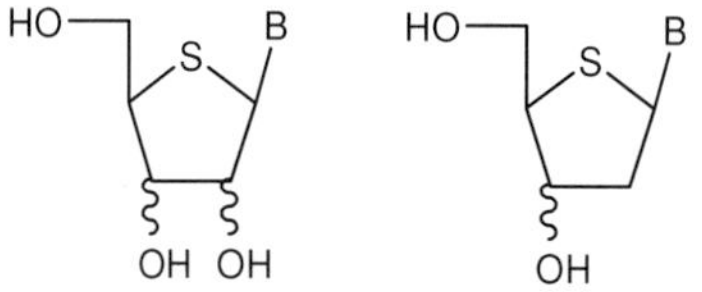

N-Thionucleosides

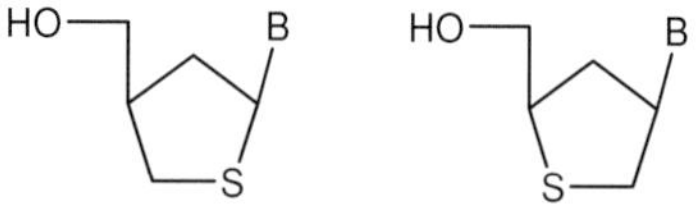

N-Isothionucleosides

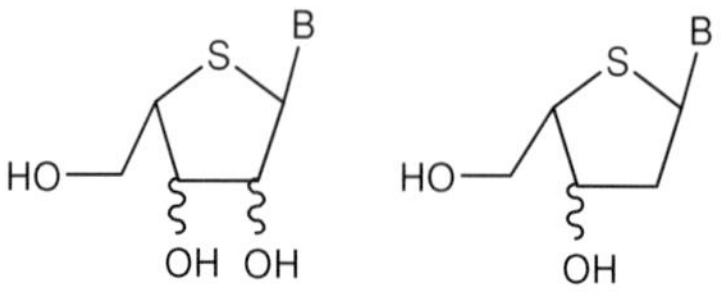

N-L-Thionucleosides

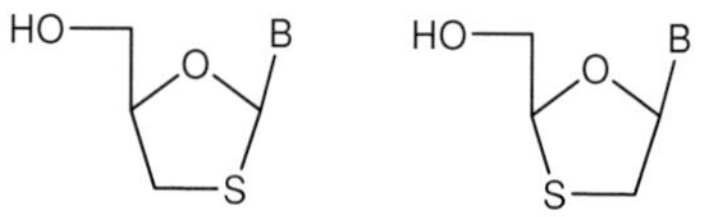

N-Thiooxonucleosides

FIGURE 4.73. Classification of N-thionucleosides.

FIGURE 4.74. Early synthesis of
thioribofuranosyl derivatives.

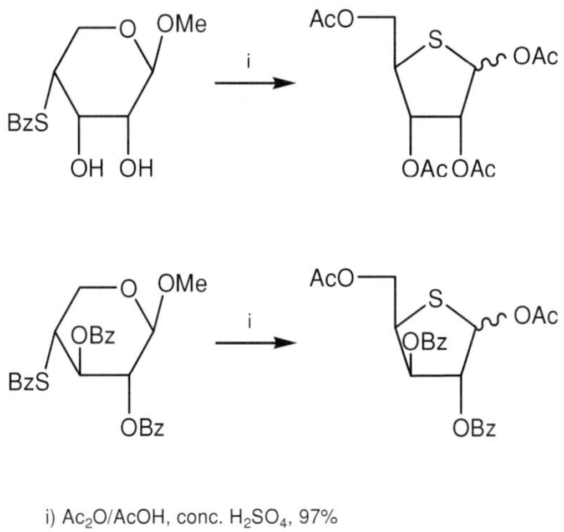

i) Ac₂O/AcOH, conc. H₂SO₄, 97%

earliest methods for preparing thioribosyl acetates was described by Reinst et
al.[122,123] involving as key steps the conversion of the 4-thiobenzoyl pyranoside
into the thioribofuranosyl acetate (Figure 4.74).

A short time later, another report introduced the use of sodium in liquid ammonia
followed by benzoylation to yield tribenzoylated thioribofuranoside as a mixture
of anomers (α:β, 1:3) (Figure 4.75).[124]

The thioribosyl derivative benzyl 3,5-di-O-benzyl-2-deoxy-1,4-dithio-D-
erythro-pentofuranoside has been prepared and used as glycosyl donor in various thionucleoside synthesis.[125,126,127] The synthesis started from 2-deoxy-ribose,
which was transformed to the methylbenzyl derivative by following a standard procedure and then treated with benzylmercaptan in acid to produce the
dithiobenzylated derivative. The next step was to invert the hydroxyl group at 4-
position by using the Mitsunobu protocol to generate the intermediate with the desired stereochemistry. Finally, tosyl protection and NaI-BaCO₃ treatment afforded
the desired thiosugar (Figure 4.76).[125]

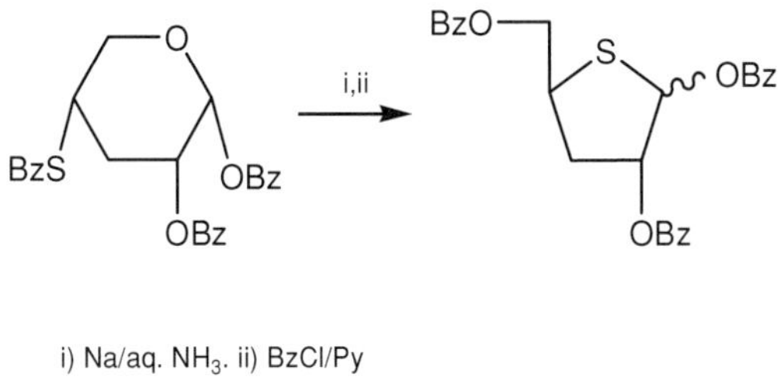

FIGURE 4.75. Preparation of benzoylated thioribofuranoside.

i) Na/aq. NH₃. ii) BzCl/Py

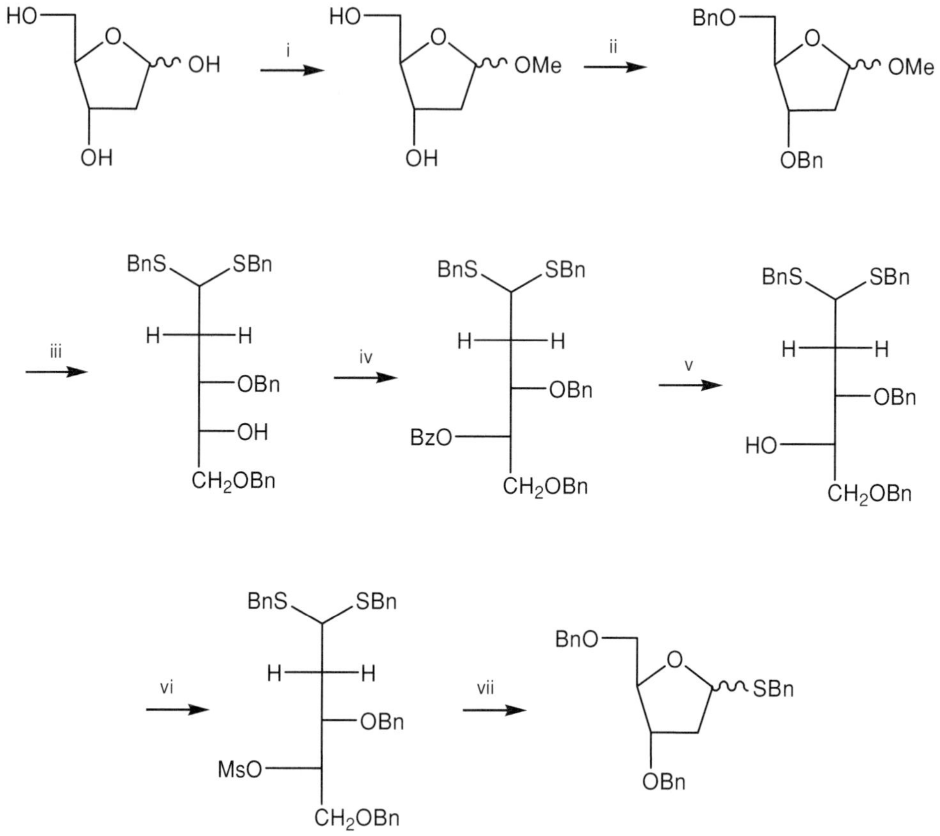

i) MeOH, HCl. ii) NaH, Bu$_4$NI, BnBr/THF. iii) BnSH, HCl. iv) PPh$_3$, PhCO$_2$H, DEAD/THF. v) NaOMe/MeOH. vi) MsCl/Py. vii) NaI, BaCO$_3$, acetone

FIGURE   4.76. Synthesis   of   benzyl   3,5-di-*O*-benzyl-2-deoxy-1,4-dithio-D-*erythro*-pentofuranoside.

4.5.2   Glycosidic Bond Formation

The general methods for preparing N-thionucleosides are similar as for N-nucleosides. However, variations from slight to significant can be found especially in the preparation of four-ring thietanocin or thiolane analogs.[126,127] Thus, according to a comprehensive review,[118] the earliest reports for N-thionucleoside formation used chloromercury salt of purine and chlorine or benzoyl thioriboside as glycosyl donor, while more recently the silyl approach has been preferred (Figure 4.77).

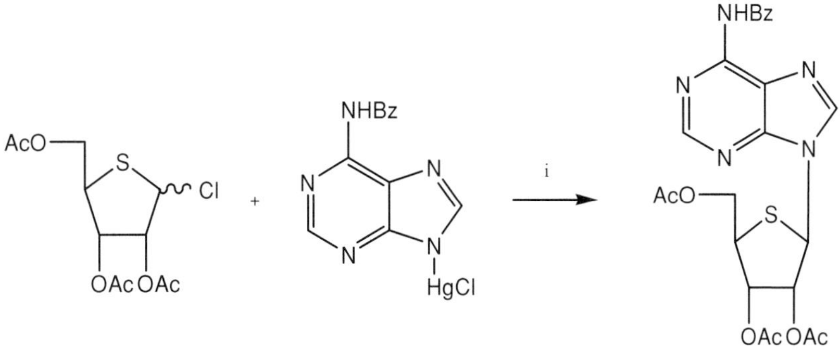

i) toluene

Ref.[122]

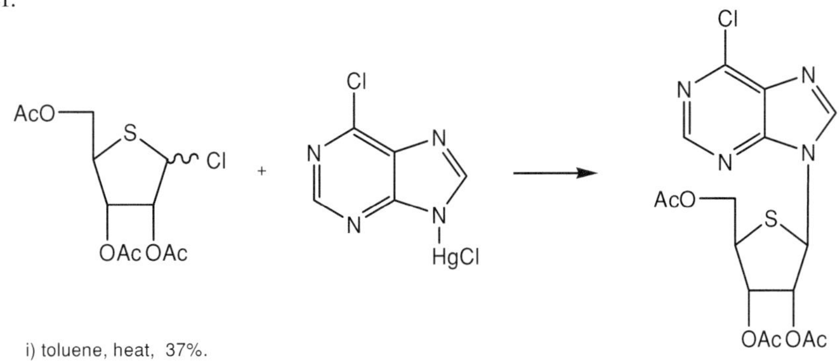

i) toluene, heat, 37%.

Ref.[128]

FIGURE 4.77. Common glycosylation reactions for the preparation of thionucleosides.

4.5.3   Chloromercuration Promoted Coupling Reactions

4.5.3.1   Silylmediated Coupling Reactions

The preparation of potential anti-HIV N-Isothionucleosides was described start-ing from glucose. The key coupling reaction proceeds in low yield between the pyrimidine base and the mesyl tetrahydrothiophene derivative under potassium conditions (Figure 4.78).[132]

N-Thioxonucleosides are another class of N-thionucleosides tested as anti-HIV agents. The conditions employed for performing the coupling reaction were TMSOTf as Lewis acid catalyst, affording a mixture of anomers (α:β, 1:2) in 64% (Figure 4.79).[133]

Thietane nucleoside was synthesized starting from the benzoyl thietane deriva-tive, which prior to the coupling reaction was treated with peroxide to produce the

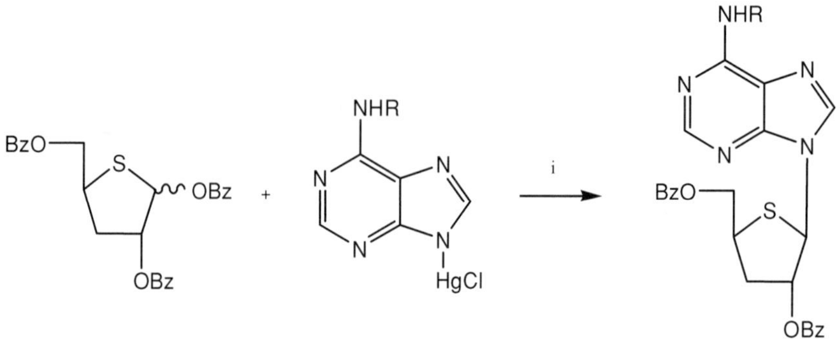

i) TiCl$_4$, EtCl$_2$, heat.

Fig.[129]

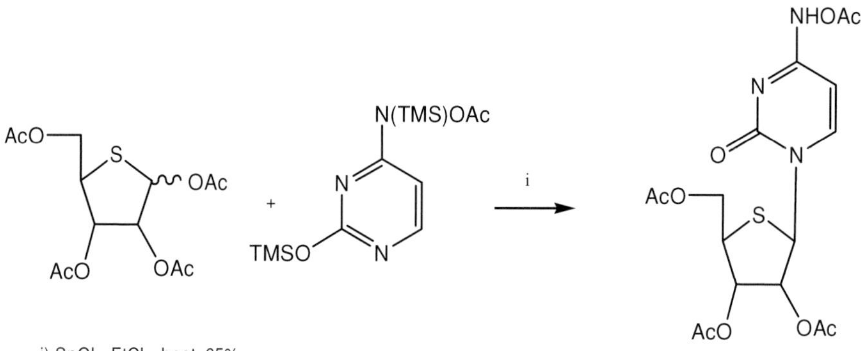

i) SnCl$_4$, EtCl$_2$, heat, 65%.

Ref.[128]

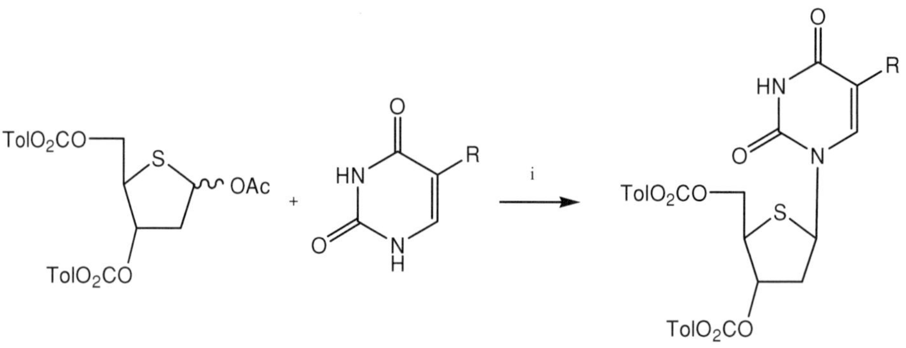

i) a) HMDS, TMSCl, MeCN. b) TfOTMS.

Ref.[130]

FIGURE 4.77. (*Continued*)

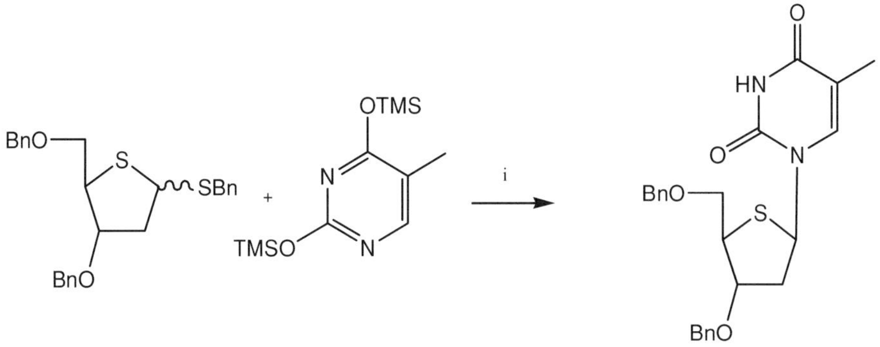

i) HgBr$_2$, CdCO$_3$, toluene.

Ref.[121]

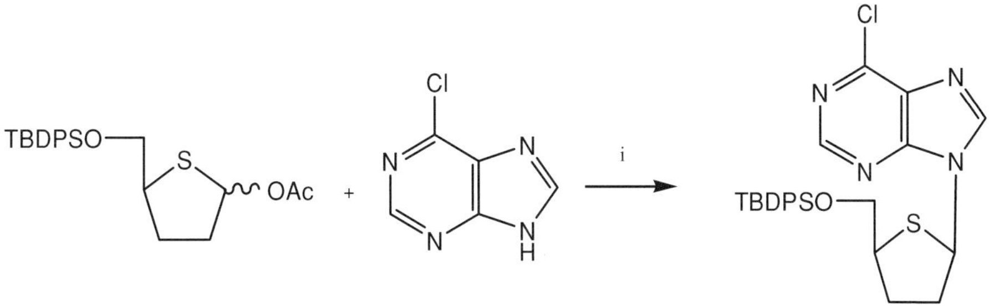

i) Et$_3$AlCl.

Ref.[131]

FIGURE 4.77. (*Continued*)

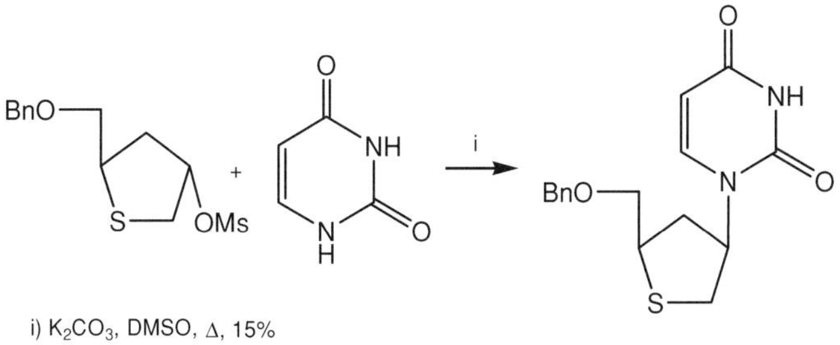

i) K$_2$CO$_3$, DMSO, Δ, 15%

FIGURE 4.78. Preparation of N-Isothionucleoside.

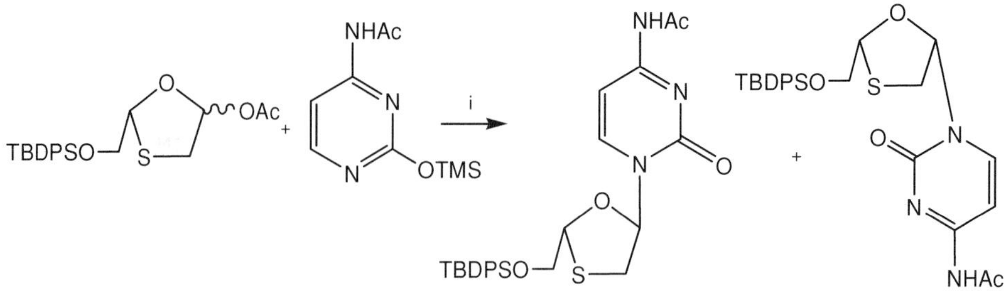

i) TMSOTf, CH$_2$Cl$_2$, 64%

FIGURE 4.79. Preparation of N-thioxonucleosides.

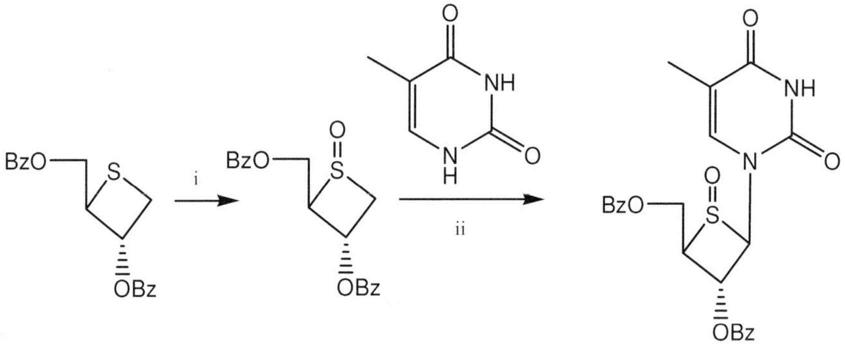

i) TMSOTf, Et$_3$N, ZnI$_2$, toluene, 30 %.

FIGURE 4.80. Synthesis of thymidine thietane nucleoside.

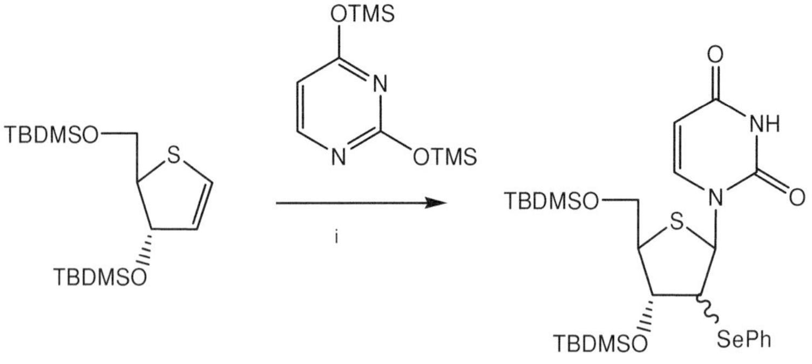

i) PhSeCl 88 %.

FIGURE 4.81. Synthesis β-4′-thionucleosides based on electrophilic glycosidation of 4-thiofuranoid glycols.

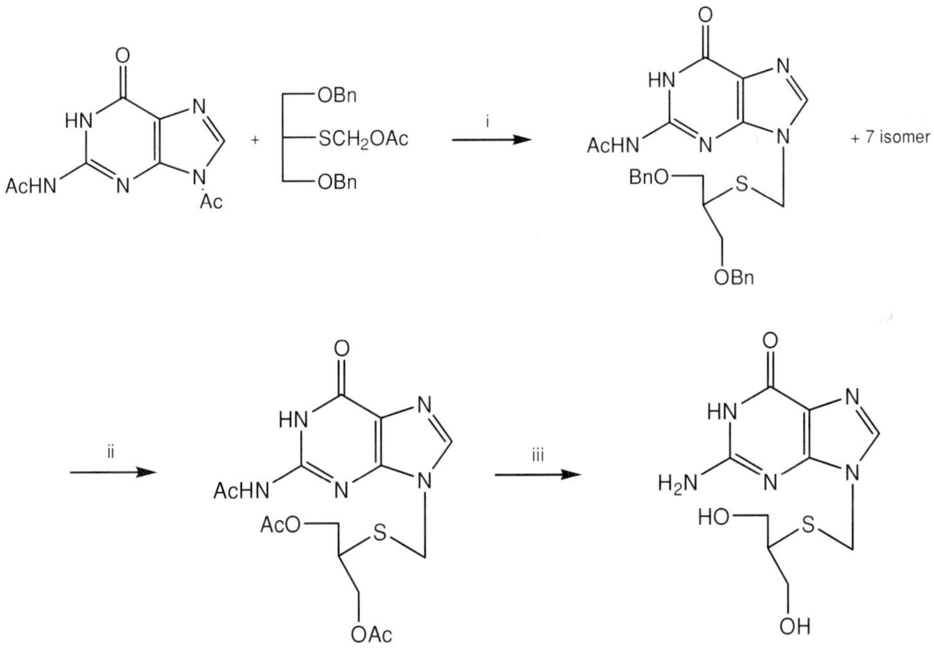

i) (p-NO$_2$C$_6$H$_4$)$_2$P(O)OH. ii) (C$_2$H$_5$)O-BF$_3$, Ac$_2$O. iii) NH$_3$.

FIGURE 4.82. Synthesis of thio analog of DHPG.

sulfoxide derivative. Then under Lewis acid conditions a Pummerer rearrangement process takes place to produce in the presence of thymine the expected thietane nucleoside (Figure 4.80).[127]

More recently the stereoselective synthesis β-4′-thionucleosides based on electrophilic glycosilation of 4-thiofuranoid glycals was described. Thus, the condensation of TBDMS-4-thioglycal with silylated uracil in the presence of PhSeCl as electrophile furnished thionucleosides in 88% as a mixture of anomers (α:β; 1:4) (Figure 4.81).[134]

The thio analog of antiviral DHPG with comparable activity to DHPG against HSV1 and human cytomegalovirus was synthesized according to the figure shown below (Figure 4.82).[110]

References

1. H. Mitsuya, R. Yarchoan, and S. Broder, *Science*, 249, 1533 (1990).
2. D.M. Huryn and M. Okabe, *Chem. Rev.*, 92, 1745 (1992).
3. H.Mitsuya and S. Broder, S. *Proc. Natl. Acad. Sci. USA*, 83, 1911 (1986).
4. C. Simons, *Nucleoside Mimetics*, Ed. Gordon and Breach Science Publishers, (2001).

5. L.A. Agrofolio, I. Guillaizeau, and Y. Saito, *Chem. Rev.*, 103, 1875 (2003).
6. G. Crisp and B.L. Flynn, *J. Org. Chem.*, 58, 6614 (1993).
7. T.S. Mansur, C.A. Evans, M. Charron, and B.E. Korba, *Bioorg. Med. Chem. Lett.*, 7, 303 (1997).
8. V. Farina and S.I. Hauck, *Synlett*, 157 (1991).
9. S.G. Rahim, N. Trivedi, M.V. BogunovicBatchelor, G.W. Hardy, G. Mills, J.W. Selway, W. Snowden, E. Littler, P.L. Coe, I. Basnak, R.F. Whale, and R.T. Walker, *J. Med. Chem.*, 39, 789 (1996).
10. (a) R.F. Heck, *J. Am. Chem. Soc.*, 90, 5518 (1968). (b) J.J. Li and G.W. Gribble, *Palladium in Heterocyclic Chemistry*; Pergamon Press: New York (2000).
11. T. Hanamoto, T. Kobayashi, and M. Kondo. *Synlett*, 2, 281 (2001).
12. G. Palmisano, and M. Santagostino, *Tetrahedron*, 49, 2553 (1993).
13. J.H. Lister, in: *The chemistry of Heterocyclic Compounds* (Eds. Weissberger, A., Taylor, E.C.), Wiley Interscience, vol. 24. (1971).
14. G. Shaw, in: *Comprehensive Heterocycle Chemistry* (Eds. Katritzky, A.R., Rees, C.W.), Pergamon Press, 5, 501 (1984).
15. M. Hocek, Eur. *J. Org. Chem.*, 245 (2003).
16. V. Nair and S.D. Chamberlain, *J. Am. Chem. Soc.*, 107, 2183 (1985).
17. V. Nair and D. Young, *J. Org. Chem.*, 49, 4340 (1984).
18. K. Tanji and T. Higashino, *Heterocycles*, 30, 435 (1990).
19. H. Vorbrügen and K. Krolikiewicz, *Angew. Chem.*, 88, 724 (1976).
20. E.C. Taylor and S.F. Martin, *J. Am. Chem. Soc.*, 96, 8095 (1974).
21. R. Mornet, N.J. Leonard, M. Theiler, and M. Doree, *J. Chem. Soc. Perkin Trans 1*, 879 (1984).
22. T.C. McKenzie and D. Glass, *J. Heterocycl. Chem.*, 24, 1551 (1987).
23. C.D. Nguyen, L. Beaucourt, and L. Pichat, *Tetrahedron Lett.*, 20, 3159 (1979).
24. K. Hirota, Y. Kitade, Y. Kanbe, and Y. Maki, *J. Org. Chem.*, 57, 5268 (1992).
25. H. Dvo_'aková, D. Dvo_'ak, and A. Hol_, *Tetrahedron Lett.*, 37, 1285 (1996).
26. L.L. Gundersen, A.K. Bakkestuen, A.J. Aasen, H. Øveras, and F. Rise, *Tetrahedron* 50, 9743 (1994).
27. A.A. Van Aerschot, P. Mamos, N.J. Weyns, S. Ikeda, E. De Clercq, and P. Herdewijn, *J. Med. Chem.*, 36, 2938 (1995).
28. S. Vottori, E. Camaioni, E. Di Francesco, R. Volpini, A. Monopoli, S. Dionisotti, E. Ongini, and G. Cristalli, *J. Med. Chem.*, 39, 4211 (1996).
29. E. Edstrom and Y. Wei, *J. Org. Chem.*, 60, 5069 (1995).
30. J. Balzarini, G.J. Kang, M. Dalal, P. Herdeweijn, E. De Clercq, S. Broder, and D.G. Johns, *Mol. Pharmacol.*, 32, 162 (1987).
31. Y. Hamamoto, H. Nakashima, T. Matsui, A. Matsuda, T. Ueda, and N. Yamamoto, *Antimicrob. Agents Chemother.*, 31, 907 (1987).
32. P.S. Manchand, P.S. Belica, M.J. Holman, T.N. Huang, H. Maehr, S.Y.K. Tam, and T. Yang, T. *J. Org. Chem.*, 57, 3473 (1992).
33. M.J. Robins, F. Hansske, N.W. Low, and J.I. Park, *Tetrahedron Lett.*, 25, 367 (1984).
34. T.S. Lin, M.Z. Luo, M.Ch. Liu, Y.L. Zhu, E. Gullen, E.G. Dutschman, and Y. Ch. Cheng, *J. Med. Chem.*, 39, 1757 (1996).
35. E.J. Corey and R.A.E. Winter, *J. Am. Chem. Soc.*, 85, 2677 (1963).
36. L.W. Dudycz, *Nucleosides, Nucleotides*, 8, 35 (1989).
37. E.J. Corey, and P.B. Hopkins, *Tetrahedron Lett.*, 23, 1979 (1982).

38. M.M. Mansuri, J.E. Jr. Starrett, J.A. Wos, D.R. Tortolani, P.R. Brodfuerhrer, H.G. Howell, and J.C. Martin, *J. Org. Chem*, 54, 4780 (1989).

39. H. Shiragami, Y. Irie, H. Yokozeki, and N. Yasuda, *J. Org. Chem.*, 53, 5170 (1988).

40. A. Rosowsky, V.C. Solan, J.G. Sodroski, and R.M. Ruprecht, *J. Med. Chem.*, 32, 1135 (1989).

41. C.H. Kim, V.E. Marquez, S. Broder, H. Mitsuya, and J.S. Driscoll, *J. Med. Chem.*, 30, 862 (1987).

42. D.H.R. Barton, D.O. Jang, and J.C. Jaszberenyi, *Tetrahedron Lett.*, 32, 2569 (1991).

43. C. Chu, U.T. Bhadti, B. Doboszowski, Z.P. Gu, Y. Kosugi, K.C. Pullaiah, and P. Van Roey, *J. Org. Chem.* 54, 2217 (1989).

44. G.W.J. Fleet, J.C. Son, and A.E. Derome, *Tetrahedron*, 44, 625 (1988).

45. W. Zhou, G. Gumina, Y. Chong, J. Wang, R.F. Schinazi, and Ch.K. Chu, *J. Med. Chem.*, 47, 3399 (2004).

46. F. Hansske and M.J. Robins, *J. Am. Chem. Soc.*, 105, 6736 (1983).

47. M.S. Motawia, J. Wendel, A.E.S. AbdelMegid, and E.B. Pedersen, *Synthesis*, 384 (1989).

48. L. Svansson, I. Kvarnström, B. Classon, and B. Samuelson, *J. Org. Chem.*, 56, 2993 (1991).

49. M. Okabe, R.C. Sun, S.Y.K. Tam, L.J. Todaro, and D.L. Coffen, *J. Org. Chem.* 53, 4780 (1988).

50. C.K. Chu, J.W. Beach, G.V. Ullas, ad Y. Kosugi, *Tetrahedron Lett.*, 29, 5349 (1988).

51. J.F. Lavallée and G. Just, *Tetrahedron Lett.*, 32, 3469 (1991).

52. J.P. Horwitz, J. Chua, and M. Noel, *J. Org. Chem.*, 29, 2076 (1964).

53. J.L. Rideout, D.W. Barry, S.N. Lehman, M.H. St. Clair, P.A.Furman, and G.A. Freeman, 3,608,606; E.P. *Chem. Abst.* 106, P38480b (1987).

54. V.E. Zaitseva, N.B. Dyatkina, A.A. Krayavskii, N.V. Skaptsova, O.V. Turina, N.V. Gnuchev, B.P. Gottikh, and A.V. Azhaev, *Bioorg. Khim.*, 10, 670 (1984).

55. J.D. Wilson, M.R. Almond, J.L. Rideout, E.P. 295, 090; *Chem. Abst.*, 111, P23914a (1989).

56. M.E. Jung and J.M. Gardiner, *J. Org. Chem.*, 113, 2614 (1991).

57. M.W. Hager and D.C. Liotta, *J. Am. Chem. Soc.*, 113, 5117 (1991).

58. G.A. Freeman, S.R. Shauer, J.L. Rideout, and S.A. Short, *Bioorg. Med. Chem.* 3, 447 (1995).

59. V.N. Barai, A.I. Zinchenko, L.A. Eroshevskaya, E.V. Zhernosek, J. Balzarini, E. De Clercq, and I.A. Mikhailopulo, *Nucleosides, Nucleotides, and Nucleic Acids* 22, 751 (2003).

60. K. Izawa, S. Takamatsu, S. Katayama, N. Hirose, S. Kosai, and T. Maruyama, *Nucleosides, Nucleotides & Nucleic Acids*, 22, 507 (2003).

61. K. Haraguchi, S. Takeda, and H. Tanaka, *Org. Lett.*, 5, 1399 (2003).

62. H. Ohrui, S. Kohgo, K. Kitano, S. Sakata, E. Kodama, K. Yoshimura, M. Matsuoka, S. Shigeta, and H. Mitsuya, *J. Med. Chem.*, 43, 4516 (2000).

63. T. Kondo, T. Ohgi, and T. Goto, *Chem Lett.*, 419 (1983).

64. H. Akimoto, E. Imamiya, T. Hitaka, and H. Nomura, *J. Chem. Soc. Perkin Trans. 1*, 1637 (1988).

65. Ch.J. Barnett and L.M. Grubb, *Tetrahedron,* 56, 9221 (2000).

66. S. Knapp and S.R. Nandan, *J. Org. Chem.*, 59, 281 (1994).

67. H. Hotoda, M. Daigo, T. Takatsu, A. Muramatsu, and M. Kaneko, *Heterocycles*, 52, 133 (2000).

68. A.G. Myers, D.Y. Gin, and D.H. Rogers, *J. Am. Chem. Soc.*, 116, 4697 (1994).
69. Ch. McGwigan, H. Barucki, S. Blewett, A. Caragio, J.T. Erichsen, G. Andrei, R. Snoock, E. De Clercq, and J. Balzarini, *J. Med. Chem.*, 43, 4993 (2000).
70. A.R. Porcari and L.B. Towsend, *Nucleosides, Nucleotides & Nucleic Acids*, 23, 31 (2004).
71. S. Hannesian and A.G. Pernet, *Adv. Carbohydr. Chem.Biochem.*, 33, 111 (1976).
72. F. De las Heras, S.Y. Tam, R.S. Klein, and J.J. Fox, *J. Org. Chem.*, 41, 84 (1976).
73. M. Bobek, J. Farkas, and F. Sorm, *Collec. Czech. Chem. Commun*, 34, 1690 (1969).
74. W.A. Asbun, and S.B. Binkley, *J. Org. Chem.*, 33, 140 (1968).
75. P.C. Sriswastava, M.V. Pickering, L.B. Allen, D.G. Streeter, M.T. Campbell, J.T. Witkowski, R.W. Sidwell, and R.K. Robins, *J. Med. Chem.*, 20, 256 (1977).
76. K.S. Ramasamy, R. Bandaru, and D. Averett, *J. Org. Chem.*, 65, 5849 (2000).
77. G. Trummlitz, and J. G. Moffat, *J. Org. Chem.*, 38, 1841 (1973).
78. U. Von Krosigk, and S.A. Benner, *Helv. Chim. Acta*, 87, 1299 (2004).
79. H.C. Zhang, G.D. Daves, *J. Org. Chem.*, 57, 4690 (1992).
80. G. Kim, and H.S. Kim, *Tetrahedron Lett.*, 41, 225 (2000).
81. J.J. Chen, J.C. Drach, and L.B. Towsend, *J. Org. Chem.*, 68, 4170 (2003).
82. S. Hannessian, S. Marcotte, Machaalani, and G. Huang, *Org. Lett.*, 2003, 5, 4277 (2003).
83. T. Yokamatsu, M. Salto, H. Abe, K. Suemune, K. Matsumoto, T. Kihara, S. Soeda, H. Shimeno, and S. Shibuya, *Tetrahedron* 53, 11297 (1997).
84. N. Katagiri, Y. Morishita, and M. Yamaguchi, *Tetrahedron Lett.*, 39, 2613 (1998).
85. Deardorff, D.R., Mattews, A.J., McKeenin, D.S., Craney, C.L. *Tetrahedron Lett.* 1986, 27, 1255.
86. Deardorff, D.R., Shambayati, S., Myles, D.C., Heerding, D. *J. Org. Chem.* 1989, 54, 3614.
87. Deardorff, D.R., Savin, K.A., Justman, C.J., Karanjawala, Z.E., Sheppeck, J.E.II., Hager, D.C., Aydin, N *J. Org. Chem.* 1996, 61, 3616.
88. P. Herdewijn, J. Balzarini, E. De Clercq, and H. Vanderhaeghe, *J. Chem. Med.*, 28, 1385 (1985).
89. D.J. Tenney, S.M. Yamanaka, C.W. Voss, C.W. Cianci, A.V. Tuomari, A.K. Sheaffer, M. Alam, and R.J. Colonno, *Antimicrob. Agents Chemother.*, 41, 2680 (1997).
90. D.H.R. Barton, and M. Ramesh, *J. Am. Chem. Soc.*, 112, 891 (1990).
91. M. Kitagawa, S. Hasegawa, S. Saito, N. Shimada, and T. Takita, *Tetrahedron Lett.*, 32, 3531 (1991).
92. M. Honjo, T. Maruyama, Y. Sato, and T. Horii, *Chem. Pharm. Bull.*, 37, 1413 (1989).
93. G.S. Bisacchi, A. Braitman, C.W. Cianci, J.M. Clark, A.K. Field, M.E. Hagen, D.R. Hockstein, M.F. Malley, T. Mitt, W.A. Slusarchyk, J.E. Sundeen, B.J. Terry, A.V. Tuomari, E.R. Weaver, M.G. Young, and R. Zahler, *J. Med. Chem.* 34, 1415 (1991).
94. Y. Ohnishi, and Y. Ichikawa, *Bioorg. Med. Chem. Lett.*, 12, 997 (2002).
95. P. Russ, P. Schelling, L. Scapozza, G. Folkers, E. De Clercq, V.E. Marquez, *J. Med. Chem.*, 46, 5045 (2003).
96. O.R. Ludek, and C. Meier, *Nucleosides, Nucleotides and Nucleic Acids*, 22, 683 (2003).
97. H.C. Zhang, and G.D. Daves, *J. Org. Chem.*, 58, 2557 (1993).
98. M.T. Crimmins, and W.J. Zuercher, *Org. Lett.*, 2, 1065 (2000).
99. T. Obara, S. Shuto, Y. Saito, R. Snoeck, G. Andrei, J. Balzarini, E. De Clercq, and A. Matsuda, *J. Med. Chem.*, 39, 3847 (1996).

100. E.A. SavilleStones, S.D. Lindell, N.S. Jennings, J.C. Head, and M.J. Ford, *J. Chem. Soc., Perkin Trans* 1, 2603 (1991).

101. L.L. Gundersen, T. Benneche, and K. Undheim, *Tetrahedron Lett.*, 33, 1085 (1992).

102. L.S. Jeong, J.G. Park, W.J. Choi, H.R. Moon, K.M. Lee, H.O. Kim, H.D. Kim, M.W. Chun, H.Y. Park, K.Kim, Y.Y. Sheng, *Nucleosides, Nucleotides and Nucleic Acids*, 22, 919 (2003),

103. R.H. Foster and D. Faulds, *Drugs*, 55, 729 (1998).

104. M.T. Crimmins and B.W. King, *J. Org. Chem.*, 61, 4192 (1996).

105. S.M. Roberts, S.J. Taylor, A.G. Sutherland, C. Lee, R. Wisdom, S. Thomas, and C. Evans, *J. Chem. Soc. Chem. Commun.*, 1120 (1990).

106. M.I. Lim and V.E. Marquez, *Tetrahedron Lett.*, 24, 5559 (1983).

107. V.E. Marquez, M.I. Lim, S.P. Treanor, J. Plowman, M.A. Priest, A. Markovac, M.S. Khan, B. Kaskar, and J.S. Driscoll, *J. Med. Chem.*, 31, 1687 (1988).

108. Y.F. Shearly, C. A. O'Dell, and G. Amett, *J. Med. Chem.*, 30,1090 (1987).

109. B.K. Chun, G.Y. Song, and Ch.K. Chu, *J. Org. Chem.*, 66, 4852 (2001).

110. Ch.K. Chu, and S. Cutler, *J. Heterocyclic Chem*, 23, 289 (1986).

111. H.J. Schaeffer, L. Beauchamp, P. de Miranda, G. Elion, D.J. Bauer, and P. Collins, *Nature*, 272, 583 (1978).

112. J.R. Barrio, J.D. Bryant, and G.E. Keyser, *J. Med. Chem*, 23, 3263 (1980).

113. G.E. Keyser, J.D. Bryant, and J.R. Barrio, *Tetrahedron Lett.*, 3263 (1979).

114. M.J. Robins and P.W. Hatfield, *Can. J. Chem.*, 60, 547 (1982).

115. L. Naesens and E. De Clercq, *Nucleotides & Nucleosides*, 16, 983 (1997).

116. Q. Dang, Y. Liu, and M.D. Erion, *Nucleotides & Nucleosides*, 17, 1445 (1998).

117. A.K. Field, M.E. Davies, C. de Witt, H.C. Perry, R. Liou, J.L. Germerhausen, J.D. Karkas, W.T. Ashton, D.B. Johnson, and R.L. Tolman, *Proc. Natl. Acd. Sci.*, 80, 4139 (1983).

118. M. Yokohama, *Synthesis*, 1637 (2000).

119. N.A. Van Drannen, G.A. Freeman, S.A. Short, R. Harvey, R. Jansen, G. Szczech, and G.W. Koszalka, *J. Med. Chem.*, 39, 538 (1996).

120. S.G. Rahim, N. Trivedi, M.V. Bogunovic, G.W. Batchelor, G. Hardy, J.W. Mills, W. Selway, E. Snowden, P.L Littler: Coe, I. Basnak, R.F. Whale, and R.T. Walker, *J. Med Chem.*, 39, 789 (1996).

121. M.R. Dyson, P.L. Coe, and R.T. Walker, *J. Med. Chem.*, 34, 2782 (1991).

122. E.J. Reist, D.E. Gueffroy, and L. Goodman, *J. Am. Chem. Soc.*, 86, 5658 (1964).

123. E.J. Reist, L.V. Fischer, and L. Goodman, *J. Org. Chem.*, 33, 189 (1968).

124. R.G.S. Ritchie, D.M. Vyals; and W.A. Szarek, *Can. J. Chem.*, 56, 794 (1978).

125. K. Haraguchi, A. Nishikawa, E. Sasakura, H. Tanaka, K. Nakamura, and T. Miyasaka, *Tetrahedron Lett.*, 39, 3713 (1998).

126. T. Naka, N. Nishizono, N. Minakawa, and A. Matsuda, *Tetrahedron Lett.*, 40, 6297 (1999).

127. N. Nishikono, N. Koike, Y. Yamagata, S. Fujii, and A. Matsuda, *Tetrahedron Lett.*, 37, 7569 (1996).

128. M. Bobek, A. Bloch, R. Parthesarathy, and R.L. Whistler, J. Am. Chem. Soc., 13, 411 (1970).

129. Ritchie et al., *J. Chem. Soc. Chem. Commun*, 1973, 86, 686.

130. M. Bobek, A. Bloch, R. Parthasarathy, and R.L. Whistler, *J. Med. Chem.*, 18, 784, (1975).

131. J.A. Secrist III, K.M. Tiwari, J.M. Riordan, and J. A. Montgomery, *J. Med. Chem.* , 34, 2361 (1991).
132. U. Niedballa and H. Vorbrüggen, *J. Org. Chem.* 1974, 39, 3654.
133. M.F. Jones, S.A. Noble, C.A. Robertson, and R. Storer, *Tetrahedron Lett.*, 32, 247 (1991).
134. J.W. Beach, L.S. Jeong, A.J. Alves, D. Pohl, H.O. Kim, C.N. Chang, S.L. Doong, R.F. Schinazi, Y.C. Cheng, C.K. Chu, *J. Org. Chem.*, 57, 2217 (1992).

5
C-Glycosides

These types of glycosides have attracted much attention, considering that many of them have demonstrated their effectiveness as therapeutic agents. The increasing significance of C-glycosides is that the conformational differences compared to O- or N-glycosides are minimal and that they are resistant to enzymatic or acidic hydrolysis since the anomeric center has been transformed from acetal to ether.[1] A glycoside is defined as C-glycoside when what it is suppose to be the anomeric carbon of a sugar is interconnected to the aglycon, generating a new C-C bond. According to Levy and Tang[2] the term C-glycoside describes those structures in which a common structural motifs the presence of carbon functionality at what would otherwise be the anomeric position of a sugar or derivative. Structurally C-glycosides can be constituted by aliphatic, or aromatic aglycon, and the sugar can be pyranose or furanose. A variety of natural product C-glycosides has been described. Examples of C-glycosides isolated from different plant genus and characterized spectroscopically are: Carminic acid, Aloin, Scoparin, Saponarin, and more recently Cucumerins (flavonoid phytoalexins)[3] and C-glucosylxanthones[4] and complex benzoquinone Altromycin B[5] among others (Figure 5.1).

Moreover, much effort and creativity have been devoted to the preparation of complex C-glycosides with potent antibiotics activity. That is the case of Aurodox[6], Lasalocid A,[7] Herbicidin,[8] and the hyperfunctionalized molecules Spongistatin[9] and Palytoxin[10] (Figure 5.2).

5.1   Synthetic Approaches for the Preparation of C-Glycosides

Accordinng to comprehensive studies,[2,11,12,13] the general strategies for C-glycosides can be overviewed as follows:

- Electrophilic glycosyl donors
- Concerted reactions
- Wittig aproximation
- Palladium mediated reactions

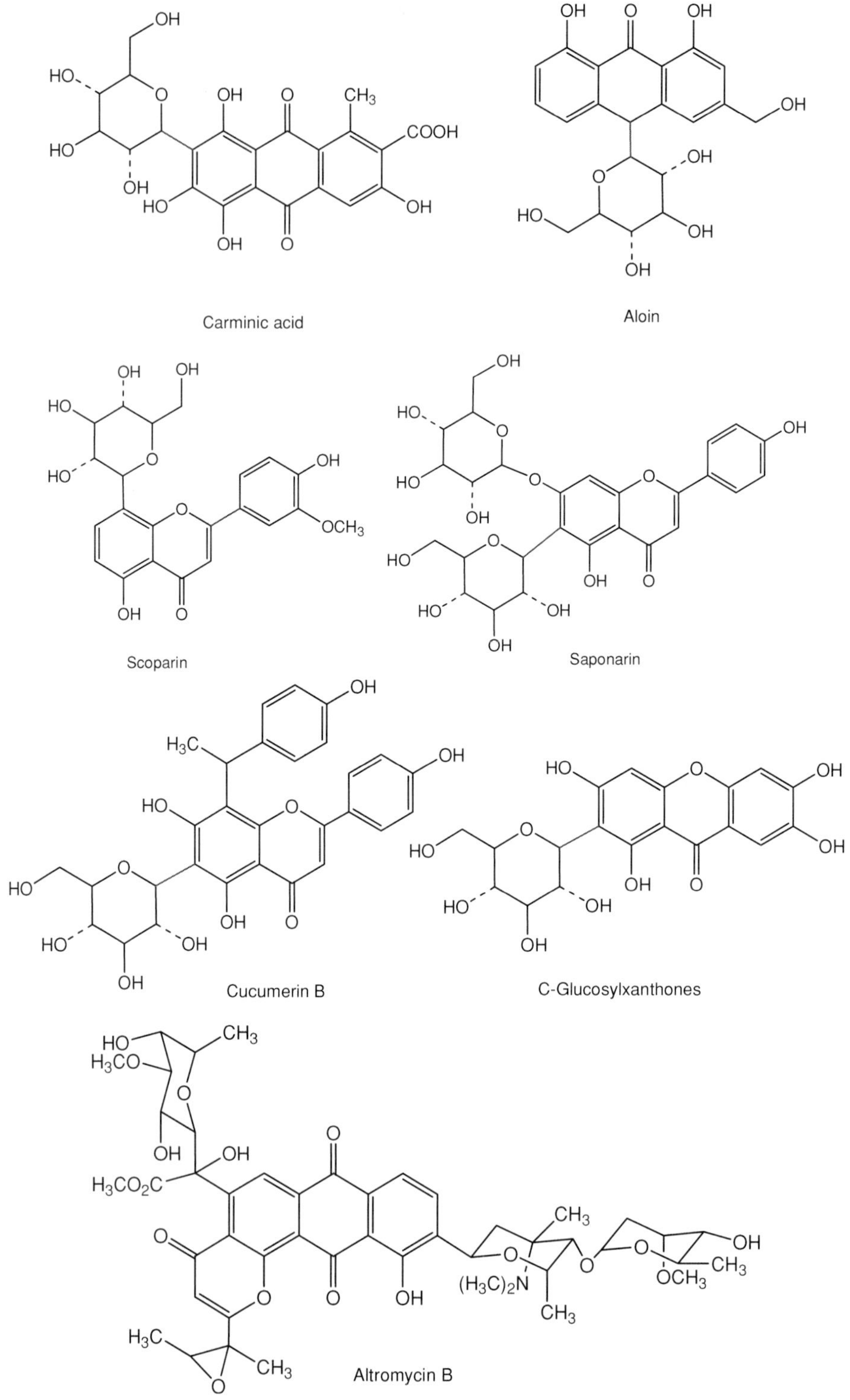

FIGURE 5.1. Some naturally ocurring *C*-glycosides.

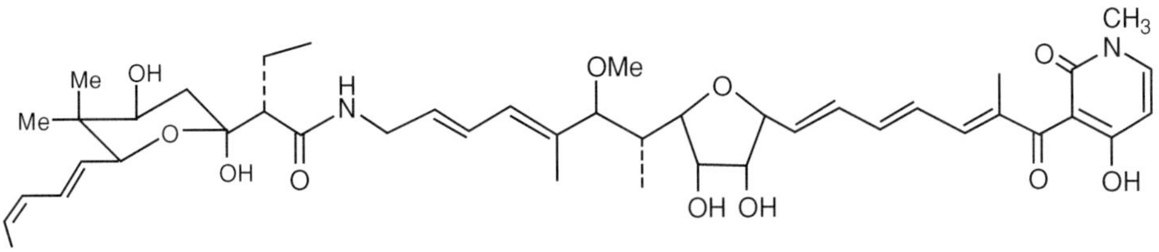

Aurodox

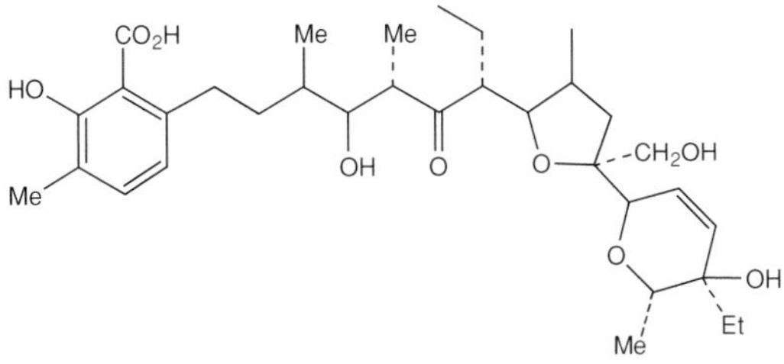

Lasalocid A

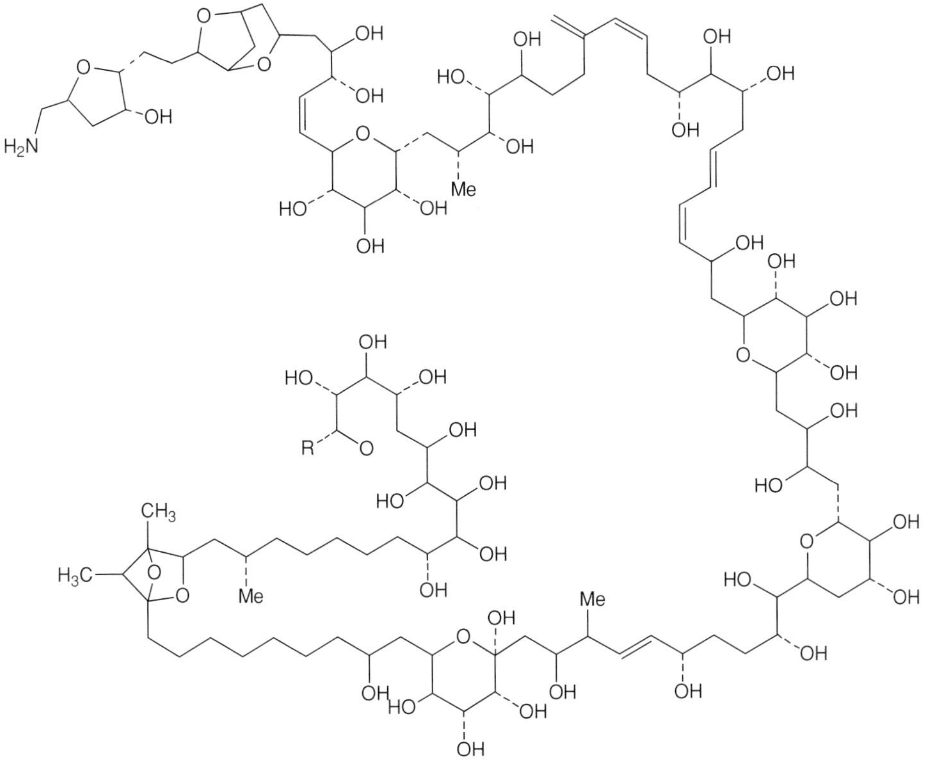

FIGURE 5.2. Complex *C*-glycoside antibiotics.

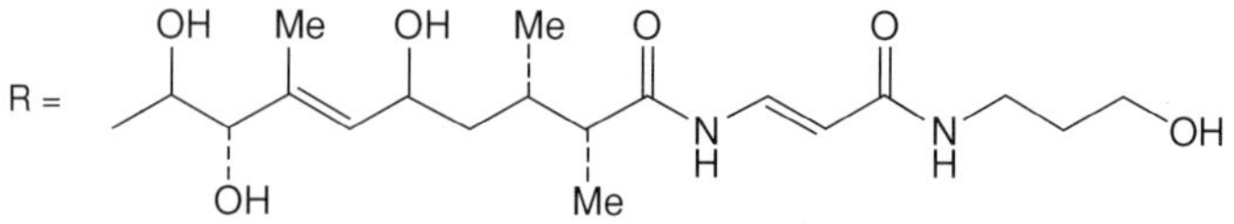

Palytoxin

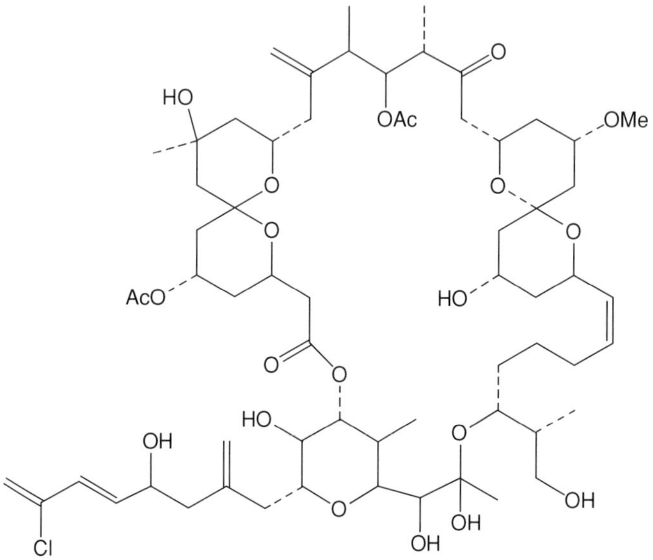

Spongistatin 1

FIGURE 5.2. (*continued*)

- Mitsunobu reaction
- Nucleophilic sugars or anomeric anions intermediates.
- Cross-metathesis reaction
- Samarium promoted reaction
- Ramberg-Bäcklund reaction
- Free radical approaches
- Exoglycals
- The tether approach

5.1.1   Electrophilic Glycosyl Donors

5.1.1.1   Glycosyl Donors Bearing Good Leaving Groups

A general approach for the *C*-glycosidic bond formation is based on nucleophilic carbon addition on the electrophilic center of a glycosyl donor. The most extensively used glycosyl donors divided in four main groups (good leaving groups, sugar lactones, glycals, and 1,2 anhydrosugars) are used as electrophilic donors

P = protecting group   M = Li, MgBr, ▬≡▬SnBu₃

X = leaving group (I, Cl, Br, imidate, acetate and Lewis acid)

to generate *C*-glycosidic bonds when reacted with organocuprates, organotin, organozinc, cyanide, allylic Grignard, vinyl silyl reagents, and activated aromatic compounds. among others.[11]

Some of the reactions described that have been used for preparing useful intermediates or C-glycosides are shown in Figure 5.3.

It is worth mentioning that for aryl C-glycosylations, there is a dependence on the electron density of the aromatic ring and the protecting groups at the glycosyl moiety.[15] Moreover, depending on the reaction conditions, there is a competing parallel process that ultimately will drive the reaction either to the *O*- or to the C-glycoside formation. This affirmation was demonstrated in the preparation of *C*- and *O*-flavonoid glycosides by Oyama et al., which treated glycosyl fluoride with flavan under Lewis acid conditions. It was observed that $BF_3.ET_2O$ and 2,6-di-tert-butyl-4-methyl pyridine (DTBMP) resulted predominantly in the formation of the 5-O-β-glycoside, while if the reaction is carried out only with BF_3-Et_2O, the C-glycoside is obtained (Figure 5.4).

5.1.1.2   Other Electrophilic Glycosyl Donors

Additionally, the introduction of other electrophilic centers at the anometic position has extended the possibilities for preparation of C-glycosides by using electrophilic sugars. Some of these electrophilic sugars are lactols, anomeric esters, glycals, anhydrides, and lactones.

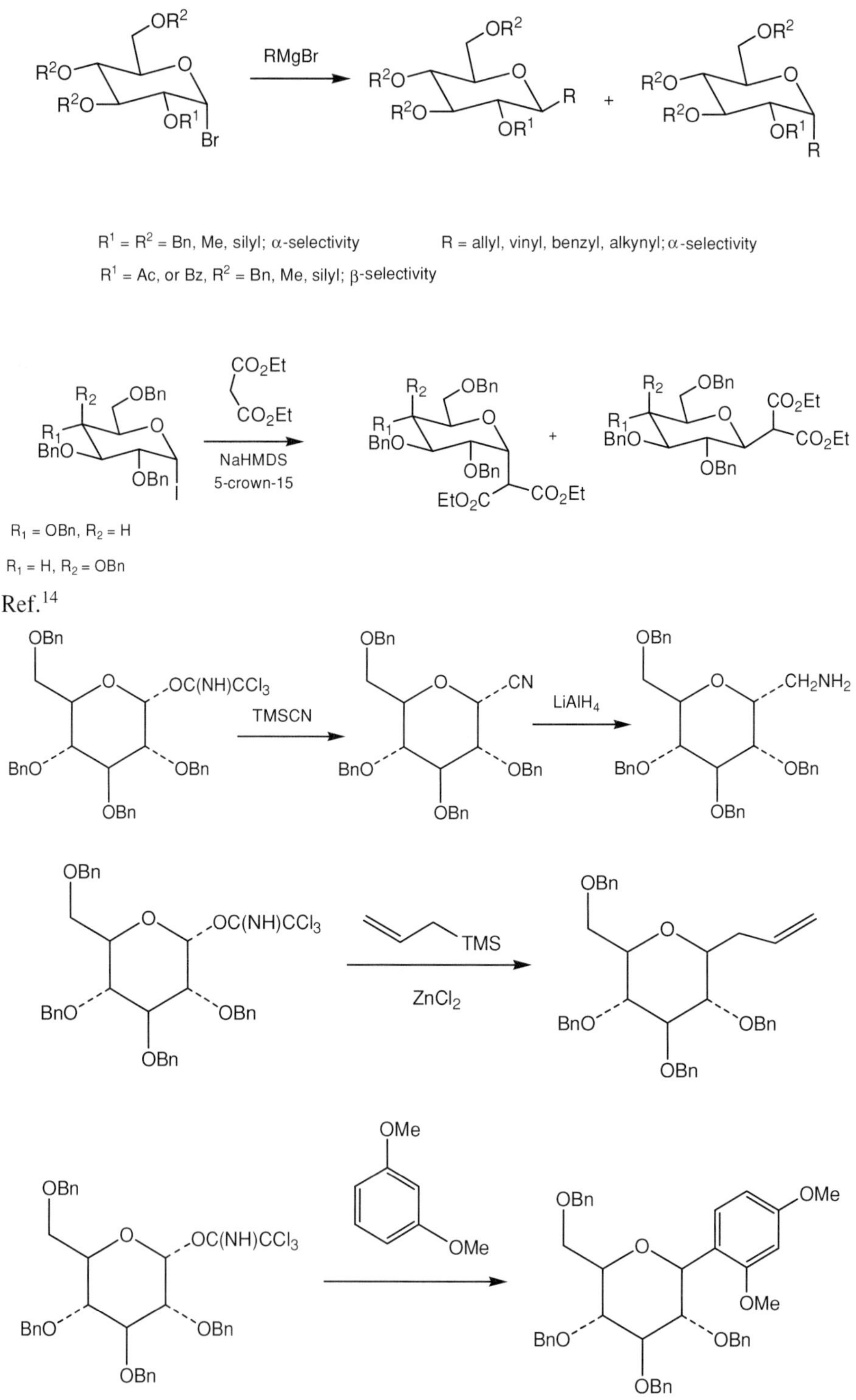
RMgBr
R¹ = R² = Bn, Me, silyl; α-selectivity
R¹ = Ac, or Bz, R² = Bn, Me, silyl; β-selectivity
R = allyl, vinyl, benzyl, alkynyl; α-selectivity
NaHMDS
5-crown-15
R₁ = OBn, R₂ = H
R₁ = H, R₂ = OBn
Ref.14
TMSCN
LiAlH₄
ZnCl₂

FIGURE 5.3. Preparation of C-glycosides or intermediates from electrophilic glycosyl donor with good leaving groups.

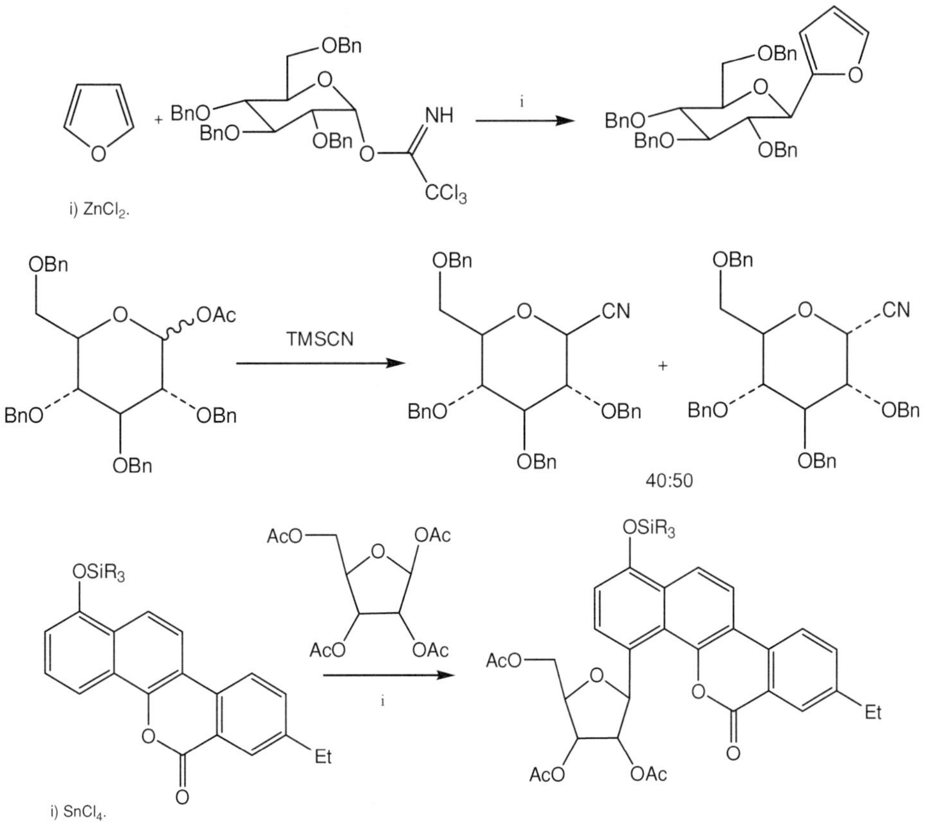

i) ZnCl₂.

40:50

i) SnCl₄.

FIGURE 5.3. (*continued*)

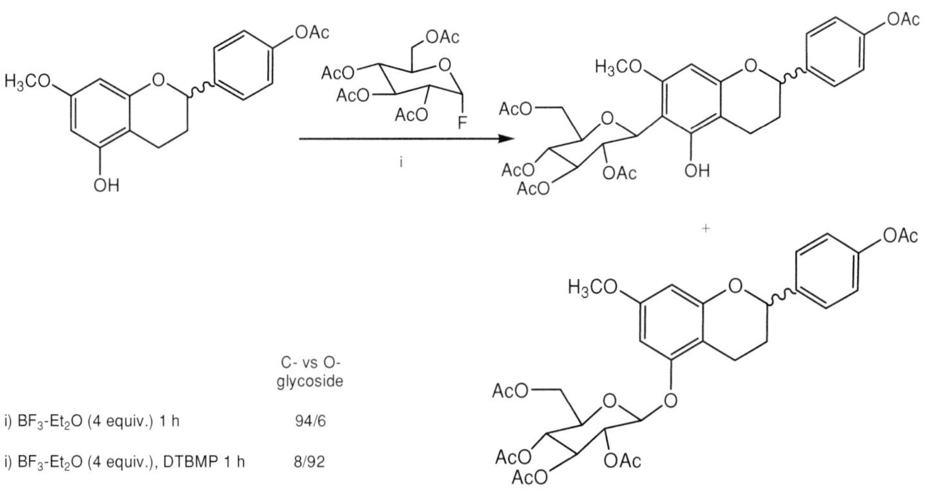

	C- vs O-glycoside
i) BF₃-Et₂O (4 equiv.) 1 h	94/6
i) BF₃-Et₂O (4 equiv.), DTBMP 1 h	8/92

FIGURE 5.4. C-glycosylations involving glycosyl donors with leaving group.

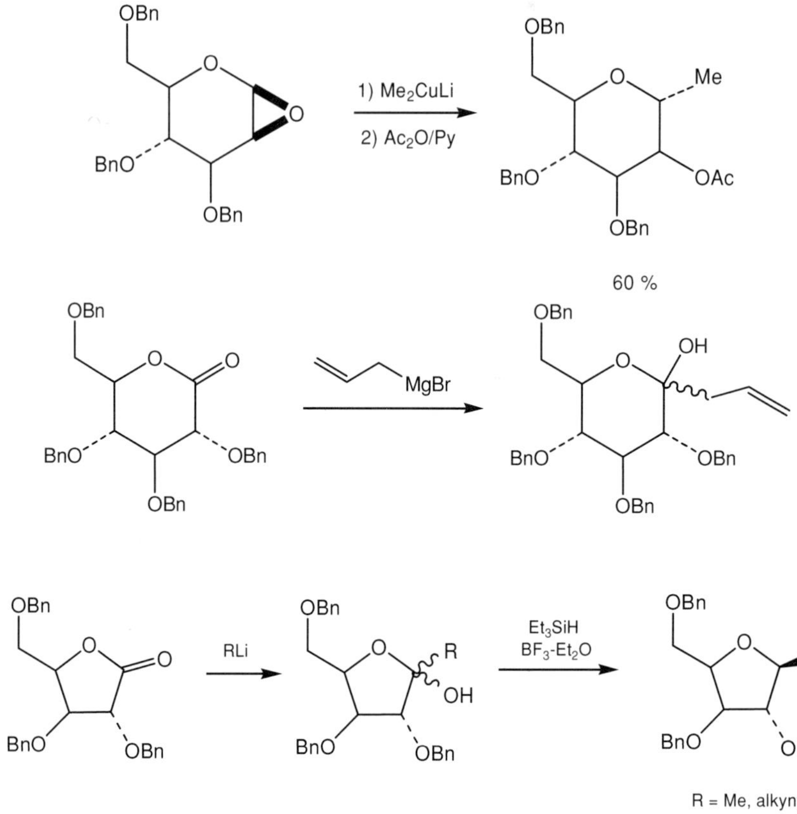

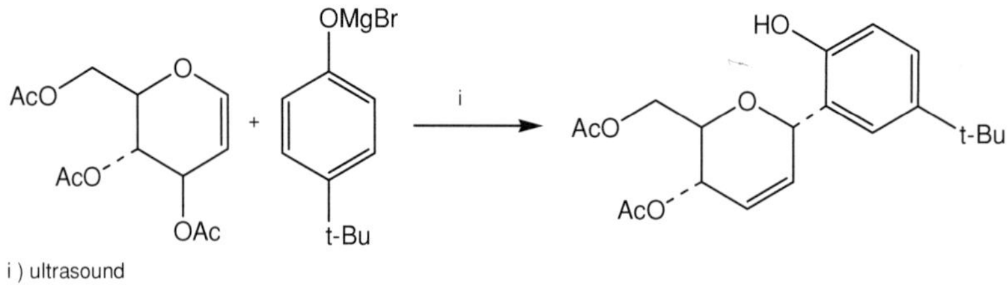

FIGURE 5.5. *C*-glycoside formation with electrophilic sugars.

Some of the reactions carried out for preparing C-glycoside intermediates involving these alternative glycosyl donors are shown in Figure 5.5. In 1,2-anhydrosugars the stereoselectivity is 1,2-trans type and involves a typical S_N2 process. On the other hand, glycals exhibit high stereoselectivity, and in glycosyl acetates the stereocontrol relies on the electronic and steric properties of the nucleophiles.

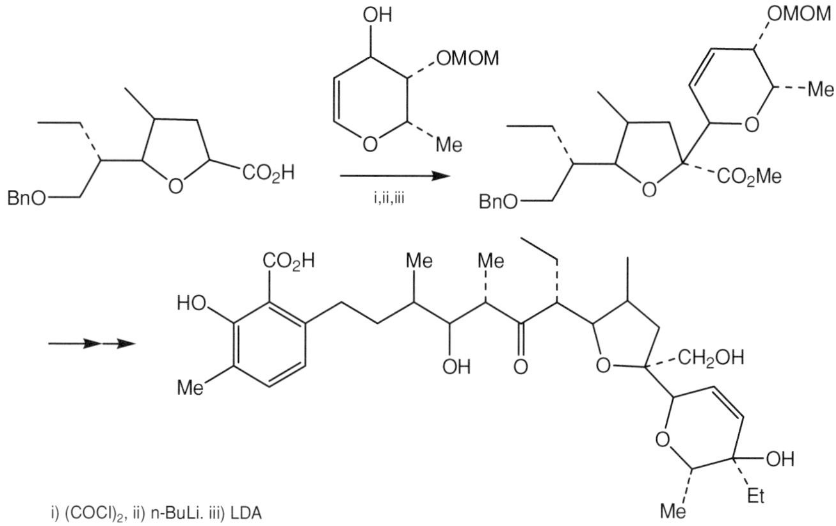

FIGURE 5.6. Synthesis of Lasalocid A.

5.1.2 Concerted Reaction and Ring Formation

This type of reactions includes sigmatropic rearrangements and cycloadition transformations. As an example of the applicability of the sigmatropic rearrangment for preparation o C-glycosydes, Ireland[7] reported the synthesis of Lasalocid A, consisting in the coupling of acid derivative with protected glycal as a result of enolate addition and Claisen rearrangement. Series of transformation of this precursor will give place to Lasalocid A (Figure 5.6).

To exemplify the effectiveness of cycloadditions for preparation of C-glycosides, Schmidt et al.[17] prepared p-methoxyphenyl 2,3,4,6-tetraacetyl C-glucopyranose, by following a Diels-Alder approach. The reaction between heterodiene and dienophile produced cycloadduct that was successively transformed to give the desired product (Figure 5.7).

Protected monosaccharide is reacted with Wittig ilide to produce a ring opening unsaturated intermediate, which was cyclized to produce a mixture of α,β C-glycosides. The α form could be converted to the β form under sodium methoxide conditions (Figure 5.8).[18]

Cation-mediated cyclization reactions of silyl enols ether-containing thioglycosides give bicyclic ketotetrahydrofurans. Treatment with sodium amalgam in buffered methanol yields the expected dihydropyran, which was transformed to the diol intermediate, and after separation converted to the bis-acetonides (Figure 5.9).[19]

Ring closure of polyalcohol has been proposed as a suitable strategy for preparing C-glycosides.[20] Condensation between iodine pyranoside intermediate with

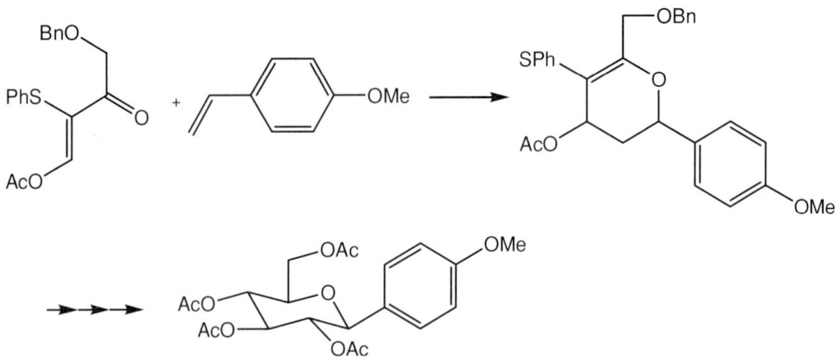
BnO
PhS
AcO
O
+
OMe
OBn
SPh
O
AcO
OMe
OAc
OMe
AcO
O
AcO
OAc

FIGURE 5.7. The Diels-Alder reaction for *C*-glycoside formation.

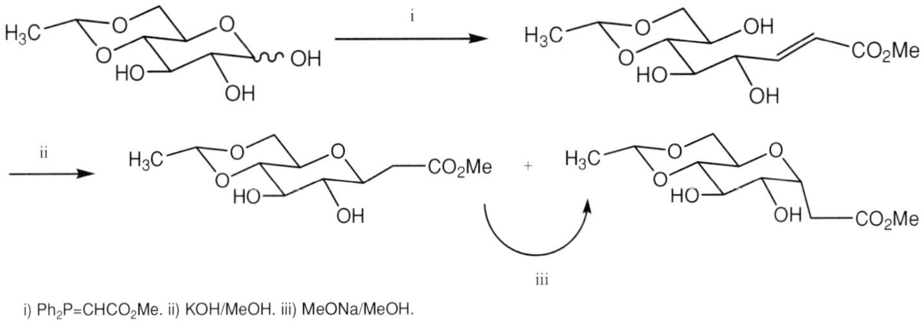
H3C
O
O
O
HO
OH
OH
i
OH
CO2Me
ii
+
iii
i) Ph2P=CHCO2Me. ii) KOH/MeOH. iii) MeONa/MeOH.

FIGURE 5.8. Wittig reaction for *C*-glycoside formation.

an aldohexose will result in the condensation product which undergoes ciclyzation to give the mixture of *C*-disaccharides shown in Figure 5.10.

5.1.3 Palladium-Mediated Reactions

Heck-type reactions have been successfully assayed for preparing interesting *C*-glycosides. Such is the case of Vineomicinone B2 prepared by palladium catalyzed condensation between TBS protected glycal with antracene derivative.[21] Further transformations will generate *C*-glycoside Vineomcinone B2 (Figure 5.11).

Other palladium-mediated coupling includes Stille (palladium-catalyzed vinyl substitution),[22] and Suzuki cross-coupling reactions. [23]

5.1.4 Mitsunobu Reaction

A Mitsunobu reaction is an additional useful reaction for preparing *C*-glycosides (Figure 5.12). When tetra-*O*-methyl glucopyranose is reacted with 1-naphtol in the

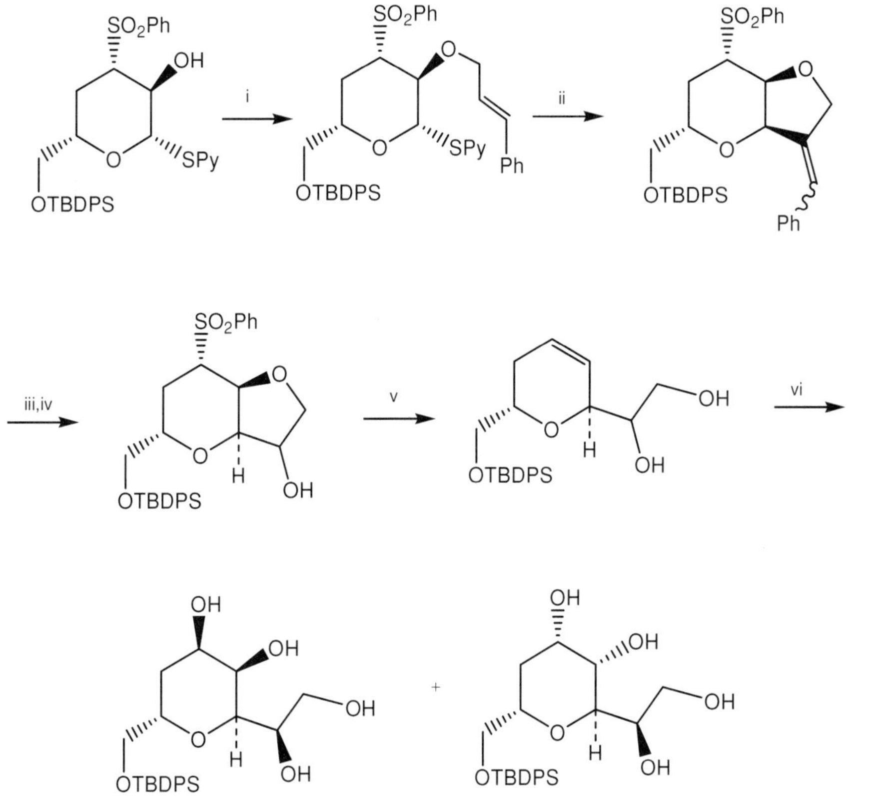

i) PhCH=CHCH$_2$Br, CH$_2$Cl$_2$, 50% aq. NaOH, r.t. ii) AgOSO$_2$CF$_3$, MS, CH$_2$Cl$_2$, r.t. then DBU. iii) O$_3$, CH$_2$Cl$_2$, −78°C, then PPh$_3$, −78°C to r.t. iv) NaBH$_4$, MeOH, 0°C. v) 6% Na(Hg), Na$_2$HPO$_4$, MeOH, 0°C. vi) OsO$_4$, NMO, 9:1 acetone-H$_2$O, r.t.

FIGURE 5.9. Formation of tetrahydrofurans and application to the synthesis of 2-octulopyranosides.

presence of Mitsunobu conditons (diethylazidodicarboxylate and triphenylphosphine), the resulting product is the *O*-glycoside, which is rearranged with BF$_3$-Et$_2$O to the corresponding *C*-glycoside.[24]

5.1.5 Nucleophilic Sugars

Anomeric carbons are considered electrophilic sites by nature; however, it is possible to invert this reactivity by using metallic bases. The resulting carbanion character is known as *umpolung* reactivity and allows the species to behave as nucleophiles. A variety of glycosyl donors have been converted to lithium or stannane glycosyl anions (Figure 5.13).[11]

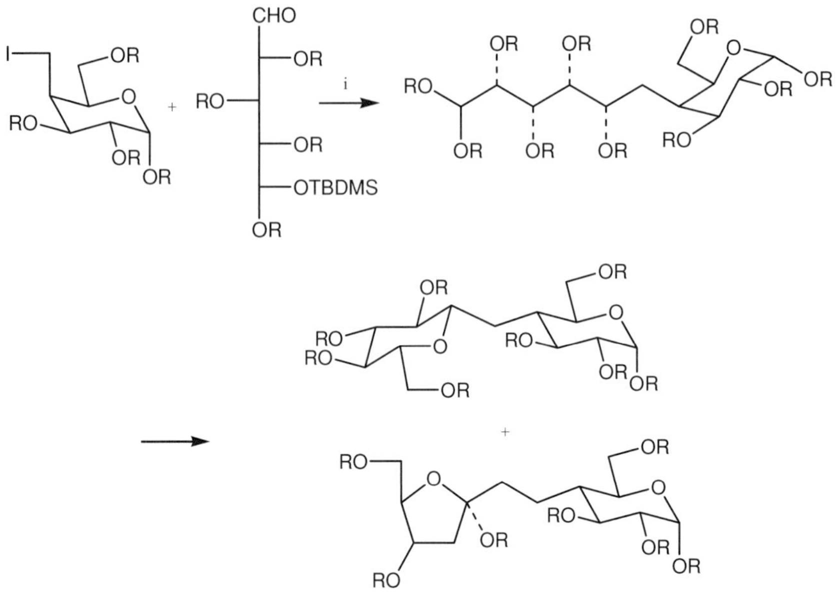

i) n-BuLi, THF

FIGURE 5.10. *C*-disaccharide formation from aldohexoses.

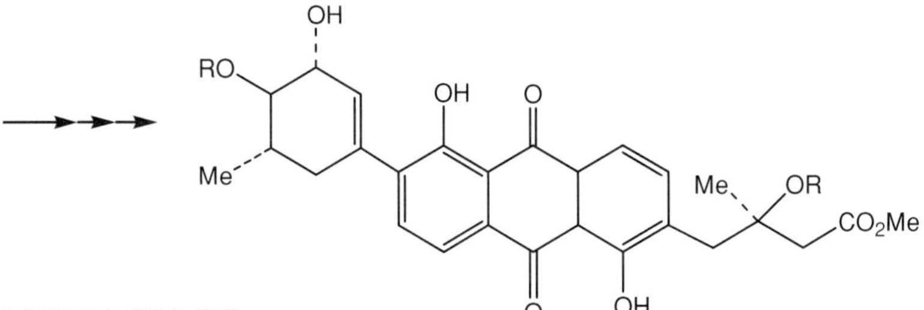

i) Pd(Ph$_3$)$_2$Cl$_2$/DIBAL/THF

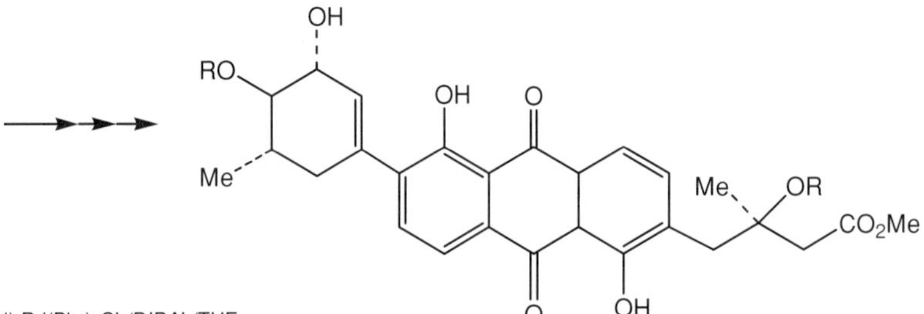

i) Pd(Ph$_3$)$_2$Cl$_2$/DIBAL/THF

FIGURE 5.11. Synthesis of Vineomicinone B2 methyl esther.

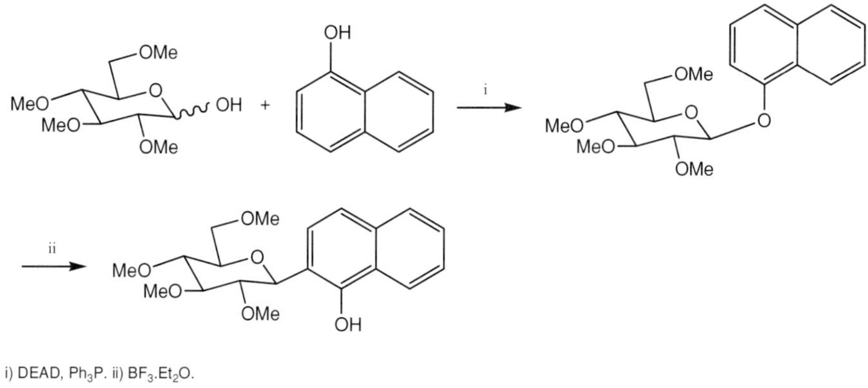

i) DEAD, Ph₃P. ii) BF₃.Et₂O.

FIGURE 5.12. Mitsunobu reaction for aromatic C-glycoside formation.

By using this possibility, the synthesis of the C-glycosyl asparagine analog has been completed by Kessler and coworkers.[25] The transformation of the stannane to the lithium donor was followed by the coupling reaction with the aldehyde glutamic acid derivative to provide the β-D-linked C-glycoside. Removal of Boc protecting group and dehydroxylation reaction under Barton-McCombie condition provided the target molecule (Figure 5.14).

Another accomplishment following this umpolung strategy was the preparation of the aromatic C-glycoside shown in Figure 5.15. Hence, lithium glycal (obtained from glycal treatment with lithium diisopropylamide) was reacted with quinolic ketal to yield addition product, which was transformed to the aromatic C-glycal.[26, 12]

Aldol condensations between glycosyl donors containing active methylene carbons with glycosyl acceptors have been also proposed as suitable approaches for preparing C-disaccharides. Martin et al,[27] described a procedure for preparing

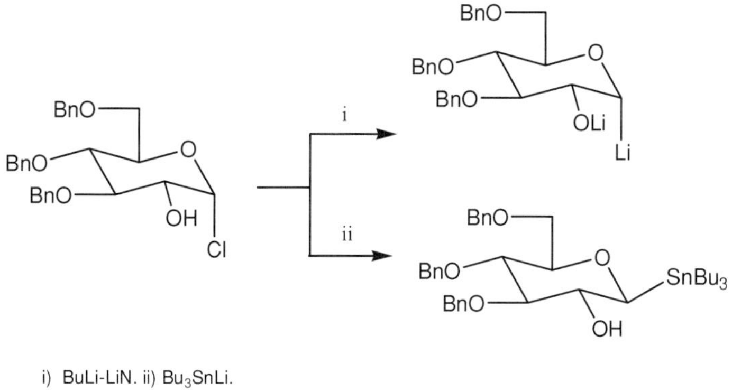

i) BuLi-LiN. ii) Bu₃SnLi.

FIGURE 5.13. Preparation of lithium and stannane glycosyl anions.

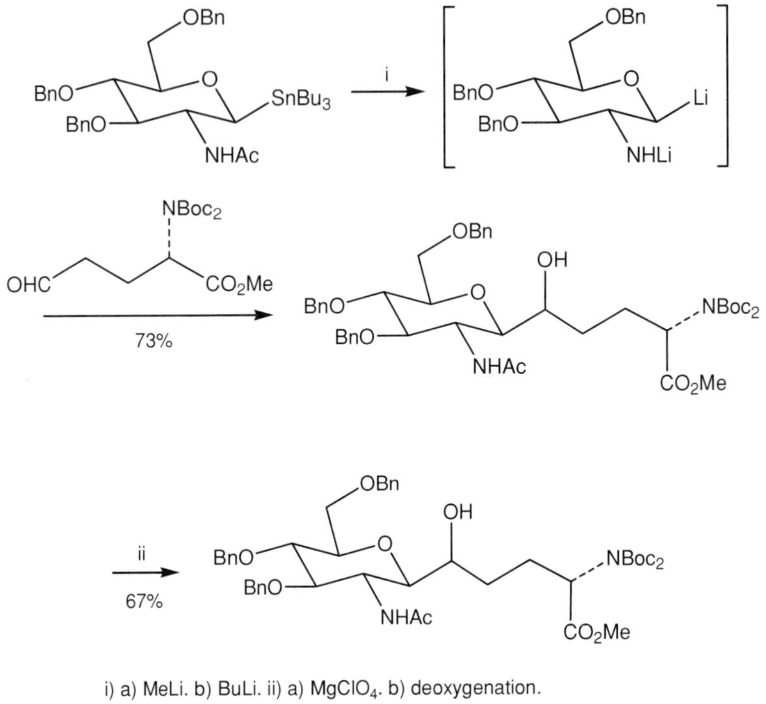

i) a) MeLi. b) BuLi. ii) a) MgClO$_4$. b) deoxygenation.

FIGURE 5.14. Preparation of C-analogs of glycosyl asparagines from anionic glycosyl donors.

(1,6)- and (1,1)-linked C-disaccharides based on the nitroaldol condensation between the glycosylnitromethane peracetate with the galactose-derived aldehyde to provide after dehydration, reduction of the double bond, and and radical denitration the desired C-disaccharide (Figure 5.16).

5.1.6   Cross-Metathesis Reaction

Cross-metathesis reaction is an emerging methodology for C-C bond formation. The air stable Grubbs ruthenium complex[28] has become an attractive catalyst for the olefin cross methatesis reactions and has been also applied successfully for the preparation of pseudosaccharides. The coupling reaction between C-allyl α-D-galactopyranoside with 4-acetoxystyrene led to the formation of the cross-metathesis product (Figure 5.17).[29]

5.1.7   Samarium Promoted Reaction

The synthesis of a C-glycoside analog of α-1,3-mannobiose has been reported via SmI$_2$-promoted C-glycosilation. The general approach is based on the Barbier-type

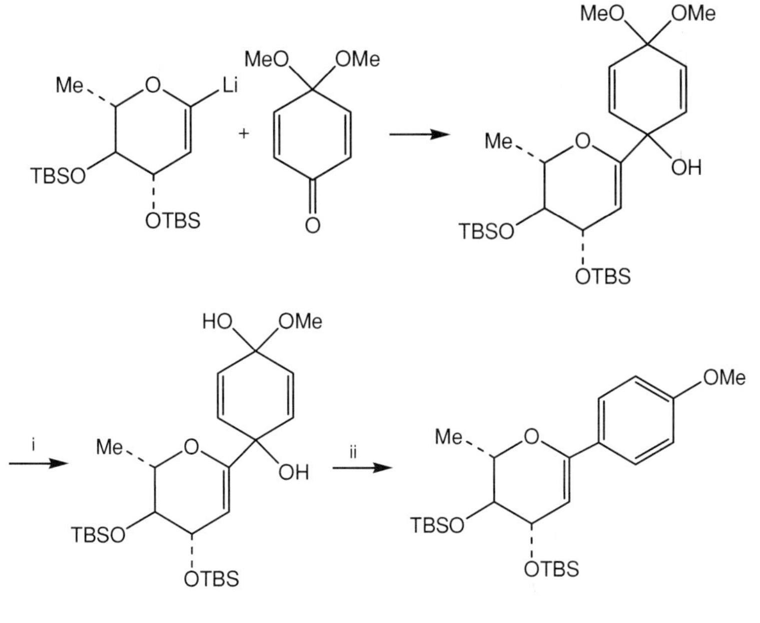

i) DIBAL/CH$_2$Cl$_2$. ii) POCl$_3$/Py.

FIGURE 5.15. C-glycoside formation using lithium glycal nucleophilic donor.

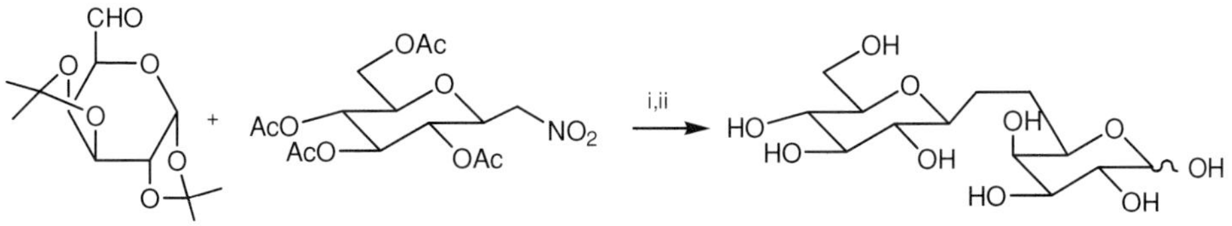

i) KF, MeCN, DCH-18-crown-6. ii) a) Ac$_2$O, Py, CHCl$_3$. b) NaBH$_4$, MeOH, CH$_2$Cl$_2$, 0°C. c) Bu$_3$SnH, AIBN, reflux. d) NaOMe, MeOH. e) H$_3^+$O.

FIGURE 5.16. C-disaccharide formation with glycosyl donors containing active methylene carbons.

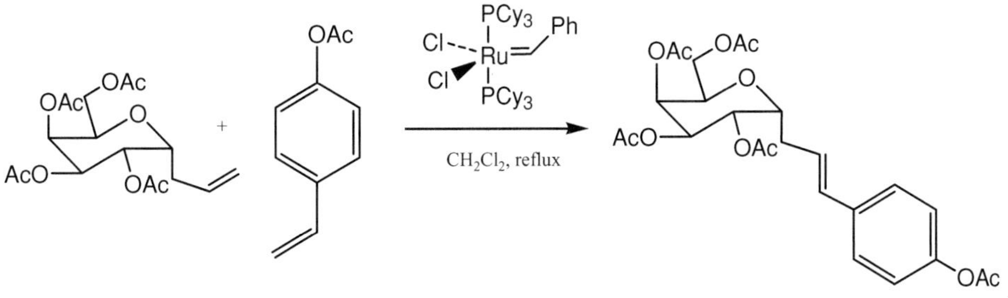

FIGURE 5.17. Cross-metathesis reaction for C-glycoside formation.

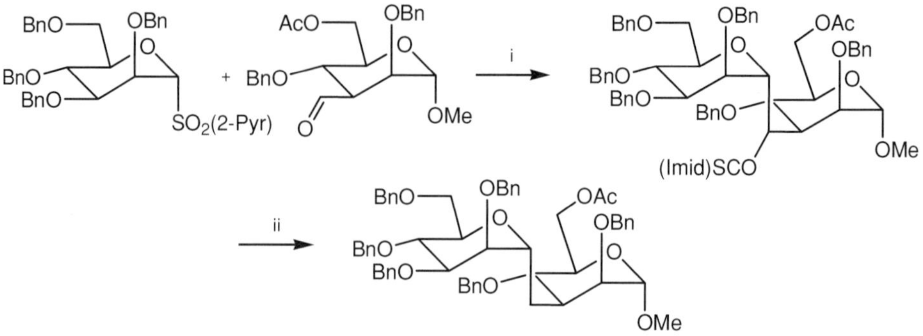

i) a) SmI₂ (2.8 eq), THF, 20°C. b) (Imid)₂CS (15 eq), CH₃CN, reflux, 35% ii) F₅PhOH, Ph₃SnH, AIBN, toluene, reflux, 65%.

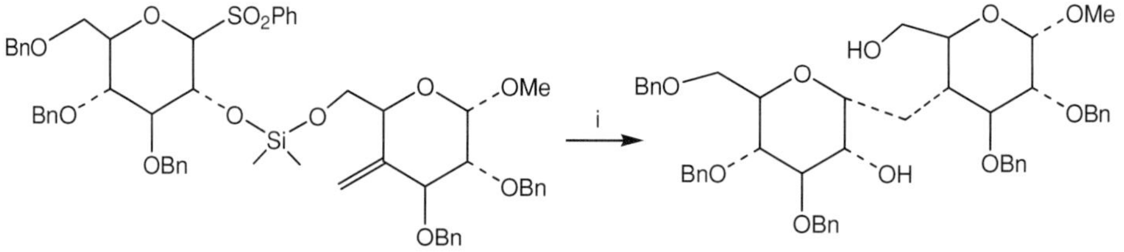

i) SmI₂, PhH, HMPA, 60°C. b) aq. HF.

FIGURE 5.18. Samarium promoted C-glycosylation.

coupling[30] and involves the use of pyridyl sulfone glycosyl donor with a sugar aldehyde in the presence of SmI₂ as catalyst. This procedure has been exploited successfully for the preparation of disaccharides under the tether approach (Figure 5.18).[31]

5.1.8   The Ramberg-Bäcklund Reaction

This novel procedure introduced by Franck et al. is becoming a practical and versatile approach for the preparation of biologically active C-glycosides such as aromatic,[5] aminoacids,[32,33] or glycerolipids.[34] The reaction sequence for C-glycoside formation consists of the initial S-glycoside formation, transformation to the sulfone derivative, Ramberg-Bäcklund rearrangement involving sulfone extrusion, and hydrogenolysis (Figure 5.19).

Another C-disaccharide was prepared by transformation of benzylated exoglycal to the iodide derivative, which in turn was coupled with the sulfur glycosyl donor. Further transformation to the sulfone and Ramberg-Bäcklund rearrangement produced the unsaturated disaccharide, which was finally reduced under Perlman conditions to provide disaccharide in 70% yield (Figure 5.20).[35]

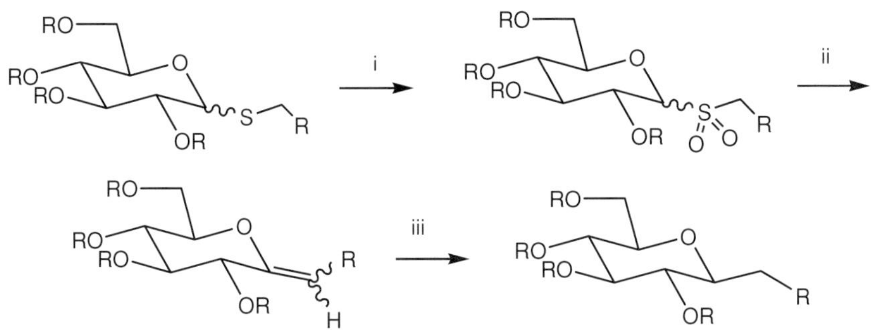

i) mCPBA, Na₂HPO₄/CH₂Cl₂, 77%. ii) KOH/Al₂O₃, CBrF₂CBrF₂/tBuOH, 50°C. iii) H₂, Pd(OH)₂/EtOAc.

FIGURE 5.19. The Ramberg-Bäcklund approach for C-glycoside formation.

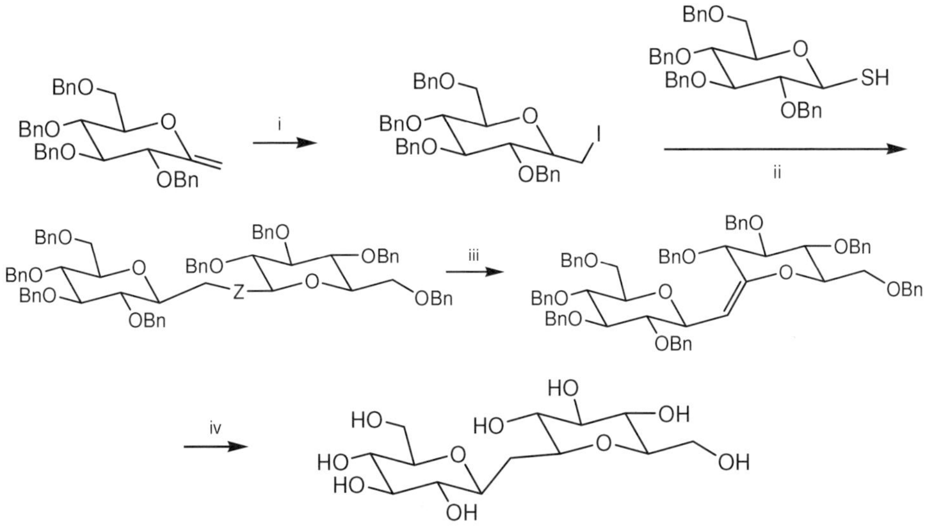

i)a) 9-BBN. b) Ph₃P, I₂, Imid. 81%. ii) K₂CO₃, then oxidation. iii) KOH, CCl₄, tBuOH, 60°C. iv) H₂, Pd(OH)₂, 70%.

FIGURE 5.20. Synthesis of C-isotrehalose.

5.1.9   Free Radical Approach

This approach is based on the generation of free radical at the anomeric carbon by using glycosyl donors, which are subjected to stannous treatment of free radical conditions that in turn will react with mainly exoglycals to produce a C-glycosidic linkage. The general methods leading to anomeric radicals formation are summarized in Figure 5.21:[36]

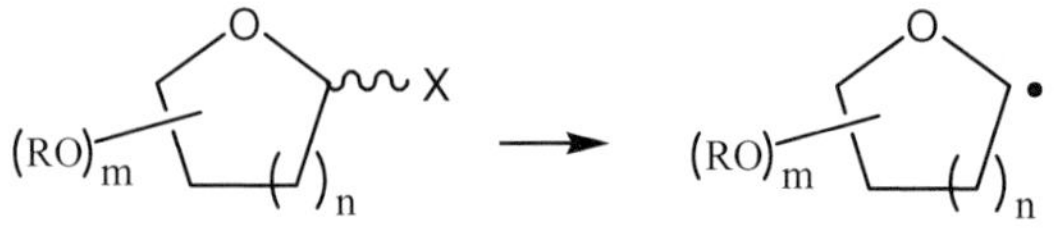

a) Bu$_3$SnH (Hg-hν or Δ)

X = NO$_2$, I, Br, Cl, SeAr, SAr

b) Bu$_3$SnSnBu$_3$H (Hg-hν)

X = NO$_2$, I, Br, Cl, SeAr, SAr

c) (C$_2$H$_5$)$_3$B (air) or N-acetoxy-2-thiopyridone (W-hν)

X = TeAr

d) Ester of N-Hydroxy-2-thiopyridone (W-hν)

X = COOH

e) (Diacyloxyiodo)arenes (Hg-hν or Δ)

X = COOH

f) Hg-hν

X = COBu-t etc

g) W-hν

X = Co(dmgH)$_2$Py

FIGURE 5.21. General methods leading to anomeric radicals formation.

The coupling reaction between acetobromoglucose and the unsaturated lactone shown in Figure 5.22 will result in the C-disaccharide formation, where a free radical mechanism promoted by a mixture of AIBN-Bu$_3$SnH is involved.[37]

Anomer radicals may also generate rearranged products as a result of 1,2-migration particularly for the case of acetoxy and phosphate groups. This feature has been exploited successfully for preparing 2-deoxy sugars from commercially available sugars Figure 5.23.[38]

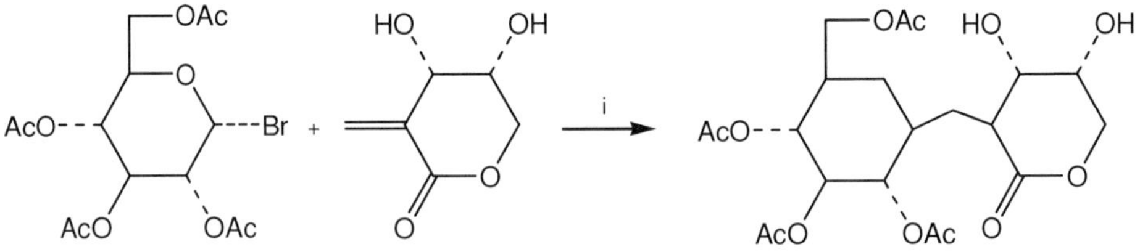

i) AIBN, Bu$_3$SnH.

FIGURE 5.22. Free radical coupling reaction.

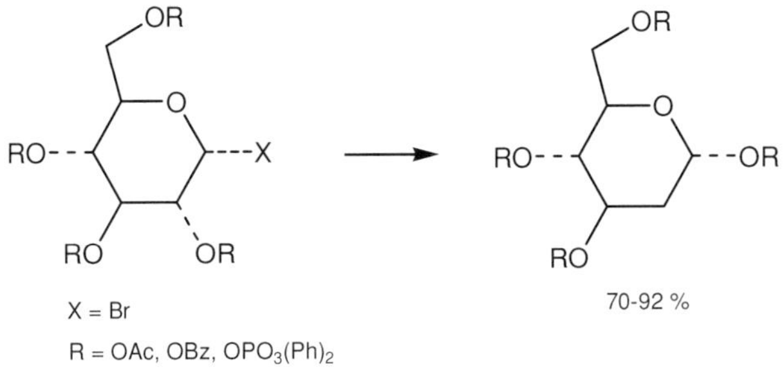

X = Br
R = OAc, OBz, OPO$_3$(Ph)$_2$

70-92 %

FIGURE 5.23. Synthesis of 2-deoxy sugars by reductive 1,2-migration of glycosyl bromide.

5.1.10 Exoglycals

Exoglycals have been described as another possibility for preparing C-glycosyl derivatives. The term exo-glycal is given to those unsaturated sugars with exocyclic double bonds. The most representative of these compounds are 1,2- and 5-6-unsaturated sugars (Figure 5.24).[39]

They were first prepared by reacting lactones with ethyl isocyanoacetate and subsequent hydrogenolysis[40,41] (Figure5.25). This reaction has not been exploited extensively due to sugar oxazole formation.

More recently two methods were reported for direct olefination of lactones. One is based on phosphorous Wittig-type reaction[42] and the other by direct methylenation using the Tebbe reagent[43] (Figure 5.26).

Alternative methods for the preparation of exoglycals includes β-elimination of halides[44,45] dehydration, Grignard nucleophilic addition, sulfone extrusion (Ramberg-Bäcklunnd olefination),[46] and tosyl hydrazones (Bamford-Stevens conditions)[47] among others (Figure 5.27).

The synthesis of several C-disaccharides by using exoglycals has been described. Such is the case of the preparation of C-disaccharide by reaction of two molecules of the C-methylene intermediate under Lewis acid conditions (Figure 5.28). The reaction was proposed to proceed via oxonium cation.[48]

A 1,3-dipolar cycloaddition of exomethylene sugar with glycosyl nitrone has been proposed as an approach for the formation of amino-C-ketosyl disaccharides (Figure 5.29).[49]

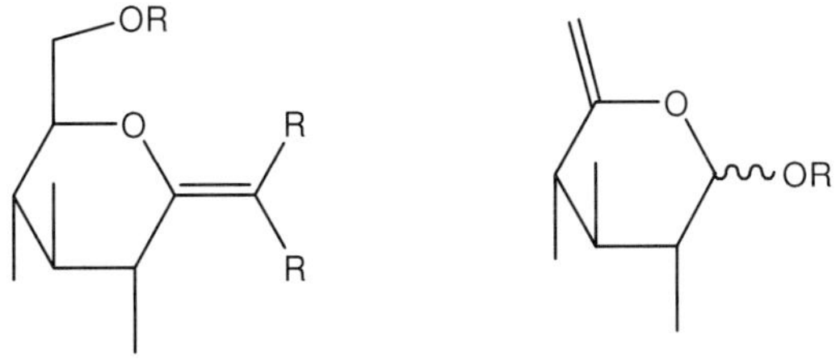

FIGURE 5.24. 1,2- and 5,6- unsaturated sugars.

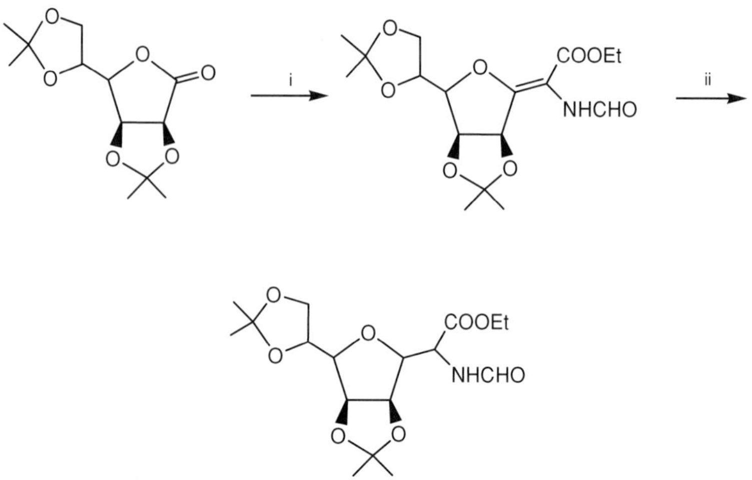

i) a) EtOOCCH₂NC, KH. b) AcOH. ii) a) H₂, Pd/C b) H₂O.

FIGURE 5.25. First synthesis of 1,2-unsaturated sugar.

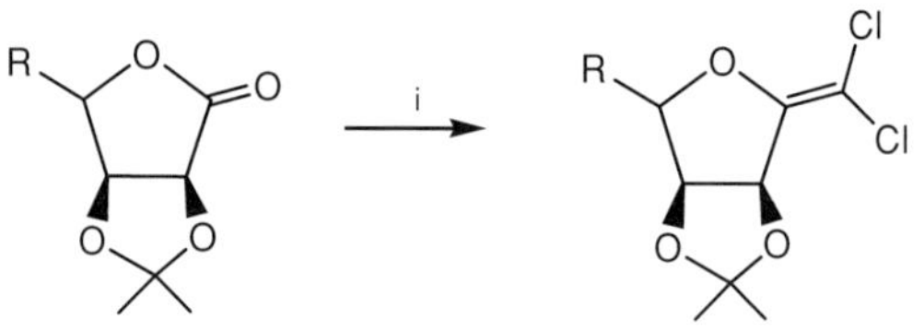

i) P(NMe₂)₃, CCl₄, −30-0°C.

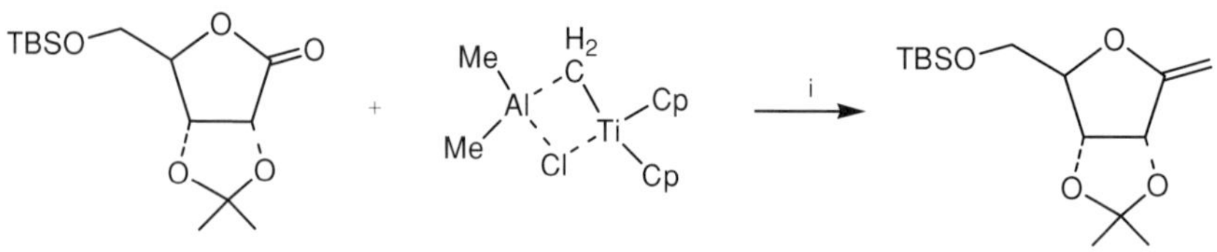

i) THF, Tol, −40°C.

FIGURE 5.26. Early methods for preparation of exoglycals.

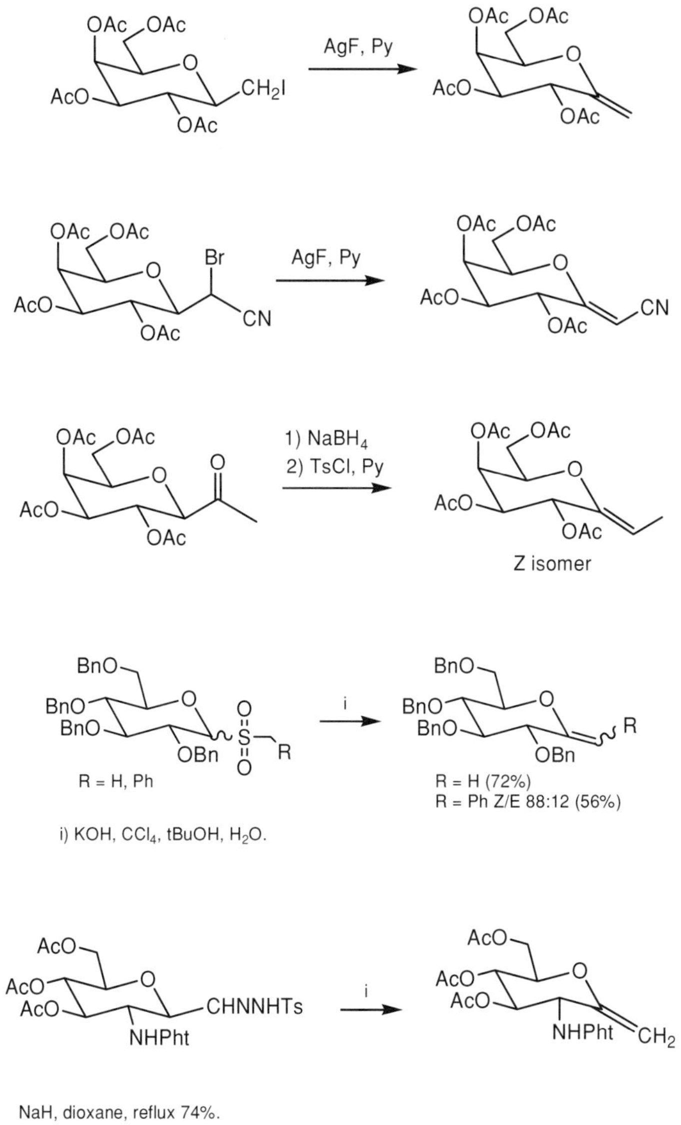

FIGURE 5.27. Alternative methods for the preparation of exoglycals.

5.1.11   The Tether Approach

Various approaches for C-glycoside construction are comprehensively reviewed, focusing mainly on the methylene formation.[50] The strategies presented are based on the concept that a nucleophilic anomeric donor is condensed with an exomethylene sugar to produce a C-disaccharide linkage.[51] According to this strategy methyl

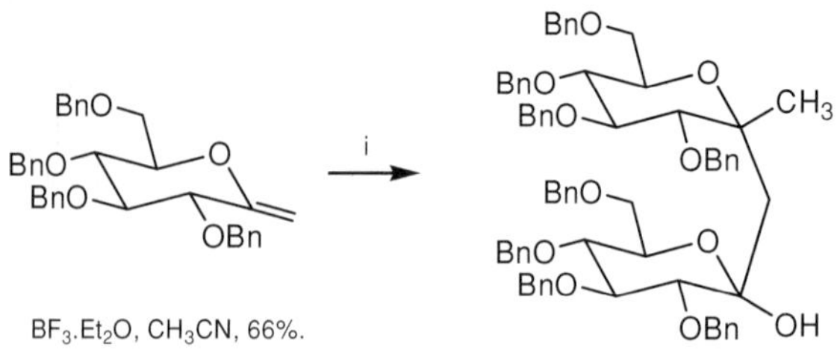

FIGURE 5.28. Preparation of C-disaccharide from exoglycals.

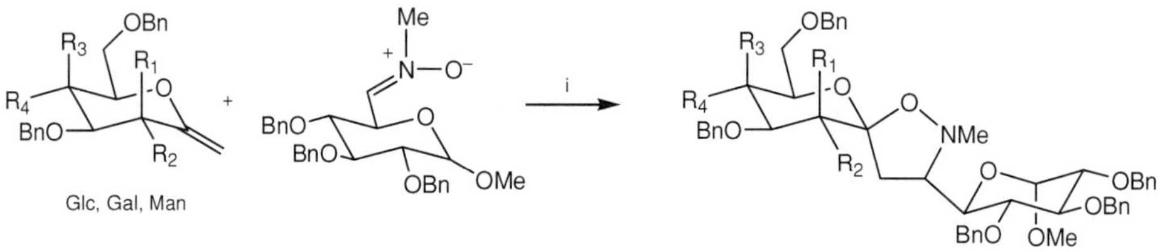

FIGURE 5.29. Dipolar cycloaddition of exoglycals.

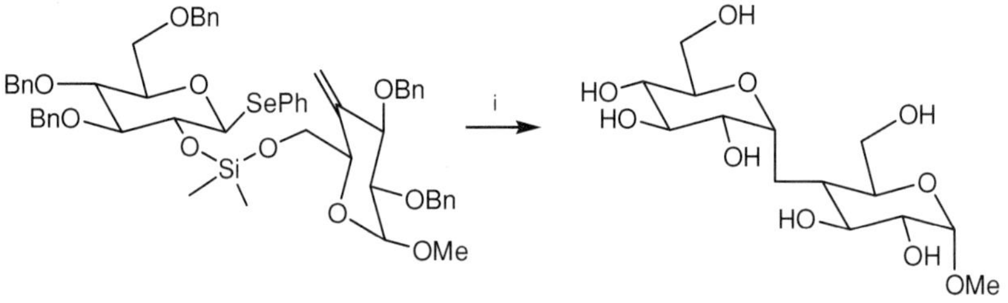

i) a) Bu₃SnH, AIBN, PhH, 60°C. b) HF, THF. c) H₂,Pd/C, MeOH, AcOEt.

FIGURE 5.30. C-glycoside construction under the tether approach.

α-C-isomaltoside was prepared from the silaketal connected precursor as shown in Figure 5.30.

The tether approach considers the preliminary formation of a temporary attachment usually involving a silyl protecting group, as tether which is cleaved after formation of the desired C-C bond. The general conditions involve the use of selenoglucopyranosides[52] or phenylsulfoxides[37,53] as glycosyl donors. An important application of this methodology can be seen in the preparation of O-C mixed sulfated trisaccharide (Figure 5.31).[13]

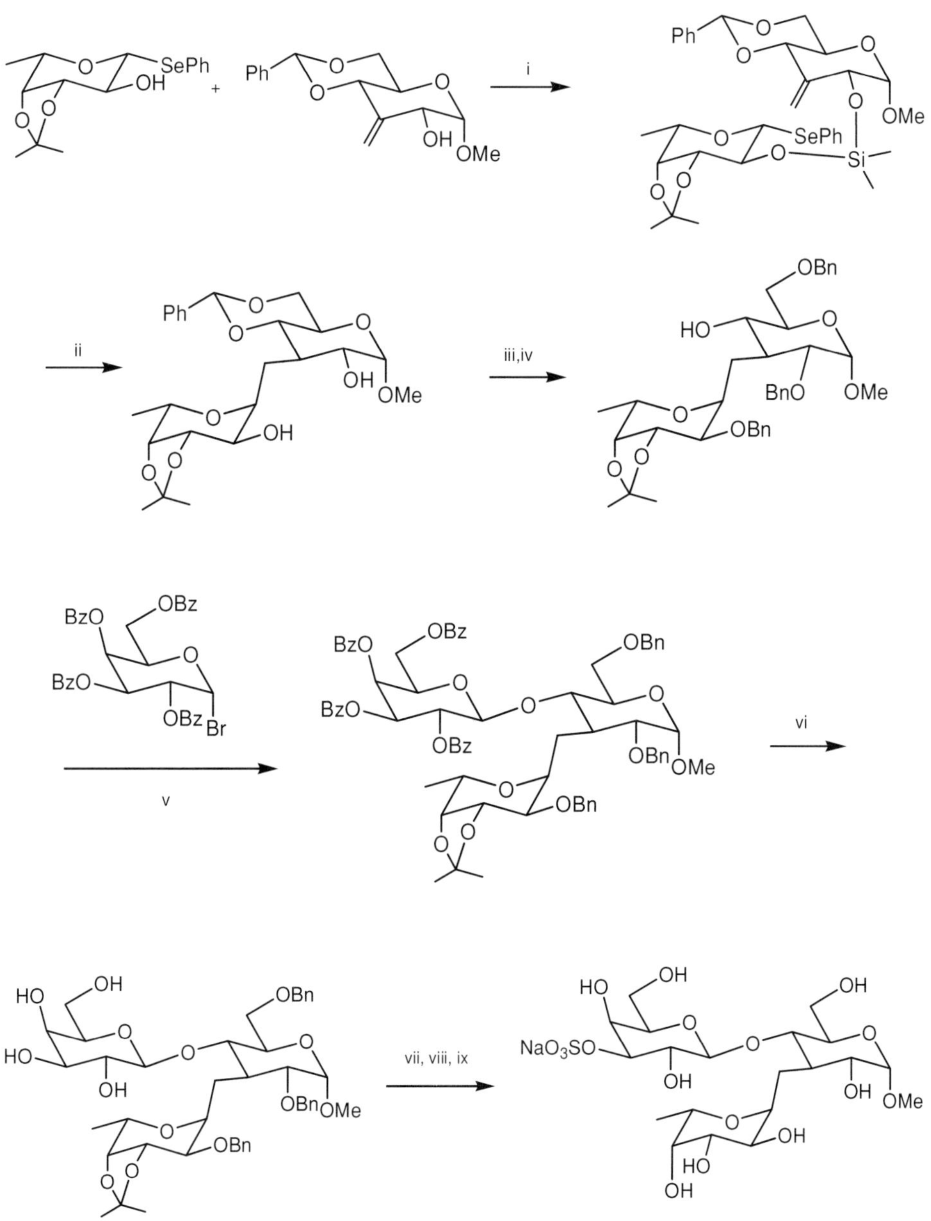

i) a) BuLi; Me$_2$SiCl$_2$ (4.4 equiv.), THF, −78°C to 20°C. b) imidazole, THF, r.t. ii) Bu$_3$SnH (2 equiv.), AIBN, PhMe, 110°C, 17 h. then Bu$_4$NF, THF, r.t., 60%. iii) BnBr, NaH, DMF, 100%. iv) NaBH$_3$CN, HCl, 70%. v) AgOTf, collidine, MS. 80%. vi) MeONa/MeOH, 97%. vii) Bu$_2$SnO, MeOH. viii) SO$_3$/Me$_3$N, 70%, 3 steps. ix) H$_2$, Pd/C, 100%.

FIGURE 5.31. Preparation of sulfated C-trisaccharide under the tether methodology.

References

1. P.S. Belica, and R.W. Franck, *Tetrahedron Lett.* **39**, 8225 (1998).
2. D.E. Levy, and C. Tang, *The Chemistry of C-Glycosides*, Pergamon Press: Oxford (1995).
3. D.J. McNally, K.V. Wurms, C. Labbé, S. Quideau, and R.R. Belanger, *J. Nat. Prod.* **66**, 1280 (2003).
4. P.M. Pauletti, I. Castro-Gamboa, D.H. Siqueira-Silva, M.C. Marx-Young, D.M. Tomazela, M.N. Eberlin, V. da Silva-Bolzani, *J. Nat. Prod.* **66**, 1384 (2003).
5. P. Pasetto and W. Franck, *J. Org. Chem.* **68**, 8042 (2003).
6. R.E. Dolle and K.C. Niclolaou, *J. Am. Chem. Soc.* **107**, 1691 (1985).
7. R.E. Ireland, R.C. Anderson, R. Badoub, B. Fitzsimmons, S. McGarvey, S. Thaissivongs, and C.S. Wicox, *J. Am. Chem. Soc.*, **105**, 1983 (1983).
8. F. Emery and P. Vogel, *Tetrahedron Lett.* 4209 (1993).
9. L. Paterson and L.E. Keown, *Tetrahedron Lett.* **38**, 5727 (1997).
10. M.D. Lewis, J.K. Cha, and Y. Kishi, *J. Am. Chem. Soc.* **104**, 4976 (1982).
11. Y. Du, R.J. Linhardt, and I. Vlahov, *Tetrahedron* **54**, 9913 (1998).
12. (a) M.H.D. Postema, *C-glycoside Synthesis*; CRC Press: Boca Ratón (1995); b) M.H.D. Postema *Tetrahedron* **48**, 8545 (1992).
13. P. Sinaÿ, *Pure & Appl. Chem.* **69**, 459 (1997).
14. J. Gervay, and M.J. Hadd, *J. Org. Chem.* **62**, 6961 (1997).
15. K. Oyama and T.J. Kondo, *Org. Chem.* **69**, 5240 (2004).
16. E. Calzada, C.A. Clarke, C. Roussin-Bouchard, and R.H. Wightman, *J. Chem. Soc. Perkin Trans 1*, 717 (1995).
17. R.W. Schmidt, B. Frick, B. Haag-Zeino and S. Apparao, *Tetrahedron Lett.* **28**, 4045 (1987).
18. J.M. Lancelin, J.R. Pougny, and P. Sinaÿ, *Carbohydr. Res.* **136**, 369 (1985).
19. D. Craig, M.W. Pennington, and P. Warner, *Tetrahedron* **55**, 13495 (1999).
20. R.R. Schmidt and P. Shorma, *Carbohydr. Res.* **14**, 1353 (1985).
21. M.A. Tius, J. Gomez-Galano, X. Gu, and J.H. Zaidi, *J. Am. Chem. Soc.* **113**, 5775 (1991).
22. A. Abas, R.L. Beddoes, J.C. Conway, P. Quayle, and C.J. Urch, *Synlett*, 1264 (1995).
23. B.A. Johns, Y.T. Pan, A.D. Elbein, and C.R. Johnson, *Carbohydr. Res.* **136**, 369 (1985).
24. T. Kometani, H. Kondo, and Y. Fujimori, *Synthesis* 1005 (1988).
25. F. Burkhart, M. Hoffmann, and H. Kessler, *Angew. Chem. Int. Ed. Engl.* **36**, 1191 (1997).
26. K.A. Parker and C.A. Coburn, *J. Am. Chem. Soc.* **113**, 8516 (1991).
27. O.R. Martin and W. Lai, *J. Org. Chem.* **58**, 176 (1993).
28. (a) P. Schawab, M.B. France, J.W. Ziller, and R.H. Grubbs, *Angew. Chem., Int, Ed. Engl.* **34**, 2039 (1995). (b) M.H.D Postema, J.L. Piper, and R.L. Betts, *J. Org. Chem.* **70**, 829 (2005).
29. R. Roy, R. Dominique, and S.K. Das, *J. Org. Chem.* **64,** 5408 (1999).
30. T. Mazéas, J.M. Skrydstrup, and J.M. Beau, *Angew. Chem,* **107**, 990 (1995).
31. S.L. Krintel, J. Jiménez-Barbero, and T. Skrydstrup, *Tetrahedron Lett.*, **40**, 7565 (1999).
32. A. Dondoni and A. Marra, *Chem. Rev.* **100**, 4395 (2000).
33. Y. Ohnishi and Y. Ichikawa, *Bioorg. Med. Chem. Lett.* **12**, 997 (2002).
34. G. Yang, R.W. Franck, R. Bittman, P. Samadder, and G. Arthur, *Org. Lett.* **3**, 197 (2001).
35. F.K. Griffin, D.E. Paterson, and R.J. Taylor, *Angew. Chem., Int. Ed.* **38**, 2939 (1999).
36. H. Togo, W. He, and Y. Waki, and M. Yokohama, *Synlett* 700 (1998).
37. A. Chenede, E. Perrin, E.D. Rekai, and P. Sinaÿ, Synlett 420 (1994).

38. B. Giese, K.S. Gröninger, T. Witzel, H.G. Korth, and R. Sustmann, *Angew. Chem. Int. Ed. Engl.*, **26**, 233 (1987).
39. C. Taillefumier and Y. Chapleur, Chem. Rev. **104**, 263 (2004).
40. K. Bischofberger, R.H. Hall, and A.J. Jordaan, *Chem. Soc. Chem. Commun.* 806 (1975).
41. R.H. Hall, K. Bischofberger, S.J. Eitelman, and A.J. Jordaan, *Chem. Soc. Perkin Trans.* 1, 743 (1977).
42. Y. Chapleur *J. Chem. Soc. Chem. Commun.* 449 (1984).
43. C.S. Wilcox, G.W. Long, and H. Suh, *Tetrahedron Lett.* **25**, 395 (1984).
44. M. Brockhaus and J. Lehmann, *Carbohydr. Res.* **53**, 21 (1977).
45. S.J. Eitelman, R.H. Hall, and A.J. Jordaan, *Chem. Soc., Chem. Commun.* 923 (1976).
46. F.K. Griffin, D.E. Paterson, P.V. Murphy, and R.J.K. Taylor, *Eur. J. Org. Chem.* 1305 (2002).
47. M. Toth and L. Somsak, *J. Chem. Soc., Perkin Trans. 1*, 942 (2001).
48. L. Lay, F. Nicotra, L. Panza, G. Russo, E. Caneva, *J. Org. Chem.* **57**, 1304 (1992).
49. X. Li, H. Takahasi, H. Ohtake, and S. Ikegami, Tetrahedron Lett, **45**, 3981 (2004).
50. G. Casiraghi, F. Zanardi, G. Rassu, and P. Spanu, *Chem. Rev.* **95**, 1677 (1995).
51. B. Vauzeilles, D. Cravo, J.-M. Mallet, and P. Sinaÿ, *Synlett*, 522 (1993).
52. A.J. Fairbanks, E. Perrin, and P. Sinaÿ, *Synlett* 679 (1996).
53. D.S.T. Mazeas, O. Doumeix, J.-M. Beau, *Angew. Chem. Int. Ed. Engl.* **33**, 1383 (1994).

6
Glycoconjugates

Carbohydrates covalently attached to proteins and lipids produce three types of glycoconjugates: proteoglycans, glycoproteins, and glycolipids. Although in the first two cases the types of linkages are the same, chemically proteoglycans behave as polysaccharides and glycoproteins having much less carbohydrate content as proteins. The third important class of glycoconjugates, constituted by a carbohydrate residue covalently attached to a lipidic component, has been classified into four types depending on the lipidic nature: glycoglycerol, glycosyl polyisoprenol pyrophosphates, fatty acid esthers, and glycosphingolipids.[1]

The most common monosaccharides residues found in glycoconjugates are D-galactose, D-mannose, N-acetyl-D-glucosamine, N-acetyl-D-galactosamine, L-fucose, D-xylose, and sialic acids (Figure 6.1).

6.1 Biological Function and Structural Information

Glycoproteins and glycolipids are major components of the outer surface of mammalian cells. The former have been implicated in several essential events such as immune defense, viral replication, cell-cell adhesion, inflammation and cell growth, while the latter in cell-cell recognition, growth, differentiation, and interaction with proteins of viral and bacterial pathogens.

It is attributed to the hepatic Gal^2/GalNAc-binding receptor as the first recognition discovery of carbohydrates as biological signals.[2] Subsequently, Man-6-phosphate receptor for lysosomal enzymes and Man-receptor from alveolar macrophages were reported and investigated.[3,4]

In the cellular immune system, some specific glycoproteins are implicated in the folding, quality control, and assembly of peptide-loaded major histocompatibility complex antigens and the T cells receptor complex. Furthermore, the oligosaccharides linked to glycoproteins provide protease protection, ER-associated retrograde transport of misfolded proteins, loading of antigenic peptides into MHC I class I, and influence the range of antigenic peptides generated in the endosomal pathway for presentation by MHC class II.[5]

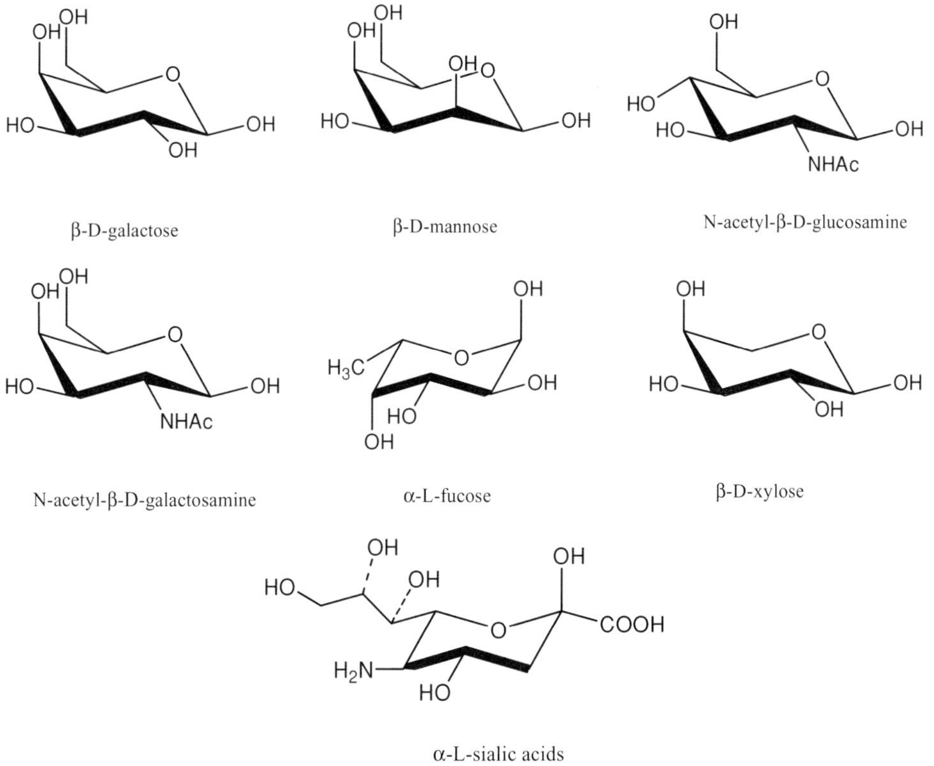

β-D-galactose   β-D-mannose   N-acetyl-β-D-glucosamine

N-acetyl-β-D-galactosamine   α-L-fucose   β-D-xylose

α-L-sialic acids

FIGURE 6.1. Monosaccharides residues of glycoproteins.

In addition, enveloped viruses such as human immunodeficiency virus (HIV) evades immune response by exploiting the host glycosylation machinery to protect potential antigenic epitopes.[6] They also use the host secretory pathway to fold and assemble their often heavily glycosylated coat proteins.

Another important fact to mention is that normal cells and tumor cells have evident differences in glycoprotein content on their cell membranes. Altered glycoproteins of the tumor membranes such as Thomsen-Friedenreich (T antigen) are tumor-associated antigen and belong to the class of O-glycoproteins.[7−9]

6.1.1 Classification of Glycocoproteins

Based on the type of the glycosidic bond formed between the sugar and the protein residues, glycoproteins are divided in N- and O-glycans. The first type involved the glycosidic linkage between asparagine and N-acetylglucosamine and the second involves an O-glycosidic linkage between the sugar residue (fucose, galactose, N-acetylgalactosamine, and N-acetylglucosamine) and the oxygen in the side chain serine, threonine, or hydroxyl lysine.

It is known that N-linked glycans contain the pentasaccharide Manα1-6(Manα1-3)Manβ1-4GlcNAcβ1-4GlcNAc as a common core, and they have been classified into four main groups on the basis of the structure and the location of glycan residues added to the trimannosyl core: oligomannose, complex, hybrid, poly-N-acetylglucosamine (Figure 6.2).[10]

First type

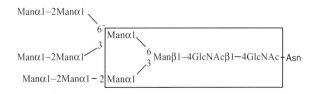

Second type

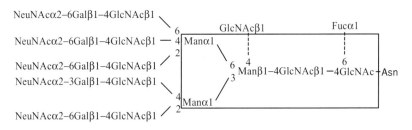

Third type

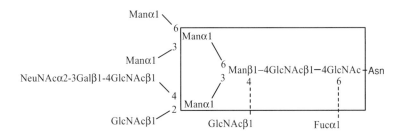

Fourth type

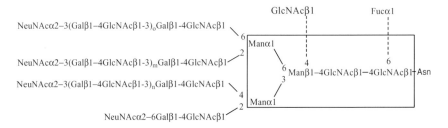

FIGURE 6.2. The four groups of N-linked glycans.

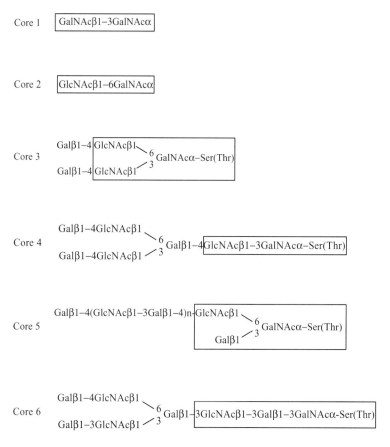

FIGURE 6.3. Core structures in O-linked glycans.

O-glycans do not present common core structures and until now they have been classified in at least six groups according to different core structures (Figure 6.3).

6.1.2 Recognition Sites

There are two main classes of glycosidic linkage depending on the type of glycosidic bond formed between the sugar residue and the protein: the *O*-linked glycans involving the amino acids serine, threonine, and hydroxyl lysine, and *N*-linked glycans involving the amino acid asparagine in the form of tripeptide with sequence AsnXSer (where X is any amino acid except proline).

Thorough studies with sugar analogs indicate that presumably the most important of the substituents is the equatorial OH-group on carbon 3. Also important is the OH-group on carbon 4 which can be either axial or equatorial depending on the glycoprotein. Regarding C-2, there is certain tolerance; however, the size of the group should not be too large. Finally, C-6 and the anomeric carbon apparently do not play significant roles in the binding (Table 6.1).[11]

TABLE 6.1. Sugar requirements for three different glycoproteins.

	Rat hepatic	Chicken hepatic	MBP-A
1		α ≈ β large substituents tolerated, negative group	
	Detrimental	Enhancing	Tolerable
2	Eq. N-Ac enhance binding	Eq. N-Ac enhance binding	No effect by N-Ac
3		Eq. OH required	
4	Axial OH required	Eq. OH required	Eq. OH required
5		Large subst. accepted	

6.1.3 Structural Information of Glycoproteins

A better understanding about the conformation of glycoproteins has been reached by using NMR, molecular dynamics (MD), and in some cases X-ray diffraction techniques. The high motion of oligosaccharides mainly across the glycosidic linkage (Figure 6.4) has limited the unambiguous conformational determinations in glycoproteins. However, the conformations from the MD simulations are in good agreement with the values from NMR studies. It has been observed that ω-angle prefers *Gauche* conformation by solvation effects with φ-angle largely determined by the anomeric effect, and the ψ-angle highly influenced by non-bonded interactions.[12]

The linkage between the sugar residue and the amino acid asparagine (N-linked glycans) is planar along the C1-NH-C=O glycosidic linkage and flexible along the CO-CH₂-CH- bonds (Figure 6.5).

Based on the considerations that N-glycosidic linkages are rigid for the amide group and flexible for the side chain angles, the conformational motion of the glycoproteins depends on the flexibility of the asparagine side chain. This flexibility will have considerable effect on the volume occupied by the sugar and the shielding effects of the carbohydrate over the protein surface.

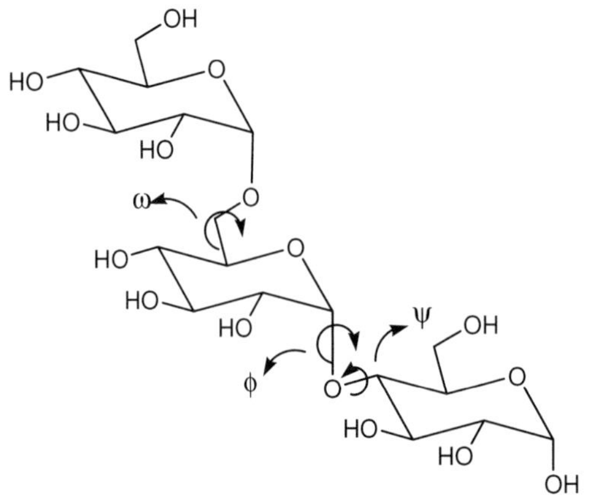

FIGURE 6.4. Angles of rotation of carbohydrates.

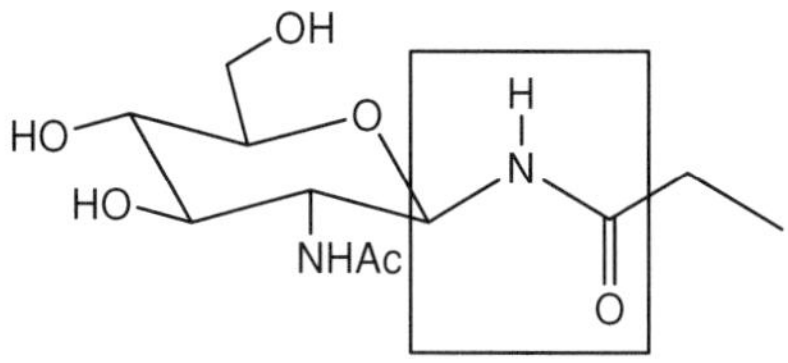

FIGURE 6.5. Planarity of the Asn-GlcNAc glycosidic linkage.

Hydrogen bond and van der Waals interactions showed for some cases stacked conformations, and distances across a carbohydrate residue (from O-1 to O-4) of 5.4 Å and for the first three residues of the core of an N-linked oligosaccharide extend to approximately 16 Å from head to tail.[12]

6.2   Carbohydrate Binding Proteins

Carbohydrate binding proteins are defined as those proteins that interact through noncovalent bonds with carbohydrates. Of particular interest are **Lectins.** which bind reversibly to mono- and oligosaccharides with high specificity, and are apparently devoid of catalytic activity.[13]

Carbohydrate binding proteins are widespread macromolecules found in virus, bacteria, plants, and animals and act as recognition determinants including clearance of glycoproteins from the circulatory system, control of intracellular traffic of glycoproteins, recruitment of leukocytes to inflammatory sites, adhesion of infectious agents to host cells, and cell interactions in the immune system in malignancy and metastasis.[13]

Depending on the affinity shown toward the type of monosaccharide, they can be classified in mannose, galactose/N-acetylgalactosamine, N-acetylglucosamine, L-fucose, and N-acetylneuraminic acid. Due to their high specificity, lectins specific for galactose do not recognize glucose or mannose, nor N-acetylglucosamine with N-acetylgalactosamine, but mannose-specific animals lectins do recognize fucose.

Lectins also exhibit high specificity for di-, tri-, and tetrasacharides and some interact only with oligosaccharides. Moreover, different lectins specific for the same oligosaccharide may recognize different regions of its surface. Some of the lectins and their affinity ligands are shown in Table 6.2.

High-resolution studies involving the protein sequence determination and three-dimensional analysis gave insight about the structure and molecular interaction between the sugar ligands and the proteins. As result of this structural analysis, it was observed on the basis of common structural features that lectins fall into three main categories:

1. simple,
2. mosaic or multidomain,
3. macromolecular assemblies.

TABLE 6.2. Lectines and affinity ligands.

Family	Lectin	Abbrev	Ligand
Legume (plant lectins)	concanavalin	ConA	MeαMan, MeαGlc Manα3(Manα6)Man
	Erithina corallodendron	EcorL	Galβ4Glc
	Fava bean	Favin	MeαMan
	Griffonia simplicifolia	GSIV	Fucα2Galβ3(Fucα4)GlcNAc
	Red kidney bean	PHA	Complex pentasaccharide
	Lathyrus ochrus	LOL I,II	Manα3Manβ4GlcNAc, complex octasaccharide
	lentil	LCL	MeαMan, MeαGlc
	pea	PSL	Manα3(Manα6)Man
	Peanut	PNA	Galβ4Glc
	Soybean	SBA	Biantennary pentasccharide
cereal	Wheat germ	WGA	NeuAc(α2-3)Galβ4Glc GlcNAcβ4GlcNAc sialoglycopeptide
Amaryllidaceae	Snow drop	GNA	MeαMann mannopentaose
Moraceae	Artocarpus integrifolia	Jacalin	MeαGal
Galectins (animal lectins)	Human heart	Galectin 1	Galβ4GlcNAc octasaccharide
	Rat liver	Galectin 2	Galβ4Glc

Simple lectins are most of known plant lectins (Legume, Cereal, Amaryllidaceae, Moraceae, Euphorbiaceae), animal lectins (galectins or formely S-lectins), and Pentraxins, and contain a nonidentical small number of subunits of molecular weight below 40 kDa.

Mosaic or multidomain include viral hemagglutinins and animal lectins C- (endocytic lectins, collectins, selectins), P-, and I type. Their molecular weight is variable and is formed by different protein domains, only one of them having the carbohydrate binding site.

Macromolecular assemblies are common in bacteria and usually present in the form of fimbriae, which are filamentous, heteropolymeric organelles present on the surface of the bacteria.[14]

Most plants lectins recognize and interact with terminal nonreducing units of oligo- and polysaccharides, glycoproteins, and glycolipids. Anomeric preference is an important finding observed for different carbohydrate binding proteins, for instance, all mannose/glucose binding lectins display great preference for the α-anomeric forms;[15] however, lectins from *Ricinus communis* bind preferentially to β-galactosidases while other lectins make no difference in binding to anomers of GalNAc and GlcNAc. A considerable amount of structural information about carbohydrate binding proteins such as the complete amino acid sequences for various lectins is available.[13,16]

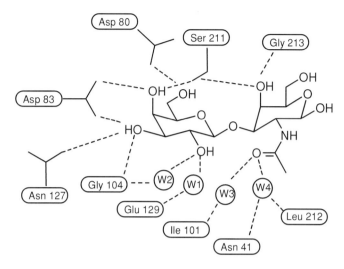

FIGURE 6.6. Gal($\beta 1 \rightarrow 3$)GalNAc in the combining site of peanut agglutinin.

6.2.1 Combining Sites

Lectins combine with carbohydrates mainly through weak forces such as hydrogen bonding, coordination with metal ions, and hydrophobic interactions. The hydrogen bridge interaction is established between the carbohydrate hydroxyl groups and the amino groups. Additionally, contacts between the carbohydrate and the protein are mediated by water bridges (Figure 6.6).[17]

Although carbohydrates are essentially polar molecules, there is a significant nonpolar or hydrophobic interaction that occurs between the N-acetyl group of amino sugars and the glycerol moiety of neuraminic acid, with the aromatic amino acids phenylalanine, tyrosine, and tryptophan. In the combining site of wheat germ agglutinin with sialyllactose, several van der Waals contacts stabilize the orientation of the sugar ring through nonpolar stacking interactions with the aromatic side chain of Tyr64, and Tyr66 that interacts through nonpolar with the glycerol tail of the N-acetyl neuraminic acid (Figure 6.7).[18]

Several classes of lectins are ion-dependent for their functional interaction with the ligands. Divalent ions such as calcium and magnesium participate in the stabilization of the amino acid positions that interact with the sugars. The Ca^{2+} ion establishes a coordination bond with the carbonyl group of asparagine and with one carboxylate oxygen of an acidic amino acid. The Mn^{2+} does not coordinate any residues that interact directly with the protein, but is involved in fixing the Ca^{2+}.position (Figure 6.8).[19,20]

In the interaction of Concanavalin A with the branched trisaccharide Man(α1-6)[Man(α1-3)]Man, several hydrogen bond contacts between the hydroxyl group of the sugar and the amino acid residues are observed. Some of these interactions are

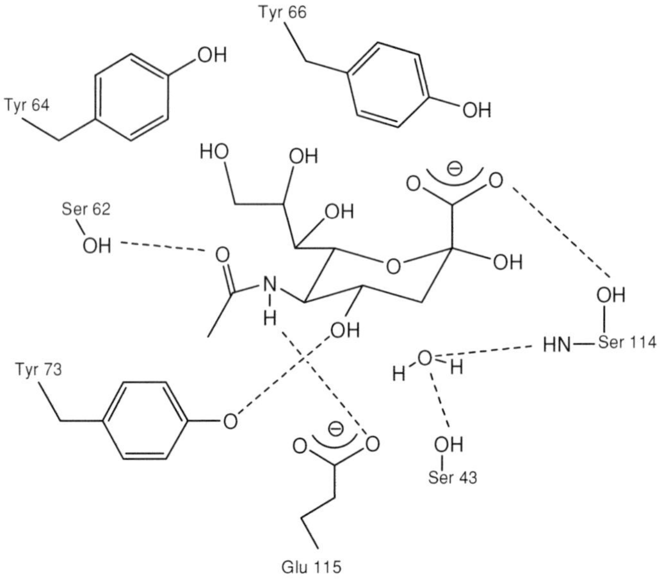

FIGURE 6.7. Sialyllactose in the combining site of wheat germ agglutinin.

bifurcated or involve water and contribute importantly in the recognition process (Figure 6.9).[13,21]

Carbohydrate binding proteins are classified in two types: calcium-dependent (C-type glycoproteins), and thiol reagent-dependent (S-type). The former are structurally more diverse (although the binding region known as carbohydrate recognition domain CRD is highly conserved) and more specific to organs and tissues, while the latter are structurally more conserved and are more widespread among the organs and tissue examined.[22] Other carbohydrate binding proteins that do not

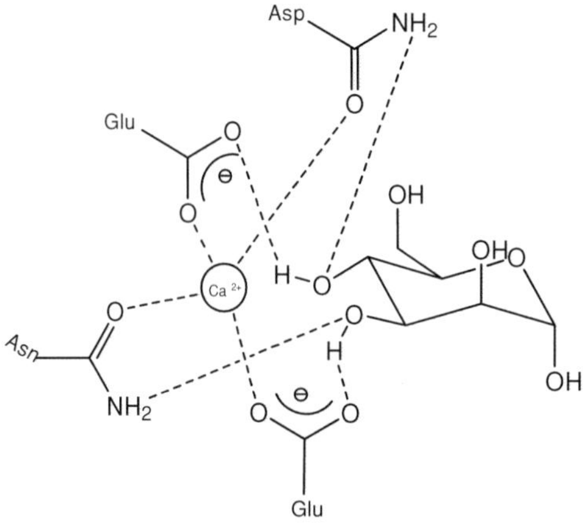

FIGURE 6.8. Mannose binding protein C with bound mannose.

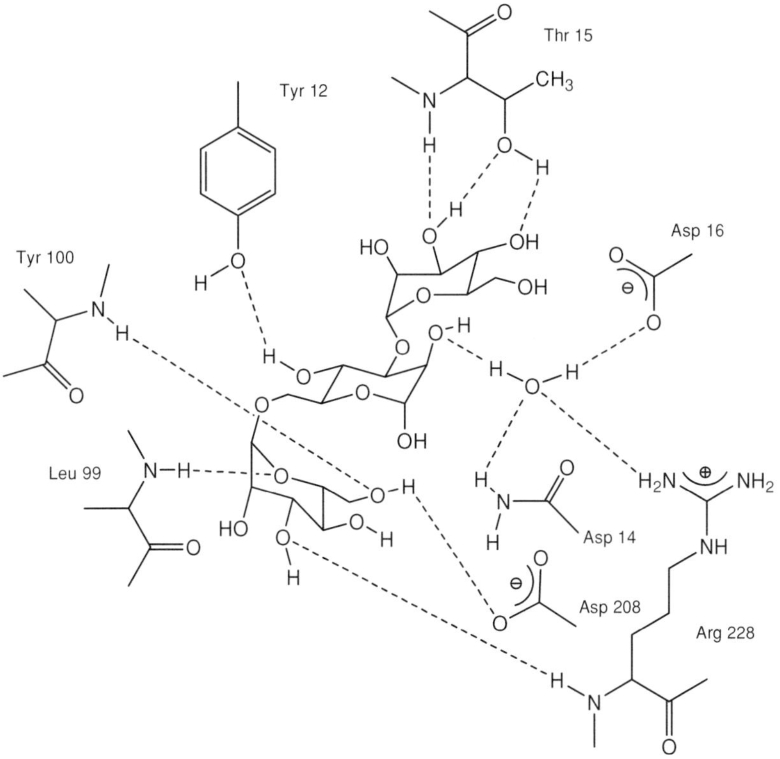

FIGURE 6.9. Trimannoside binding site of Concanavalin A.

fall into this two categories are fibronectin and laminin, serum immunoglobulins, mannose-phosphate receptor, viral hemagglutinins, and serum amyloid protein.

Another important class of carbohydrate binding proteins are known as **Selectins** (classified as E-, P-, and L-selectins) and are defined as nonenzymatic and nonimmune proteins involved in the leukocyte recruitment to sites of inflammation.[23][24] It has been found that the tetrasaccharide sialyl Lewis x is the recognition molecule and the use of synthetic sialyl Lewis x confirmed the hypothesis that sulfation increases the affinity for L-selectins.[25]

6.3   Glycopeptide Synthesis

The design of glycopeptides requires a combination of sugar and peptide chemistry, being a substantial part the installation of the O- or N-glycosidic bond[26][27] The synthetic approach is in principle designed on the basis of the glycosidic bond required. Thus, while in the case of O-glycopeptides, the synthetic methods relies on the common strategies for the preparation of O-glycosides, for the preparation

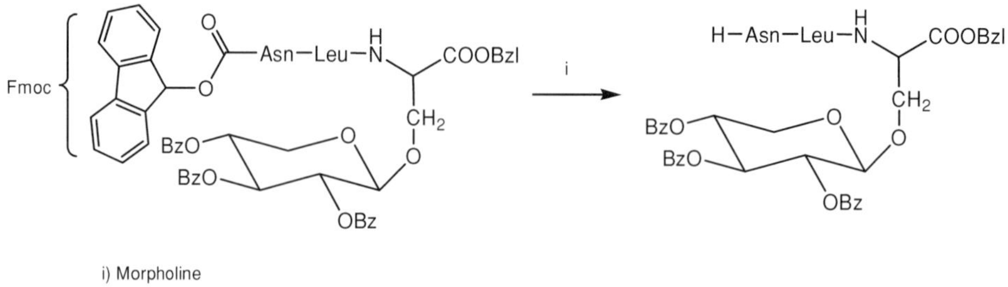

i) Morpholine

FIGURE 6.10. Peptide protecting group Fmoc and removal conditions.

of N-glycopeptides the strategy of choice involves the coupling between the amino glycosyl donor with aspartate in the presence of a condensing agent or by enzymatic catalysis.

Compatibility between the protecting groups and the glycosidic bond when they are subjected to different reaction conditions such as acid or base conditions is a sensitive issue. For instance, it is known that the glycosidic bond in acetals is acid–sensitive; however, in the case of O-glycosyl serine and threonine, they conversely present base-sensitivity. The introduction of selective protecting groups for amino acid functionalities that can be cleaved under mild conditions without affecting the glycoside bond or protecting groups attached to the sugar moiety is a feasible approach. Widely employed protecting groups used for this purpose is the Fmoc protecting group (9-fluorenyl)methoxycarbonyl), Pyroc (2-(pyridyl)ethoxycarbonyl), and Aloc (allyloxycarbonyl) for the peptide and Mpm (4-methoxy-benzyl ether) for the sugar region. The conditions needed for the cleavage of the mentioned protecting group in the presence of other functionalities are indicated in Figure 6.10.[28,29]

The synthesis of N-α-FMOC amino acid glycosides was carried out from O'Donnell Schiff bases or from N-α-FMOC amino protected serine or threonine and the appropriate glycosyl bromides under Koenigs-Knorr modified conditions.[30] The α-FMOC-protected glycosides were incorporated into 22 encephalin glycopeptides analogs (Figure 6.11).

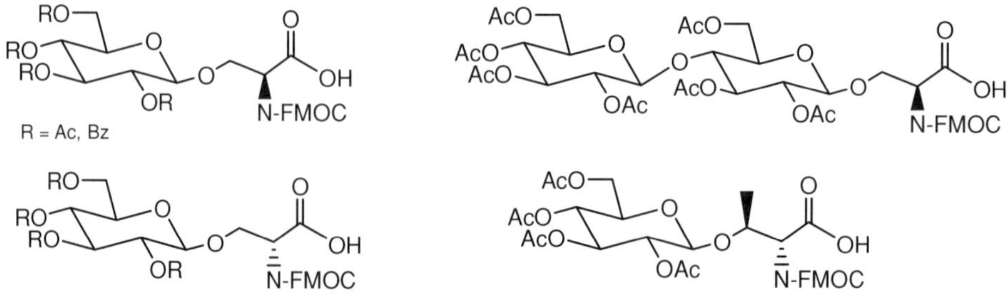

FIGURE 6.11. N-α-FMOC-amino acid glycosides.

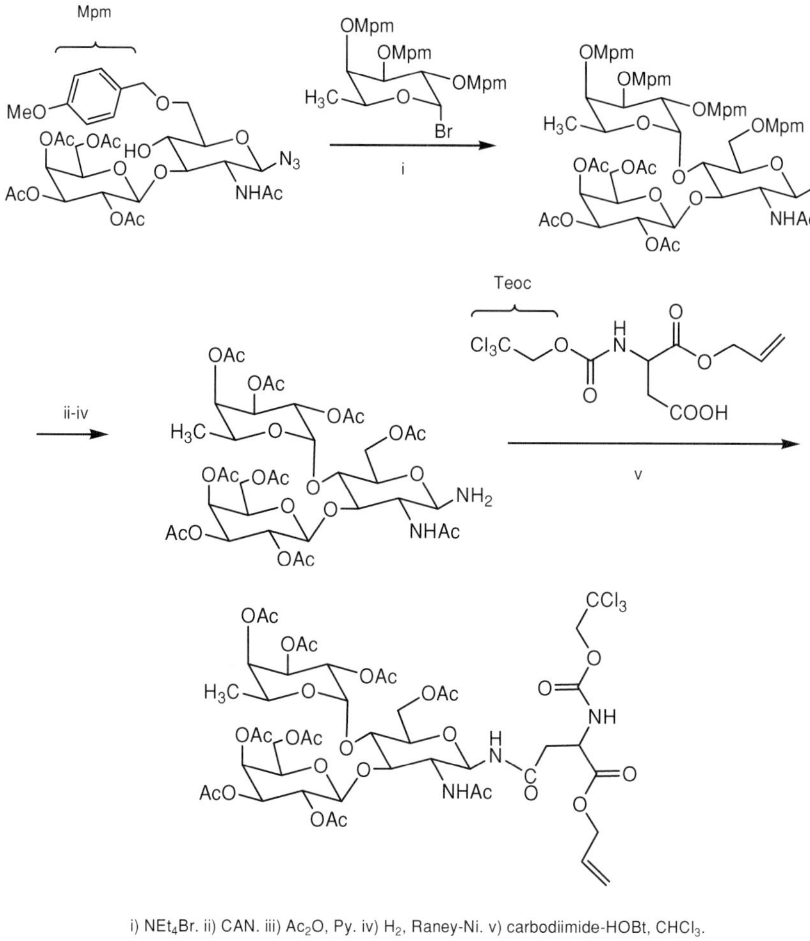
Mpm
MeO
OAc
OAc
HO
O
N3
AcO
NHAc
OMpm
H3C
Br
i
Teoc
Cl3C
H
N
O
COOH
ii-iv
NH2
v
CCl3
NH
i) NEt4Br. ii) CAN. iii) Ac2O, Py. iv) H2, Raney-Ni. v) carbodiimide-HOBt, CHCl3.

FIGURE 6.12. Synthesis of glycopeptide Lewis^a.

Pyroc is another protecting group useful in peptide chemistry. It is stable to acids, bases, and hydrogenolysis, but sensitive to morpholine. The allylic protecting group Aloc is also stable to acid, base, and can be removed under of Pd(0) catalysis or weak base as morpholine.[29]

A tumor-associated antigen Lewis^a was synthesized by applying a combination of compatible sugar and peptide protecting groups. For this method the azide group was used as anomeric amine precursor (Figure 6.12).[30]

Enzymes has been useful for peptide elongation using an engineered *subtilisin* and disaccharide bond formation with glycosyltransferase as shown in Figure 6.13.[31]

A novel chemoenzymatic synthesis of Eel Calcitonine glycopeptide analog having natural N-linked oligosaccharides such as disialo biantennary complex

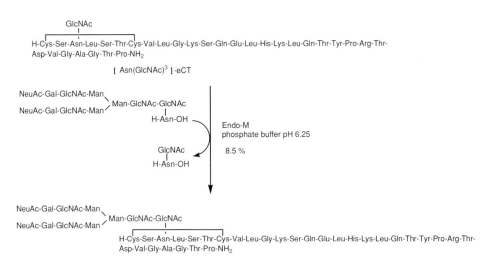

i) Thiosubtilisin mutant, pH = 9, 50°C. ii) H_2, Pd/C. iii) UDP-Gal β1-4-GalTase.

FIGURE 6.13. Chemo-enzymatic synthesis of glycopeptide.

type as model compound for glycoprotein has been described. Natural oligosaccharides were next added by a transglycosylation reaction using endo-β-N-
acetylglucosaminidase from *Mucor hiemalis* (Figure 6.14).[32]

According to Sears & Wong[33] there are three basic approaches for preparing
glycopeptides with complex glycans: (1) a converged method consisting in the
independent preparation of the sugar and peptide components, and final assemble,
(2) the preparation of the sugar attached to an amino acid using glycopeptide
chemistry and simultaneous peptide-linked to a glycal, (3) solid-phase synthesis
of the glycopeptide and chemoenzymatic elaboration of the glycal (Figure 6.15).

GlcNAc
|
H-Cys-Ser-Asn-Leu-Ser-Thr-Cys-Val-Leu-Gly-Lys-Ser-Gln-Glu-Leu-His-Lys-Leu-Gln-Thr-Tyr-Pro-Arg-Thr-
Asp-Val-Gly-Ala-Gly-Thr-Pro-NH₂

[Asn(GlcNAc)³]-eCT

NeuAc-Gal-GlcNAc-Man
                      \
                       Man-GlcNAc-GlcNAc
                      / |
NeuAc-Gal-GlcNAc-Man   H-Asn-OH
                                          Endo-M
                                          phosphate buffer pH 6.25
                      GlcNAc
                        | 8.5 %
                      H-Asn-OH

NeuAc-Gal-GlcNAc-Man
                      \
                       Man-GlcNAc-GlcNAc
                      / |
NeuAc-Gal-GlcNAc-Man   H-Cys-Ser-Asn-Leu-Ser-Thr-Cys-Val-Leu-Gly-Lys-Ser-Gln-Glu-Leu-His-Lys-Leu-Gln-Thr-Tyr-Pro-Arg-Thr-
                       Asp-Val-Gly-Ala-Gly-Thr-Pro-NH₂

FIGURE 6.14. A transglycosylation reaction for preparation of glycopeptide.

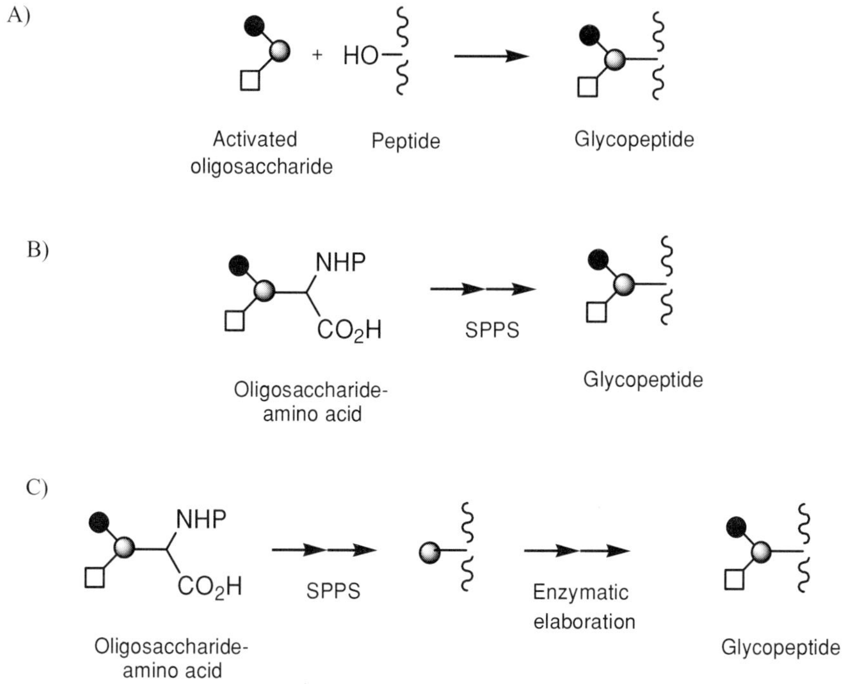

FIGURE 6.15. Proposed general approaches for glycopeptide synthesis.

6.4 Glycoprotein Synthesis

Glycoproteins are essential macromolecules involved in a wide range of functions related to cellular recognition processes. Natural glycoproteins usually exist as a mixture of glycoforms and are found difficult to isolate for their structural characterization and for understanding more about their function.[34,35]

As mentioned, glycoproteins can be obtained by fermentation process, however this natural approximation produce a population of many different glycoforms as result of the participation of many glycosidases and transferases for a given protein, although the mixture can be useful for preparing homogeneous core which in turn might be reelaborated enzymatically.[33] The synthetic preparation of glycoproteins can be considered to some extend glycopeptide chemistry, although the complexity is undoubtedly superior. The synthesis of glycoproteins has received considerable attention, most of it involving a combination of chemical and enzymatic methods.[35,36−41]

A general strategy proposed by Duus et al.[42] considers the assembly of glycosylated amino acid building blocks in solid-phase peptide synthesis according to the general figure shown in Figure 6.16.

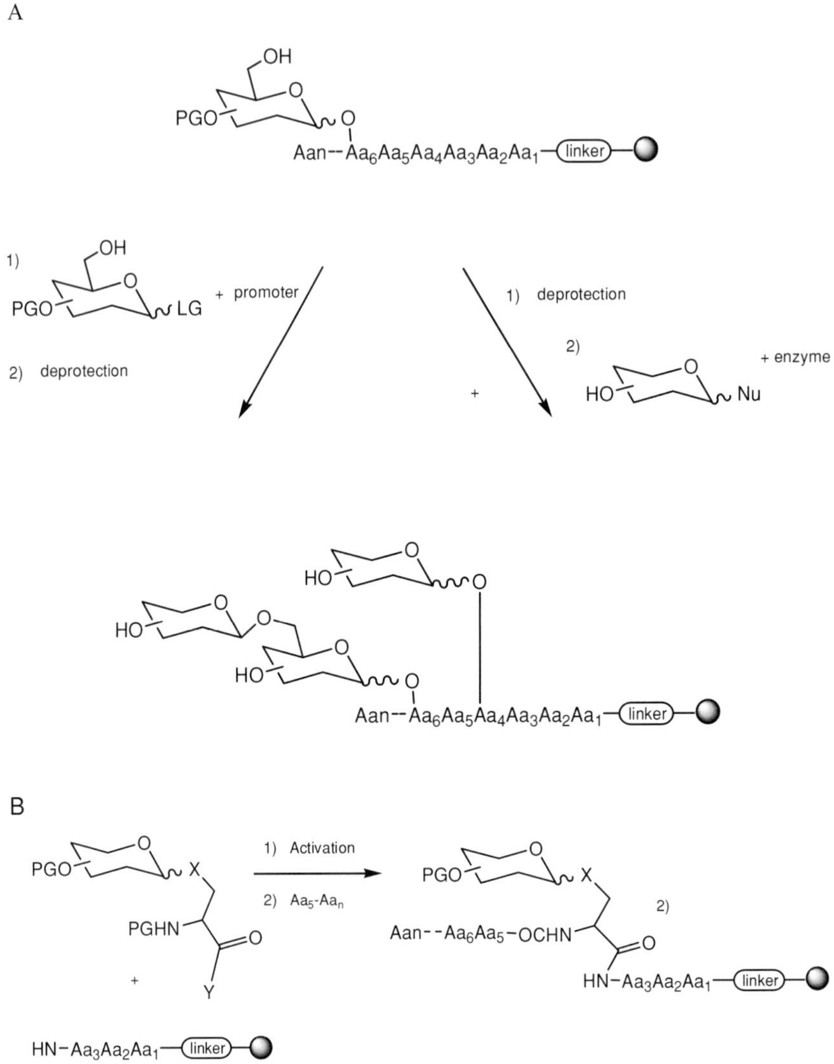

A

B

FIGURE 6.16. Strategies for glycopeptide synthesis.

According to a comprehensive review the strategies described so far for chemical glycoprotein synthesis are (a) indiscriminate glycosylation, (b) chemoselective and site-specific glycosylation (c) site-selective glycosylation.[43]

6.4.1 Indiscriminate Glycosylation

This nonselective approach consists of the preparation of sugars bearing functionalities that under proper conditions may react with a protein. Some of the sugar derivatives used for this purpose are shown in Figure 6.17.

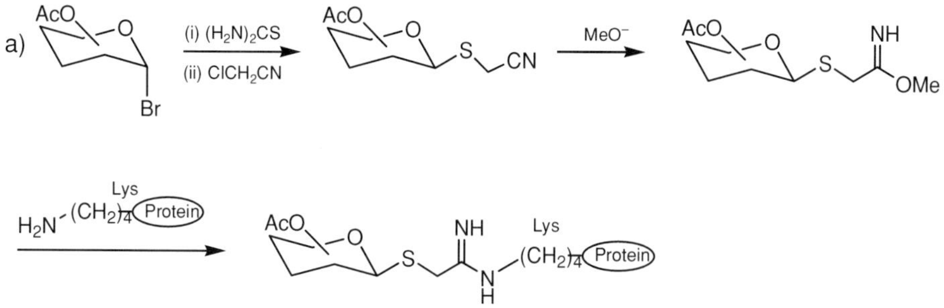

a)

Ref.[44]

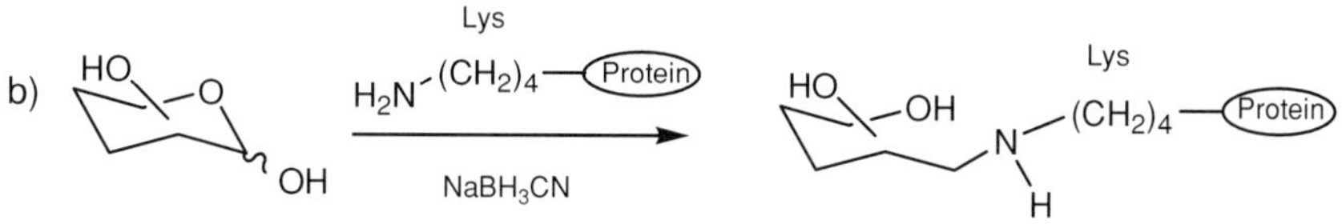

b)

Ref.[45]

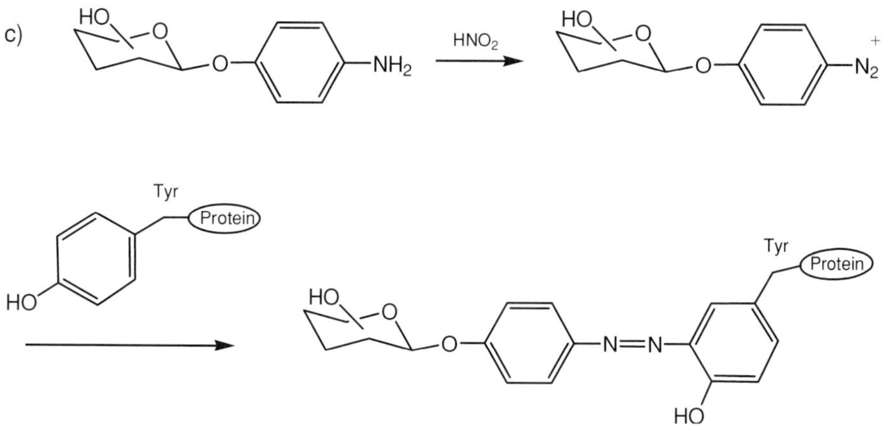

c)

Ref.[46]

d)

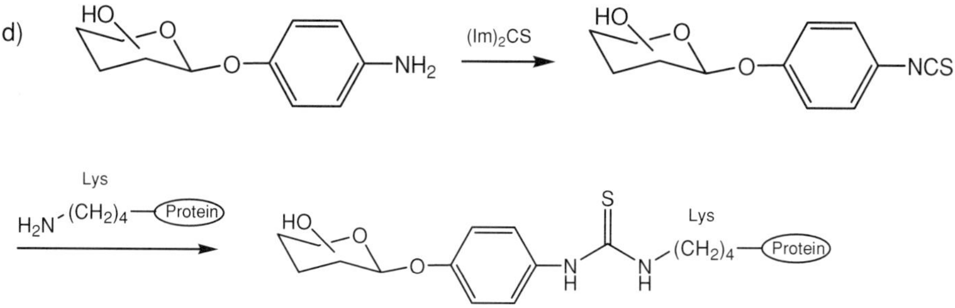

Ref.[47]

FIGURE 6.17. Indiscriminate glycoprotein syntheses.

e)

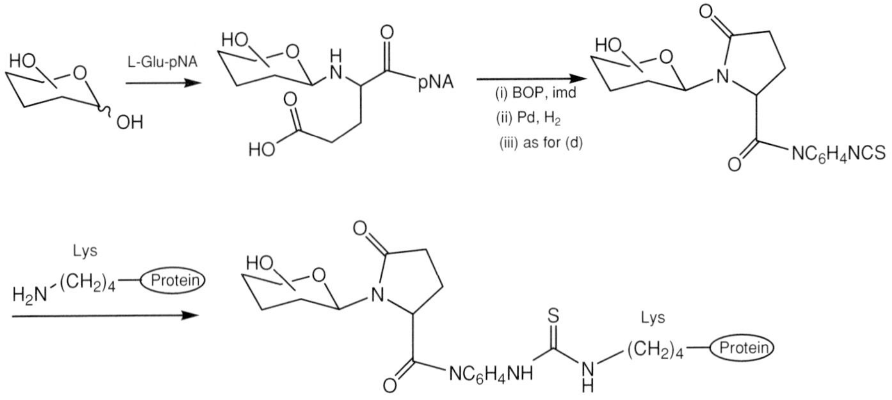

Ref.[48]

f)

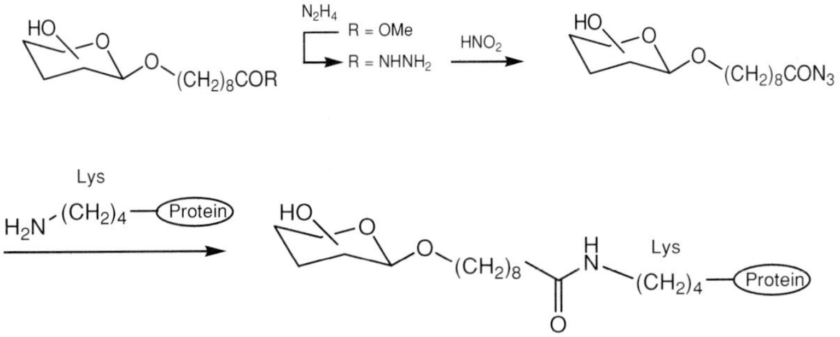

Ref.[49]

g)

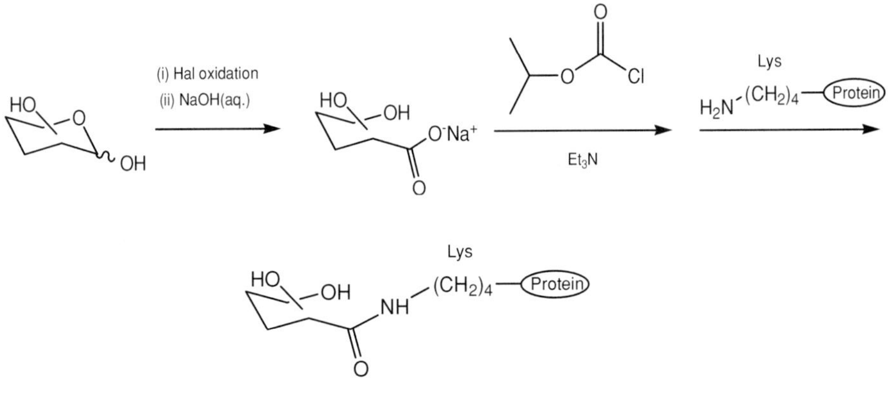

h)

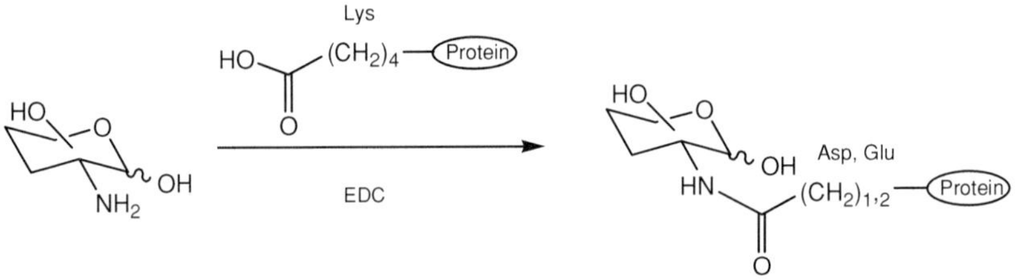

Ref.[50,51]

FIGURE 6.17. (*Continued*)

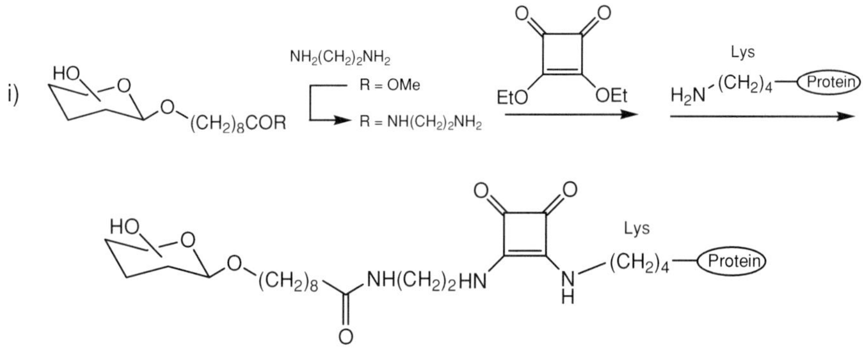

Ref.[52]

FIGURE 6.17. (*Continued*)

6.4.2 Chemoselective and Site-Specific Glycosylation

This approach intends to direct selectively the glycosidic linkage by using chemical and enzymatic tools. Such selectivity has been attempted under a strategy coined with the term chemoselective ligation, and some enzymes involved in this strategy are galactose oxidase,[53] horseradish peroxidase Examples of these step reactions are indicated in Figure 6.18.

6.4.3 Site-Selective Glycosylation

This possibility implies the choice of site selectivity on the glycan. In order to reach this goal a combined site-directed mutagenesis and chemical modification has been performed.[62,63] This strategy involves the introduction of cysteine as chemoselective tag at preselected positions within a given protein and then reaction of its thiol group with glycomethanethiosulfonate (Figure 6.19).

6.4.4 Enzymatic Synthesis

Three basic strategies are considered for obtaining glycoproteins follow-ing an enzymatic approach: the elaboration of glycans through the use of glycosyltransferases,[64-67] the trimming of glycans by purification of glycoform mixtures through selective enzymatic degradation,[68] and alteration of glycans or glycoprotein remodelation, consisting in combined trimming of existing glycan structures followed by elaboration to alternative ones. Theses methods were used for preparing an unnatural glycoform of ribonuclease B by using endoH degra-dation and elaboration with galactosyltransferase, fucosyltransferase, and sialyl-transferase system to construct an sLex glycoform.[69] Other approaches for the assembling of peptides are the "native peptide ligation"[70] and endoglycosidase-catalyzed transglycosylation.[32]

Recent advances on glycoprotein synthesis proposes and in vitro approaches in-volve the following sequential steps: (a) remodeling of recombinant glycoproteins

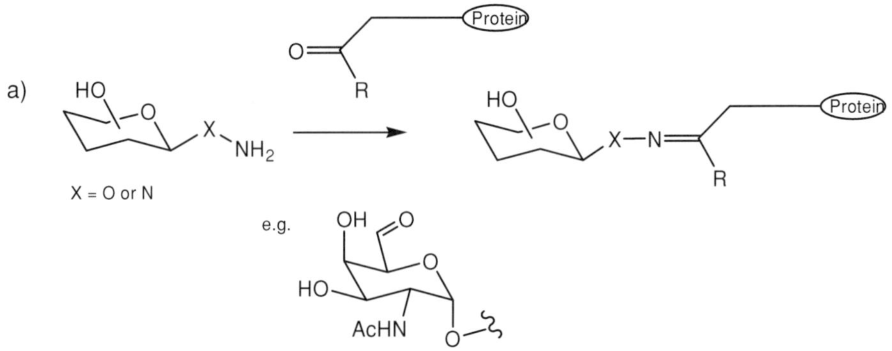

Ref.[54-56]

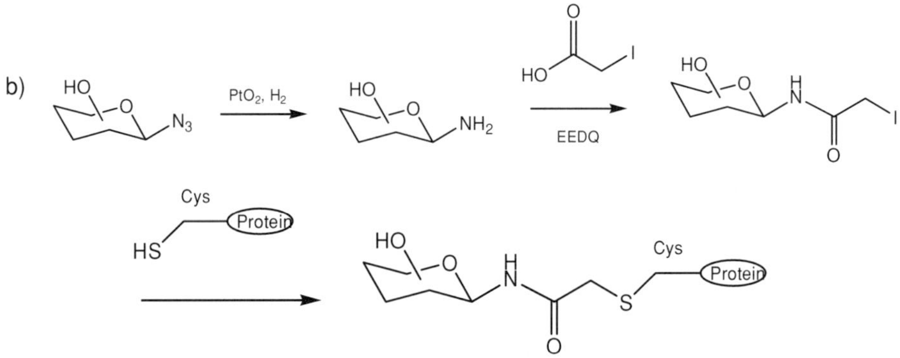

Ref.[57,58]

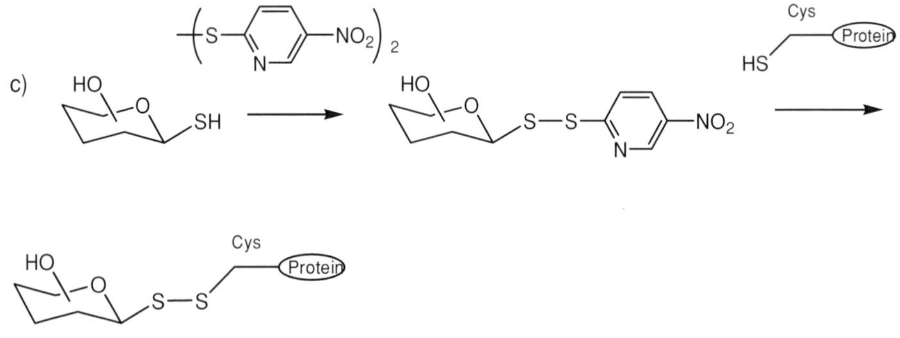

Ref.[59]

FIGURE 6.18. Chemoselective and site-specific glycoprotein syntheses.

by using glycosidases and glycosyltransferases: (b) ligation of synthetic glycopeptides by enzymatic or chemical methods: (c) intein-mediated coupling of glycopeptides to larger proteins expressed as intein-fusion proteins: (d) ligation of glycopeptides to larger proteins containing N-terminal cysteine expressed as TEV

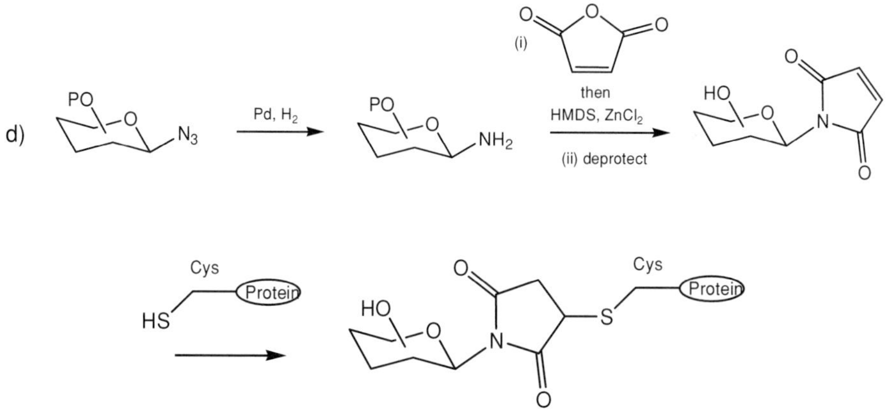

Ref.[60]

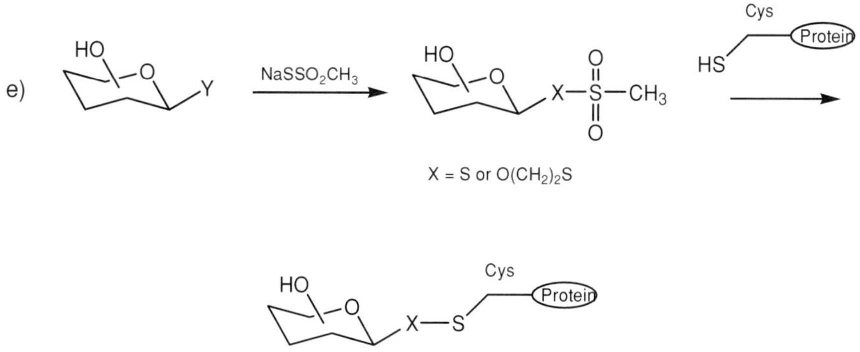

$X = S$ or $O(CH_2)_2S$

Ref.[61-62]

FIGURE 6.18. (*Continued*)

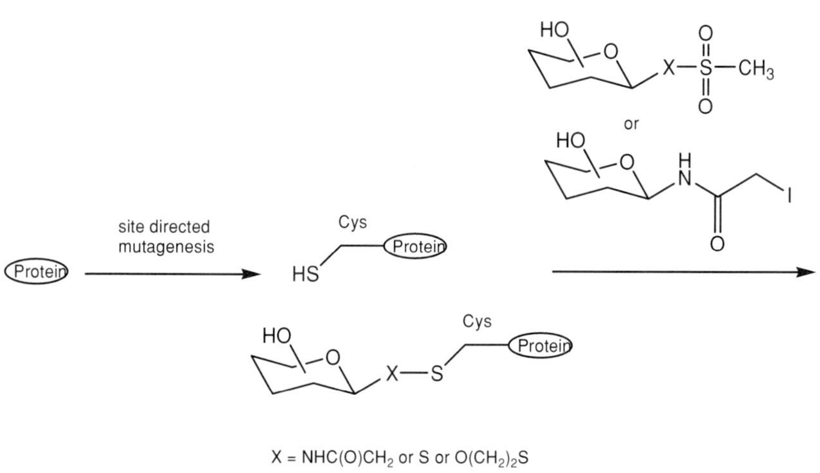

$X = NHC(O)CH_2$ or S or $O(CH_2)_2S$

FIGURE 6.19. Site-selective glycoprotein syntheses.

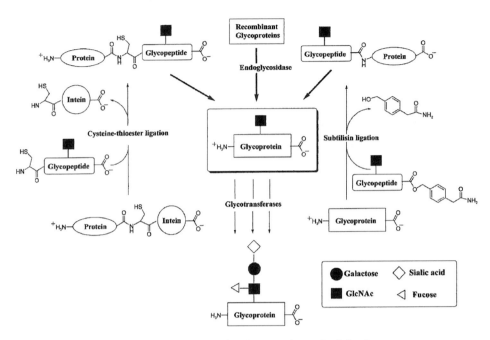

FIGURE 6.20. Strategies for glycoprotein synthesis in vitro.

protease cleavable fusion proteins: (e) in vitro translation; and (f) pathway re-engineering in yeast system to produce human–type N-lynked glycoforms (Figure 6.20).[71]

6.5   Synthesis of Antigenic Glycoconjugates

The preparation of complex glycoconjugates has been a current strategy for the design of synthetic vaccines and usually involves the preparation of the oligosaccharide moiety, which provides the immune specificity by chemical or enzymatic methods, and further attachment through a linker with an immunogenic protein. There has been a continuous effort for developing glycoconjugates containing antigens such as MBr1 antigen Globo-H, the blood group determinant and ovarian cancer antigen Lewis^y, N3 antigens associated with gastrointestinal cancer, the adenocarcinoma antigen KH-1, and the small cell lung carcinoma antigen fucosyl GM1, among others (Figure 6.21), as promising alternatives to develop potentially useful carbohydrate-based anticancer vaccines accessible for clinical program. The synthetic approach becomes justified if we consider that cancer and normal cell growing in tissue culture generally show minimal level of expression of such antigens.[72]

The chemical synthesis of most of these complex oligosaccharides represents a formidable challenge and requires a convenient combination of strategies that

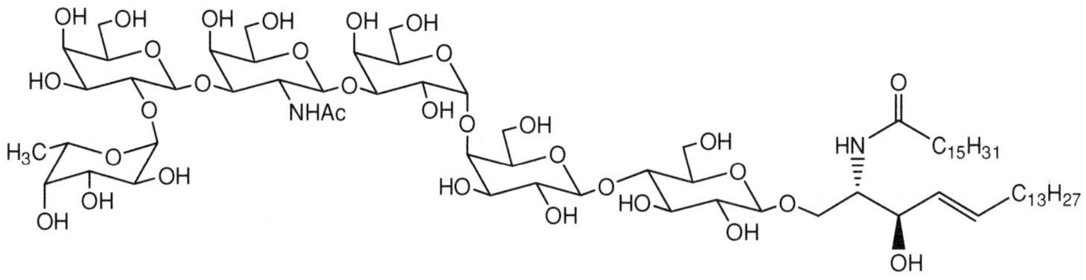

MBr1 antigen / Globo-H

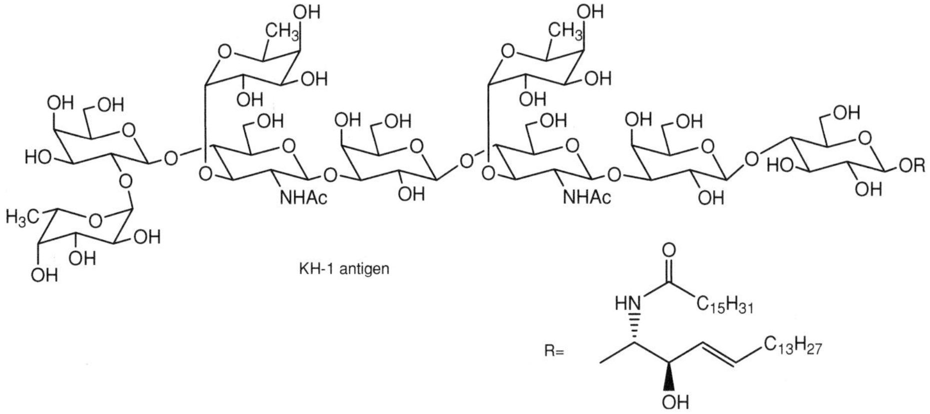

KH-1 antigen

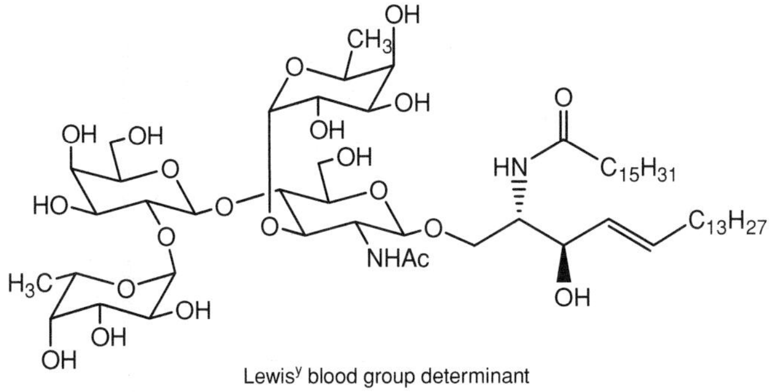

R=

Lewis^y blood group determinant

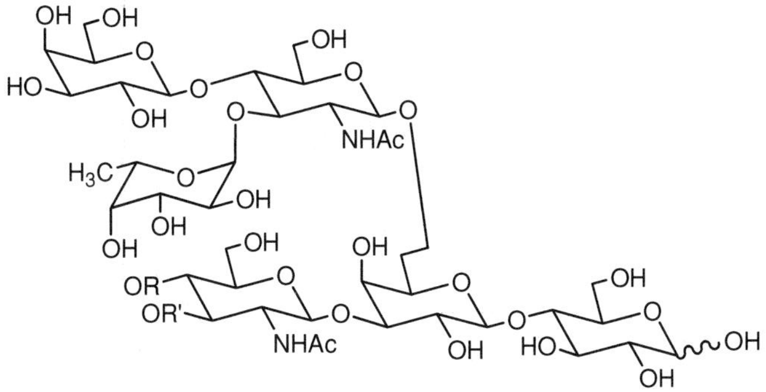

major N3 antigen (R = α-fucose, R' = β-Gal)
minor N3 antigen (R' = α-fucose, R = β-Gal)

FIGURE 6.21. Carbohydrate structures of tumor-associated antigens.

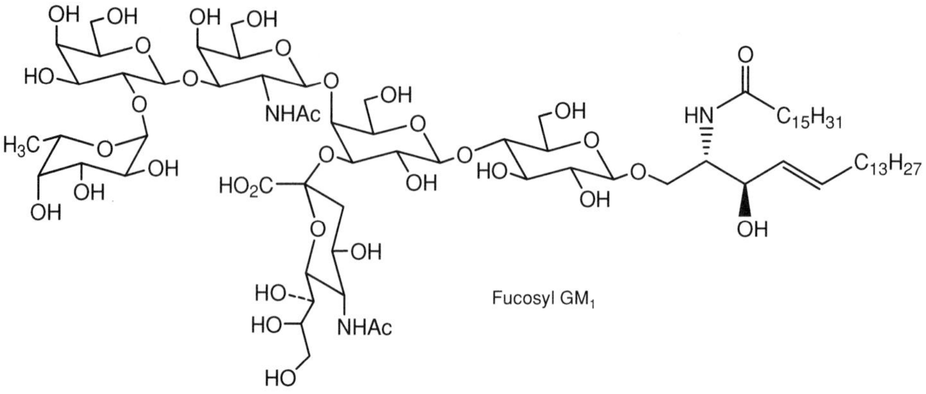

FIGURE 6.21. (*Continued*)

allowed suitable manipulations using appropriate protecting groups, glycosyl donors, acceptors, and coupling reactions conditions.

For instance, the synthesis of glycolipid KH-1 was achieved by Desphande et al.[72] based on the glycal methodology (Figure 6.22).

Likewise, the synthesis of the water-soluble galactosphingolipid analog that binds specifically to recombinant gp 120 was prepared by condensation of *C*-glucosyl aldehyde with Wittig reagent affording the oxazolidone which was transformed into the *C*-glycosylamino acid. By following a subsequent standard protocol represented in Figure 6.23 the target glycolipid was constructed.[73]

Glycoside ceramides are important molecules involved in apoptosis or active cell death. In leukemia cell lines C2 ceramide induces apoptosis via the sphingomyelin pathway. It has been observed that α-galactosylceramides having more than 10 carbons in fatty acid chain have immune stimulatory activities. Thus, the α-Gal-C2 was synthesized by direct glycosylation of C2-Cer with galactosyl fluoride donor in the presence of silver perchlorate as condensing agent (Figure 6.24).[74]

The convergent synthesis is a procedure consisting in the parallel preparation of fragments or building block that will be connected through a coupling reaction, prior deprotection. This procedure was applied successfully for preparation of glycosylphosphatydil inositols (GPI), which are involved in the attachment of glycoproteins with eukaryotic cells (Figure 6.25).[75]

The potential of carbohydrates as antibiotics, antiviral, and anticancer substances has been established.[76,77] Besides, it has been demonstrated their involvement in fertilization, embryogenesis, regulation of the immune system tissue repair, neuronal development, intracellular pathways, and cancer transformation, among others.[12] There is an increasing understanding of how carbohydrates behave biologically between normal and disease states and with this accurate information, novel carbohydrates and therapeutic approaches are developed.[78] For instance, novel glycoside sulfates have been reported as novel potentially useful drugs (Figure 6.26).[76,77]

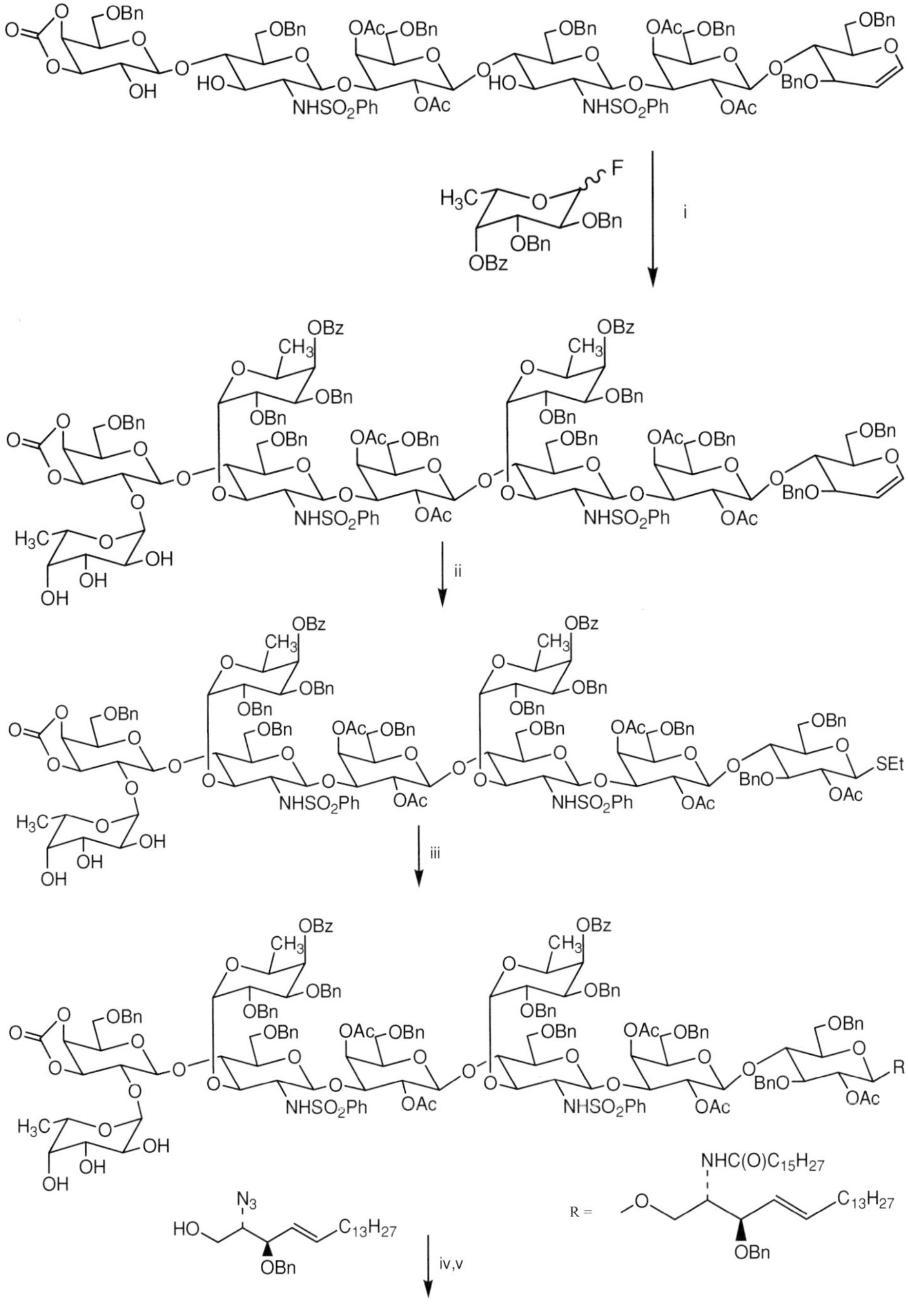

i) Sn(OTf), PhMe/THF (10:1), 4Å MS. ii) a) DMDO, CH$_2$Cl$_2$. b) EtSH, CH$_2$Cl$_2$, H$^+$ (cat.).
c) Ac$_2$O, Py, CH$_2$Cl$_2$. iii) MeOTf, Et$_2$O/CH$_2$Cl$_2$ (2:1) 4 Å M.S. iv) a) H$_2$/Pd-CaCO$_3$, palmitic anh.
EtOAc. v) a) Na/NH$_3$, THF, then MeOH. b) Ac$_2$O, Et$_3$N, DMAP, CH$_2$Cl$_2$. c) MeONa, MeOH.

FIGURE 6.22. Synthesis of KH-1 antigen.

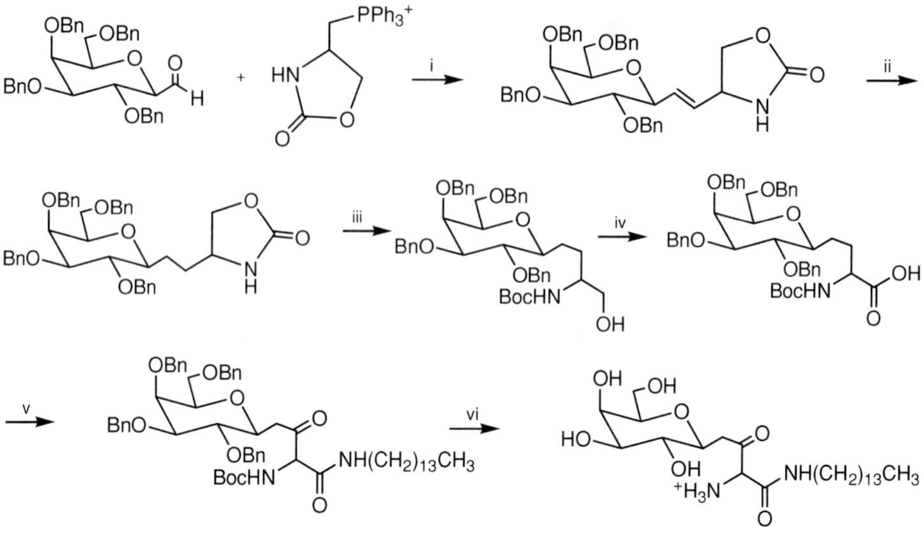

i) BULi, THF. ii) TsNHNH, NaOAc, DME, H₂/cat. iii) a) Boc₂O, Et₃N, DMAP.
b) C_SCO₃, MeOH. iv) Jones. v) EDC, HOBt, tetradecylamine. vi) a) TFA. b) H₂/Pd-C.

FIGURE 6.23. Synthesis of galactosphingolipid analog.

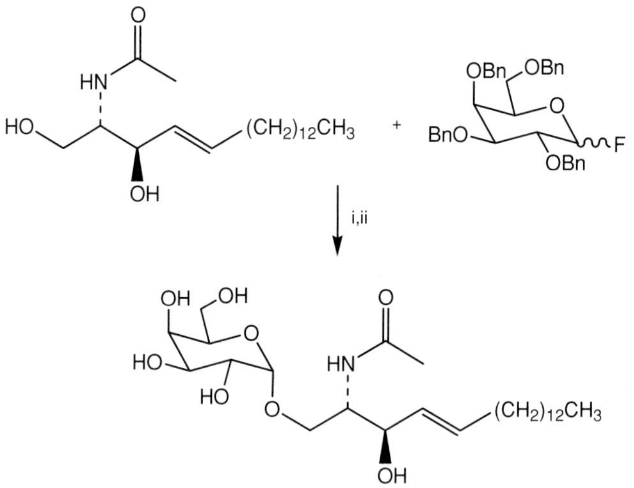

i) AgClO₄, THF, MS, 2h. ii) Na/NH₃, 2h.

FIGURE 6.24. Synthesis of glycosylceramide.

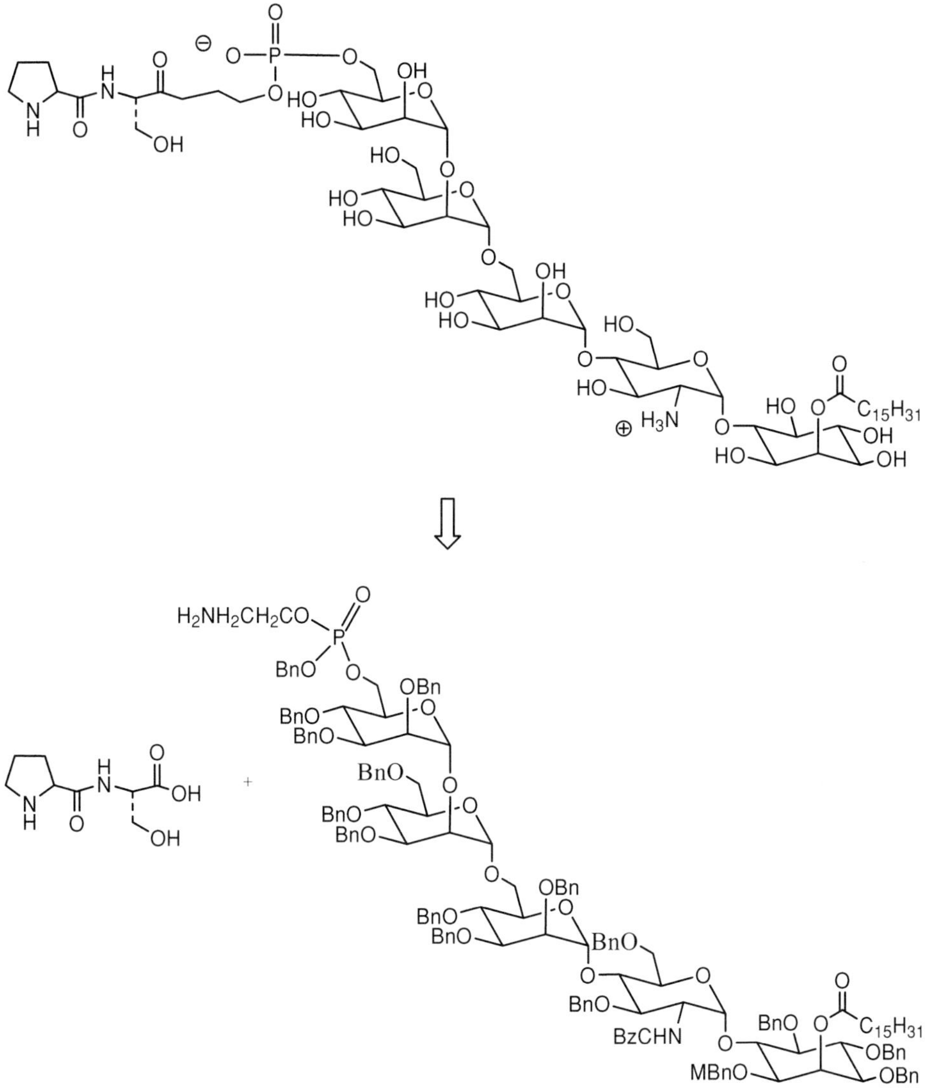

FIGURE 6.25. Retrosynthesis for the preparation of GPI-anchored peptide using convergent synthesis.

It has been mentioned that carbohydrate-based agents such as glycoproteins and polysaccharides obtained from synthetic routes is an emerging and promising strategy for the preparation of vaccines.[79-81] This possibility has become available due the remarkable progress for the chemical and enzymatic preparation of oligosaccharides.

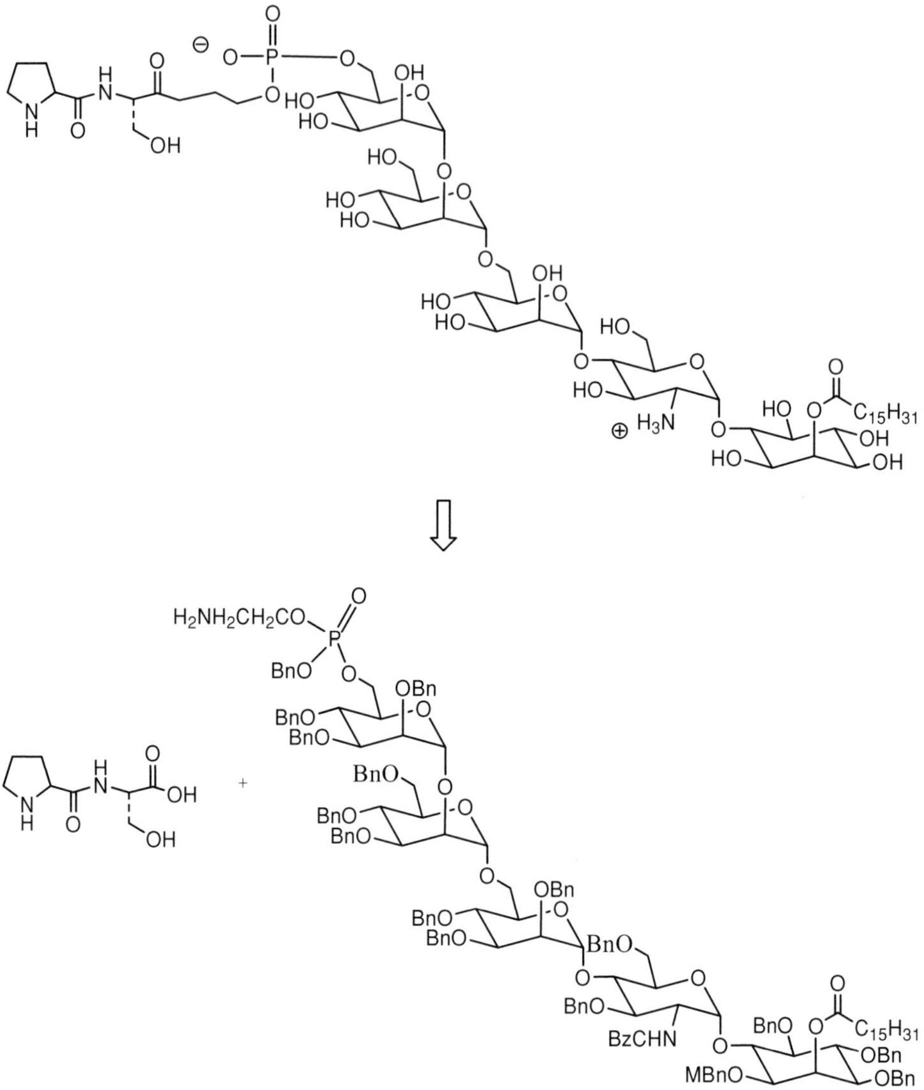

FIGURE 6.25. (*Continued*)

Recent developments on carbohydrate chemistry made possible the design and escalation of new immunogenic carbohydrates. A newly developed synthetic carbohydrate attached to a protein carrier was reported by Verez-Bencomo and Fernández-Santana, and currently administrated against *Haemophilus influenzae* type b disease. The chemical synthesis leading to the oligomeric polyribosylribitol phosphate is described in Figure 6.27.[82]

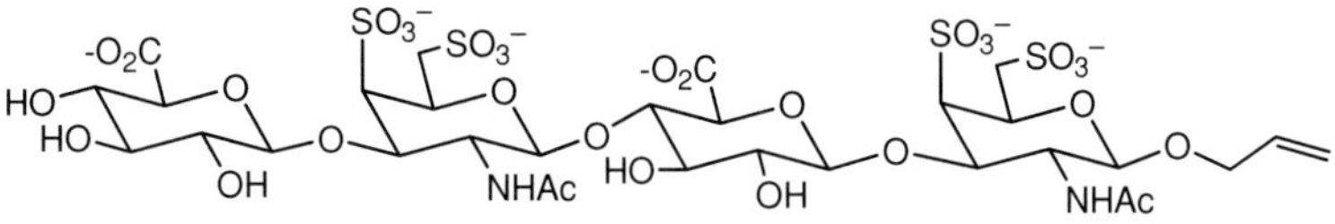

Chondroitin sulfates

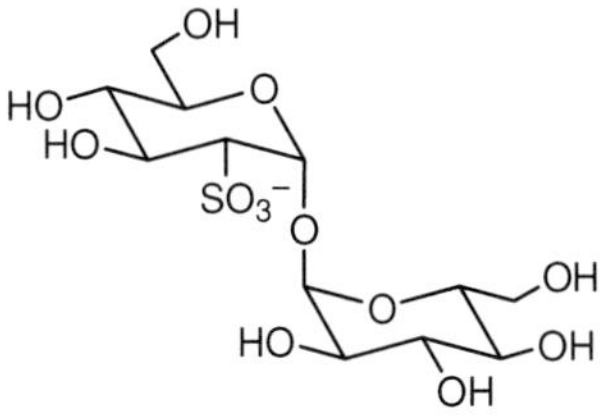

tetrahalose-2-sulfate

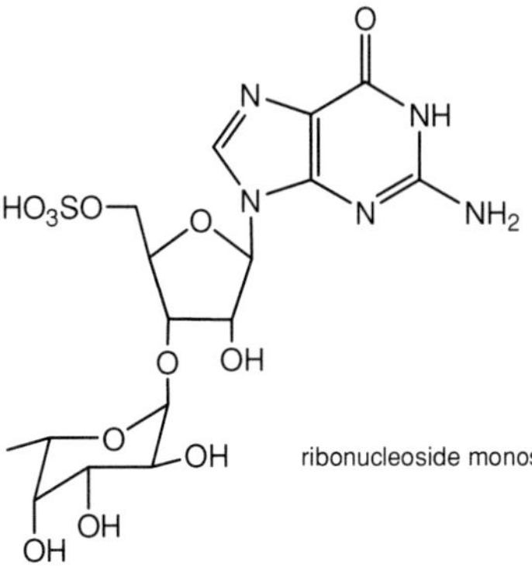

ribonucleoside monosulfate

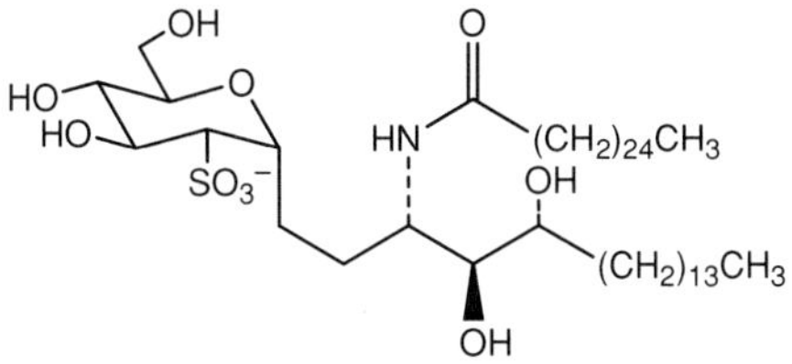

C-glycoside KRN7000

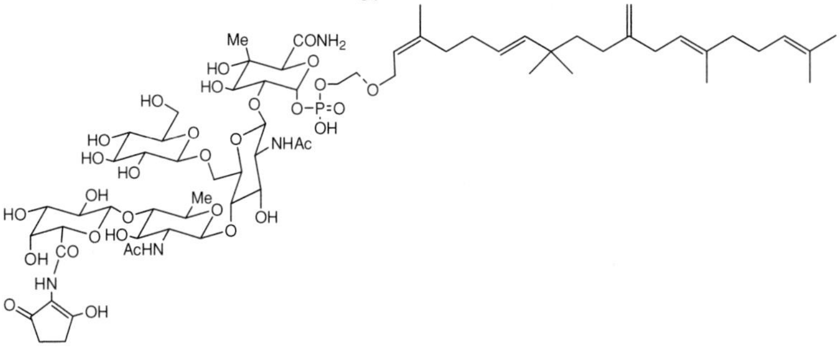

Menomycin A

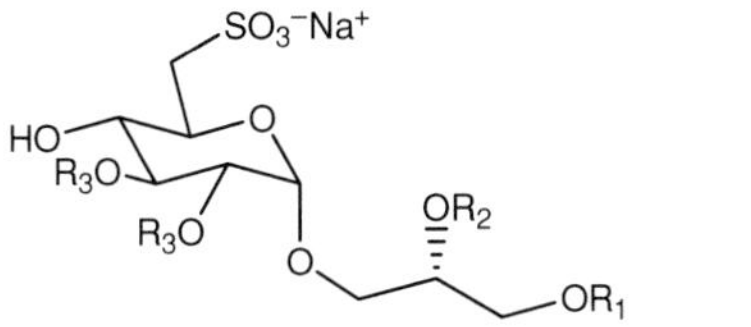

R_1 = oleoyl R_2 = palmitoyl R_3 = H
R_1 = linoleoyl R_2 = palmitoyl R_3 = H
R_1 = palmitoyl R_2 = palmitoyl R_3 = H
R_1 = linoleoyl R_2 = palmitoyl R_3 = palmitoyl

FIGURE 6.26. Novel glycoside sulfates and phosphates as potential drugs.

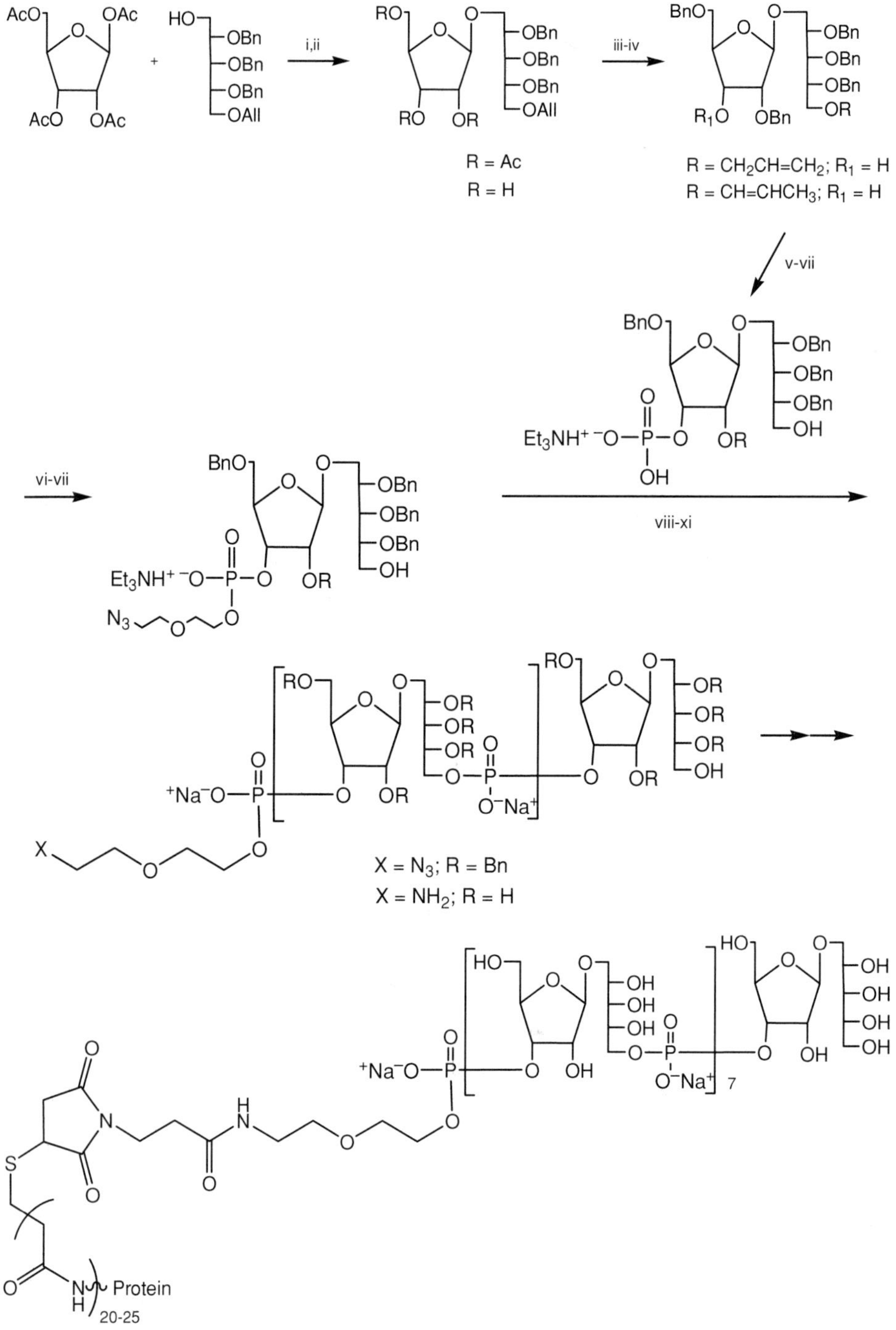

i) BF$_3$Et$_2$O, CH$_2$Cl$_2$. ii) CH$_3$ONa, CH$_3$OH. iii) BnCl, BuSnO, NaH, Bu$_4$NI. iv) tBuOK, DMSO, 100°C.
v) PCl$_3$, imidazole, CH$_3$CN. vi) N$_3$(CH$_2$)$_2$O(CH$_2$)$_2$OH, I$_2$. vii) AcOH-H$_2$O, 80°C. viii) PivCl, Py.
ix) Py-H$_2$O. x) H$_2$, Pd-C, EtOH-H$_2$O-EtOAc-AcOH, 1.5 atm. xi) cation exchange resin on Sephadex SP-C25.

FIGURE 6.27. Synthetic carbohydrate conjugate vaccine Quimi-Hib.

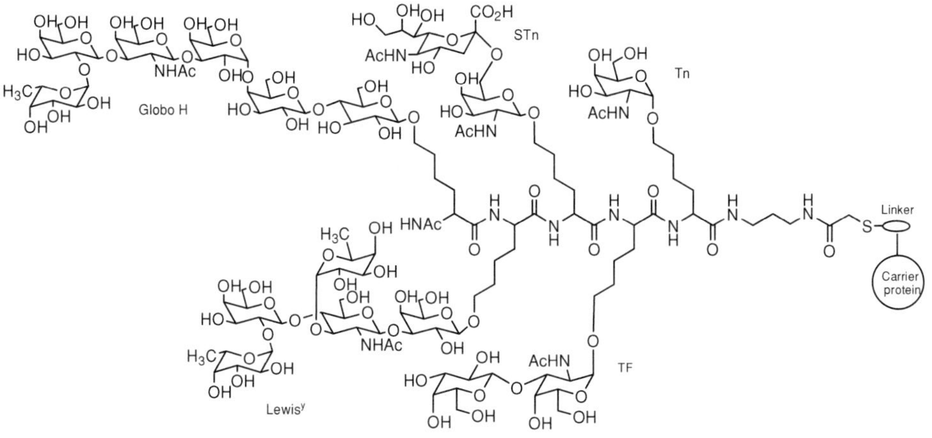

FIGURE 6.28. Various tumor antigenic agents coupled to a linker developed as potential synthetic vaccine.

Another alternative therapeutic strategy for inducing immune response through the use of synthetic carbohydrate vaccines has been proposed by Danishefsky et al., involving the attachment of different tumor antigenic agents (Globo H, STn, Tn, Lewis^y), coupled to a linker and this to a protein carrier (Figure 6.28).[83]

References

1. J.F. Robyt, *Essentials of Carbohydrate Chemistry,* Springer-Verlag, New York, Inc. **1998**, 279.
2. A. Morell, R.A. Irvine, I. Sternliev, I.H. Scheinberg, and G. Ashwell, G. *J. Biol. Chem.* **1968**, 243, 155.
3. H.D. Fischer, A. Gonzalez-Noriega, W.S. Sly, and D.J. Morre, *J. Biol. Chem.* **1980**, 255, 9608.
4. P. Stahl, P.H. Schlesinger, E. Sigardson, J. Rodman, Y.C. Lee, *Cell* **1980**, 19, 207.
5. P.M. Rudd, T. Elliot, P. Cresswell, I.A. Wilson, and R.A. Dwek, *Science* **2001**, 291, 370.
6. I.N. Reitter, R.E. Means, and R.C. Desrosiers, *Nature Med.* **1998**, 4, 679.
7. G.F. Springer, *Science* **1984**, 224, 1198.
8. J. Samuel, A.A. Noujaim, G.D. MacLean, M.R. Sureshand, and B.M. Longenecker, *Cancer Res.* **1990**, 50, 4801.
9. M. Fukada, *Biochim. Biophys. Acta,* **1985**, 780, 119.
10. R. Kornfeld, and S. Kornfeld, *Annu. Rev. Biochem.* **1985**, 54, 631.
11. Y.C. Lee, *FASEB J.* **1992**, 6, 3193–3200.
12. R. Dwek, *Chem. Rev.* **1996**, 683.
13. H. Lis and N. Sharon, *Chem. Rev.* **1998**, 98, 637.
14. W. Gaastra and A.-M. Svennerholm, *Trends. Microbiol.* **1996**, 4, 444.
15. I.J. Goldstein, H.C. Winter, R.D. Poretz, *In Glycoproteins,* Elsevier **1997**, 403–474.
16. W.I. Weis and K. Drickamer, *Annu. Rev. Biochem* **1996**, 65, 441.

17. R. Ravishankar, N. Ravindran, A. Suguna, A. Surolia and M. Vijayan, *Curr. Science*, **1997**, 72, 855.
18. N. Sharon, *Trends Biochem. Sci.* **1993**, 18, 221.
19. W.I. Weis, and K. Drickamer, *Annu. Rev. Biochem.* **1996**, 65, 441.
20. N. Sharon, and H. Lis, *Essays Biochem.* **1995**, 30, 59.
21. J.H. Naismith and R.A. Field, *J. Biol. Chem.* **1996**, 271, 972.
22. K. Drickamer, *J. Biol. Chem.* **1988**, 263, 9575-9560.
23. F. Fukumori, N. Takeuchi, T. Hagiwara, H. Ohbayashi, T. Endo, N. Kochibe, Y. Nagata, and A. Kobata, *J. Biochem.* **1990**, 107, 190.
24. L.A. Lasky, *Annu. Rev. Biochem.* **1995**, 64, 113.
25. S. Hemmerich, H. Leffler, and S.D. Rosen *J. Biol. Chem.* **1995**, 270, 12035.
26. H. Kunz, *Angew. Chem. Int. Ed. Engl.* **1987**, 26, 294.
27. H. Garg and R.W. Jeanloz, *Adv. Carbohydr. Chem. Biochem.* **1985**, 43, 135.
28. H. Kunz, *Pure & Appl. Chem.* **1993**, 65, 1223.
29. H. Kunz and C. Unverzagt, *Angew. Chem. Int. Ed. Engl.* **1984**, 23, 436.
30. J. März and H. Kunz, *Synlett* **1992**, 589.
31. C.H. Wong, M. Schuster, P. Wang, and P. Sears, *J. Am. Chem. Soc.* **1993**, 115, 5893.
32. M. Mizuno, K. Haneda, R. Iguchi, I. Muramoto, T. Kawakami, S. Aimoto, K. Yamamoto, and T. Inazu, *J. Am. Chem. Soc.* **1999**, 121, 284.
33. P. Sears and C.-H. Wong, *Science* **2001**, 291, 2344.
34. P.P. Deshpande, H.M. Kim, A. Zatorski. T.K. Park, G. Raguphathi, P.O. Livingston, D. Live and S.J. Danishefsky, *J. Am. Chem. Soc.* **1998**, 120, 1600.
35. C.R. Bertozzi, D.G. Cook, W.R. Kobertz, F. Gonzalez-Scarano, M.D. Bednarski, *J. Am. Chem. Soc.*, **1992**, 114, 10639.
36. L.M. Obei, C.M. Linardic, L.A. Karolak, and Y.A. Hannun, *Science* **1993**, 259, 1769.
37. J. Xue, N. Shao, and Z. Guo, *J. Org. Chem.* **2003**, 68, 4020-4029.
38. U. Kempin, L. Henning, D. Knoll, P. Welzel, D. Müller, and J. Markus, van Heijenoort, *Tetrahedron*, **1997**, 53, 17669.
39. S. Loya, V. Reshef, E. Mizrachi, C., Silbertein, Y. Rachamim, S. Carmeli and A. Hizi, *J. Nat. Prod.* **1998**, 61, 891.
40. A. Persidis, *Nature Biotechnology* **1997**, 15, 479.
41. T. Buskas, Y. Li and G.-J. Boons, *Chem. Eur. J.* **2004**, 10, 3517.
42. J.Ø Duus, P.M.St. Hilaire M .Meldal, and K. Bock, *Pure. Appl. Chem.* **1999**, 71, 755.
43. (a) B. G. Davis, *Chem. Rev.* **2002**, 102, 579. (b) C.R. Bertozzi and L.L. Kiessling, *Science* **2001**, 23, 2357.
44. Y.C. Lee, C.P. Stowell, and M.J. Krantz, *Biochemistry* **1976**, 15, 3956.
45. G.R. Gray, *Arch. Biochem. Biophys.* **1974**, 163, 426.
46. C.R. McBroom, C.H. Samanen, and I.J. Goldstein, *Methods Enzymol.* **1972**, 28, 212.
47. D.H. Buss and I.J. Goldstein, *J. Chem. Soc. C* **1968**, 1457.
48. C. Quétard, S. Bourgerie, N. Normand-Sdiqui, R. Mayer, G. Strecker, P. Midoux, A.C. Roche, and M. Monsigny, *Bioconjugate Chem.* **1998**, 9, 268.
49. R.U. Lemieux, D.R. Bindle, and D. A. Baker, *J. Am. Chem. Soc.* **1975**, 97, 4076.
50. W.O. Baek and M.A. Vijayalaksmi, *Biochim. Biophys. Acta* **1997**, 1336, 394.
51. K.-Y. Jiang, O. Pitiot, M. Anissimova, H. Adenier,and M.A. Vijayalakshmi, *Biochim, Biophys. Acta.* **1999**, 1433, 198.
52. V.P. Kamath, P. Diedrich, and O. Hindsgaul, *Glycoconjugate J.* **1996**, 13, 315.
53. G.A. Lemieux and C.R. Bertozzi, *TIBTECH* **1998**, 16, 506.
54. S.E. Cervigni, P. Dumy, and M. Mutter, *Angew. Chem. Int. Ed. Engl.* **1996**, 35, 1230.
55. Y. Zhao, S.B.H. Kent, and B.T. Chait, *Proc. Natl. Acad. Sci. USA*, **1997**, 94, 1629.

56. P. Durieux, J. Fernandez-Carneado, and G. Tuchscherer, *Tetrahedron Lett.* **2001**, 42, 2297.

57. N.J. Davis and S.L., Flitsch, *Tetrahedron Lett.* **1991**, 32, 6793.

58. L.A. Marcaurelle and C.R. Bertozzi, *J. Am. Chem. Soc.* **2001**, 123, 1587.

59. W.M. Macindoe, A.H. van Oijen, and G.-J. Boons, *Chem. Commun.* **1998**, 847.

60. I. Shin, H.-J. Jung, and M.R. Lee, *Tetrahedron Lett.* **2001**, 42, 1325.

61. B.J. Davis, R.C. Lloyd, and J.B. Jones, *J. Org. Chem.* **1998**, 63, 9614.

62. B.G. Davis, M.A.T. Maughan, M.P. Green, and A. Ullman, *Tetrahedron: Asymmetry* **2000**, 11, 245.

63. B.G. Davis, R.C. Lloyd, and J.B. Jones, *Bioorg. Med. Chem.* **2000**, 8, 1527.

64. J.C. Paulson, R.L. Hill, T. Tanabe, and G. Ashwell, *J. Biol. Chem.* **1997**, 252, 8624.

65. S. Tsuboi, Y. Isogai, N. Hada, J.K. King, O. Hindsgaul, and M. Fukuda, *J. Biol. Chem.* **1996**, 271, 27213.

66. C. Unversagt, *Tetrahedron Lett.* **1997**, 32, 5627.

67. R.A. Geremia, E.A. Petroni, L. Ielpi, and B. Herissat, *Biochem. J.* **1996**, 318, 133.

68. B. Friedman, S.C. Hubbard, and J.R. Rasmussen, *Glycoconjugate, J.* **1993**, 10, 257.

69. K. Witte, P. Sears, R. Martin, and C.-H. Wong, *J. Am. Chem. Soc.* **1997**, 119, 2114.

70. G.G. Kochendoerfer and S.B.H. Kent, *Curr. Opin. Chem. Biol.* **1999**, 3, 665.

71. C.-H. Wong, *J. Org. Chem.* 70, 4221 (2005).

72. P.P. Deshpande, H.M. Kim, A. Zatorski, T.K. Park, G. Raguphathi, P.O. Livingston, D. Live, and S.J. Danishefsky, *J. Am. Chem. Soc.* **1998**, 120, 1600.

73. C.R. Bertozzi, D.G. Cook, W.R. Kobertz, F. Gonzalez-Scarano, and M.D. Bednarski, *J. Am. Chem. Soc.* **1992**, 114, 10639.

74. L.M. Obei, C.M. Linardic, L.A. Karolak, and Y.A. Hannun, *Science* **1993**, 259, 1769.

75. J. Xue, N. Shao, and Z. Guo, *J. Org. Chem.* **2003**, 68, 4020–4029.

76. U. Kempin, L. Henning, D. Knoll, P. Welzel, D. Müller, J. Markus, van Heijenoort, *Tetrahedron*, **1997**, 53, 17669.

77. S. Loya, V. Reshef, E. Mizrachi, C. Silbertein, Y. Rachamim, S. Carmeli, and A. Hizi, *J. Nat. Prod.* **1998**, 61, 891.

78. A. Persidis, *Nature Biotechnology* **1997**, 15, 479.

79. T. Buskas, Y. Li, and G.-J. Boons, *Chem. Eur. J.* **2004**, 10, 3517.

80. P.H. Seeberger, R.L. Soucy, Y.-U. Kwon, D.A. Snyder, and T. Konemitsu, *Chem. Commun.* **2004**, 1706.

81. S. Bay, V. Huteau, M.-L. Zarantonelli, R. Pires, J. Ughetto-Monfrin, M.-K. Taha, P. England, and P. Lafaye, *J. Med. Chem.* **2004**, 47, 3916.

82. V. Verez-Bencomo, et al., *Science* **2004**, 305, 522.

83. S.J. Danishefsky and J.R. Allen, *Angew. Chem. Int. Ed.* **2000**, 39, 836–863.

7
Hydrolysis of Glycosides

The glycosidic bond might be degraded by chemical and/or enzymatic agents. Comparative studies revealed that chemical hydrolysis is nonspecific and on the other hand the enzymatic is regio and stereospecific. The glycosides are chemically susceptible to acid conditions, and only in some cases to basic conditions. In general, the acid sensitivity is attributed to the sugar moiety and the basic nonstability to the aglycon nature.

7.1   Acidic Hydrolysis

When a glycoside is subjected to acid conditions, a process called *acetolysis* takes place. This phenomenon is more clearly seen on O-glycosides, where even weak acid conditions can be sufficient for O-glycoside breakage. Some simple glycosides such as β methyl-2,3,4,6-tetra-O-methyl-D-glucopyranose are hydrolyzed under diluted HCl conditions to yield a hydroxy-2,3,4,6-tetra-O-methyl-D-glucopyranose. Likewise β ethyl-glucopyranose is hydrolyzed to a mixture of anomers (Figure 7.1).

In general, S-glycosides are more resistant than their counterparts O-glycosides to an acidic medium; however, the former can be hydrolyzed under the conditions described in Figure 7.2.

Disaccharides can be readily hydrolyzed under weak acidic conditions, producing their constitutive monomers in equivalent quantities (Table 7.1).

Depending on the strength of the hydrolytic conditions, polysaccharides undergo fragmentation, producing oligosaccharides, disaccharides, and monomers. The degradation degree relies on acid concentration, branching, and solubility. Thus, cellulose, being the most abundant natural polisaccharide in nature, requires high acidic concentrations in order to be fully degraded to glucose. On the contrary, some other polysaccharides at lower concentrations produce dimers and monomers (Table 7.2).

Partial hydrolysis is important in certain cases in which disaccharides are not either affordable materials or easily obtained ones through synthetic means. Such

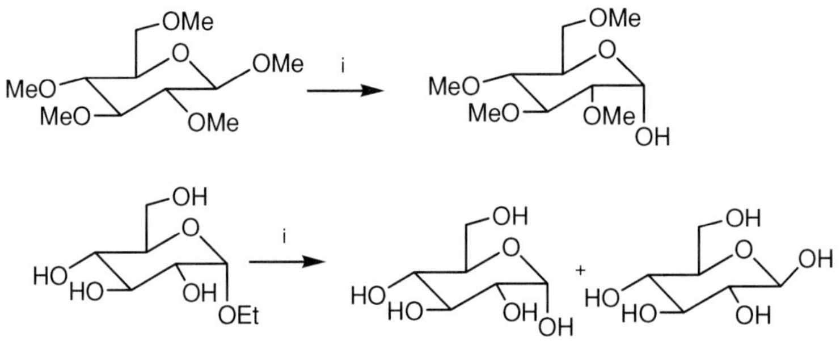

i) HCl dil.

FIGURE 7.1. Acid hydrolysis of simple *O*-glucosides.

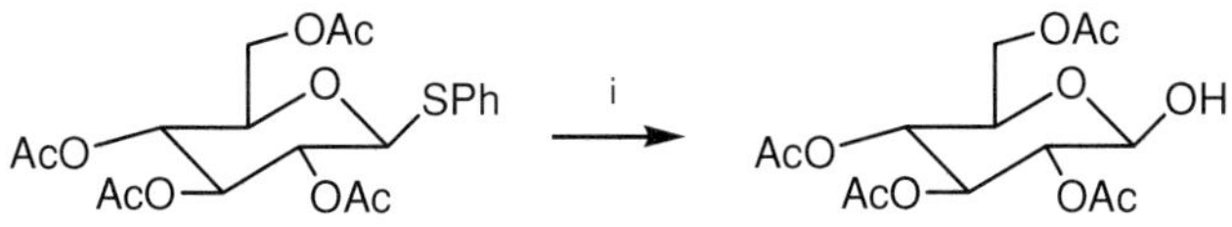

i) NBS/acetone-H$_2$O.

FIGURE 7.2. Hydrolysis of thioglycosides.

TABLE 7.1. Acid hydrolysis of disaccharides.

Disaccharide	Hydrolysis product
(+)-sucrose	D-(+)-glucose D-(−)-fructose
(+)-lactose	D-(+)-glucose D-(+)-galactose
(+)-cellobiose	D-(+)-glucose D-(+)-glucose

TABLE 7.2. Acid hydrolysis of polysaccharides.

Polysaccharide	Partial hydrolysis	Total hydrolysis
cellulose	1,4-cellobiose	D-glucose
laminarin	1,3-laminaribiose	D-glucose
curdlan	1,3-laminaribiose	D-glucose
quitine	1,4-N-acetyl glucosamine	2-amino-2-deoxy-D-glucose
manan	1,4-mannobiose	D-glucose
pululan	1,4-maltotriose	D-glucose

is the case of 1,3-laminaribiose synthetically obtained in poor yields (9.5%),[1] but readily available from polysaccharide curdlan.[2]

7.2 Basic Hydrolysis

Some glycosides have been shown to be partially sensitive against basic conditions, besides their naturally high acid sensitivity. It is been experimentally found that three classes of O-glycosides might be subject to basic hydrolysis:[3]

1. phenolic glycosides
2. enolic glycosides
3. β-substituted alcohol glycosides

7.2.1 Phenolic Glycosides

A typical example of phenolic glycoside decomposition under basic conditions is observed in the treatment of salicin with barium hydroxide giving as result a cyclic acetal and the release of the aglycon (Figure 7.3).

7.2.2 Enolic Glycosides

Within this type of glycosides, the three varieties to be considered are (1) 4-hydroxycoumarins, (2) purine and pirimidin glycosides,and (3) simple enols (Figure 7.4).

7.2.3 β-Susbstituted Alcohol Glycosides

Glycoside picrocine is hydrolyzed in diluted potassium hydroxide solution, through a mechanism that involves an intermediate carbanion formation to give a conjugated unsaturated product and glucose as breakage product (Figure 7.5).

Contrary to acid hydrolysis of disaccharides where degradation products are their constitutive units, in most of the cases for basic conditions, nonsugar derivatives are produced as result (Table 7.3).

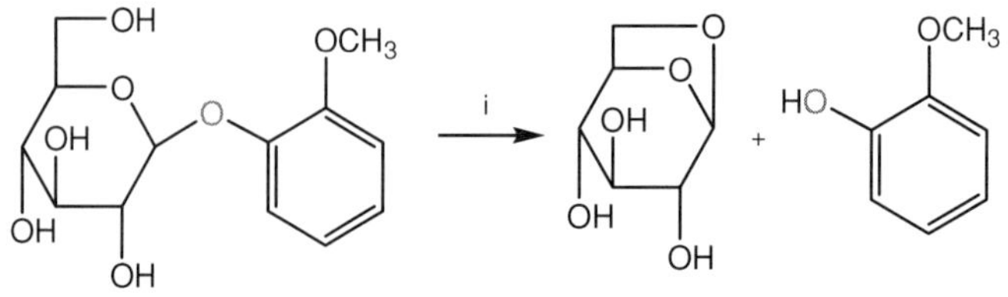

i) Ba(OH)$_2$.

FIGURE 7.3. Basic hydrolysis of phenolic glycosides.

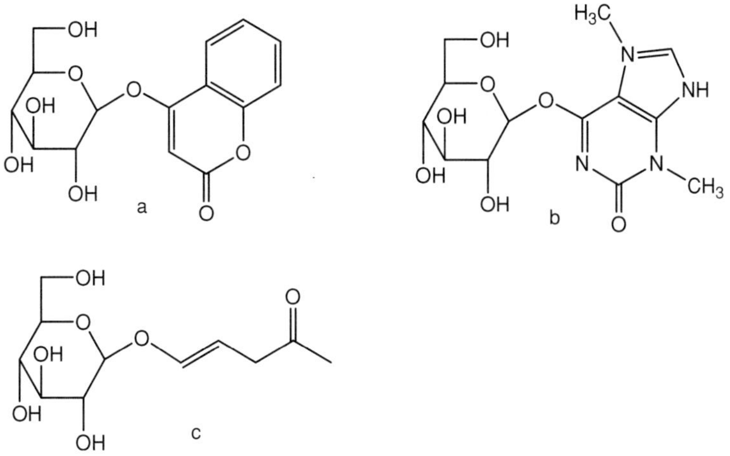

FIGURE 7.4. Basic hydrolysis of enolic glycosides.

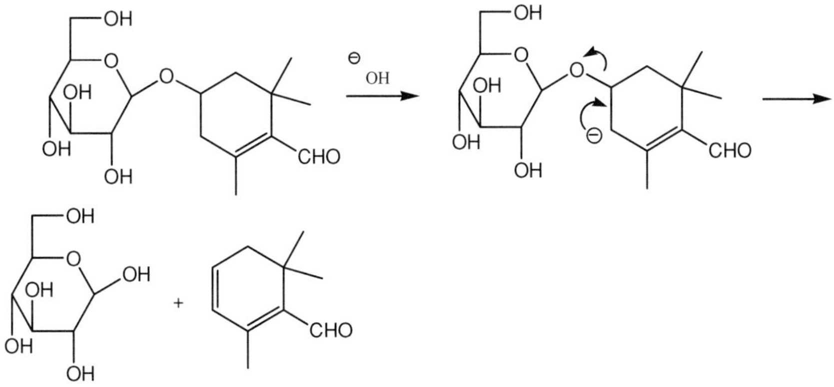

FIGURE 7.5. Basic hydrolysis of β-substituted alcohol glycosides.

TABLE 7.3. Degradation products of disaccharides under basic conditions.

Disaccharide	Hydrolysis conditions (KOH)	Temperature (°C)	Product
cellobiose	1.5 N	50	lactic acid
gentobiose	2 N	50	lactic acid
lactose	0.2 N	100	D-galactose
maltose	0.15 N	25	fenilhydrazone of D-mannose

7.3   Enzymatic Hydrolysis

β-Glycosides are the natural substrates for hydrolytic enzymes known as β-glycosidases. So far, at the biochemical level, the rule of most of glycosidases is not totally well understood. However, some of them have been related to feeding, detoxification processes, or even as a defense mechanism against herbivorous pathogens through releasing of tiocyanates, cianides, and phytohormones. It is been established that there is an specific glycosidases for each aldopyranose, being the sugar composition responsible for the recognition pattern. Some of the best-studied hydrolyses are the β-glycosidases and among them β-glucosidases, β-glucuronidases, β-cellulase, β-glucanasas, β-quitinases, all of them with important biological and economical implications.[4]

7.3.1   β-Glucosidases

There is strong evidence indicating that their action is mainly directed toward the defense mechanism and growth regulation. For instance, cyanogenic glycosides are hydrolyzed, for the releasing of cyanide ions as a defense mechanism against animals. In humans the equivalent of β-glucosidase is called glucocerebrosidase (with low genomic homology to the plant counterpart) and catalyzes the degradation of glucosylceramide inside lysosome. The lack or deficiency of this enzyme produces the Gaucher disease characterized by accumulation of esphyngosylglucosides and glucosylceramides.

7.3.2   β-Glucanasas, β-Quitinases

The natural substrates for these oligosaccharide hydrolytic enzymes are laminarin and quitine, respectively, being present in fungi, yeast, and insects. Some of the processes related to the activity of these enzymes are seed degradation, cellular elongation control, growth regulation, pollen growth regulation, digestion, and fertilization. Moreover, within the context of the defense mechanisms, these enzymes can be able to digest the fungi cellular wall and also to release oligosaccharides that induce the production of antimycotic substances called phytolexines.

7.3.3   β-Cellulase

Cellulose is the most abundant natural polisaccharide on earth. Cellulitic enzymes, particularily cellobiohydrolases CBHI, CBHII, EGI and EG II found in fungi *Trichoderma reesei,* have been thoroughly studied for determing the three-dimensional structure, the genomic sequence, receptors, and substrate specificity.

7.3.4   β-Glucuronidase

In animals this enzyme is responsible for the detoxification processes, coupling mainly aromatic compounds and eliminating them as glucuronides. In plants there is not detectable β-glucuronidase activity; however, the development of the

CBH I and CBH II

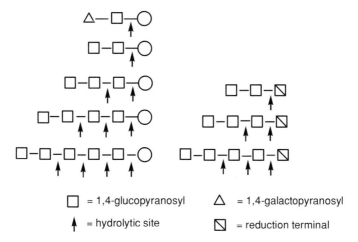

FIGURE 7.6. Enzymatic specificity on low–molecular-weight substrates.

GUS gene fusion containing *E. coli* β-glucuronidase has been widely used as a gene marker.[5] Transgenic plants containing exogenic information fused to the β-glucuronidase gene marker can be conveniently monitored by using fluorogenic o histochemical glucuronides.

7.3.5 *Glycosidase Enzymatic Activity Detection*

Detection can be achieved not only qualitatively, but also quantitatively; for doing so, high- and low-molecular-weight substrates have been designed. Claeyssens[6] demonstrated hydrolytic specificity of cellulases CBH I and CBH II through the use of synthetic fruorogenic substrates containing the highly fluoroescent coumarin umbelliferone or *p*-nitrophenol, in the form of *O*-glucosides. The cleavage of the glycoside releases the chromophore, which can be easily measured in a fluorometer or spectrophotometer. The synthetic design of mono-, di-, tri-, and tetrasaccharides attached to the mentioned chromophores have been of great advantage to determine the specificity during enzymatic cleavage (Figure 7.6).

7.3.6 *Regarding β-1,4-Glucanases (EG)*

The utilization of polysaccharides covalently attached to dyes has been reported. The complex Ostatin Brilliant Redhydroxyethylcellulose (OBR-HEC) is applied as a specific substrate for EG, Remazol Brilliant Blue-xylan (RBB-X) the specific substrate for β-1,4-xylanases.

   Likewise, β-1,3-glucanases are detected by using an electrophoresis technique on polyacrilamide gels utilizing laminarin as substrate. The generated fragments are reacted further with azoic stain 2,3,5-triphenyltetrazolium to produce a color

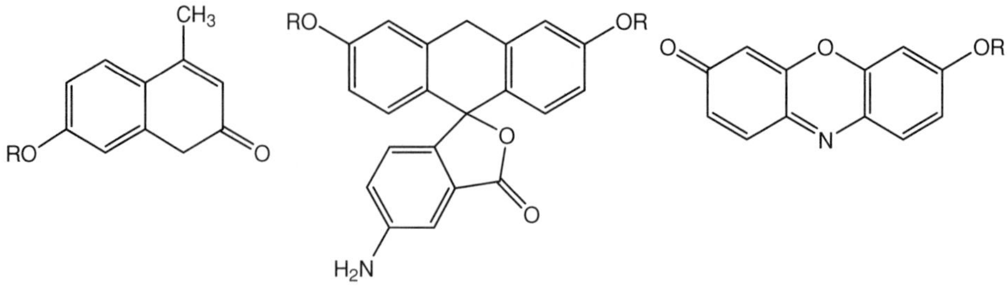

R = glucose, galactose, glucucronic acid, N-acetylglucosamine.

FIGURE 7.7. Fluorescent *O*-glycosides.

complex.[7] Despite their high sensitivity, this method cannot distinguish between endo and exo glucanase.

7.3.7   Fluorescent O-Glycosides

As mentioned before, fluorogenic aglycons are very useful molecules to monitor enzymatic activity. In principle, the fluorescent compound does not exhibit fluorescence in the glycoside form and exerts its fluorescence when released as a result of the enzymatic activity (Figure 7.7). Some of the fluorescent compound widely used for enzymatic detections are umbelliferone, fluoresceine, and resorufin, having been coupled to most of the biologically important sugars as *O*-glycosides.

The generated fluorescence is quantified in fluorometers constituted basically by a radiation source and two monocromatic mirrors (f1 and f2). The first one selects the light for producing fluorescence activation, and the second transmits selectively fluorescence emission. A detector will measure the intensity of the fluorescence generated (Figure 7.8).

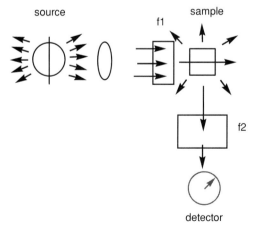

FIGURE 7.8. Basic diagram of fluorometer.

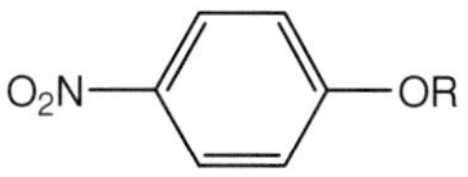

FIGURE 7.9. Absorption glycosides.

R = glucose, galactose, glucuronic acid, N-acetylglucosamine.

7.3.8   O-Glycosides Measured by Absorption

Quantification of enzymatic activity following absorption detection is based on the use of synthetic *p*-nitrophenol in the form of *O*-glycosides as substrate (Figure 7.9). The release of the aglycon from the sugar moiety produces slight yellow color measured as absorbance.

7.3.9   Histochemical O-Glycosides

Generally, a histochemical substrate is to be considered as a good candidate, should be such that in the form of *O*-glycosides it is water-soluble and when the enzyme hydrolyzes the glycosidic bond releases the aglycon, which precipitates immediately. A compound that closely fulfills these requirements is 5-bromo-4-chloro-N-acetyl-3-indoxyl, which has been attached to most of the biologically important monosaccharides. Although this is commercially available, it is highly sensitive, producing an easily detectable blue precipitate. It shows some diffusion before the monomers undergo dimerization in the presence of oxygen, to produce the blue indigo precipitate (Figure 7.10).

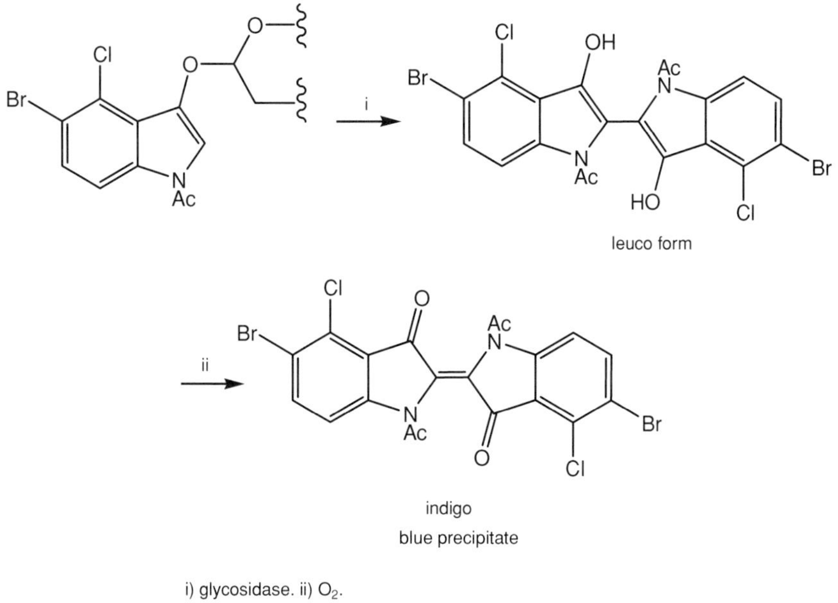

leuco form

indigo
blue precipitate

i) glycosidase. ii) O₂.

FIGURE 7.10. 5-Bromo-4-chloro-indoxyl aldopyranose hydrolysis.

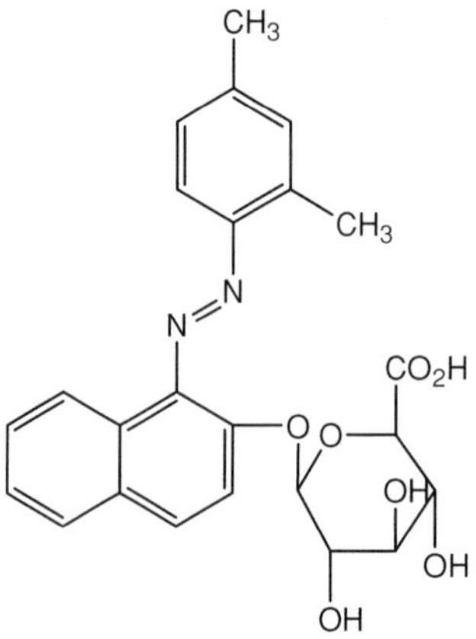

FIGURE 7.11. Phenylazonaphtol glucuronides as histochemical substrates.

Alternatively, phenylazonaphtols O-glycosides (Figure 7.11) known as Sudan glucuronide has been tested as histochemical substrate for enzymatic detection of gene marker β-glucuronidase in transgenic plants.[8,9] The water-soluble Sudan glucuronide releases the fenilazo naphtol stain after enzymatic hydrolysis, which can be seen in the sites of enzymatic activity as red crystals.

The mechanism and stereochemistry of enzymatic hydrolysis may occur with either inversion or retention of the configuration at the anomeric center. The first type of hydrolysis is carried out by the called inverting glycosidases, and the second by retaining glycosidases, being the vast majority of β-glucosidases of the latter type. This has been proved through NMR studies, by measuring the chemical shift and magnitude of the coupling constant of the anomeric carbon. The most accepted

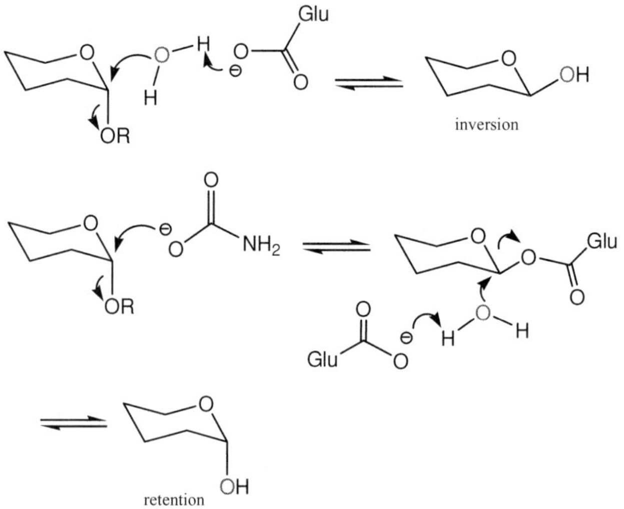

FIGURE 7.12. Schematic representation of retention and inversion hydrolysis mechanism.

mechanism involves protonation of substrate, carboxylate participation attached to enzyme, intermediate formation glycoside-enzyme, and displacement as shown in Figure 7.12.[4]

References

1. P. Bächli,and E.G. Percival, *J. Chem. Soc.* **1952**, 1243.
2. L.-X. Wang, N. Sakairi, and H. Kuzuhara, *J. Carbohydr. Chem.*, **1991**, 10, 349.
3. C.E. Ballou, *Adv. Carbohydr. Chem.*, **1954**, 9, 59.
4. A. Esen, β-Glucuronidase, *Biochemistry and Molecular Biology*, ACS, **1993**, 7.
5. R.A. Jefferson, S.M. Burges, and D. Hirsh, *Proc. Natl. Acad. Sci. USA*, **1986**, 83, 8447.
6. M. Clayssens, H. Van Tildeburgh, P. Tomme, T., Wood, and S. McRae, *Biochem. J.*, **1989**, 261, 819.
7. S.Q. Pan, X.S. Ye, and J. Kuc, *Anal. Biochem.*, **1989**, 136, 182.
8. N. Terryn, M. Brito-Arias, G. Engler, C. Tire, R. Villarroel, M. van Montagu, and D. Inze, *Plant Cell*, **1993**, 5, 1761.
9. E. Van der Eycken, N. Terryn, J.L. Goemans, G. Carlens, W. Nerinckx, M. Claeyssens, J. Van der Eycken, M. Van Montagu, M. Brito-Arias, and G. Engler, *Plant Cell Report* **2000**, 19, 966.

8
Nuclear Magnetic Resonance of Glycosides

8.1 NMR of O-Glycosides

Nuclear magnetic resonance (^{1}H, ^{13}C NMR), X-ray diffraction, and mass spectrometry are considered among the most important analytical methods for structural elucidation. Characterization by means of ^{1}H, ^{13}C NMR, mono- and bidimensional spectroscopy is a powerful tool for structural assignment of simple and complex glycosides. Pioneering studies[1-4] on simple monosaccharides were essential for understanding through the chemical shifts and coupling constants the conformational behavior of sugars.

Some basic considerations derived from the referred studies mentioned above that apply to simple saccharides are

Pyranoside rings of the D-series generally prefer to assume conformation $^4C_1$ and those of the L-series the conformation $_1C^4$. The anomeric proton usually resonate at lower field than methine protons, whereas methylene protons resonate at somewhat higher fields.

In D-pyranoses with $^4C_1$ conformation, the α-anomer resonance is downfield compared to the β-anomer, and the value of the coupling constant between H-1 and H-2 at three bond distance $^3J_{1-2}$ determine if the anomeric proton is equatorial or axial, and therefore if the glycoside is α or β. Usually for axial-axial interactions, the observed values are 8-10 Hz and for axial-equatorial or equatorial-equatorial 2-3 Hz. Thus for β-glucose, $^3J_{1,2} = 8$ Hz, $^3J_{2,3} = {}^3J_{3,4} = {}^3J_{4,5} = 10$ Hz, H-1 appears as doublet, and H-2, H-3, H-4 appear as 10 Hz triplets, and H-5 as double double doublet as it is coupled to the two H6s.

The α-galactose presents $^3J_{1,2} = 3$ Hz, $^3J_{2,3} = 10$ Hz, $^3J_{3,4} = 4$ Hz, $^3J_{4,5} < 1$ Hz. H-1 appears as 3 Hz doublet, H-2 and H-3 as double of doublets, H-4 as doublet, and H-5 as triplet for coupling with two H6s. The different possible arrangements are for better understanding represented in Newman proyection (Figure 8.1).

Equatorial protons are positioned at lower field than chemically equivalent axial protons except in those cases were there is a carbonyl group adjacent to H-equatorial, or when there is a synaxial interaction with H-axial, in which a deshielding effect is observed.[1c]

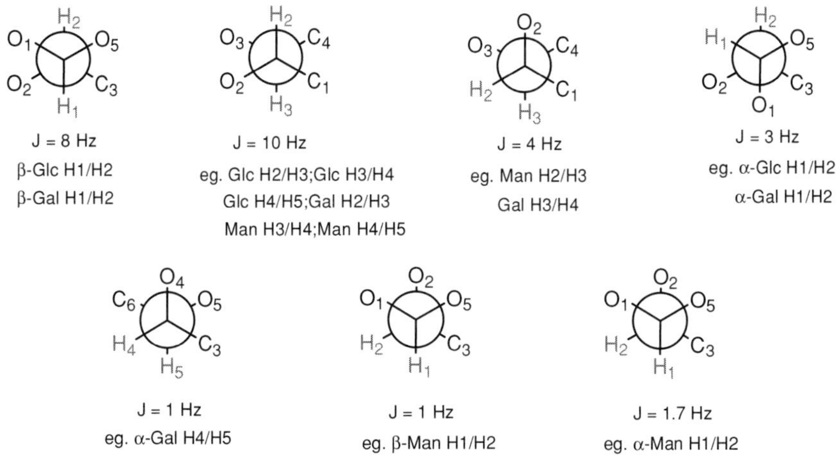

FIGURE 8.1. Newman projections showing the arrengements of hydrogens in $^4C_1$ and $_4C^1$ chair conformations and the expected coupling constants.

The magnitude of coupling constant $^3J_{H-H}$ besides torsion angle dependence may be affected by other factors such as substituent electronegativity, bond length, and bond distance. Solvent effects on $^3J_{(HH)}$ appear to be relatively minor, except in cases where solvent-induced conformational changes occur.[5]

The ^{13}C chemical shift may also reveals along with de ^1H NMR the anomeric configuration, but the one bond ^{13}C-^1H coupling constants can be remarkably useful to determine the anomeric configuration in pyranoses. For instance, the $^1J_{CH}$ for the α-anomer is 170 Hz and for the β-anomer 160 Hz, being for the L-isomer the reverse.[6]

The chemical shift values of the ring protons are dependent of the groups attached to the hydroxyl groups. For instance, a characteristic shift of ring-proton resonances to lower field occurs when the hydroxyl group is esterified with acetyl, sulfate, or phosphate where normally downfield shifts of ~0.2-0.5 ppm are observed. If the protecting group is acetate, for nonaromatic solvents C-6 resonates at lower field, followed by C-2, C-4, and at highest field the 3-acetoxyl signal.[4] The proton magnetic resonance of 4,6-O-benzylidene pyranosides have been measured and the values of the coupling constants $J_{1,2}$, $J_{1,3}$, $J_{2,3}$, and $J_{3,4}$ support the assignment of the chair conformation to the pyranoid ring.[2]

The coupling constants $^3J_{H-H}$ values on saturated systems can be predicted by applying the Karplus equation,[7] which correlates the dihedral angle θ values with the magnitude of the coupling constant $^3J_{H-H}$.

$$^3J_{H-H} = A + B\cos\theta + C\cos2\theta$$

where θ is the dihedral angle between H1-C1-C2-H2, A = 4.22, B = −0.5, and C = 4.5 Hz for C-C bond distance 1.543 Å.

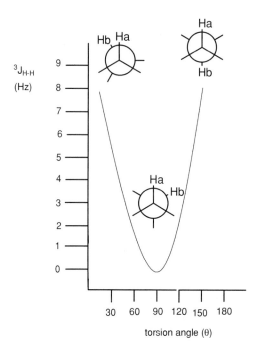

FIGURE 8.2. Relationship between coupling constant and torsion angle.

Coupling constants for vicinal protons at three bond distances are two or three times bigger when they are eclipsed or antieclipsed (0° or 180°) to each other than when they are synclinal or gauche (60°) (Figure 8.2).

Karplus analysis is more accurate when comparative studies are performed between structurally similar compounds. For the study of conformational differences between structurally similar molecules the Karplus equation adopts the form of

$$^3J_{H-H} = K \cos2\theta$$

where K is dependent on H1-C1-C2-H2 fragment, when θ is having values between 50 and 70°, or 110 and 130°, slight variations are observed, while for values close to 0, 90 and 180°, no observable changes are detected.

The effect of the relative orientation and electronegativity of substituents on the magnitude of $^3J(aa)$, $^3J(ae)$, and $^3J(ee)$ has been predicted by a simple set of additivity constants. The step followed in the derivation of the additivity constants considers that antiperiplanar substituents exert a negative and gauche substituents a positive effect on J. The resulting data were fitted equation $^3J = {}^3J^0 + \Sigma\Delta J$ (x), where $^3J^0$ represents the reference value. Some of the additivity constants ΔJ (x) for a given substituent are given in Table 8.1.[5]

More recently, a computer program known as ALTONA was developed for the calculation of dihedral angles from ^1H NMR. This program calculates plots of H-C-C-H diedral angles from proton-proton NMR vicinal coupling constants using an empirically generalized Karplus-type equation, which takes into account the

TABLE 8.1. Additivity constants ΔJ (x) for a substituent X.

X	$\Sigma \Delta J$ (ae)(x) or $\Sigma \Delta J$ (ee)(x) X anti	X gauche	$\Sigma \Delta J$ (aa)(x) X gauche
H,C	0.0	0.0	0.0
I,S	−0.3	+0.1	−0.3
Br	−0.9	+0.3	−0.7
N	−1.1	+0.3	−0.6
N_3	−1.4	+0.4	−1.1
Cl	−1.2	+0.4	−1.0
O	−1.8	+0.5	−1.4
F	−2.5	+0.7	−2.0

electronegativity and the orientation of the substituents attached to the considered fragment.[8]

The NMR spectra of fully acetylated 1-thioaldopyranosides having the configurations β-D-xylo, α-L-arabino, β-D-ribo, β-D-gluco, and β-D-galacto were determined in different solvents, observing that the H-1 signal in these derivatives appears ~0.35 ppm to higher field than its position in the 1-oxygenated analogs.[9]

Also, detailed studies of ^{1}H NMR spectra of a series of hexopyranosyl halides have been accomplished. The first-order assignments revealed several stereospecific dependencies, mainly upon the orientation of the halogen substituent with respect to the pyranose ring and the relative orientation of other substituents attached to the ring.[3b,10,11]

^{1}H and ^{13}C chemical shifts and J-coupling patterns for common D-aldohexoses, D-aldopentoses, and some methyl monosaccharides are described in Tables 8.2 and 8.3.[12,13]

A wide number and variety of O- and to a lesser extent C-glycosides islolated from natural sources have been reported and their NMR analysis described. The chemical shifts and coupling constants of some of them are described just as representative examples in Table 8.4.

Nuclear Overhauser effects (NOE) is a dipole-dipole relaxation experiment and has been one of the most useful experiments for the structural assignments of glycosides on the basis of shielding and deshielding effects.[19] Glycosylation sites can be identified by comparison of ^{1}H NMR spectral data of the peracetylated and the nonprotected sugar, since free OH groups causes significant downfield shift (in the range of 1 to 0.5 ppm). The approach known as "structural-reporter-group" has been introduced to identify individual sugars or sequences of residues and can be used to identify structural motifs or specific sugars and linkage compositions found in relevant databases.[13]

For complex molecules the interpretation is often problematic, especially due the presence of internal motion. Some of the difficulties encountered for NMR structural assignment for oligosaccharides are[20]

The limited number of C,H dipolar couplings measured across a single bond
The distribution of C,H bond vectors is not isotropic due to the geometry of the pyranose ring.

TABLE 8.2. ^{1}H chemical shifts and couplings ($^3J_{H-H}$) of D-aldohexoses and aldopentoses measured at 400 MHz in D$_2$O.

Compound	H-1	H-2	H-3	H-4	H-5	H-6a	H-6b
α-glucose	5.09	3.41	3.61	3.29	3.72	3.72	3.63
	3.6	9.5	9.5	9.5		2.8	5.7, 12.8
β-glucose	4.51	3.13	3.37	3.30	3.35	3.75	3.60
	7.8	9.5	9.5	9.5		2.8	5.7, 12.8
α-galactose	5.16	3.72	3.77	3.90	4.00	3.70	3.62
	3.8	10.0	3.8	1.0		6.4	6.4
β-galactose	4.48	3.41	3.56	3.84	3.61	3.70	3.62
	8.0	10.0	3.8	1.0		3.8	7.8
α-mannose	5.05	3.79	3.72	3.52	3.70	3.74	3.63
	1.8	3.8	10.0	9.8		2.8	6.8, 12.2
β-mannose	4.77	3.85	3.53	3.44	3.25	3.74	3.60
	1.5	3.8	10.0	9.8		2.8	6.8, 12.2

^{1}H chemical shifts and couplings ($^3J_{H-H}$) of D-aldopentoses measured at 400 MHz in D$_2$O.

Compound	H-1	H-2	H-3	H-4	H-5a	H-5b
α-xylose	5.09	3.42	3.48	3.52	3.58	3.57
	3.6	9.0	9.0		7.5	7.5
β-xylose	4.47	3.14	3.33	3.51	3.82	3.22
	7.8	9.2	9.0		5.6	10.5, 11.4
α-arabinose	4.40	3.40	3.55	3.83	3.78	3.57
	7.8	9.8	3.6		1.8	1.3, 13.0
β-arabinose	5.12	3.70	3.77	3.89	3.54	3.91
	3.6	9.3	9.8		2.5	1.7, 13.5
α-ribose	4.75	3.71	3.83	3.77	3.82	3.50
	2.1	3.0	3.0		5.3	2.6, 12.4
β-ribose	4.81	3.41	3.98	3.77	3.72	3.57
	6.5	3.3	3.2		4.4	8.8, 11.4
α-lyxose	4.89	3.69	3.78	3.73	3.71	3.58
	4.9	3.6	7.8		3.8	7.2, 12.1
β-lyxose	4.74	3.81	3.53	3.73	3.84	3.15
	1.1	2.7	8.5		5.1	9.1, 11.7

Due to the flexibility of the glycosidic bond that connects the different sugars moieties, different alignment tensors can be observed.

More recently the use of a novel procedure known as "residual dipolar coupling" has been introduced by Tian & Prestegard as an alternative approach for studing the conformational and the motional properties of oligosaccharides.[21] The approach is based on the solution for each ring of an order matrix that combines different types of couplings, $^1D_{CH}$, $^2D_{CH}$, and D_{HH}.

Dipolar couplings arise from through space spin-spin interactions and are dependent of both internuclear distance (r) and an angle between the magnetic field and the internuclear vector (θ) as described by the equation

$$D_{ij} = \xi_{ij} \left(\frac{(3\cos^2\theta - 1)}{2} \right) (1/r^3)$$

TABLE 8.3. ^{13}C chemical shifts of some aldoses

Compound	C-1	C-2	C-3	C-4	C-5	C-6
α-glucose	92.9	72.5	73.8	70.6	72.3	61.6
β-glucose	96.7	75.1	76.7	70.6	76.8	61.7
α-galactose	93.2	69.4	70.2	70.3	71.4	62.2
β-galactose	97.3	72.9	73.8	69.7	76.0	62.0
α-mannose	95.0	71.7	71.3	68.0	73.4	62.1
β-mannose	94.6	72.3	74.1	67.8	77.2	62.1
α-arabinose	101.9	82.3	76.5	83.8	62.0	
β-arabinose	96.0	77.1	75.1	82.2	62.0	
α-ribose	97.1	71.7	70.8	83.8	62.1	
β-ribose	101.7	76.0	71.2	83.3	63.3	

where ξij is a constant that depends on the properties of nuclei i and j.

Direct measurements of dipolar interactions can be achieved by dissolving molecules in oriented media such as crystals composed of bicelles or phage. Despite the fact that molecular tumbling remains fast in these media, the sampling of orientations are no longer isotropic, and consequently the dipolar couplings do not average to zero and splittings are observed between the dipolar coupled spin pairs.

The knowledge of the molecular geometry of a fragment and the measurement of five or more interdependent residual couplings from the fragment allows the determination of the Saupe order matrix elements (Sij) from a set of linear equations relating dipolar couplings to the known geometry factors and the unknown order tensor elements.

$$D_{resid} \propto \sum_{ij} S_{ij} \cos \theta_i \cos \theta_j$$

where θij are the angles between the internuclear vectors.

Determination of the Saupe order matrices for individual rigid fragments of a molecule allows both structural characterization and assestments of internal motions between fragments.[10]

NMR studies carried out by De Bruyn,[22] using as models series of disaccharides, provided valuable information about conformational behavior from the chemical shifts and the torsion angles present around the glycosidic bond (Figure 8.3). Also it has been reported that the ^{13}C chemical shifts for the glycoside and the aglycone carbon can be directly correlated with one of the torsion angles psi (ψ) defined by the bonds C(1)-O(1)-C(4)-H(4).[19]

The sign of θ and ψ has been previously calculated through the method known as hard sphere exoanomeric effect[23] which predicted the relative stability of the different conformers around the torsion angles, considering the bond length, bond angle, and atomic size. It has been observed that for a number of disaccharides there is a variation of the chemical shifts as a function of ψ, compared with the values of their corresponding monosaccharides (Table 8.5).

TABLE 8.4. ^1H chemical shifts and couplings ($^3J_{H-H}$) of natural O- and C-glycosides

Natural glycoside	H-1	H-2	H-3	H-4	H-5	H-6, H-6'	Ref
Glc	4.63 d(8.0)	3.17 dd (8.0, 9.0)	3.36 t (9.0)	3.26 t (9.0)	3.30 m	3.66 dd (6.0, 12.0); 3.89 dd (2.0, 12.0)	14
Glc	5.04 d (7.5)	3.50 dd (9.0, 7.5)	3.44 m	3.38 t (9.0)	3.44 m	3.71 dd (5.5, 12.3), 3.91 dd (2.0, 12.3)	15
Fuc	4.32 d (8.0)	3.59 dd (8.0, 10.0)	3.54 br dd (3.0, 10.0)	3.63 br d (3.0)	3.58 dq (1.0, 6.0)	1.21 d	16
Glc	4.88 d (10.0)	5.35 t (10.0)	3.49 m	3.49 m	3.33 m	3.58 dd (5.5, 12.5), 3.78 m	17 Zou et al
	3.47 dd (1.9, 9.2)	3.66, t (9.2)	3.76 dd (2.8, 9.2)	5.21 br s	3.37 d (13.1), 3.88 dd (1.8, 13.1)		18 Diaz et al

TABLE 8.5. Chemical shifts of disaccharides and anomers of glucopyranose.

	H1′	H2′	H3′	H4′	H5′	H6a	H6b	H1	H2	H3	H4	H5	H6a	H6b
α-1	5.10	3.56	3.80	3.47	3.86	3.85	3.78	5.45	3.64	3.83	3.47	4.03	3.85	3.76
β-1	5.41	3.55	3.80	3.47	3.86	3.78	3.78	4.81	3.39	3.59	—	—	3.90	3.56
α-2	5.38	3.56	3.77	3.47	4.02	3.82	3.82	5.24	3.63	3.86	3.67	3.48	—	—
β-2	5.36	3.57	3.76	3.45	4.02	3.82	3.82	4.67	3.36	3.64	3.64	3.93	3.82	3.82
α-3	5.41	3.59	3.68	3.42	3.72	3.74	3.74	5.23	3.58	3.97	3.64	3.60	3.92	3.77
β-3	5.41	3.58	3.69	3.42	3.74	3.77	3.77	4.66	3.28	3.77	3.62	3.87	3.84	3.77
α-4	4.63	3.37	3.52	3.42	3.46	3.75	3.75	5.45	3.65	3.87	3.47	3.88	3.85	3.79
α-5	4.73	3.37	3.54	3.42	3.49	3.73	3.73	5.23	3.73	3.92	3.52	3.49	3.90	3.74
β-5	4.63	3.37	3.54	3.42	3.49	3.73	3.73	4.74	3.44	3.74	3.49	3.58	3.97	3.82
β-6	4.51	3.32	3.52	3.42	3.50	3.75	3.75	4.67	3.29	3.60	3.65	—	—	—
α-7	—	—	—	—	—	—	—	5.23	3.50	3.71	3.36	3.82	—	—
β-7	—	—	—	—	—	—	—	4.67	3.27	3.45	3.35	3.82	—	—

1 = kojibiose; 2 = nigerose; 3 = maltose; 4 = soforose; 5 = laminaribiose; 6 = cellobiose; 7 = glucose

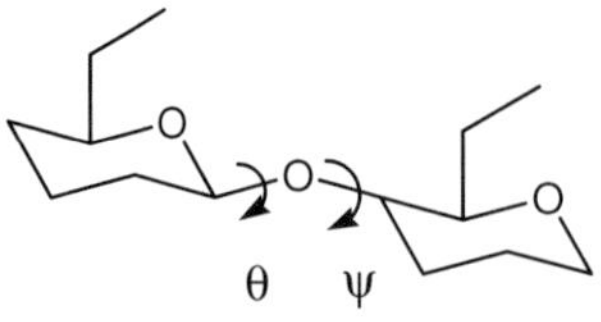

FIGURE 8.3. Torsion angles around the glycosidic bond.

The development of Karplus relationship for three-bond C-O-C-C spin-coupling constants by Bose et al.[24] suggests that $^3J_{COCC}$ obeys a Karplus relationship similar to that observed for $^3J_{HH}, ^3J_{HC}$, and other vicinal spin-coupling constants. However, the precise form of this relationship that is the shape and amplitude of the Karplus curve is unknown. Also in this work, $^3J_{COCH}$ values have been measured to asses the phi (ϕ) and the psi (ψ) torsion angles (Figure 8.4) and Karplus relationships have been reported for this vicinal coupling.[25]

Comparative conformational studies using a combination of NMR spectroscopy and molecular mecanics of lactose disaccharide (βGal[1-4]Glc) and its C-analog showed that for the former the population in solution is about 90% *syn* and 10% *anti*, while for the latter the conformation is more flexible in the forms 55%, 40%, and 5% *syn, anti, gauche-gauche,* respectively.[26]

^{1}H NMR spectra of oligosaccharides follows in many cases complex patterns due to extensive overlap within the region δ 3.0-4.2. However, the use of pyridine-d_5 improves the signal dispersion, increasing the resolution especially in overcrowded regions. The localization of anomeric protons is a valuable tool for recognizing the number of monosaccharide residues.

A number of one- and two-dimensional methods provide thorough information to assert the complete assignment unambigiously. One-dimensional NMR analysis provides useful information about the chemical shifts and scalar couplings of well-resolved signals such as anomeric protons (δ 4.4-5.6) and methyl groups for 6-deoxy monosaccharides (fucose, quinovose, ramnose) at (δ 1.1-1.3). The effect on the proton chemical shift of glycosylation is a typical deshielding of the proton across the glycosidic bond and the two neighboring positions of the aglycone. This behavior is due to repulsion between hydrogens and to the effect of the lone pair of the oxygen to the hydrogens.[27]

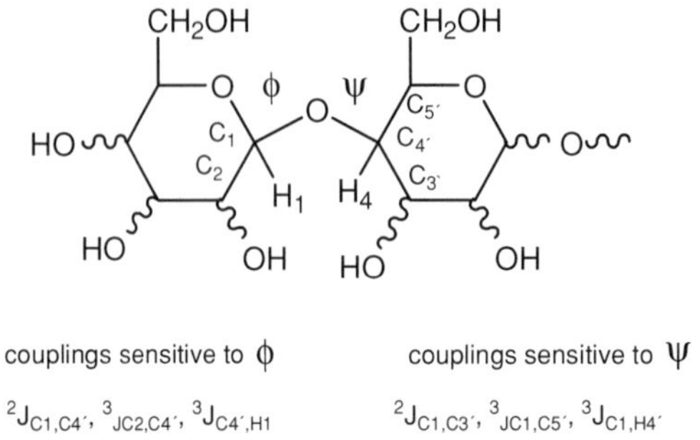

couplings sensitive to ϕ

$^2J_{C1,C4'}, ^3J_{C2,C4'}, ^3J_{C4',H1}$

couplings sensitive to ψ

$^2J_{C1,C3'}, ^3J_{C1,C5'}, ^3J_{C1,H4'}$

FIGURE 8.4. Couplings sensitive to ϕ and ψ torsion angles.

Conformational analysis on more complex glycosides is based mainly on the inter-residue ^{1}H-^{1}H Nuclear Overhauser effects (NOE).[28] and also ^{13}C-^{1}H long-range coupling constants across the glycosidic linkage for studying the preferred conformation of oligosaccharides in solution. Selective irradiation of the anomeric proton reveals inter-residual contacts with aglycone protons. In this way 1→2, 1→3, 1→4, and 1→6 combinations as well as −α and −β linkages may be determined.[29] Long-range ^{1}H-^{1}H couplings involving four bonds between anomeric and aglycone protons ($^4J_{HCOCH}$) are usually very small that could be detected but not measured.[30]

Two-dimensional NMR is a reliable method for determining inter-ring connectivity. Through space dipolar interactions between the anomeric and the trans glycosidic proton can be detected in the form of NOE signals and represent the basis for linkage and sequence analysis.[31] Also, the interglycosidic connectivities are established on the basis of long-range ($^3J_{CH}$) by HMBC studies.[27] The usefulness of this method has been later demonstrated in a number of structural elucidations.[32]

Bidimensional homonuclear techniques such as the TOCSY experiment have been useful for the NMR characterization of the naturally occurring complex glycosides such as glycoresin tricolorin E,[33] allowing the total assignment of the sugar region, including the anomeric protons for each of the 4 monosaccharides established (Figure 8.5).

Likewise, the complete ^{1}H and ^{13}C assignments of a synthetic octasaccharide fragment of the *O*-specific polysaccharide of *Shigella dysenteriae* type 1 by using 2D TOCSY at 600 MHz was described. In the contour plot it is possible to observe the connectivity between the sugar units and the detailed assignment of the protons.[34] Moreover, a 2D selective TOCSY-DQFCOSY experiment for identification of individual sugar components in oligosaccharides is described, assuming that unambiguous sequential assignment of the proton signals for individual components is reached.[35]

High-resolution ^{1}H NMR spectroscopy has been applied in the structural analysis of glycoproteins. The initial efforts to assign all the anomeric and nonanomeric protons were done by using spin decoupling and nuclear Overhauser spectroscopy.[36]

Nuclear magnetic resonance of carbohydrate related to glycoconjugates has been analyzed. One of the first high-resolution studies was reported back in 1973 on intact glycolipids in a 220-MHz magnet.[37] Subsequent studies on underivatized and permethylated glycosphingolipids in dimethylsulfoxide-d$_6$ and chloroform, respectively, allowed one to assign all the anomeric protons and a number of nonanomeric proton resonances.[38]

Early studies on high-resolution NMR. spectra of glycans chain in D$_2$O allowed one to assign the anomeric and nonanomeric protons as well as the coupling constants of sugar residues found in glycoproteins.[36,39] More recently, the complete resolution of acetyl protected sialic acid glycopeptides was achieved by using NOESY and DQF-COSY technique.[6]

For the NMR-analysis of carbohydrate-protein complexes the transfer nuclear Overhauser effect (trNOE) experiment seems to be a promising alternative.[40]

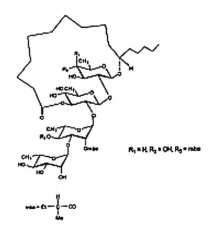

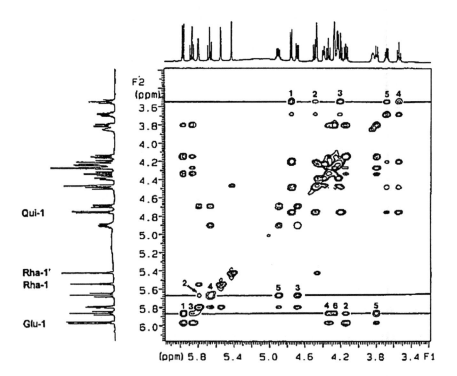

FIGURE 8.5. Expanded region of TOCSY spectrum for characterization of tricolorin E.

Recent advances on conformational analysis of oligosaccharides allows to determine the inter-residue interactions based on the dihedral angles ϕ and ψ along the interglycosidic linkage.[41] In this connection, recent conformational advances on E-selectin-sialyl Lewis^x complex have led to the determination of the bioactive conformation of the silayl Lewis^x tetrasaccharide.[42]

NMR spectroscopy of glycoproteins has been achieved by using a combination of homo- and heteronuclear experiments at natural abundance.[43] Increased refinement is possible when a ^{15}N-labeled sample was used and the mobility of the glycan chain could be assessed by the measurements of ^{13}C line widths obtained from the high-resolution HSQC spectra.[41]

8.2 *N*-Glycosides

The conformational analysis of N-glycosides has been extensively studied on the basis of chemical shifts and coupling constant determinations mainly around the C-N linkage. Torsion angles symbolized as χ for furanosides rings are also dependent on the Karplus equation, and similarly plays an important rule for the conformational analysis of 5-member rings.[44] For purines the angle χ is formed between O4′-C1-N9-C4 atoms, and O4′-C1-N1-C2 for pyrimidines. When torsion angles O4′-C1 N9-C4 for purines, and O4′-C1-N1-C2 for pyrimidines are eclipsed, then $\chi = 0°$. Positive angles of χ occurs for rotation clockwise for N9-C4 for purines and N1-C2 for pyrimidines. The conformation *syn* in nucleosides corresponds to the angle $\chi = 0 \pm 90°$, and *anti* to $180 \pm 90°$ (Figure 8.6).

Regarding furanoside rings, there are different nonplanar conformations possibly assumed in terms of five endocyclic torsion angles symbolized as v_0, v_1, v_2, v_3, v_4, corresponding to the bonds O4′-C1′, C1′-C2′, C2′-C3′, C3′-C4′, and C4′-O4′. The two most common conformations founded are the envelope (E), referring to 4 atoms on the plane, and twist (T) for 3 atoms on the plane. The puckering of the furanoside rings of nucleosides is explained by Sorenssen et al.[45] Unmodified nucleosides are present as an equilibrium between the C-3′-endo conformation, located around P = 18°, and the C-2′-endo conformation centered around P = 162° (Figure 8.7).

Besides the torsion angle described for the N-glycosidic bond, there are for the case of oligosaccharides, additional torsion angles symbolized as $\omega,\omega',\phi,\phi',\psi,$

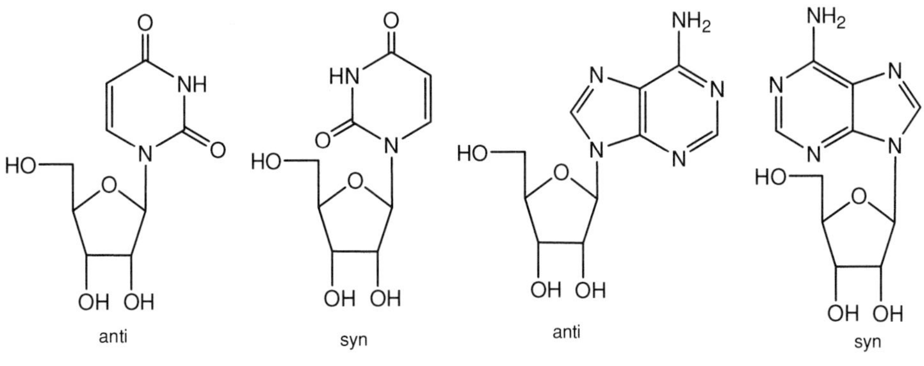

FIGURE 8.6. *Syn-anti* conformations for purines and pyrimidines.

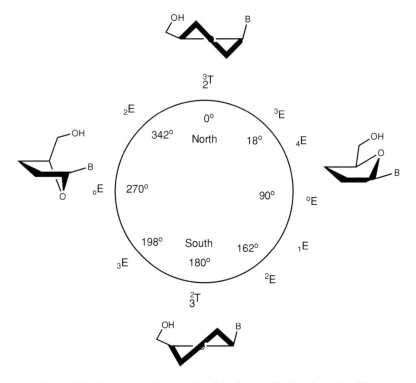

FIGURE 8.7. Pseudorotacional cycle of the furanoside ring in nucleosides.

and ψ' corresponding to the bonds P-O5, P-O3′, O5′-C5′, O3′-C3′, C5′-C4′, C4′-O3′, respectively (Figure 8.8).

The vicinal coupling constants at 3 bond distance are dependent of the dihedral angle θ, and the relationship determined by the Karplus equation.

$$^3J_{HH} = A\cos 2\theta - B\cos \theta + C$$

where A, B, and C are constants and their values are given in Table 8.6.

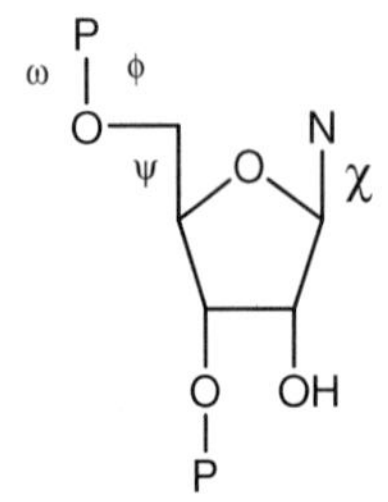

FIGURE 8.8. Torsion angles for oligosaccharides.

TABLE 8.6. Karplus A, B, C constant values for nucleotide molecular fragments.

	torsion angle	J (Hz)	A	B	C
HO-CH-CH-OH	sugar ring	J1′2′, J2′3′, J3′4′	10.2	0.8	0
H-C-C-H	ψ	J4′5′, J4′5″	9.7	1.8	0
H-C-O-H	ϕ'	J2′OH2′	10.4	1.5	0.2
	ϕ	J3′OH3′			
		J5′OH5′			
H-C-O-P	ϕ'	J3′P	18.1	4.8	0
	ϕ	J5′, J5″			

TABLE 8.7. $^3J_{2'3'}$ (Hz) values for furanoside ring in nucleosides.

Nucleoside	$^3J_{1'2'}+^3_{J3'4'}$ (Hz)	$^3J_{2'3'}$ (Hz)
β-D-ribonucleosides		
pyrimidine (anti)	9.9 ($\pm$ 0.2)	5.3 ($\pm$ 0.2)
pyrimidine (syn)	10.3 ($\pm$ 0.2)	6.2 ($\pm$ 0.2)
purine (anti)	9.7 ($\pm$ 0.3)	5.2 ($\pm$ 0.1)
purine (syn)	9.6 ($\pm$ 0.3)	5.5 ($\pm$ 0.2)
β-D-deoxyribonucleosides		
pyrimidine (anti)	10.6 ($\pm$ 0.2)	6.7 ($\pm$ 0.2)
pyrimidine (syn)	11.0 ($\pm$ 0.1)	8.0 ($\pm$ 0.1)
purine (anti-syn)	6.3 ($\pm$ 0.1)	6.3 ($\pm$ 0.1)
C-nucleoside pyrimidic	10.6 ($\pm$ 0.3)	5.2 ($\pm$ 0.2)
C-nucleoside puric	10.1 ($\pm$ 0.3)	5.0 ($\pm$ 0.2)

The exchange of –OH for -OPO$_3$ does not affect sensibly the Karplus relationship; therefore, the values are valid for both nucleosides or oligonucleosides. However, as mentioned, ^3J there is a dependence of other factors such as bond length, bond angle, electronegativity, and substituent orientation. Some of the values reported for ribose and deoxyribose are described in Table 8.7.

The analysis of the *C*-Nucleosides β-pseudouridine (β–ψ) and α-pseurouridine (α–ψ) in aqueous solution have been described and the observed coupling constants given in Table 8.8.[1b]

TABLE 8.8. Coupling constant (Hz) for β- and α-pseudouridine at 30°

Coupling constant	β–ψ	α–ψ
J$_{61'}$	0.8	1.3
J$_{1'2'}$	5.0	3.3
J$_{2'3'}$	5.0	4.2
J$_{3'4'}$	5.2	7.9
J$_{4'5'B}$	3.2	2.4
J$_{4'5'C}$	4.6	5.7
J$_{5'B5'C}$	−12.7	−12.7

References

1. (a) R.U. Lemieux and A.R. Morgan, *Can. J. Chem.*, **43**, 2199 (1965). (b) G. Kotowycz and R.U. Lemieux, *Chem. Rev.*, **73**, 669 (1973). (c) R.U. Lemieux and J.D. Stevens, *Can. J. Chem.*, **43**, 2059 (1965).
2. B. Coxon, *Tetrahedron*, **21**, 3481 (1965).
3. (a) L.D. Hall, *Adv. Carbohydr. Chem.*, **19**, 51 (1964). (b) L.D. Hall, J.F. Manville, and N.S. Bhacca, *Can. J. Chem.*, **47**, 1 (1969).
4. D. Horton, and J.H. Lauterbach, *Carbohydr. Res.*, **43**, 9 (1975).
5. C. Altona, and C.A.G. Haasnoot, *Organic Magnetic Resonance*, **13**, 417 (1980).
6. E. Breitmaier, *Structure Elucidation by NMR I Organic Chemistry* 3th Ed. John Wiley & Sons pp 46 (2002).
7. M. Karplus, *Chem. Phys.*, **11**, 30 (1959).
8. C.M. Cerda-García-Rojas, L. G. Zepeda, and P. Joseph-Nathan, *Tetrahedron Computer Methodology*, 3, 113, (1990).
9. C.V. Holland, D. Horton, M.J. Miller, and N.S. Bhacca, *J. Org. Chem.*, **32**, 3077 (1967).
10. G. Hajdukovic, M.L. Martin, P. Sinaÿ, and J.R. Pougny, *Organic Magnetic Resonance* **7**, 366 (1975).
11. D. Horton and W. N. Turner, *J. Org. Chem.* **30**, 3387 (1965).
12. K. Bock and H. Thøgerson, *Annual Reports on NMR Spectroscopy* Ed by G.A Webb, Academic Press **13**, 37 (1982).
13. J. Ø Duus, C.H. Gotffredsen, and K. Bock, *Chem. Rev.*, **100**, 4589 (2000).
14. A. Delazar, M. Byres, S. Gibbons, Y. Kumarasamy, M. Modarresi, L. Nahar, M. Shoeb, and S. D. Sarker, *J. Nat. Prod.* **67**, 1584 (2004).
15. G. Bohr, C. Gerhäuser, J. Knauft, J. Zapp, and H. Becker, *J. Nat. Prod.* **68**, 1545 (2005).
16. S.O. Lee, S. Z. Choi, S. U. Choi, K. C. Lee, Y. W. Chin, J. Kim, Y. C. Kim, and K.R. Lee, *J. Nat. Prod.* **68**, 1471 (2005).
17. J-H Zou, J.S. Yang, and L. Zhou, *J. Nat. Prod.* **67**, 664 (2004).
18. F.Diaz, H-B Chai, Q.Mi, B-N.Su, J.S.Vigo, J.G. Graham, F.Cabieses, N.R. Farnsworth, G.A. Cordell, J.M. Pezzuto, S.M. Swanson, and A. D.Kinghorn, *J. Nat. Prod.* **67**, 352 (2004).
19. P. Manitto, D. Monti, and G. Speranza, *J. Chem. Soc. Perkin Trans. 1*, 1297 (1990).
20. H. Neubauer, J. Meiler, W. Peti, and C. Griesinger, *Helv. Chim. Acta*, **84**, 243 (2001).
21. F. Tian, H.M. Al-Hashimi, J.L. Craighead, and J.H. Prestegard, *J. Am. Chem. Soc.*, **123**, 485 (2001).
22. A. De Bruyn, *J. Carbohydr. Chem.*, **10**, 159 (1991).
23. R.U. Lemieux and S. Koto, *Tetrahedron*, **30**, 1933 (1974).
24. B. Bose, S. Zhao, R. Stenutz, F. Cloran, P.B. Bondo, G. Bondo, B. Hertz, I. Carmichael, and A.S. Serianni, *J. Am. Chem. Soc.*, **120**, 11158 (1998).
25. K. Bock, A. Brignole., and B.W. Sigurskjold, *J. Chem. Soc. Perkin Trans. II*, 1711 (1996).
26. A. Imberty, *Current Opinion in Structural Biology*, **7**, 617 (1997).
27. P.K. Agrawal and A.K. Pathak, *Phytochemical Analysis*, **7**, 113 (1996).
28. I. Tvaroska, M. Hricovini, and E. Petrakova, *Carbohydr. Res.*, **189**, 359 (1989).
29. (a) Y.S. Bae, J.F.W. Burger, J.P. Steynberg, D. Ferreira, and R.W. Heminway, *Phytochemistry*,
30. (a) G. Batta and A. Liptak, *J. Am. Chem. Soc.*, **106**, 248 (1984). (b) G. Masiiot, C. Lavaud, C. Delaude, G.V Binst, S.P.F. Miller, and H.M. Fales, *Phytochemistry*, **29**, 3291 (1990).

31. J.H. Prestegard, T.A.W. Koerner, P.C. Demou, and R.K. Yu. *J. Am. Chem. Soc.*, **104**, 4993 (1982).

32. (a) N.M. Duc, R. Kasai, K. Ohtani, A. Ito, N.T. Nham, K. Yamasaki, and O. Tanaka, *Chem. Pharm. Bull.*, **42**, 634 (1994). (b) T. Nakamura, T. Takedo, and Y. Ogihara, *Chem. Pharm. Bull.*, **42**, 1111 (1994).
(c) B. Razanamahefa, C. Demetzos, A.-L Skaltsounis, M. Andriantisiferana, and F. Tillequin, *Heterocycles*, 38, 357 (1994). (d) T. Nakanishi, K. Tanaka, H. Murata, M. Somekawa, and A. Inada, *Chem. Pharm. Bull*, 41, 183 (1994). (e) S.T. Thulborg, S.B Christensen, C. Cornett, C.E. Olsen, and E. Lemmich, *Phytochemistry*, 36, 753 (1994).

33. (a) M. Bah and R. Pereda-Miranda, *Tetrahedron*, **41**, 13063 (1996). (b) R. Pereda-Miranda and M. Bah, *Current Topics in Medicinal Chemistry*, **3**, 1 (2003).

34. B. Coxon, N. Sari, G. Batta, and G. Pozsgay, *Carbohydr. Res.*, **324**, 53 (2000).

35. H. Sato, and Y. Kajihara, *J. Carbohydr. Chem.*, **22**, 339 (2003).

36. F.G.J. Vliegenthart, L. Dorland, and H. van Halbeek, *Adv. Carbohydr. Chem. Biochem.*, **41**, 209 (1983).

37. M. Martin-Lomas and D. Chapman, *Chem. Phys, Lipids*, **10**, 152 (1973).

38. (a) J. Dabrowski, P. Handfland, and H. Egge, *Biochemistry*, **19**, 5652 (1980). (b) K.E. Falk, K.-A. Karlsson, and B.E. Samuelson, *Arch. Biochem. Biophys.*, **192**, 164 (1979).

39. L.S. Wolfe, R.G. Senior, and N.M.K. Ng Yin Kin, *J. Biol. Chem.*, **249**, 1838 (1974).

40. (a) F. Ni, *Prog. Nucl. Magn. Reson Spectros.*, **26**, 517 (1994). (b) F. Casset, T. Peters, M. Etzler. E. Korchagina, S. Nifantev, Perez, and A. Imberty, *Eur. J. Biochem.*, **239**, 710 (1996).

41. (a) T. Peters, and B.M. Pinto, *Current Opinion in Structural Biology*, **6**, 710 (1996). (b) T. Weimar, S.L. Harris, .J.B. Pitnar, K. Bock, and B.M. Pinto, *Biochemistry*, **34**, 13672 (1995).

42. K. Scheffler, B. Ernst, A. Katopodis, J.L. Magnani, W.T. Wang, R. Weisemann, and T. Peters, *Angew. Chem. Int. Ed.*, **34**, 1841 (1995).

43. (a) D.F. Wyss, and J.S. Choi, and G. Wagner, *Biochemistry*, **34**, 1622 (1995). (b) D.F. Wyss, J.S. Choi, J. Li, M.H. Knoppers, K.J. Willis, A.R.N. Arulandaman, A. Smolyar, E.L. Reinherz, and G. Wagner, *Science*, **269**, 1273 (1995). (c) R. Liang, A.H. Androtti, and D. Kahne, *J. Am. Chem. Soc.*, **117**, 10395 (1995).

44. D. Davies, *Progress in NMR Spectroscopy*, **12**, 140 (1978).

45. M.H. Sorenssen, C. Nielsen, and P. Nielsen, *J. Org. Chem.*, **66**, 4878 (2001).

9
X-Ray Diffraction of Glycosides

X-ray crystallography is a powerful tool for obtaining molecular information regarding bond lengths, bond angles, hydrogen bond interactions, and torsion angles, which are necessary elements for understanding the conformation of glycosides. Improved diffractometers, faster computational processors, and newer mathematical programs have made possible the structural resolution of simple and complex substances of glycosidic nature particularly those with noncentrosymmetric space groups.

Early studies on simple glycosides allowed to confirm that the sugar residue is pyranoid (and not acyclic), assuming two possible chair conformations ($^4C_1$ and $_4C^1$), usually orienting the substituent to the equatorial position.[1]

In hydrogen bond interactions on pyranoids some facts that almost occur invariably are (a) the ring-oxygen atom is always a hydrogen bond acceptor, (b) each hydroxyl group is associated with two hydrogen bonds, one as donor and one as acceptor, (c) in disaccharides there might be intramolecular hydrogen bonding between two residues, (d) the hydrogen bond O-O distance has values around 2.68 to 3.04 Å.

Crystallographic observations on the anomeric effect demonstrated that the bond shortening and preferred *gauche* conformation of the glycosidic bonds in pyranoses was a consequence of an electronic distribution in the hemiacetal and acetal moiety of these molecules.[2]

On the other hand, the primarily alcohols can be present in three staggered orientations, defined as gg, gt, and tg, refering to torsion angles O5-C5-C6-O6 and the second to C4-C5-C6-O6 [g ± 60°, t 180°]. An alternative nomenclature refers O5-C5-C6-O6 as +g = gt, −g = gg, t = tg.

The general standard molecular dimensions for pyranosides are described in Table 9.1, being the C-C bond length in the range of 1.523–1.526, C-C-C angles 110.4–110.5°, and usually shorter glycosidic bond 1.398 for axial and 1.385 for equatorial disposition (Table 9.1).[3]

The distortion degree from the ideal chair conformation has been studied by Cremer and Pople,[4] who, by following a mathematical approximation, were able to propose three puckering parameters described as spherical polar set Q (total puckering amplitude), and the angles θ and φ, describing the distortion suffered

TABLE 9.1. Standard molecular dimensions for $^4C_1$ chair conformations in pyranosides.

Bond type	Bond lengths (Å)	Bond lengths (Å)	Angle type	Bond angle (°)	Bond angle (°)	4 atoms ring	Torsion angles (°)
C-C ring	1.526	1.523	C-C-C ring	110.4	110.5		
C-C exo	1.516	1.514	C-C-C exo	112.5	112.7		
C-O exo	1.420	1.426	C-C-O ring	110.0	110.0		
			C-C-O exo	109.7	109.6		
C5-O5 axial	1.434	1.436	C5-O5-C1	114.0	114.0	C-C-C-C	53
C1-O5 axial	1.419	1.419	O5-C1-O1	112.1	111.6	C-C-C-O	56
C1-O1 axial	1.398	1.415	C5-O5-C1	112.0	112.0	C-C-O-C	60
C5-O5 eq.	1.426	1.436	O5-C1-O1	108.0	107.3	C-C-C-C	53
C1-O5 eq.	1.428	1.429				C-C-C-O	57

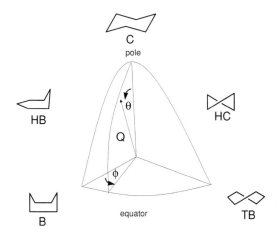

FIGURE 9.1. One octant of the sphere on which the conformations of six-membered rings can be mapped for a constant Q.

by six-member rings from the ideal chair conformation. The chair corresponds to $\theta = 0°$, $\phi = 0°$; boat for $\theta = 90°$, $\phi = 0°$; and twist boat for $\theta = 90°$, $\phi = 90°$ (Figure 9.1). The pyranoside ring varies slightly and in terms of Cremer and Pople puckering parameters, the range of values is $Q = 0.55–0.58$ Å with θ within 5° of 0 or 180°.[3]

9.1   X-Ray Diffraction of O-Glycosides

One of the pioneering studies about sugar X-ray analysis was presented by Levy and Brown[5] reporting the structure of sucrose and sucrose NaBr.H_2O. Through these studies it was observed that although they were energetically equivalent, the chair conformation was different, due to slight hydrogen bridge interactions on the furanoside moiety (Figure 9.2).

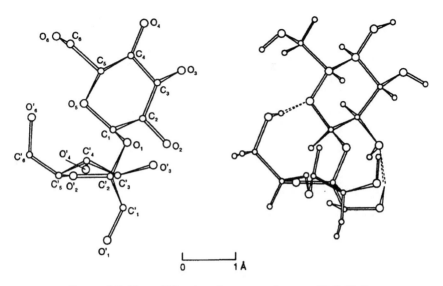

FIGURE 9.2. X-ray diffraction of sucrose and sucrose NaBr.H₂O.

Another disaccharide characterized by X-ray crystallography was octa-O-acetyl-β-D-cellobiose, which presents space group $P2_1,2_1,2_1$, with both pyranoside residues in $^4C_1$ chair conformation slightly more distorted in comparison with cellobiose. Moreover, the torsion angles determined were for O5-C1-C4′ −77°, and for C1-O1-C4-C5 104° (Figure 9.3). The sign value indicates according with the Klyne & Prelog notation to the right if positive and to the left if negative.[6]

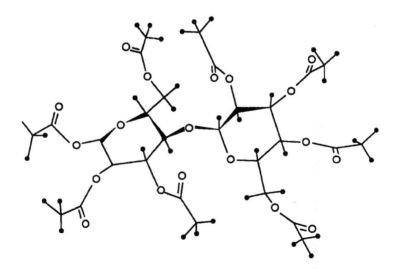

FIGURE 9.3. Chair conformation for octa-O-acetyl-β-D-cellobiose.

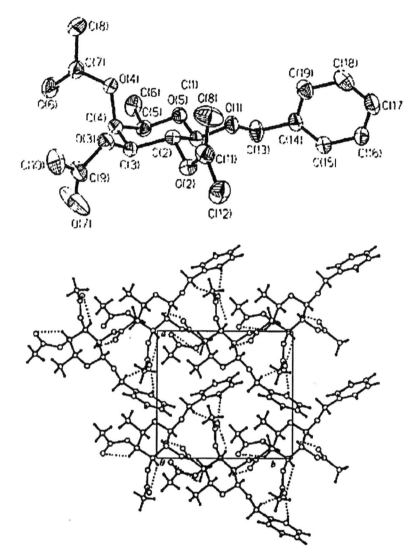

FIGURE 9.4. Thermal ellipsoid drawing and packing diagram showing the hydrogen bonding along [001] of phenyl methyl 2,3,4,-tri-O-acetyl-b-D-fucopyranoside.

The crystal structure of benzyl 2,3,4-tri-O-acetyl-β-D-fucopyranoside is described,[7] presenting a monoclinic system, space group P2$_1$, with bond distances C-O 1.423 Å, C-C 1.513 Å, and shorter C-O 1.380 Å for equatorially anomeric bond. The angle disposition for the endocyclic bond C1-O5-C5 is of 112.4 (3)°, being this value typical for chair conformation ^{4}C$_1$ in pyranoside with substituents positioned at equatorial positions. The perspective view of the molecule shows equatorial disposition for all substituents except position 4 that remains axial (Figure 9.4).

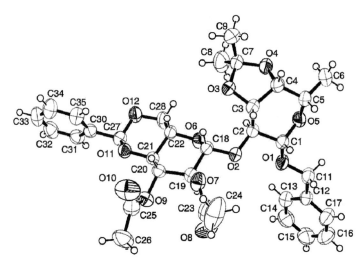

FIGURE 9.5. Perspective Ortep view of phenylmethyl glucosyl fucopyranosyl derivative showing the distortion degree between 5- and 6-member fused rings on chair conformation.

Disaccharide phenylmethyl-O-(2,3-di-O-acetyl-4,6-O-benzylidene-β-D-glucopyranosyl)-(1→2)-3,4-O-isopropylidene-β-D-fucopyranoside shows for fucopyranoside moiety a distorted chair due the five-member ring acetonide at O3 and O4 positions, with Cremer and Pople puckering parameters of Q = 0.556 (3), θ = 159.9 (3)°, and φ = 220.8 (8)°. In contrast for the glucopyranosyl moiety with a six-member ring benzylidene ring attached at positions O4 and O6, the chair conformation is less distorted with Cremer and Pople puckering parameters of Q = 0.597 (3), θ = 170.5 (3)°, and φ = 156.0 (16)° (Figure 9.5).[8]

The solid-state crystal structure of glycoresin tricolorin A was solved by using an intense syncroton radiation to collect data. The cystals belongs as usual to the P2₁ having cell dimensions a = 14.025(1), b = 33.337(1), c = 25.512(1) Å, β = 91.07(1)°. The energy maps were calculated as a function of two glycosidic linkage torsion angles defined as φ = Θ (O5-C1-O1-Cx) and ψ = Θ (C1-O1-Cx-C(x+1), indicating a higher level of conformational freedom along the ψ-axis.

The size of the crystal unit cell demonstrate the presence of four independent tricolorin A molecules per asymmetric unit and the refined structure showed the presence of 18 water molecules forming a channel along the hydrophilic region (Figure 9.6).[9]

Other selected pyranosides analyzed by X-ray diffraction and their parameters determined are shown in Table 9.2.

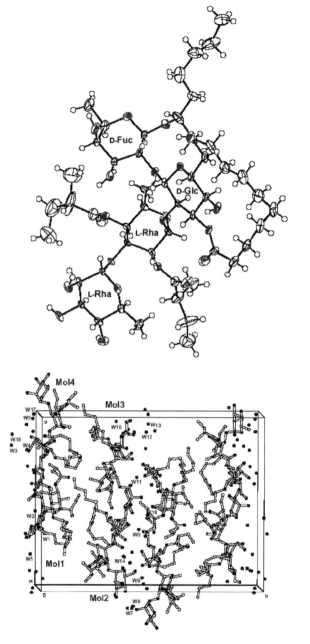

FIGURE 9.6. ORTEP representation of tricolorin A and graphical representation of the unit cell.

TABLE 9.2. X-ray diffraction parameters of some selected pyranosides.

Structure	Symmetry cell	Symmetry space	Conformation	Ref
	orthorhombic	P 2_1 2_1 2_1	$^4C_1$	10
	monoclinic	P 2_1	$^4C_1$	11
	monoclinic	P2_1	chair for α-anomer and boat β-anomer	12
	orthorhombic	P 2_1 2_1 2_1	$^4C_1$	13
	monoclinic	C 2	$_4C^1$	14
	orthorhombic	P 2_1 2_1 2_1	$^4C_1$	15
	monoclinic	C 2	$_4C^1$ $^4C_1$	16
	monoclinic	P2_1	$_4C^1$ $^4C_1$	17

9.2   X-Ray Diffraction of Nucleosides

A number of N-glycosides and C-glycosides has been solved by X-ray analysis, presenting as common features space group $P2_12_12_1$ or $P2_1$, the furanoside ring in the twist conformation, and symmetric system monoclinic or orthorhombic.

For instance, the hypermodified nucleoside queuosine presents a space group $P2_12_12_1$, cell dimensions a = 26.895, b = 7.0707, c = 23.883 Å, and symmetric system orthorhombic (Figure 9.7). The three-dimensional structure determined by X-ray has been also helpful to understand the recognition process at the tRNA level. Thus, based on this information it was possible to determine that the bulky group cyclopentenediol due to the trans disposition assumed is not involved in any interaction codon-anticodon, therefore suggesting that another type of interaction was taken place.[18]

The unusual conformation of α-D anomer of 5-aza-7-deaza-2'-deoxyguanosine has been reported by Seela et al.[19] In this work it is described that the title compound adopts a high-anticonformation with the C1'-C2' and N9-C8 bonds nearly eclipsed with torsion angle C1'-C2'-N9-C8 = 30.3 (4)°. It can be also observed that for 2'-deoxy-α-D-ribonucleosides the C2' endo sugar puckering with either a half chair or envelope conformation is preferred (Figure 9.8).

The solid-state conformation of constrained carbocyclic nucleosides (N)-methano-carba-AZT and N-(S)-methano-carba-AZT was determined by X-ray

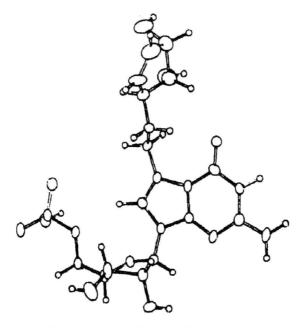

FIGURE 9.7. Perspective view of hypermodified nucleoside queuosine.

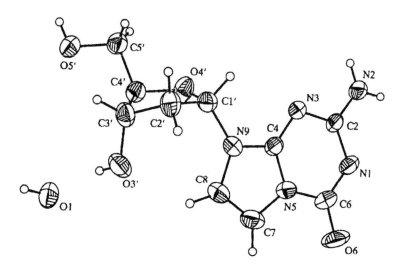

FIGURE 9.8. Perspective view of α-D anomer of 5-aza-7-deaza-2′-deoxyguanosine.

diffraction. As expected with the prediction, their thermal ellipsoid presented a rigid pseudoboat conformation for the bicycle [3.1.0] hexane system, which makes them assume in nearly perfect $_2$E and $_3$E envelope conformations in the pseudorotational cycle (Figure 9.9).[20]

Other selected N-nucleosides that have been analyzed by X-ray diffraction and their parameters determined are shown in Table 9.3.

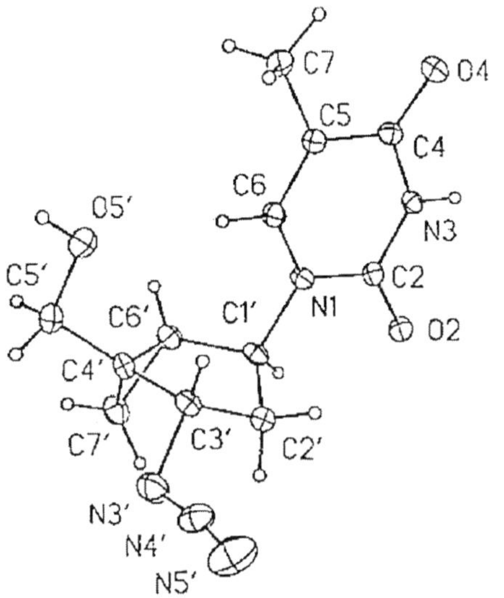

FIGURE 9.9. X-ray structure of (N)-*methano*-carba-AZT.

TABLE 9.3. N-Glycosides and their X-ray diffraction parameters.

Structure	Symmetry cell	Symmetry space	Sugar puckering	Conformation	Ref
	Monoclinic	P 2_1	$^3T_2$	anti [χ = -125.37 (13)$^\circ$]	21
	Orthorhombic	P 2_1 2_1 2_1	unsymmetrical twist	anti	22
	Orthorhombic	P2_1 2_1 2_1	$^2T_3$	anti and high anti [χ = -101.1 (3)$^\circ$]	23
	Orthorhombic	P2_1 2_1 2_1	$^3T_2$	anti and high-anti [χ = -101.8 (5)$^\circ$]	24
	Orthorhombic	P2_1 2_1 2_1	$^3T_4$	anti [χ = -106.5 (3)$^\circ$]	25
	Orthorhombic	P2_1 2_1 2_1	$^3T_2$	anti	26
	Orthorhombic	P2_1 2_1 2_1	$_3T^2$	anti and high anti [χ = -103.5 (3)$^\circ$].	27

TABLE 9.3. (*Continued*)

	Monoclinic	P 2₁	$^2T_3$	*anti* [χ = – 117.1 (5)°]	28
	orthorhombic	P 2₁ 2₁ 2₁	C1'-*exo*, C2'-*endo* twist and C2'-*endo* envelope	*anti*	29
	Orthorhombic	P2₁ 2₁ 2₁	S-type	*anti* [torsion angle = – 105.3 (2)°]	30

References

1. (a) S. Furberg, and C.S. Petersen, *Acta Chem. Scan.* **1962**, 16, 1539. (b) Reeves, R.E. *J. Am. Chem. Soc.* **1950**, 72, 1499.
2. G.A. Jeffrey, J.A. Pople, J.S. Binkley, and S. Vishveshwara, *J. Am. Chem. Soc.* **1978**, 100, 373.
3. G.A. Jeffrey, *Acta Cryst.* **1990**, B46, 89.
4. D. Cremer and J.A. Pople, *J. Am. Chem. Soc.* **1975**, 97, 1354.
5. G.M. Brown and H.A. Levy, *Science,* **1963**, 141, 921.
6. F. Leung, H.D. Chanzy, S. Pérez,and H., Marchessault, *Can J. Chem.* **1976**, 54, 1365.
7. M.A. Brito-Arias, E.V.García-Baez, E. Durán-Páramo, and S. Rojas-Lima, *J. Chem. Crystallogr.*, **2002**, 32, 237.
8. M.A. Brito-Arias, E. Duran-Paramo, I. Mata, and E. Molins, *Acta Cryst.*, **2002**, C58, o537.
9. A. Rencurosi, E.P. Mitchell, G. Cioci, S. Péres, R. Pereda-Miranda, and A. Imberty, *Angew. Chem. Int. Ed.* **2004**, 4, 2.
10. I. Matijasic, G. Pavlovic, and R. Trojko Jr, *Acta Cryst.*, C59, o184 (2003).
11. O. Renaudet, P. Dumy and C. Philouze, *Acta Cryst.*, C57, 309 (2001).
12. C.G. Suresh, B. Ravindran, K.N. Rao, and T. Pathak, *Acta Cryst.*, C56, 1030 (2000).
13. Z.-Z. Qiu, X.-P. Hui, P.-F. Xu, *Acta Cryst*, C61, O475 (2005).
14. N. Low, C. Garcia, M. Melguizo, J. Cobo, M. Nogueras, A. Sánchez, M.D. López, and M.E. Light, *Acta Cryst.*, C57, 222, (2001).

15. Z.-H. Cheng, T. Wu, S.W.A. Bligh, A. Bashall, and B.-Y. Yu, *J. Nat. Prod.* **2004**, 67, 1761.
16. L. Eriksson, R. Stenutz, and G. Widmalm, *Acta Cryst.*, C**56**, 702 (2000)
17. R. Stenutz, M. Shang, and A. S. Serianni, *Acta Cryst.*, C**55**, 1719 (1999).
18. S. Yokohama, T. Miyazawa, Y., Litaka, Z. Yamaizumi, H. Kasai, and S. Nishimura, *Nature*, **1979**, 107, 282.
19. F. Seela, H. Rosemeyer, A. Melenewski, E.-M. Heithoff, H. Eickmeier, and H. Reuter, *Acta Cryst.*, C**58**, O142 (2002).
20. V.E. Marquez, A. Ezzitouni, P. Russ, M.A. Siddiqui, H. Ford, Jr., R.J. Feldman, H. Mitsuya, C. Goerge, J.J. Barchi, Jr., *J. Am. Chem. Soc.* **1998**, 120, 2780.
21. F. Seela, P. Chittepu, J. He, and H. Eickmeier, *Acta Cryst.*, C**60**, O884 (2004).
22. F. Seela, Y. Zhang, K. Xu, and H. Eickmeier, *Acta Cryst.*, C**61**, O60 (2005).
23. F. Seela, K. Xu, and H. Eickmeier, *Acta Cryst.*, C**61**, O408 (2005).
24. W. Lin, K. Xu, H. Eickmeier, and F. Seela, *Acta Cryst.*, C**61**, O195 (2005).
25. F. Seela, K.I. Shaikh, and H. Eickmeier, *Acta Cryst.*, C**61**, O151 (2005).
26. F. Seela, V.R. Sirivolu, J. He, and H. Eickmeier, *Acta Cryst.*, C**61**, O67 (2005).
27. W. Lin, F. Seela, H. Eickmeier, and H. Reuter, *Acta Cryst.*, C**60**, O566 (2004).
28. F. Seela, K.I. Shaikh, and H. Eickmeier, *Acta Cryst.*, C**60**, O489 (2004).
29. J. W. Bats, J. Parsch, and J. W. Engels, *Acta Cryst.*, C**56**, 201 (2000).
30. F. Seela, A.M. Jawalekar, and H. Eickmeier, *Acta Cryst.*, C**60**, O387 (2004).

10
Mass Spectrometry of Glycosides

High-resolution mass spectrometry has become another valuable tool for characterization of simple and complex glycosides. The method is based on the collision of a high-energy electron against a sample under study producing as result a cation radical fragment known as the molecular ion, which should match with the molecular weight of the molecule. The mass spectrum also registers a number of fragments being the most intense the base peak assigned a relative intensity of 100. Mass spectrometry can be applied as high and low ionization experiments, being for the former the most suitable for glycosides electron impact and for the latter fast atom bombardment (FAB) and electrospray ionization the routine experiments for characterization of glycosides. In terms of sensitivity of the measurement this instrumental method requires small amount of sample, even in the order of nanogram quantities.

The fragmentation patterns of acetyl-protected pentoses and hexoses was studied and their main m/z fragment established. For instance, for methyl-β-D-xylopyranoside triacetate the main fragment follows the two alternative routes shown in Figure 10.1.[1]

High ionization experiments such as electron impact has been found to be a suitable approach for the determination of the molecular weight through their corresponding molecular ion of protected glycosides such as peracetylated O-glycosides of low molecular weight. For instance, by using electron impact it was possible to determine the molecular weight, and the common fragmentation patterns, of m/e 331 and 169 (100) of the phenylazonaphtol-β-D-glucopyranoside pentaacetate (Figure 10.2).[2]

However, for most of nonprotected glycosides, high ionization does not provide reliable information and commonly decomposition is observed due to thermal unstability. The introduction of shoft ionization techniques such as fast atom bombardment (FAB) and electrospray ionization has produced great progress for the structural characterization of simple and complex glycosides. This important analytical procedure is especially useful for determining the molecular weight through detection of the molecular ion, as well as sugar sequence. The choice of the matrix and the solubility of the sample are essential aspects to consider for obtaining

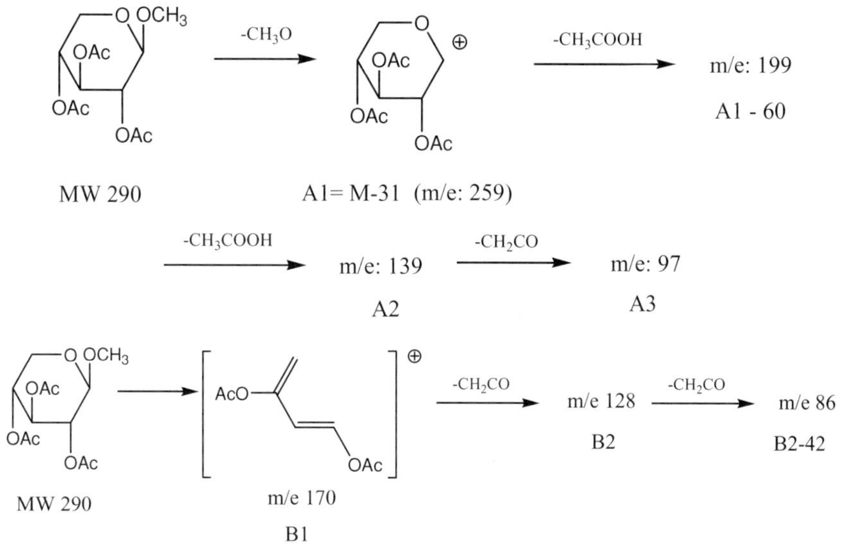

FIGURE 10.1. Fragmentation pattern of methyl-β-D-xylopyranoside.

the best resolution. Glycerol is the matrix most commonly used and it is the best choice for underivatized carbohydrates and glycopeptides. Some other matrices used alternatively for hydrophobic samples are thioglycerol, tetraethyleneglycol, and triethanolamine.[3]

The use of derivatives also plays an important rule and may improve the spectral interpretation and the sensitivity. The most commonly used derivatives are the per-*O*-acetyl and the per-*O*-methyl. Usually for the former the fragmentation pathways are less specific and furnish more information, although the spectrum is more difficult to interpret.

For the assignment of the molecular ion it is important to recognize the pseudomolecular ions produced during a FAB experiment, which can be positive-ion and negative-ion mode. In the positive-ion mode the usually present signals are $[M+H]^+$, $[M+NH_4]^+$, $[M+Na]^+$, and $[M+K]^+$, and for the negative $[M-H]^-$, and for those molecules that cannot lose a proton $[M+Cl]^-$, or $[M+SCN]^-$.

Some of the most common fragmentation pathway produced by polysaccharides and glycoconjugates are represented in Figure 10.3:[4]

Likewise, application of different ionization techniques in the study of natural glycosides has been performed and derived. From this it has been possible to assign the main potential fragmentation sites in *O*- and *C*-glycosides (Figure 10.4).[5]

Negative ion FAB-MS in triethanolamine of synthetically prepared glycoresin composed by fucose, glucose, and quinovose attached to jalapinolic acid (Figure 10.5) shows $[M-H]^-$ peak (m/z 1216) in agreement with the expected molecular weight.[6]

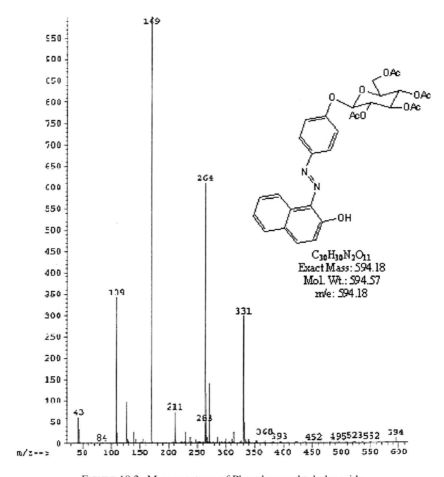

FIGURE 10.2. Mass spectrum of Phenylazonaphtol glucoside.

Mass spectrometry has been also applied successfully for glycoprotein structural determination of primary structure. The first glycoprotein primary structure was determined through electron impact and chemical ionization;[7] however, soft ionization methods of fast atom bombardment (FAB), electrospray (ES), or matrix-assisted laser desorption ionization (MALDI) are conducting most of the glycoprotein structural determinations.

FAB is particularly useful for analyzing the permethyl derivatives of oligosaccharides released from glycoproteins by chemical or enzymatic methods. When the atom or ion beam collides with the matrix, a substantial number of sample molecules are ionized producing positively charged species called quasimolecular ions $[M+H]^+$ and $[M+Na]^+$.[8]

In order to optimize fragment ion information of glycoproteins, three approaches are currently being used: inducing fragmentation by collisional activation,

Pathway A

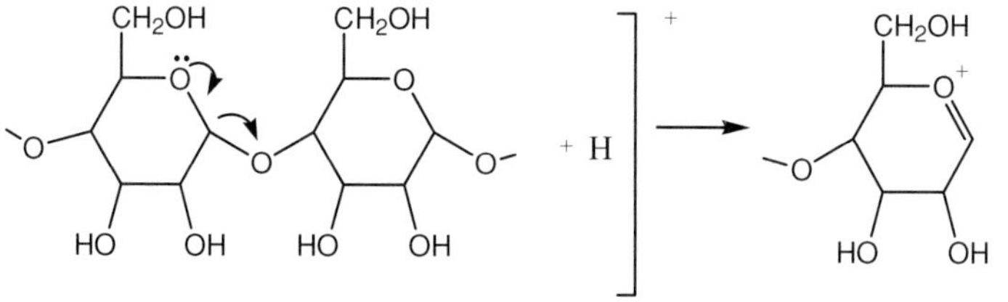

Pathway B

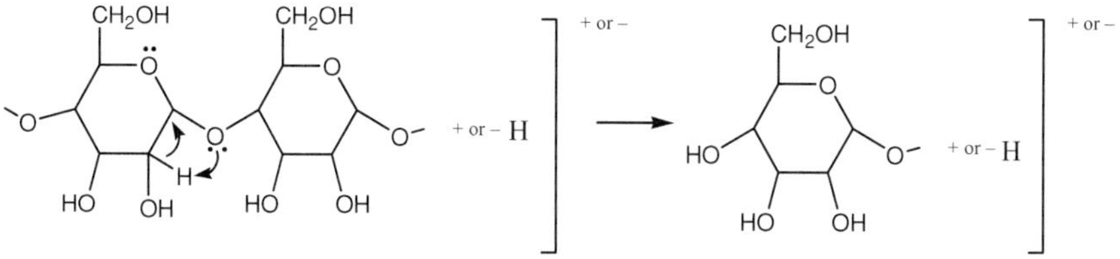

Pathway C

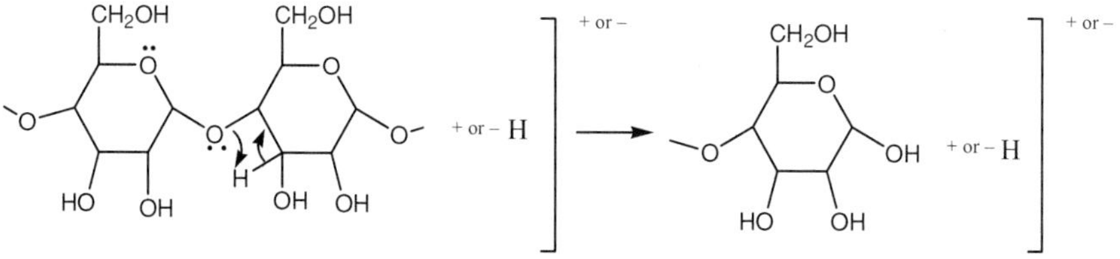

Pathway D

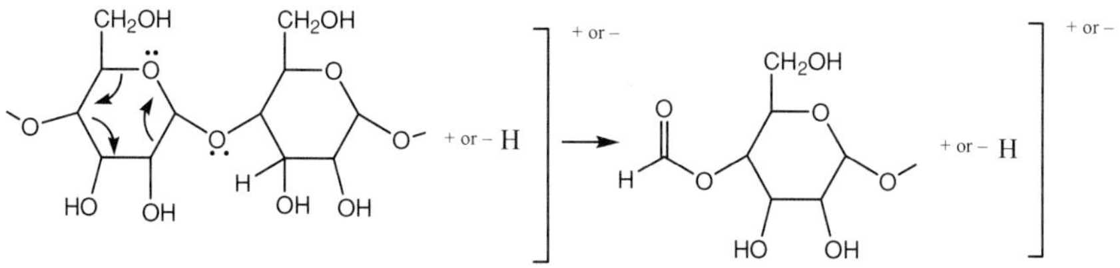

FIGURE 10.3. The most common fragmentation pathways.

Pathway E

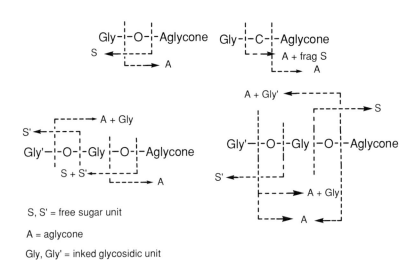

FIGURE 10.3. (*continued*)

monitoring natural ionization-induced fragmentation, and selecting derivatives that enhance and direct fragmentation.

During collisional activation of collected fractions from an enzymatic digest, the first step is to identify in the MS mode the fractions containing sugar fragment-ions. Then switching to the MS/MS mode of a doubly or triply charged ion a composite spectrum containing fragmentation of saccharide and peptide is obtained. Since glycosidic bonds are weaker than peptide bonds, the basic oligosaccharide sequence is determined.[9]

The natural fragmentation approach relies on the fragmentation created by internal energy transfer to the ion during the ionization process, and now is becoming most limited in use than the previous one.[10]

Gly–|–O–|–Aglycone   Gly–|–C–|–Aglycone
S ◄─|────┘   └─► A + frag S
        └──► A   └──► A

                                    A + Gly' ◄─ ─ ─ ─ ─ ┐
                                                        ┌─ ─ ─ ─ ─► S
          ┌─ ─ ─ ─► A + Gly
S' ◄─|─ ─ ┐
Gly'–|–O–|–Gly–|–O–|–Aglycone   Gly'–|–O–|–Gly–|–O–|–Aglycone
      S + S' ◄─ ─ ─ ─ ┘
              └─ ─ ─► A   S' ◄─ ─|─ ─ ┘
                                            ┌─ ─ ─► A + Gly
                                            └─► A ◄─┘

S, S' = free sugar unit

A = aglycone

Gly, Gly' = inked glycosidic unit

FIGURE 10.4. The main potential fragmentation sites in *O*- and *C*-glycosides.

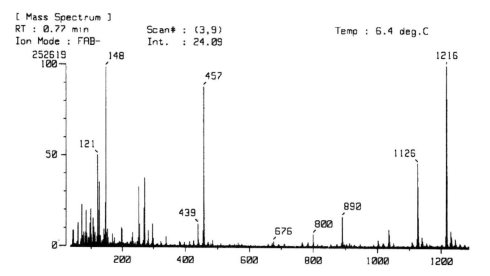

FIGURE 10.5. FAB-MS negative mode of synthetically prepared protected glycoresin.

Derivatization methods are likewise divided into tagging of reducing ends and protection of most of all of the functional groups. The first type facilitates chromatographic purifications and enhances the formation of reducing end fragment ions. The second type involves primarily the permethylation, which form abundant fragment ions arising from cleavage on the reducing side of each HexNAc residue.

The permethylation of Tamm-Horsfall glycoprotein was effected and the FAB mass spectrum obtained, showing molecular ions for core 2-type structures carrying up to three sialyl Lex moieties (Figure 10.6).[11]

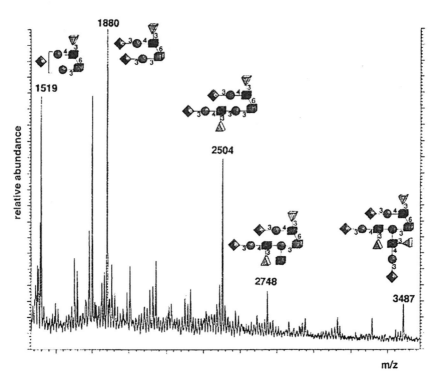

FIGURE 10.6. Partial FAB mass spectrum of permethylated *O*-oligosaccharides from gly-coprotein uromodulin.

References

 1. K. Bieman, *J. Am. Chem. Soc.* **85**, 2289 (1963).
 2. M. Brito-Arias, unpublished results.
 3. K. Harada, M. Suzuki, and H. Kambara, *Org. Mass. Spectrom*, 1982, 17, 386.
 4. M. Dell, *Advances in Carbohydrate Chemistry and Biochemistry*, 1987, 45, 19.
 5. J.-L. Wolfender, M. Maillard, A. Marston, and K., Hostettman, *Pytochemical Analysis*, 1992, 3, 193.
 6. M.A. Brito-Arias, R. Pereda-Miranda, and C.H. Heathcock, *J. Org. Chem.*, 2004.
 7. Morris, et al., *J. Biol. Chem. 1978*, 253, 5155.
 8. A. Dell and H.R. Morris, *Science*, 2001, 291, 2351.
 9. Teng-umnuay et al., *J. Biol. Chem.* 1998, 273, 18242.
10. A. Dell, *Adv. Carbohydr. Chem. Biochem.* 1987, 45, 19.
11. R.L. Easton, M.S. Patankar, G.F. Clark, H.R. Morris, and A. Dell, *J. Biol. Chem.* 2000, 275, 21928.

Index

Printed in the United States of America